W0041980

Trait-Mediated Indirect Interactions

Ecological and Evolutionary Perspectives

There is increasing evidence that the structure and functioning of ecological communities and ecosystems are strongly influenced by flexible traits of individuals within species. A deep understanding of how trait flexibility alters direct and indirect species interactions is crucial for addressing key issues in basic and applied ecology. This book provides an integrated perspective on the ecological and evolutionary consequences of interactions mediated by flexible species traits across a wide range of systems. It is the first volume synthesizing the rapidly expanding research field of trait-mediated indirect effects, and highlights how the conceptual framework of these effects can aid the understanding of evolutionary processes, population dynamics, community structure and stability, and ecosystem function. It not only brings out the importance of this emerging field for basic ecological questions, but also explores the implications of trait-mediated interactions for the conservation of biodiversity and the response of ecosystems to anthropogenic environmental changes.

TAKAYUKI OHGUSHI is a Professor at the Center for Ecological Research at Kyoto University. His research focuses on the population biology of insect herbivores, plant–herbivore interactions, multitrophic interactions and the linkage from gene to ecosystem. In particular, he is interested in how trait-mediated indirect effects create ecological communities and biodiversity.

OSWALD J. SCHMITZ is the Oastler Professor of Population and Community Ecology in the Yale University School of Forestry and Environmental Studies. He studies the linkage between two important components of natural systems: biodiversity and ecosystem services, using field experimentation guided by formal mathematical theory of trait-based species interactions.

ROBERT D. HOLT is Arthur R. Marshall Jr. Chair in Ecology and Eminent Scholar in the Department of Biology at the University of Florida. He is an evolutionary and community ecologist whose contributions are principally theoretical, but always tied to concrete processes in the natural world. He has received the International Ecology Institute Prize in Terrestrial Ecology and the Sewall Wright Award from the American Society of Naturalists.

Ecological Reviews

SERIES EDITOR Hefin Jones *Cardiff University*, UK
SERIES EDITORIAL BOARD
Mark Bradford *University of Georgia*, USA
David Burslem *University of Aberdeen*, UK
Alan Gray *CEH Wallingford*, UK
Catherine Hill *British Ecological Society*, UK
Sue Hartley *University of York*, UK
Mark Hunter *University of Michigan*, USA
Heikki Setala *University of Helsinki*, Finland
Phillip Warren *University of Sheffield*, UK

Ecological Reviews publishes books at the cutting edge of modern ecology, providing a forum for volumes that discuss topics that are focal points of current activity and likely to be of long-term importance to the progress of the field. The series is an invaluable source of ideas and inspiration for ecologists at all levels from graduate students to more established researchers and professionals. The series has been developed jointly by the British Ecological Society and Cambridge University Press and encompasses the Society's Symposia as appropriate.

Biotic Interactions in the Tropics: Their Role in the Maintenance of Species Diversity
Edited by David F. R. P. Burslem, Michelle A. Pinard and Sue E. Hartley

Biological Diversity and Function in Soils
Edited by Richard Bardgett, Michael Usher and David Hopkins

Island Colonization: The Origin and Development of Island Communities
By Ian Thornton
Edited by Tim New

Scaling Biodiversity
Edited by David Storch, Pablo Margnet and James Brown

Body Size: The Structure and Function of Aquatic Ecosystems
Edited by Alan G. Hildrew, David G. Raffaelli and Ronni Edmonds-Brown

Speciation and Patterns of Diversity
Edited by Roger Butlin, Jon Bridle and Dolph Schluter

Ecology of Industrial Pollution
Edited by Lesley C. Batty and Kevin B. Hallberg

Ecosystem Ecology: A New Synthesis
Edited by David G. Raffaelli and Christopher L. J. Frid

Urban Ecology
Edited by Kevin J. Gaston

The Ecology of Plant Secondary Metabolites: From Genes to Global Processes
Edited by Glenn R. Iason, Marcel Dicke and Susan E. Hartley

Trait-Mediated Indirect Interactions

Ecological and Evolutionary Perspectives

Edited by

TAKAYUKI OHGUSHI
Kyoto University, Japan

OSWALD J. SCHMITZ
Yale University, USA

ROBERT D. HOLT
University of Florida, USA

CAMBRIDGE
UNIVERSITY PRESS

CAMBRIDGE
UNIVERSITY PRESS

Shaftesbury Road, Cambridge CB2 8EA, United Kingdom

One Liberty Plaza, 20th Floor, New York, NY 10006, USA

477 Williamstown Road, Port Melbourne, VIC 3207, Australia

314–321, 3rd Floor, Plot 3, Splendor Forum, Jasola District Centre, New Delhi – 110025, India

103 Penang Road, #05–06/07, Visioncrest Commercial, Singapore 238467

Cambridge University Press is part of Cambridge University Press & Assessment,
a department of the University of Cambridge.

We share the University's mission to contribute to society through the pursuit of
education, learning and research at the highest international levels of excellence.

www.cambridge.org
Information on this title: www.cambridge.org/9780521173131

© Cambridge University Press & Assessment 2012

This publication is in copyright. Subject to statutory exception and to the provisions
of relevant collective licensing agreements, no reproduction of any part may take
place without the written permission of Cambridge University Press & Assessment.

First published 2012

A catalogue record for this publication is available from the British Library

Library of Congress Cataloging-in-Publication data
Trait-mediated indirect interactions: ecological and evolutionary
perspectives / edited by Takayuki Ohgushi, Oswald J.
Schmitz, Robert D. Holt.
   p.   cm. – (Ecological reviews)
Includes index.
ISBN 978-1-107-00183-1
1. Coevolution.   2. Adaptation (Biology)   3. Biotic communities.   I. Ohgushi,
Takayuki.   II. Schmitz, Oswald J.   III. Holt, Robert D.
QH372.E26   2012
576.8′7–dc23

                                                                2012015492

ISBN   978-1-107-00183-1   Hardback
ISBN   978-0-521-17313-1   Paperback

Additional resources for this publication at www.cambridge.org/9780521173131

Cambridge University Press & Assessment has no responsibility for the persistence
or accuracy of URLs for external or third-party internet websites referred to in this
publication and does not guarantee that any content on such websites is, or will
remain, accurate or appropriate.

Contents

List of contributors   *page* viii
Foreword
Peter W. Price   xiii
Preface   xv

1   Introduction
     Takayuki Ohgushi, Oswald J. Schmitz and Robert D. Holt   1

     PART I   COMMUNITY   7
2   Perspective: kinds of trait-mediated indirect effects in
     ecological communities. A synthesis
     Thomas W. Schoener and David A. Spiller   9
3   Consequences of trait changes in host–parasitoid
     interactions in insect communities
     F. J. Frank van Veen and H. Charles Godfray   28
4   The impact of trait-mediated indirect interactions in marine
     communities
     Jeremy D. Long and Mark E. Hay   47
5   Trait-mediated indirect interactions in size-structured
     populations: causes and consequences for species
     interactions and community dynamics
     Volker H. W. Rudolf   69
6   Trait-mediated effects, density dependence and the dynamic
     stability of ecological systems
     Robert D. Holt and Michael Barfield   89
7   Plant effects on herbivore–enemy interactions in natural
     systems
     Kailen A. Mooney and Michael S. Singer   107

8   The implications of adaptive prey behaviour for ecological
    communities: a review of current theory
    Scott D. Peacor and Clayton E. Cressler   131

9   Community consequences of phenotypic plasticity of
    terrestrial plants: herbivore-initiated bottom-up trophic
    cascades
    Takayuki Ohgushi   161

10 Model-based, response-surface approaches to quantifying
    indirect interactions
    Toshinori Okuyama and Benjamin M. Bolker   186

PART II   COEVOLUTION   205

11 Perspective: trait-mediated indirect interactions and the
    coevolutionary process
    Benjamin J. Ridenhour and Scott L. Nuismer   207

12 Evolutionary indirect effects: examples from introduced
    plant and herbivore interactions
    Jennifer A. Lau   221

13 Indirect evolutionary interactions in a multitrophic system
    Timothy P. Craig, Joanne K. Itami, Michael Dixon and Terry R. Hams   244

14 The role of trait-mediated indirect interactions for
    multispecies plant–animal mutualisms
    Rebecca E. Irwin   257

15 Consequences of trait evolution in a multispecies system
    Craig W. Benkman, Adam M. Siepielski and Julie W. Smith   278

PART III   ECOSYSTEM   293

16 Perspective: interspecific indirect genetic effects (IIGEs).
    Linking genetics and genomics to community ecology and
    ecosystem processes
    *Gerard J. Allan, Stephen M. Shuster, Scott Woolbright, Faith Walker,
    Nashelly Meneses, Arthur Keith, Joseph K. Bailey and Thomas
    G. Whitham*   295

17 Species functional traits, trophic control and the ecosystem
    consequences of adaptive foraging in the middle of food
    chains
    Geoffrey C. Trussell and Oswald J. Schmitz   324

18 Effects of herbivores on terrestrial ecosystem processes: the
    role of trait-mediated indirect effects
    *Mark D. Hunter, Barbara C. Reynolds, Myra C. Hall and Christopher
    J. Frost*   339

19 Functional and heritable consequences of plant genotype on community composition and ecosystem processes
Jennifer A. Schweitzer, Joseph K. Bailey, Dylan G. Fischer, Carri J. LeRoy, Thomas G. Whitham and Stephen C. Hart　371

20 Microbial mutualists and biodiversity in ecosystems
Jennifer A. Rudgers and Keith Clay　391

21 Integrating trait-mediated effects and non-trophic interactions in the study of biodiversity and ecosystem functioning
Alexandra Goudard and Michel Loreau　414

PART IV APPLIED ECOLOGY　433

22 Perspective: consequences of trait-mediated indirect interactions for biological control of plant pests
Maurice W. Sabelis, Arne Janssen and Izabela Lesna　435

23 Natural enemy functional identity, trait-mediated interactions and biological control
Tobin D. Northfield, David W. Crowder, Randa Jabbour and William E. Snyder　450

24 Trait-mediated effects modify patch-size density relationships in insect herbivores and parasitoids
Peter A. Hambäck, Petter Andersson and Tibor Bukovinszky　466

25 Plasticity and trait-mediated indirect interactions among plants
Erik T. Aschehoug and Ragan M. Callaway　489

26 Climate change, phenology and the nature of consumer–resource interactions: advancing the match/mismatch hypothesis
Jeffrey T. Kerby, Christopher C. Wilmers and Eric Post　508

27 Coda
Takayuki Ohgushi, Oswald J. Schmitz and Robert D. Holt　526

Index　530

The colour plates are to be found between pages 432 and 433.

Contributors

GERARD J. ALLAN
Department of Biological Sciences,
Northern Arizona University,
Flagstaff, USA
gery.allan@nau.edu

PETTER ANDERSSON
Department of Botany, Stockholm
University, Stockholm, Sweden
petter.andersson@botan.su.se

ERIK T. ASCHEHOUG
Division of Biological Sciences,
University of Montana, Missoula, USA
erik.aschehoug@mso.umt.edu

JOSEPH K. BAILEY
Department of Ecology and
Evolutionary Biology, University of
Tennessee, Knoxville, USA
Joe.Bailey@utk.edu

MICHAEL BARFIELD
Department of Biology, University of
Florida, Gainesville, USA
mjb01@ufl.edu

CRAIG W. BENKMAN
Department of Zoology and
Physiology, University of Wyoming,
Laramie, USA
cbenkman@uwyo.edu

BENJAMIN M. BOLKER
Department of Mathematics and
Statistics and Department of
Biology, McMaster University,
Hamilton, Canada
bolker@mcmaster.ca

TIBOR BUKOVINSZKY
Department of Terrestrial
Ecology, Netherlands Institute of
Ecology, Wageningen,
The Netherlands
t.bukovinszky@nioo.knaw.nl

RAGAN M. CALLAWAY
Division of Biological Sciences,
University of Montana, Missoula,
USA
ray.callaway@mso.umt.edu

KEITH CLAY
Department of Biology,
Indiana University,
Bloomington, USA
clay@indiana.edu

TIMOTHY P. CRAIG
Department of Biology,
University of Minnesota, Duluth,
USA
tcraig@d.umn.edu

CLAYTON E. CRESSLER
Department of Ecology and
Evolutionary Biology, University
of Michigan, Ann Arbor, USA
cressler@umich.edu

DAVID W. CROWDER
Department of Entomology,
Washington State University,
Pullman, USA
dcrowder@wsu.edu

MICHAEL DIXON
US Fish and Wildlife Service,
Lakewood, USA
Michael_D_Dixon@fws.gov

DYLAN G. FISCHER
Environmental Studies Program, The
Evergreen State College, Olympia,
USA
fischerd@evergreen.edu

CHRISTOPHER J. FROST
Warnell School of Forest Resources,
University of Georgia, Athens, USA
cfrost@warnell.uga.edu

H. CHARLES J. GODFRAY
Department of Zoology, University of
Oxford, Oxford, UK
charles.godfray@zoo.ox.ac.uk

ALEXANDRA GOUDARD
Lycée Champollion, Grenoble,
France
alexandra.goudard@ac-grenoble.fr

MYRA C. HALL
Georgia Perimeter College, Decatur,
USA
myra.hall@gpc.edu

PETER A. HAMBÄCK
Department of Botany, Stockholm
University, Stockholm, Sweden
Peter.Hamback@botan.su.se

TERRY R. HAMS
Golder Associates, Saskatoon, Canada
hams0005@d.umn.edu

STEPHEN C. HART
School of Natural Sciences and Sierra
Nevada Research Institute, University
of California, Merced, USA
shart4@ucmerced.edu

MARK E. HAY
School of Biology, Georgia Institute of
Technology, Atlanta, USA
mark.hay@biology.gatech.edu

ROBERT D. HOLT
Department of Biology, University of
Florida, Gainesville, USA
rdholt@ufl.edu

MARK D. HUNTER
Department of Ecology and
Evolutionary Biology, University of
Michigan, Ann Arbor, USA
mdhunter@umich.edu

REBECCA E. IRWIN
Biology Department,
Dartmouth College, Hanover,
USA
Rebecca.E.Irwin@Dartmouth.edu

JOANNE K. ITAMI
Department of Biology,
University
of Minnesota-Duluth, USA
jitami@d.umn.edu

RANDA JABBOUR
Department of Plant, Soil, and
Environmental Sciences,
University of Maine,
Orono, USA
Randa.jabbour@maine.edu

ARNE JANSSEN
Section Population Biology, Institute
for Biodiversity and Ecosystem
Dynamics, University of Amsterdam,
Amsterdam, The Netherlands
A.R.M.Janssen@uva.nl

ARTHUR KEITH
Department of Biological Sciences,
Northern Arizona University,
Flagstaff, USA
Arthur.Keith@nau.edu

JEFFREY T. KERBY
Department of Biology,
Pennsylvania State University,
University Park, USA
jtkerb@gmail.com

JENNIFER A. LAU
W.K. Kellogg Biological Station and
Department of Plant Biology,
Michigan State University, East
Lansing, USA
jenlau@msu.edu

CARRI J. LEROY
Environmental Studies Program,
The Evergreen State College, Olympia,
USA
leroyc@evergreen.edu

IZABELA LESNA
Section Population Biology, Institute
for Biodiversity and Ecosystem

Dynamics, University of Amsterdam,
Amsterdam, The Netherlands
i.k.a.lesna@uva.nl

JEREMY D. LONG
Department of Biology and Coastal
Marine Institute Laboratory, San
Diego State University, San Diego,
USA
jlong@sciences.sdsu.edu

MICHEL LOREAU
Station d'Ecologie Expérimentale du
CNRS, Moulis, France
michel.loreau@ecoex-moulis.cnrs.fr

NASHELLY MENESES
Department of Biological Sciences,
Northern Arizona University,
Flagstaff, USA
nm49@nau.edu

KAILEN A. MOONEY
Department of Ecology and
Evolutionary Biology, University of
California, Irvine, USA
mooneyk@uci.edu

TOBIN D. NORTHFIELD
Department of Zoology, University of
Wisconsin, Madison, USA
northfield@wisc.edu

SCOTT L. NUISMER
Department of Biological Sciences,
University of Idaho, Moscow, USA
snuismer@uidaho.edu

TAKAYUKI OHGUSHI
Center for Ecological Research, Kyoto
University, Otsu, Japan
ohgushi@ecology.kyoto-u.ac.jp

TOSHINORI OKUYAMA
Department of Entomology,
National Taiwan University, Taipei,
Taiwan
okuyama@ntu.edu.tw

SCOTT D. PEACOR
Department of Fisheries and
Wildlife, Michigan State University,
East Lansing, USA
peacor@msu.edu

ERIC POST
Department of Biology,
Pennsylvania State University,
University Park, USA
esp10@psu.edu

PETER W. PRICE
Department of Biological Sciences,
Northern Arizona University,
Flagstaff, USA
peter.price@nau.edu

BARBARA C. REYNOLDS
Department of Environmental
Studies, University of North
Carolina-Asheville, Asheville, USA
kreynolds@unca.edu

BENJAMIN J. RIDENHOUR
Department of Biological Sciences,
University of Notre Dame, South
Bend, USA
Benjamin.Ridenhour.1@nd.edu

JENNIFER A. RUDGERS
Department of Biology, University of
New Mexico, Albuquerque,
USA
jrudgers@unm.edu

VOLKER H. W. RUDOLF
Department of Ecology and
Evolutionary Biology, Rice
University, Houston, USA
volker.rudolf@rice.edu

MAURICE W. SABELIS
Section Population Biology,
Institute for Biodiversity and
Ecosystem Dynamics, University of
Amsterdam, Amsterdam, The
Netherlands
M.W.Sabelis@uva.nl

OSWALD J. SCHMITZ
School of Forestry and
Environmental Studies,
Yale University, New Haven,
USA
oswald.schmitz@yale.edu

THOMAS W. SCHOENER
Department of Evolution and
Ecology and Center for
Population Biology,
University of California,
Davis, USA
twschoener@ucdavis.edu

JENNIFER A. SCHWEITZER
Department of Ecology and
Evolutionary Biology,
University of Tennessee,
Knoxville, USA
Jen.Schweitzer@utk.edu

STEPHEN M. SHUSTER
Department of Biological Sciences,
Northern Arizona University,
Flagstaff, USA
stephen.shuster@nau.edu

ADAM M. SIEPIELSKI
Department of Biology,
University of San Diego, San Diego,
USA
adamsiepielski@sandiego.edu

MICHAEL S. SINGER
Department of Biology,
Wesleyan University, Middletown,
USA
msinger@wesleyan.edu

JULIE W. SMITH
Department of Biology,
Pacific Lutheran University,
Tacoma, USA
smith@plu.edu

WILLIAM E. SNYDER
Department of Entomology,
Washington State University,
Pullman, USA
wesnyder@wsu.edu

DAVID A. SPILLER
Department of Evolution and
Ecology and Center for Population
Biology, University of California,
Davis, USA
daspiller@ucdavis.edu

GEOFFREY C. TRUSSELL
Marine Science Center and
Department of Biology,

Northeastern University,
Boston, USA
g.trussell@neu.edu

F. J. FRANK VAN VEEN
Centre for Ecology and Conservation,
College of Life and Environmental
Sciences, University of Exeter,
Penryn, UK
f.j.f.van-veen@exeter.ac.uk

FAITH WALKER
Department of Biological Sciences,
Northern Arizona University,
Flagstaff, USA
Faith.Walker@nau.edu

THOMAS G. WHITHAM
Department of Biological Sciences,
Northern Arizona University,
Flagstaff, USA
thomas.whitham@nau.edu

CHRISTOPHER C. WILMERS
Environmental Studies Department,
University of California,
Santa Cruz, USA
cwilmers@ucsc.edu

SCOTT WOOLBRIGHT
The Institute for Genomic Biology,
University of Illinois,
Urbana, USA
sawg@illinois.edu

Foreword

PETER W. PRICE

The word 'ubiquitous' crops up repeatedly in this book in relation to trait-mediated indirect interactions (TMIIs) in ecology, where one species alters the interchange between two other species. Indeed, the subject of this book has a pervasive relationship with all of ecology, providing a theme and a concept that enmeshes all levels of organization, from individuals to populations, communities, ecosystems and global phenomena. Another term used frequently is 'strong effect' for the influence of such indirect interactions in communities and ecosystems. If we are dealing with ever-present and major impacts in ecology, then certainly these kinds of interactions deserve much attention. In fact, the volume of literature in this field appears to be undergoing exponential growth. Therefore, this volume is a timely reminder that the field is expanding rapidly, and a guide on how it can grow along new routes of research and application with time. TMIIs as a category of interchanges in ecology are worthy of attention from all ecologists and those in related fields such as agriculture, forestry, conservation, epidemiology, parasitology and animal husbandry.

We should not be surprised by the pervasive and robust influences of TMIIs. One example: as with humans, all plants and animals produce body odours, but more so than with humans, the body odours in nature have strong effects on other species, both direct and indirect. A phytochemical may have a direct impact on a herbivore, as well as an indirect effect by acting as an attractant to the enemy of the herbivore. In a community of plants, herbivores, parasitoids and predators there is therefore a rich blend of aromas wafting around, mediating the interactions of a multitude of species. These are not necessarily feeding links that would enter into a food web, but they would become important components of an interaction web, far richer than the conventional food web. Similar increases in interaction richness and complexity are observed when we compare direct feeding links on plants versus indirect interactions emanating from herbivores altering plant traits, which provide new resources for other species. Interactions may more than triple in number when trait-mediated interchanges are recognized.

In the equation resulting in great biodiversity there is also the role of plant genotype, which influences all trophic levels above and below ground, community structure and ecosystem processes. Therefore, genotypic diversity within populations and species, and their hybrids, has wide-ranging indirect effects, many of them surprisingly strong.

Then we should reflect upon the effects of geographic adjustment of species ranges resulting from climate change, invasive species and species introductions by humans, say, for biological control purposes. All involve animals and plants which influence residents, including in many indirect ways, creating non-analogue communities – never seen before – and a mismatch between consumers and resources. The cohesiveness of communities is disrupted in many indirect ways, such as in diffuse coevolutionary processes.

Part of the wonder and amazement that we enjoy in this literature is derived from our naturalist's fascination with nature. The subject of this book depends on the keen observer, a witness to behavioural interactions; the accomplished chemist who can unravel unseen, cryptic relationships and the formerly mysterious world of molecular processes; and the experimenter who can support or dispel notions and hypotheses on the natural world.

All these topics and many more are discussed in this book, which has become a compendium of the large literature on the vast array of interactions under the umbrella of trait-mediated indirect effects. Theory, concept, hypothesis, modelling, empirical results, predictions and applications to practical environmental problems are blended into this volume. The recognition of indirect effects enriches ecology by an order of magnitude.

Preface

Ecologists have long found the complexity of ecological systems, brought about by the diversity of species and the myriad direct and indirect interactions among them, to be a source of awe. Such complexity represents a formidable scientific challenge as ecologists struggle to identify the fundamental mechanisms that make sense of the web of interactions that drive ecological functioning and sustainability. The most fundamental mechanism driving complexity is the process of evolution by natural selection, a process that requires variation in phenotypic traits among members of a species. One dimension of this variation is the propensity for organisms to show adaptive shifts in phenotypes to meet challenges imposed by changing environmental conditions. But this fundamental evolutionary mechanism seems somewhat forgotten in modern analyses of community and ecosystem dynamics, where conceptualizations and empirical approaches typically treat organisms (and entire species) as having a fixed set of phenotypic traits, invariant to changes in environmental contexts. Ecologists have largely ignored the community and ecosystem consequences of phenotypic adjustment to environmental conditions, and thereby effectively have overlooked the potential to explain much of context-dependency (and hence variation) in species interactions. Taking into account such flexibility in individual traits will, we believe, enhance the power of our field to explain ecological patterns and to predict the consequences of environmental change.

This volume grew from a symposium entitled 'Trait-mediated indirect effects in insect communities', held in Durban, South Africa, at the International Congress of Entomology meeting in July 2008. The symposium brought together a collection of international contributors uniquely qualified to evaluate and expand our understanding of how trait-mediated indirect effects structure insect communities. To offer a broader view of the field, we invited additional authors working in a wide range of ecological systems to contribute chapters. We intend to provide ecologists with insight into the 'state-of-the-art' of research focused on trait-based effects, an approach which can link in novel ways individual, population, community and ecosystem ecology. Our goal is to foster research on trait-mediated interactions and thereby stimulate ecologists to address more systematically how trait-based effects can modify expectations based on classical approaches that view traits as invariant across environments. This volume reveals in many ways how the conceptual framework of trait-mediated indirect interactions can greatly improve our understanding of evolutionary processes, population dynamics, community structure and stability, and ecosystem properties and functions.

We are very grateful to the British Ecological Society for including this volume in the series titled *Ecological Reviews*. We thank the authors for their hard work in helping us put this volume together, and the many colleagues who kindly reviewed the chapters. Finally, we particularly thank commissioning editor Dominic Lewis, assistant editor Lynette Talbot and production editor Caroline Mowatt for their assistance with the production of this volume. It has been greatly rewarding for us to be involved in crafting a book focused on this emerging, important research area.

CHAPTER ONE

Introduction

TAKAYUKI OHGUSHI

Center for Ecological Research, Kyoto University

OSWALD J. SCHMITZ

School of Forestry and Environmental Studies, Yale University

and

ROBERT D. HOLT

Department of Biology, University of Florida

Community ecology is experiencing a resurgence, driven in part by its central importance in addressing critical applied problems, ranging from the control of pest and invasive species, to the wise harvest of natural resources, to projecting the impact of global climate change. A fundamental tenet of community ecology is that species do not exist in isolation: they are directly, and more importantly, indirectly interconnected with myriad other species. The essential 'glue' that holds communities together and that makes them more than the haphazard sum of individual species is the nexus of indirect interactions among three or more species that emerges from direct interactions such as predation, competition and mutualism between pairs of species. The recognition of indirect effects has triggered a rapid growth of empirical and theoretical research that aims to predict community-level dynamics under different contexts.

Indirect effects occur when the impacts of one species on another are influenced by one or more intermediate species. Indirect effects are diverse, but can be classified broadly as either (1) density-mediated or (2) trait-mediated. Density-mediated indirect effects (DMIEs) result from numerical responses of species to each other. For instance, a fox may kill rabbits, reducing rabbit population size, and so relaxing herbivory upon herbaceous plants. Hence, the fox's indirect effect on plants is mediated by density changes in rabbits; this is known as a trophic cascade. DMIEs, such as depicted by this trophic cascade, and other mechanisms such as apparent competition between prey mediated by the numerical response of a shared predator, have been well studied and have contributed to our understanding of community organization and ecosystem functioning in both terrestrial and aquatic systems (Holt and Lawton 1994; Polis *et al.* 2000; Terborgh and Estes 2010). However, the fox may not only kill rabbits, it may alter their behaviour and other traits. For instance, rabbits may hide more in the presence of foxes and so have less opportunity to feed on herbs. This

Trait-Mediated Indirect Interactions: Ecological and Evolutionary Perspectives, eds. Takayuki Ohgushi, Oswald J. Schmitz and Robert D. Holt. Published by Cambridge University Press. © Cambridge University Press 2012.

indirect effect of the fox upon the plants could be strong, even though rabbit abundance remains high in the presence of foxes. In this case, it is a change in a rabbit's trait – altered behaviour to avoid predation – that determines the outcome of the predator–prey direct interaction and the nature of the indirect effect of foxes on plants. In this same food chain, consumption by the rabbit could change plant architecture and biomass, altering the exposure of the rabbit to capture. In principle, changes in traits of any species in a community could shift interactions among other species, as well as altering the interactions involving that species itself. The objective of this volume is to display the rich variety of ways such trait effects arise, and to examine the consequences of such trait effects for a wide range of ecological issues.

What do we mean by *trait*? Basically, we suggest that anything that can be measured about an individual organism or a strategy (sensu Vincent and Brown 2005) can be considered a trait. If we can measure it, then in principle other organisms can also measure it, and respond to this metric. As evolutionary biologists, the traits we most care about are typically those that are heritable and markedly affect fitness. In some cases, traits may be relatively independent of the environment, and of the sort that can be measured in museum specimens (e.g., body size and shape). Such invariance is indeed assumed in classical models in community ecology (in effect: seen one rabbit, seen 'em all). But in other cases, traits are highly plastic, varying in accord with changes in abiotic and biotic factors. Such traits can only be measured in a specific environmental context (for example life history traits, metabolic rates, or per capita attack rates by a predator upon its prey), and ideally would be described as functions of trait values against environmental variables, i.e., norms of reaction. If trait plasticity is an evolved state, it is reasonable to presume the trait in question is important in determining fitness, and that relative rankings of phenotypes by fitness vary with environmental circumstances.

Plasticity has long been of interest in evolutionary biology (e.g., Scheiner and Lyman 1991), but now community ecologists are increasingly aware that species traits and plasticity in traits can have multiple consequential effects for other species, not just for the species itself and its immediate interactors. These effects mean that the interactions determining community structure and dynamics can be much more pervasive than suggested by simple pairwise metrics such as niche overlap (permitting resource competition) and trophic transfers (comprising the links in food webs). Because trait effects are often large, trait plasticity implies that individuals in a given species have different direct and indirect effects in different situations. To the extent that these context-dependent effects are predictable, incorporating an understanding of them is essential to developing a predictive theory of communities. Such a trait-based approach also fosters the integration of evolutionary and community ecology.

Traditionally, theoretical models of ecology portraying the outcomes of interspecific interactions assumed that species have fixed properties.

The variables at the core of community dynamics were then naturally the densities of each interacting species, possibly expanded to include different age classes. In recent years, it has become clear that theories based solely on density interactions provide an insufficient foundation for community theory. Parameters that govern interactions between a pair of species may depend not only on the traits of these species, but also on the traits and abundance of third (or more) parties. Because traits can be plastic or evolutionarily labile, they may themselves change on timescales commensurate with changes in density (Abrams 1995). If species B plastically changes a key trait in response to species A, this may influence the strength and even qualitative sign of interactions between species B and C. There is thus an emergent indirect interaction – a *trait-mediated indirect interaction* – between species A and C. The word 'emergent' reflects the fact that interacting entities can produce phenomena – higher order structures and functionalities – that are more than can be captured by a simple averaging of the properties of the entities considered apart (Page 2011). In other words, a trait-mediated indirect interaction ('TMII') occurs when species A indirectly influences species C by inducing a modification in the traits, such as behaviour, morphology and/or life history of species B. Two conditions must be satisfied for a TMII: (1) a species affects a trait of another species; (2) the latter species affects a third species, completing the indirect effect (Werner and Peacor 2003).

In general, a wide range of trait-mediated indirect effects (TMIEs) is possible via plastic shifts in behaviour, morphology, physiology and life histories of affected species. The recognition of TMIEs came to prominence first in predator–prey systems. For example, as in our fox–rabbit example, prey can avoid predation risk by altering their behaviours, morphologies and life histories in the presence of predators, leading to trophic cascades with enhanced primary production (Schmitz *et al.* 2004). More recently, TMIEs have received considerable attention in plant–herbivore systems because of the prevalence of herbivore-induced changes in plant traits (Ohgushi 2005). A change in leaf morphology, stem architecture or phenology of leaf production following herbivory can potentially alter the competitive efficacy of plants for light, and shift interactions with herbivores, pollinators, or seed dispersers, or even predators (Ohgushi *et al.* 2007). Such TMIEs can match or even outweigh classical DMIEs in determining community dynamics (Preisser *et al.* 2005; Trussell *et al.* 2006; Werner and Peacor 2006). The chapters in this volume demonstrate that ecological communities are replete with TMIIs, and that a deep understanding of such interactions is crucial for addressing many issues in both basic and applied ecology.

The richness of trait-mediated effects results in a far more reticulate pattern of linkages among species than do classical food web interactions, leading to emergent properties of communities that are hard to predict from analyses of individual species or species pairs (Ings *et al.* 2009; Beckerman *et al.* 2010). Phenotypic plasticity produces a complex tapestry of impacts on organisms, cascading upward and downward through trophic levels, and surprising chain reactions across entire communities. We believe the time is ripe for a synthesis of this new development in population and community ecology, leading to a fresh understanding of a wide array of ecological processes (including ecosystem dynamics), and this volume is a step towards such a synthesis.

The authors in this book emphasize conceptual issues and provide illustrative empirical and theoretical studies that highlight the central importance of TMIIs for ecological processes ranging from diffuse coevolution, to population dynamics, to community organization, to ecosystem functions. The chapters, we believe, collectively provide crucial steps towards an integrated perspective on the ecological and evolutionary consequences of TMIEs in a wide range of ecological systems, consequences linking evolution, community and ecosystem ecology. This perspective enhances our ability to answer basic ecological questions, foster the conservation of biodiversity, and project how ecosystems may respond to anthropogenic environmental changes.

References

Abrams, P. A. (1995) Implications of dynamically variable traits for identifying, classifying, and measuring direct and indirect effects in ecological communities. *American Naturalist*, **146**, 112–134.

Beckerman, A. P., Petchey, O. L. and Morin, P. J. (2010) Adaptive foragers and community ecology: linking individuals to communities and ecosystems. *Functional Ecology*, **24**, 1–6.

Holt, R. D. and Lawton, J. H. (1994) The ecological consequences of shared natural enemies. *Annual Review of Ecology and Systematics*, **25**, 495–520.

Ings, T. C., Montoya, J. M., Bascompte, J. *et al.* (2009) Ecological networks – beyond food webs. *Journal of Animal Ecology*, **78**, 253–269.

Ohgushi, T. (2005) Indirect interaction webs: herbivore-induced effects through trait change in plants. *Annual Review of Ecology, Evolution, and Systematics*, **36**, 81–105.

Ohgushi, T., Craig, T. P. and Price, P. W. (2007) *Ecological Communities: Plant Mediation in Indirect Interaction Webs.* Cambridge: Cambridge University Press.

Page, S. E. (2011) *Diversity and Complexity.* Princeton, NJ: Princeton University Press.

Preisser, E. L., Bolnick, D. I. and Benard, M. F. (2005) Scared to death? The effects of intimidation and consumption in predator–prey interactions. *Ecology*, **86**, 501–509.

Polis, G. A., Sears, A. L. W., Huxel, G. R., Strong, D. R. and Maron, J. (2000) When is a trophic cascade a trophic cascade? *Trends in Ecology and Evolution*, **15**, 473–475.

Scheiner, S. M. and Lyman, R. F. (1991) The genetics of phenotypic plasticity. II. Response to selection. *Journal of Evolutionary Biology*, **4**, 23–50.

Schmitz, O. J., Krivan, V. and Ovadia, O. (2004) Trophic cascades: the primacy of trait-mediated indirect interactions. *Ecology Letters*, **7**, 153–163.

Terborgh, J. and Estes, J. A. (2010) *Trophic Cascades: Predators, Prey, and the Changing Dynamics of Nature.* Washington, DC: Island Press.

Trussell, G. C., Ewanchuk, P. J. and Matassa, C. M. (2006) Habitat effects on the relative importance of trait- and density-mediated indirect interactions. *Ecology Letters*, **9**, 1245–1252.

Vincent, T. L. and Brown, J. S. (2005) *Evolutionary Game Theory, Natural Selection and Darwinian Dynamics*. Cambridge: Cambridge University Press.

Werner, E. E. and Peacor, S. D. (2003) A review of trait-mediated indirect interactions in ecological communities. *Ecology*, **84**, 1083–1100.

Werner, E. E. and Peacor, S. D. (2006) Lethal and nonlethal predator effects on an herbivore guild mediated by system productivity. *Ecology*, **87**, 347–361.

PART I

Community

Perspective: kinds of trait-mediated indirect effects in ecological communities. A synthesis

THOMAS W. SCHOENER and DAVID A. SPILLER

Department of Evolution and Ecology and Center for Population Biology,
University of California – Davis

Introduction

Our assessment of the importance of trait-mediated indirect effects (TMIEs) relative to density-mediated indirect effects (DMIEs) in ecological communities continues to rise. It seems only yesterday that the landmark experiment of Beckerman *et al.* (1997) was published, showing that grasshoppers, in the presence of 'disabled' spiders incapable of consumption, were intimidated enough to reduce their feeding activity and thereby maintain strong trophic cascades to the producer level. The first comprehensive review, by Werner and Peacor, appeared 6 years later and summarized numerous studies of TMIEs, including some that were able to calculate the relative importance of trait versus density effects; they concluded 'trait effects are often as strong or stronger than density effects'. A short time before, the same authors published an experiment (Peacor and Werner 2001) on a community consisting of an odonate predator and two competing anuran prey species; the predator consumed and intimidated one of the prey species, reducing its competitive effect on the second anuran species. The impact via the trait effect was 76–86%, as compared to 14–24% via the density effect. A second review by Peacor and Werner (2004) concluded that the effect of predators on prey traits modifies the magnitude of an effect farther along the pathway by 20–90%. In the same year, Schmitz *et al.* (2004) published a review of TMIEs and DMIEs in trophic cascades, whose title gives the conclusion 'Trophic cascades: the primacy of trait-mediated indirect interactions'. Shortly after, Preisser *et al.* (2005) did a meta-analysis of trait and density effects, finding for simple predator–prey interactions that certain trait effects (which they call 'intimidation') were at least as strong as density effects (which they call 'direct consumption'), being 63% versus 51%, respectively; the density effect actually declined along tritrophic cascades, while the trait effect climbed to 85% of the total effect. More recently again, Creel and Christianson (2008) reviewed risk

Trait-Mediated Indirect Interactions: Ecological and Evolutionary Perspectives, eds. Takayuki Ohgushi, Oswald J. Schmitz and Robert D. Holt. Published by Cambridge University Press. © Cambridge University Press 2012.

effects and concluded that they 'can be large, sometimes substantially larger than direct effects'. One of their examples was an elegant experiment by Pangle *et al.* (2007) showing that the risk effects of predatory water fleas on the population growth of their zooplankton prey were more than seven times larger than the consumptive (direct predation) effects. The only damper on the increasingly rosy assessment of TMIEs is from Abrams (2010) who writes 'empirical techniques used to compare the magnitudes of behavioural and non-behavioural responses to predation are likely to have overestimated the behavioural component'.

It is interesting to ask why trait effects might be expected in theory to be so large in comparison to density effects. Werner and Peacor (2003) argue especially for behaviour that trait changes are rapid so exert their influence over most of the time course of the interaction; density changes gradually, so its impact is typically weak at the beginning and is transmitted only in proportion to the individuals removed, not the entire population. This reasoning is so apparent a posteriori that it makes us wonder why the possible importance of trait effects was not recognized longer ago than it was. We think in part the answer is that the importance *was* recognized, but with respect to different kinds of traits than behavioural ones. One kind of trait change comprises the (nonlethal) effects of parasites or pathogens on their hosts, and these are well known to be frequent and often severe (e.g., Long and Hay, this volume). Such non-density changes must often have a major impact farther down the effect pathway. Indeed, expansion of the domain of traits beyond the behavioural allows inclusion of many other kinds of trait effects. The many cases of induction of plant defences by herbivores are all candidates for food-web effects (e.g., Ohgushi, this volume). More generally, Miner *et al.* (2005) document the breadth and commonness of plasticity, which they define as 'the production of multiple phenotypes from a single genotype' in ecological systems. However, in our enthusiasm to add to the catalogue of trait effects, we may sometimes go too far. For example, Werner and Peacor (2003) include as a kind of habitat or space-use trait change (see below for terminology) the subtidal study on apparent competition by Schmitt (1987), in which experimentally introducing a preferred prey caused a decrease in a non-preferred prey because of the increased immigration of predators – is this a trait change or a density change or something in between? Whatever the finer points, we have now accumulated in one place or another numerous examples of trait changes and their ecological effects. This sets the stage to ask three questions: (1) what are the axes along which trait-mediated effects can be classified? (2) what is the relative commonness and importance of the various kinds of effects? and (3) what kinds of effects are well covered by theory versus stand as grist for the theoretician's mill (sensu Paine 1988)? Our short synthetic chapter cannot come close to providing complete or even systematic answers to these questions; rather, we sketch out a framework taxonomy and hang some examples on it, many from this volume, and then give a few implications.

Table 2.1 *Three-way classification of trait-mediated indirect effects*

Kinds of traits	Trait-change mechanisms	Interaction web topologies
Feeding	Behavioural plasticity	Tritrophic cascades (Fig. 2.1)
Space use/habitat selection	Developmental plasticity	Consumptive competition/apparent competition (Fig. 2.2)
Physiology	Within-generation phenotype selection	Three-species webs with non-trophic links (Fig. 2.3)
Morphology	Evolution	Webs with four or more species (Fig. 2.4)
Life history		

We would like to distinguish three axes (Table 2.1) for the purposes of discussing the classification of TMIEs: (1) kinds of traits; (2) trait-change mechanisms; and (3) interaction-web topologies through which the effects of trait and density changes are propagated.

Our list of kinds of traits is a variation on the taxonomies of Beckerman *et al.* (2010). The list begins with the behavioural traits of feeding and space use/ habitat selection. Next is physiology, such as changes in growth rate; induced chemical plant defences are included here, although they can accompany induced morphological defences and so might go into the next category. This is followed by morphology, such as body size and defensive structures. Life history completes the list; an example is opportunistic changes in the proportion of dispersing versus sedentary forms (van Veen *et al.*, this volume).

Problematic is what to do with genetic 'traits'. A variety of studies (see Mooney and Singer, this volume) show that genotype variation influences the transmission of effects in food webs. Such differences may signify phenotypic differences; however, Mooney and Singer write 'for most plant taxa, the traits varying among genetic types that are responsible for such effects are unclear'.

Abrams in his pioneering 1995 paper gave four trait-change mechanisms: (1) behavioural plasticity, (2) developmental plasticity (this presumably included immediate homeostatic adjustments as well as longer term developmental processes, although the former could be considered a separate mechanism), (3) within-generation phenotypic selection, and (4) evolution (Table 2.1). The two kinds of behavioural traits, feeding and space use, can of course change because of behavioural plasticity, and indeed this is probably vastly commoner than the other possibilities. But behavioural traits can change as a result of the somewhat slower developmental adjustments as well as through evolution and within-generation selection. The same can be said of all the other trait kinds listed in Table 2.1 (although for genetic 'traits' only evolution can change them, true by definition if evolution is a change in gene frequency). But while the latter three trait-change mechanisms completely cross with trait kinds, behavioural mechanisms do not. One of Abrams' (1995) categories, within-generation phenotypic selection, has not lasted in the literature; indeed, the chapter of Abrams et al. in the Polis and Winemiller 1995 book on food webs does not mention this category. The apparent reason for its existence is that selection may occur without resulting in evolution. Within-generation episodes of such selection can be very strong and result in substantial changes in a population's distribution of phenotypes (e.g., Losos et al. 2006) that could have major direct and indirect ecological effects. But we think such changes if not codified by evolutionary change are likely to be relatively unimportant. We were able to find no case of a TMIE being ascribed to this trait-change mechanism. Likewise, to our knowledge there are no known real-time studies (as mechanisms 'in action') of evolutionary TMIEs, although our own research group (Kolbe, Leal, Losos, Schoener, Spiller) is working on this now with lizards. The obvious reason for their lack is that, despite the many instances of rapid evolution (Reznick and Ghalambor 2001), evolution is still not as rapid as behaviour or development; it takes a matter of generations, so for higher organisms, years, to study the possible existence of evolutionarily caused TMIEs.

The final axis is interaction-web topologies. We use this term after Ohgushi (2005, this volume) to refer to networks of species that can have both trophic and non-trophic (e.g., Borer et al. 2005) links. This distinguishes such webs from food webs, which only have trophic links (who eats whom). Figures 2.1–2.4 give the kinds of interaction webs we have found so far with reasonable empirical support. Many of these cases are in the first such compilation, Werner and Peacor's (2003) Figure 1, and we have added some others. Our goal is not to give an exhaustive list but rather to discuss the material in this volume with respect to interaction-web topology; we have included selected other examples where appropriate. Counts of the frequency of studies of the various types, as well as comparisons of the magnitudes of the respective effects, are left to a future consortium of graduate students.

Are there other interesting axes than these three? One might be kind of system, habitat or place. For example, Long and Hay (this volume) cover the marine system, and state that most studies were done in shallow intertidal habitats, probably more a matter of feasibility than anything else. The terrestrial system is well covered in this volume (Ohgushi, Mooney and Singer, van Veen), but the freshwater system is not. It is simply impossible in a finite volume to cover everything, even well-studied systems such as lentic and lotic ones. To some extent, coverage of system is perhaps not critical: Schmitz et al. (2004) tabulate TMIE studies in cascades and say that the distribution among systems and subsystems (lotic, lentic, rocky intertidal, terrestrial old fields, tropical forest, agricultural) is about as broad as that covered by food-web studies as a whole. However, Preisser *et al.* (2005) found for feeding and habitat that marine and freshwater trait effects were stronger than terrestrial ones; more such differences may emerge as more studies are done. And even within systems, studies of TMIEs may be restricted to a small number of species: Mooney and Singer (this volume) say that 'only 10 species have been investigated for intraspecific variation in enemy density and herbivore–enemy interactions, and four of these are within a single plant family'. Doubtless as more species are studied, more axes, e.g., body size (Rudolf, this volume), or rarity versus commonness, will emerge as interesting.

Kinds of traits
We now discuss the five kinds of traits listed in Table 2.1.

Feeding
Most studies of TMIEs in which an intermediate species' effect on a lower level species is reduced by predators involve a kind of intimidation, in which the intermediate species spends more time hiding or otherwise avoiding the predator and less time feeding. The classic experiment of Beckerman *et al.* (1997) mentioned above is of this type. So are many of the examples in the first review of TMIEs (Werner and Peacor 2003) and about half (9 of 20) of Schmitz *et al.*'s (2004) cases of TMIEs in trophic cascades involve feeding behaviour. The consequences of 'fear' are highlighted in the review of Preisser *et al.* (2005). Other cases are mentioned by van Veen *et al.* (this volume).

A second important class of studies in which feeding rates are reduced involves parasites. For example, parasitized periwinkles (*Littorina littorea*) have their fitness reduced because of lower grazing rates on their algal food (Long and Hay, this volume; the parasites also consume their gonads).

Space use/habitat selection
Another kind of trait commonly implicated in TMIEs involves behavioural shifts in space use. Shifts from one kind of foraging habitat to another are especially common: half of Schmitz *et al.*'s (2004) trophic cascade cases (10 of 20) involved habitat shift. While such shifts may have deleterious effects on the newly consumed food species, the species originally consumed will have their populations increased (barring certain loops in the web). A recent example involving habitat shift is that of elk in Yellowstone National Park; upon reintroduction of wolves, elk shifted away from stands of young aspen because those areas were preferred by wolves (Ripple *et al.* 2010). The result was faster growth in the young aspen, and the TMIEs even extended to properties of nutrient cycling: another example of an ecosystem TMIE, change in nitrogen mineralization, resulted from a shift in grasshopper habitat in the presence of spiders (Schmitz *et al.* 2010).

The above examples all involved linear, tritrophic webs (Fig. 2.1a), but an interesting case (Soluk and Collins 1988) involves consumptive (exploitative) competition between two predators for their mayfly prey. When each predator was alone, neither sculpin nor stoneflies had much of an effect on their mayfly prey. However, when together, predation by stoneflies caused the mayflies to increase their presence on the tops of rocks, making them vulnerable to sculpin. This kind of nonadditive effect of several predators is fairly commonly observed in space, but it may also occur with respect to time periods (Piovia-Scott et al. 2011); in both kinds, there is no escape once the number of predator species is sufficiently high.

A second class of examples of change in space use, one that does not necessarily involve selecting a different kind of habitat on the part of the prey, is the increase of dispersal likelihood in the presence of a predator. In a complex interaction web reviewed by van Veen et al. (this volume), primary parasitoids increased the dispersal rate of their aphid prey (in part by inducing a life-history change). The web also contained secondary parasitoids, and their presence was hypothesized to increase the dispersal rate of primary parasitoids. As mentioned in relation to the Schmitt (1987) study of apparent competition, his second class may not constitute pure examples of TMIEs, because in the end, what changes is the local density of one or more species. There are two ways to change density, one being by births and deaths and the other by immigration and emigration. Although in a sense it doesn't make much difference what one labels a phenomenon, if one is counting it may be somewhat unfair to classify the second way purely as a trait change rather than both a trait and density change.

Physiology

The chemical composition of plants is probably the most studied example of this kind of trait. First, and at the simplest level, herbivores can consume plant tissue and sequester some chemical that will reduce the likelihood of consumption by their predators. A famous example is the monarch butterfly receiving cardiac glycosides from the milkweed plant (Brower et al. 1967), but similar examples exist in which alkaloids from endophytes are transmitted to the plant and then to herbivores and predators, both of which are negatively affected (de Sassi et al. 2006; Finkes et al. 2006). For the most part, however, herbivores begin the chain of effects via induction of the defence in their plant food, and such effects can propagate through the community web. Second, Mooney and Singer (this volume) discuss the slow-growth/high-mortality hypothesis, in which resistance in a plant species slows the growth of its herbivores, in turn increasing their susceptibility to predators. Other examples of this nature are given by Ohgushi (this volume). In these possibilities, plants affect herbivores which in turn affect predators, constituting one case of Mooney and Singer's two cases of TMIEs. Their second type of TMIE is that plants can affect predators which then affect herbivores. Mooney and Singer cite the study of Ness *et al.*, in which carbohydrates provided to ants by nectaries increase aggression of the latter toward herbivores.

Changes in growth rate leading to a TMIE are not limited to the terrestrial system. A result of intimidation (see 'feeding' section above) can be a slower growth rate, as in Stallings *et al.*'s study of groupers discussed by Long and Hay (this volume): in the presence of the Nassau grouper, smaller grouper species hid more and grew more slowly. Slower growth may then affect the time spent feeding on certain food species.

Morphology

Changes in morphological traits per se appear not nearly as commonly involved in TMIEs as are any of the first three trait kinds just discussed, although morphological changes, defensive or otherwise, can accompany changes in chemical defences in plants. These may lead to TMIEs: for example, Ohgushi (this volume) relates the case of a stem-boring moth of willow that induces growth of lateral shoots, which have greater nitrogen and water content, thereby positively affecting other herbivores as well as their predators.

Raimondi *et al.* (2000) say they have reported the first example of a TMIE via induction of a polymorphism: a predatory whelk causes a developmental conversion in barnacles from a vulnerable (conic) to a predator-resistant (bent) form. The indirect effect on encrusting algae was relief from competition for space, in part because of the diminishment of conic forms.

A second kind of morphological trait that may change is body size. Van Veen *et al.* (this volume) review the work of Bukovinsky *et al.* on a web involving aphids, a primary parasitoid, and a secondary parasitoid. The genotype of the plant at the base of this web affected the body size of the aphids, which in turn affected the body size of the primary parasite. The chain of effects proceeded to the top of the web: secondary parasitoid species increased their emigration with smaller primary parasitoids if they were generalized enough to find alternative prey. Volker (this volume) explores a variety of theoretical expectations for size-structured populations.

Life history

Examples of life-history trait changes involved in TMIEs appear rare. Aphids have the ability to develop into a winged form, which is obviously adapted for dispersal. In the presence of parasitoids this can happen, ultimately resulting in a reduction of the local density of aphids and consequently a reduction in their effect on plants (van Veen, this volume).

Trait-change mechanisms

Of the four kinds of trait-change mechanisms listed in Table 2.1, only the first two have been studied in real time. Selection within a generation is completely unstudied with regard to TMIEs, even though it is relatively fast. The second unstudied mechanism, evolution, can be fast (Schoener 2011) but not nearly as fast as the others. Its product, genetic differences between populations subjected to different histories, has many retrospective (Losos 1994) studies with regard to the kinds and magnitudes of various TMIEs (e.g., Mooney and Singer, Shuster et al., this volume).

Both behavioural and developmental trait-change mechanisms can be considered types of plasticity. So ignored are the evolution and selection mechanisms that Peacor and Cressler (this volume) state, 'phenotypic plasticity in traits … is the underlying mechanism for [TMIEs]', repeating an attitude held in a previous paper (Werner and Peacor 2003).

It is difficult to compare the two plasticity mechanisms – behaviour and development – as agents for trait changes involved in TMIEs, because their domains are so different. Miner et al. (2005) claim that 'to date, most research on indirect interactions via plasticity has examined anti-predator behaviour'. Certainly there are a lot of studies that involve behavioural trait-change mechanisms. Such studies can be as simple as adding to a medium chemicals from predators; fright responses are almost instantaneous (Long and Hay, this volume), and rapidly attainable publishable research is a powerful desideratum for professors and students alike. The classic study of spiders and grasshoppers by Beckerman et al. (1997) involved a behavioural mechanism. Studies in which TMIEs are implicated in trophic cascades (compiled by Schmitz et al. 2004) all involve behavioural mechanisms. To repeat in a different way what we said above, no study of feeding or space use/habitat selection involves anything other than a behavioural mechanism. This observation is not a tautology, however: changes in behavioural traits can come about through developmental or evolutionary mechanisms, but such changes have not been documented in relation to TMIEs.

The second trait-change mechanism involving plasticity, development, gives behaviour a fair run for its money in terms of commonness, largely because of the many studies of induction of defences in plants by herbivores. Examples are not only found in studies of cascades such as those reviewed above, they also can commonly occur in other kinds of food webs. Long and Hay's (this volume) discussion of 'trait-mediated grazer–grazer interactions' points out that Denno and Kaplan's (2007) review finds that over half the examples of competition between herbivorous insects are mediated by inducible plant traits. Such interactions can sometimes be asymmetrical, as is true of density-mediated competitive interactions (Persson 1985). For example, prior feeding on cordgrass by a planthopper delayed development time, reduced survivorship and reduced adult size of a second planthopper species. Mediation by induced change can also occur in marine systems: in studies reviewed by Long and Hay, the seaweed *Fucus* had its palatability to other species reduced by one species of grazing periwinkle but not another. Induction by a particular herbivore species can also have positive effects on other species. An example, discussed above, is the case of a stem-boring moth of willow that induces growth of lateral shoots having greater nitrogen and water content, which then has positive effects up the food web.

While induced plant defences are by far the most commonly studied, other kinds of developmental trait-change mechanisms involved in TMIEs are occasionally documented. The most famous is the case discussed above of whelk predation on barnacles causing the development of a more predator-resistant morph, which then has consequences for space competitors of the barnacles. The case of aphids attacked by parasitoids, also discussed above, in which a winged form increases in frequency and migrates away from the local site, is a second of these rare examples. Such developmental changes may be genuinely relatively rare or an artefact of study because they are slower than behavioural ones on average.

Finally, immediate homeostatic adjustments can change aspects of an individual's physiology and morphology (e.g., body temperature, body colour) over periods of time comparable to behavioural changes. These could conceivably have food-web consequences. For now, however, immediate homeostatic changes are considered in the developmental category, as no one to our knowledge has come up with an example relating them to a TMIE.

Interaction-web topologies
Tritrophic Cascades

Figure 2.1 gives four kinds of tritrophic cascades (the top species eats the middle species which eats the bottom species) according to where a single

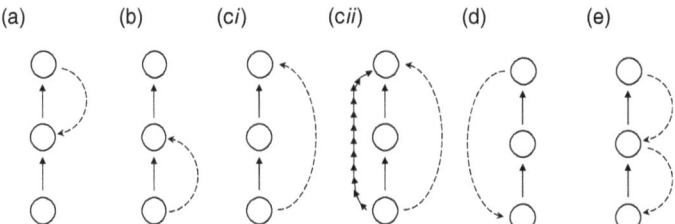

Figure 2.1 Tritrophic cascades having different kinds of trait-mediated indirect effects. Trophic links are indicated by solid arrows (the arrow points toward the consumer). Trait links are indicated by dashed arrows from the effecting to the affected species. Non-trophic density effects are indicated by carets pointing towards the affected species.

trait-mediated effect is located. There are four possibilities: the top species affects the middle species (a), the bottom species affects the middle species (b), the bottom species affects the top species (c), and the top species affects the bottom species (d).

The most abundantly and universally documented case is Fig. 2.1a, the top species affects a trait of the middle species. The classical Beckerman et al. (1997) experiment discussed above is of this type, in which the top predator affects the middle species' feeding activity. A second such relatively early example is Peacor and Werner's (2001) study in which odonate predators caused their tadpole prey to forage less, thereby increasing abundance of the tadpole's algal food species. The three-species cases in Schmitz et al.'s (2004) review fall here; they involve both feeding and space-use/habitat traits. Ripple et al.'s (2010) example of elk shifting habitat in the presence of wolves is another case of the latter. In the present volume, Long and Hay, Mooney and Singer and van Veen et al. give evidence for this type of TMIE; see also Werner and Peacor (2003).

In Fig. 2.1b, a trait effect on the middle species from the bottom species in turn affects the top species. The acquisition of chemical defences by a herbivore from its food species, which then affects the top species, illustrates this case, and we discussed several examples above. A somewhat less obvious example is from Anholt and Werner (1995): lowering the food supply available to anuran larvae increased mortality from their odonate predator due to the anuran's increased feeding activity.

In Fig. 2.1ci, the bottom species affects the top species, which in turn modifies its predatory effect on the middle species. Holt et al. (2010) model how plants can influence the foraging of predators, and this in turn affects the herbivores, for example, lions forage better in tall grass, thereby causing wildebeest abundance there to decline (in turn causing the grass to increase). A second example is given by Mooney and Singer and applied to their Case 4 (this volume): Ness et al. found that plant nectaries increased the carbohydrates available to mutualistic ants, increasing ant aggressiveness toward herbivores.

A variation on this topology (Fig. 2.1cii) involves a so far uncovered effect, one that affects density but is not trophic (Pearson 2010). An invasive plant provides substrata for spiders, increasing their densities and allowing larger webs (a trait change). The spiders in turn affect a galling dipteran of the plant. In contrast to most cases, the trait effect here is much larger than the density effect (measured on both the dipteran and plant), perhaps unsurprising for these 'ecosystem engineers'.

We know of no examples of Fig. 2.1d, the top species affecting a trait of the bottom species, although Fig. 2.4h has an example of the top species affecting the second-to-bottom species.

Finally, several different effects on traits may occur. Fig. 2.1e gives an example of what Long and Hay (this volume) call a 'trait cascade'. They cite Reynolds and Sotka's data showing that amphipods reduce feeding on *Sargassum* in the presence of cues from their predatory pinfish, and that this in turn affects *Sargassum* palatability; amphipods preferred seaweed exposed to pinfish.

Trait cascades bring up the question of when a trait effect can be said to be involved in a TMIE. In Werner and Peacor's (2003) original taxonomy, two conditions had to be satisfied for a TMIE: first a species affects a trait of another species, and then that latter species affects a third species, completing the indirect effect. Implicit in this scheme seems to be that the second effect is on density, but it could also be on a trait as in the example above. Whatever the case, such a second trait effect might travel farther and affect other species densities or traits. However, not satisfying Werner and Peacor's second requirement (nor Abrams *et al.*'s 1995 definition) would be that the trait changes in Fig. 2.1a and b could go in the other direction than that shown, because this would not lead to a third species (except through further changes in the intermediate species). In contrast, the cases shown in Fig. 2.1c and d are in fact identical topologies except that the trait effects go in opposite directions.

Consumptive (exploitative) competition and apparent competition

Figure 2.2 illustrates two other kinds of three-species webs having TMIEs, consumptive competition (Schoener 1983) and its 'inverse', apparent competition (Holt 1977).

As reviewed for inducible plant defences, cases in which the interaction between two consumers is affected by a trait change are common in consumptive competition: trait effects can be both negative and positive (Werner and Peacor 2003). Peacor and Werner (1997) showed that in the presence of a caged odonate predator, a larval anuran reduced activity, in turn reducing the predation rates of a second odonate species. The 'no escape' example for mayflies in the presence of two kinds of predators (discussed above) fits here.

TMIEs in apparent competition seem to be less common than in consumptive competition, as apparently is the former interaction in general. Werner and Peacor (2003) point out that the two prey species can have negative effects or

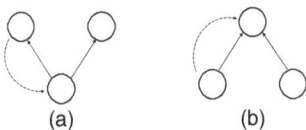

(a)   (b)

Figure 2.2 Consumptive competition and apparent competition having trait-mediated indirect effects. Trophic links are indicated by solid arrows (the arrow points toward the consumer). Trait changes are indicated by dashed arrows from the effecting to the affected species.

positive effects on one another when trait change is involved. Negative effects are expected to be more typical. An example of a negative effect is the study of Harmon *et al.* (2000), who showed that dandelions, a pollen source for cocci-nellids, increased predation upon aphids by increasing the coccinellids' resi-dency in areas with aphids. An example of a positive effect is the experiment of Huang and Sih (1990), who showed that fish predation on a salamander was weaker in the presence of an isopod – an alternative prey – than in its absence. The mechanism was that fish were more active when isopods were available, reducing the emergence rate of the salamanders from refuges and thereby lowering their vulnerability.

Three-species webs with non-trophic links

Figure 2.3 gives three kinds of topologies for three-species interaction webs in which not all species are connected by trophic (food-web) links. When not so connected, a species pair can be linked by either a trait-mediated effect, or a non-trophic interaction (which in the cases here are types of competition other than consumptive), or both.

In Fig. 2.3a, a predator–prey pair is also linked by a trait-mediated effect from predator to prey; in addition, the prey is linked by space competition to another species. An example of this is the study of Raimondi *et al.* (2000, and above): a predatory whelk causes a developmental shift in morphology of its barnacle prey, which in turn affects its space competition with encrusting algae. The case is similar to one cited by Rudolf (this volume), in which small perch shift habitat to avoid cannibalism by large perch, and this increases dietary overlap (and presumably consumptive competition) with species in the new habitat (note that we are not diagramming as separate cases webs in which interactions occur between classes within a species).

In Fig. 2.3b, a predator–prey pair has its interaction affected by a non-prey species (for cases where the third species is also a prey species, see apparent competition above). Long and Hay (this volume; see also van Veen *et al.*) cite

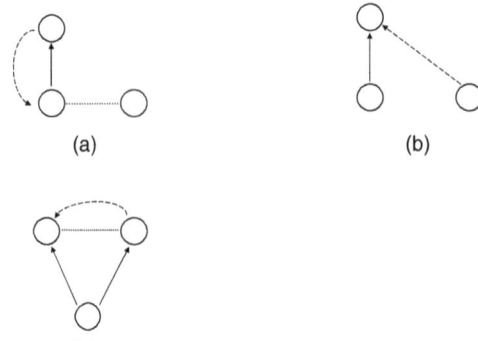

(a)

(b)

(c)

Figure 2.3 Three-species webs with non-trophic links having trait-mediated indirect effects. Trophic links are indicated by solid arrows (the arrow points toward the consumer). Trait links are indicated by dashed arrows from the effecting to the affected species. Non-trophic links are indicated by dotted lines.

Griffen and Byers' study of an alga decreasing mortality of a prey crab from a predatory crab by more than 50%.

In Fig. 2.3c, two predators of a common prey interfere with one another in addition to competing consumptively. This is case 'g' of Werner and Peacor's (2003) Figure 1. They cite as a possible example the study of Stelzer and Lamberti (1999), in which interference from a fish reduces the impact of crayfish on invertebrate prey (presumably by altering the foraging behaviour of the latter). In this case, the interference has a behavioural effect, so initiates the TMIE rather than terminates it as in Fig. 2.3a. However, it is possible in this example as well as others that interference can occur without a trait-change component, for example mortality from encounter or even from competing for space (called pre-emptive in Schoener 1983 and having components of both exploitative and interference competition). We have illustrated the dual nature here with two kinds of links between the competitors.

Webs with four or more species
Figure 2.4 illustrates a variety of webs with four species, as well as one web with five species.

The simplest is the four-species cascade (Fig. 2.4a). In their review, Schmitz *et al.* (2004) were able to find two such examples, both involving habitat shifts of

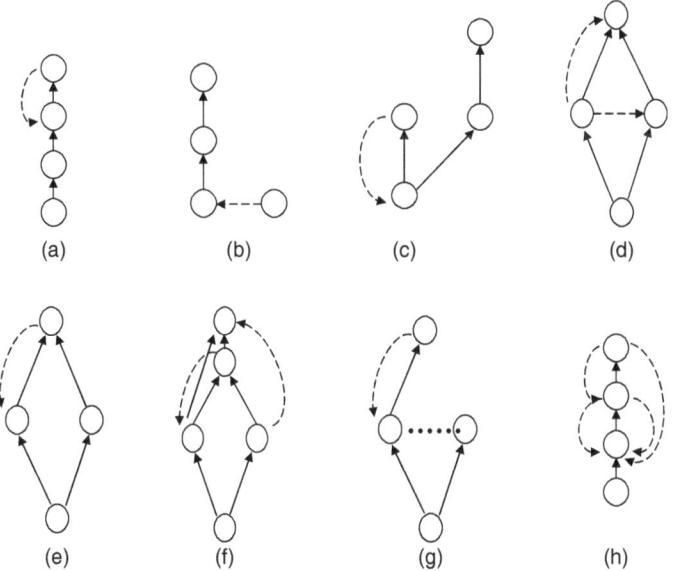

Figure 2.4 Webs with four or more species having trait-mediated indirect effects. Trophic links are indicated by solid arrows (the arrow points toward the consumer). Trait links are indicated by dashed arrows from the effecting to the affected species. Non-trophic links are indicated by dotted lines.

the trophically higher intermediate species (now there are two intermediate species). In Carpenter *et al.*'s (1987) study, adult bass removal resulted in strong bass recruitment, and the younger bass caused minnows to shift habitat to shallower water, thereby releasing their zooplankton prey from predation and decreasing algal abundance. Note that the effect on algae is the opposite of that expected were this a simple density-mediated chain of adult bass removal.

Another class of a simple linear chain involves endophytic bacteria, but the effect is bottom-up, and it begins with a non-trophic link (Fig. 2.4b). Ohgushi (this volume) reports that nitrogen-fixing soil bacteria transmit their positive effects to the plant, and this then has positive effects on herbivores and in turn their predators.

More complicated pathways exist in which bottom-up effects occur, one of which is illustrated in Fig. 2.4c. Ohgushi (this volume) reports the case of a stem-boring moth on willow, whose herbivory induces the willow to produce more palatable shoots (with higher nitrogen and water). This trait-mediated effect is then transmitted as a positive input to other herbivores and from them to predators. Thus in this example the induction does not stop with the moth–willow interaction, but goes on to other species, producing the indirect effects. Ohgushi summarizes similar examples, in some of which plant quality is increased for other herbivores and in others plant quality is decreased.

Ohgushi (this volume) gives another four-species web, that of Martinsen *et al.*, involving two kinds of herbivores, but here there are two trait-mediated effects (Fig. 2.4d). A leaf-rolling moth provides shelter for both predators and other herbivores of a plant species. This gives two trait-mediated arrows; both predators and prey are increased in the presence of the moth (although this need not be the case for other webs having this topology, since if the predators are increased enough, a net decrease in the herbivores may occur).

In their review, Werner and Peacor (2003) discuss cases of 'diamond-shaped' webs, in which one top predator feeds on two intermediate predators which compete for a common resource. One such case with two trait effects has just been discussed (Fig. 2.4d). In their own study (Fig. 2.4e), Peacor and Werner (1997) included two anurans, one of which (larval bullfrogs) becomes a better competitor than the other (larval green frogs) in the presence of the top predator (an odonate). The mechanism is a trait change: bullfrogs have their activity reduced less (only one trait link is shown in Fig. 2.4e). In a very different system, Abramsky *et al.* (1998) showed that owls caused a smaller species of gerbil to shift away from places where the owls occurred; this in turn influenced competitive interactions between the smaller gerbil species and a larger gerbil species.

Four-species 'diamonds' can be made into five-species webs by adding an additional predator at the top (Fig. 2.4f), and again Werner and Peacor (2003) give possible applications. For the four-species aquatic example just discussed, adding a fish predator changes effects on lower-level species; moreover, because fish feed on the invertebrate predators, this adds yet another trophic link to the web.

Another example of a 'diamond' web, but in this case an imperfect one, is also provided by Werner and Peacor (2003). As illustrated (Fig. 2.4g), a phorid fly is a parasitoid on one species but not the other species of ant, and this changes the dominance hierarchy during (interference) confrontations at baits. When the dominant species was a fire ant, presence of its parasitic fly caused fire ants to go under objects or into the ground, leading to dominance by other ants (Orr *et al.* 1995).

The final, most complicated example is illustrated in Fig. 2.4h and has three trophic links and four trait-mediated links (according to the authors, van Veen *et al.*, this volume). Aphids eat plants and are parasitized by a primary parasitoid, which in turn is parasitized by a secondary parasitoid. In the presence of the primary parasitoid, the aphid decreases feeding. In the presence of the secondary parasitoid, the aphid increases reproductive rate. It is speculated that volatiles from secondary parasitoids increased the feeding rate of the aphids, perhaps indicating a relatively favourable site because a dispersing aphid has to pay the cost of establishing a new feeding site by inserting their mouthparts into the host plant, a process that takes time. In addition, the secondary parasitoid causes increased emigration out of the system on the part of the primary parasitoid, and the primary parasitoid has the same effect on aphids (so two trait links from the second to the third species here). However, as noted above, increasing emigration may be better considered a density effect.

Future research directions: what do we not know?

The previous sections reviewed what is known empirically about TMIEs with respect to three axes: trait type, mechanism of trait change and interaction-web topology. Such a survey, however incomplete and specifically directed toward the work reviewed in this volume, does set something of a stage to ask what kinds of empirical data are missing, i.e., where are the gaps along the axes? Of course, the gaps may represent a real lack of representation in nature rather than a paucity of studies directed toward them. With that caveat, here are some empirical lacunae.

Three kinds of traits – feeding, space use/habitat selection and physiological – dominate real time studies of TMIEs. The first two are about equally common, whereas the third is itself dominated by studies of herbivore-induced defences. Few such studies exist on morphological or life-history traits. We suspect this imbalance is real; empirical studies tend to cover only short time periods, but morphological and life-history traits can be developmental, only somewhat longer than the shorter times for feeding or space-use trait change. Developmental traits are therefore quite feasible for study, and indeed physiological trait changes are commonly studied.

Of the four kinds of trait-change mechanisms, only two – behavioural and developmental – are represented at all in the literature; the evolutionary/ selection mechanisms for TMIEs are entirely missing. Because the behavioural mechanism is by far the commoner of the first two, we suspect that time period largely determines study frequency here. Whether the less represented mechanisms will turn out to be important remains to be seen, although we suspect (see above) that the within-generation selection mechanism will not.

Representation of the various kinds of web topologies in studies of TMIEs probably parallels representation of those webs in the literature as a whole. Most conspicuously, the larger the web, the fewer examples it is likely to have in the literature. This must in part be a matter of the difficulty of ascertaining the major pathways, the larger the web (we will return to this below). Webs with more species have more possible configurations as well as connections, and this may also be a factor. The trophic cascade, especially for three species, seems the commonest example, and there are no examples of studies of TMIEs for cascades of more than four species; in addition to the complexity of validating pathways, food webs tend to be surprisingly short (Briand and Cohen 1987; Schoener 1989). Webs with small numbers of species not in a cascade arrangement are studied much less. This is surprising with respect to consumptive (exploitative) competition, but perhaps not so surprising with respect to apparent competition, as the first experiment on this interaction was only performed a little over 20 years ago (Schmitt 1987).

From the theoretician's perspective, extensively documented kinds of traits, trait-change mechanisms and web topologies would constitute at least potential areas of investigation. Should theory be lacking for some common kind of empirical phenomenology, the theoretician is at least assured that some empiricists will pay attention to their efforts. From the other side, lack of data about certain kinds of phenomena – gaps along the axes – could be due either to such kinds being genuinely rare, or just that they are not yet much studied. Here the theoretician must be more cautious: at the worst, theory could be produced that will never apply to any data, and at best, theory may be produced on difficult-to-study systems whose eventual application lies far in the future, perhaps well beyond the theoretician's lifetime, giving no earthly (e.g., funding) reward. Peacor and Cressler (this volume) highlight several areas for which theory is incomplete or absent. To their list can be added Rudolf's (this volume) desiderata for size-stage-structured species populations.

Peacor and Cressler (this volume) point out that the dimension (number of species and links) needs to be expanded upwards. Doing so, however, could be a very daunting task. To get our heads around just how daunting, it is useful to consider the theoretical treatment of food webs using loop analysis (Yodzis 1988; Schoener 1993; Holt *et al.* this volume). In loop analysis, minus the inverse of the community matrix (the S × S matrix of partial derivatives of the population-growth differential equation of each species with respect to each species) specifies the net effects of species on each other. Yodzis concluded that the direction and magnitude of effects in real food webs were often indeterminate. In some ways Yodzis's conclusion was too pessimistic (Schoener 1993; Schmitz 1997), yet his analysis contained a great simplification: the only non-zero terms in the community matrix were those for self effects (the main diagonal) and those for predator–prey effects (the adjacent two diagonals). If trait-mediated effects occur, however, other non-zero terms will exist, and this can further (and often vastly) complicate the expression for the inverse's elements, as the inverse matrix terms are composed of determinants (of order S or S – 1) of the community matrix elements. Peacor and Werner (2004) give a specific example for a three-species cascade in which the intermediate species has its feeding rate dependent on predator density. Because the coefficient for the term describing the consumption of the bottom species by the intermediate species is a function of the top species' density, the derivative of the bottom species' growth equation with respect to the top species' density is non-zero. This looming pessimism is somewhat countered by the more mixed message of Peacor and Cressler (this volume) that theoretical evidence for the need to include trait modifications is equivocal. Whatever the case, theory has been almost exclusively restricted to three-species systems, and even there mostly to cascades. What is the extent of theoretical possibility for the four- and five-species webs of Fig. 2.4? Furthermore, webs in which not all the species are trophically linked seem entirely ignored by theory. There is plenty of grist for the theoretician's mill among interaction webs with TMIEs that are known to exist in nature, enough to support a number of cottage industries.

Apparently both for theory and data there is plenty of work yet to be done.

References

Abrams, P. (1995) Implications of dynamically variable traits for identifying, classifying, and measuring direct and indirect effects in ecological communities. *American Naturalist*, **146**, 112–134.

Abrams, P. A. (2010) Implications of flexible foraging for interspecific interactions: lessons from simple models. *Functional Ecology*, **24**, 7–17.

Abrams, P. A., Menge, B. A., Mittelbach, G. G., Spiller, D. A. and Yodzis, P. (1996) The role of indirect effects in food webs. In G. A. Polis, and K. O. Winemiller, eds., *Food Webs: Integration of Pattern and Dynamics*. New York: Chapman and Hall, pp. 371–395.

Abramsky, Z., Rosenzweig, M. L. and Subach, A. (1998) Do gerbils care more about competition or predation? *Oikos*, **83**, 75–84.

Anholt, B. R. and Werner, E. E. (1995) Interaction between food availability and predation mortality mediated by adaptive behavior. *Ecology*, **76**, 2230–2234.

Beckerman, A. P., Uriarte, M. and Schmitz, O. J. (1997) Experimental evidence for a behavior-mediated trophic cascade in a terrestrial food chain. *Proceedings of the National Academy of Sciences of the United States of America*, **94**, 10735–10738.

Beckerman, A., Petchey, O. L. and Morin, P. J. (2010) Adaptive foragers and community ecology: linking individuals to communities and ecosystems. *Functional Ecology*, **24**, 1–6.

Borer, E. T. Seabloom, E. W., Shurin, J. B. *et al.* (2005) What determines the strength of a trophic cascade? *Ecology*, **86**, 528–537.

Briand, F. and Cohen, J. E. (1987) Environmental correlates of food-chain length. *Science*, **238**, 956–960.

Brower, L. P., Van Zandt Brower, J. and Corvino, J. M. (1967). Plant poisons in a terrestrial food chain. *Proceedings of the National Academy of Sciences of the United States of America*, **57**, 893–898.

Carpenter, S. R., Kitchell, J. F., Hodgson, J. R. *et al.* (1987) Regulation of lake primary productivity by food-web structure. *Ecology*, **68**, 1863–1876.

Creel, S. and Christianson, D. (2008) Relationships between direct predation and risk effects. *Trends in Ecology and Evolution*, **23**, 194–201.

de Sassi, C., Müller, C. B. and Krauss, J. (2006) Fungal plant endosymbionts alter life history and reproductive success of aphid predators. *Proceedings of the Royal Society of London, Series B*, **273**, 1301–1306.

Denno, R. F. and Kaplan, I. (2007) Plant-mediated interactions in herbivorous insects: mechanisms, symmetry, and challenging the paradigms of competition past. In T. Ohgushi, T. P. Craig and P. W. Price, eds., *Ecological Communities: Plant Mediation in Indirect Interaction Webs*. Cambridge: Cambridge University Press, pp. 19–50.

Finkes, L. K., Cady, A. B., Mulroy, J. C., Clay, K. and Rudgers, J. A. (2006) Plant-fungus mutualism affects spider composition in successional fields. *Ecology Letters*, **9**, 344–353.

Harmon, J. P. Ives, A. R., Losey, J. E., Olson, A. C. and Rauwald, K. S. (2000) *Coleomegilla maculata* (Coleoptera: Coccinellidae) predation on pea aphids promoted by proximity to dandelions. *Oecologia*, **125**, 543–548.

Holt, R. D. (1977) Predation, apparent competition, and the structure of prey communities. *Theoretical Population Biology*, **12**, 197–229.

Holt, R. D., Holdo, R. M. and van Veen, F. J. F. (2010) Theoretical perspectives on trophic cascades: Current trends and future directions. In J. Terborgh and J. A. Estes, eds., *Trophic Cascades*. Washington DC: Island Press, pp. 301–318.

Huang, C. F. and Sih, A. (1990) Experimental studies on behaviorally mediated, indirect interactions through a shared predator. *Ecology*, **71**, 1515–1522.

Losos, J. B. (1994) Integrative approaches to evolutionary ecology: *Anolis* lizards as model systems. *Annual Review Ecology Systematics*, **25**, 467–493.

Losos, J. B., Schoener, T. W., Langerhans, R. B. and D. A. Spiller (2006) Rapid temporal reversal in predator-driven natural selection. *Science*, **314**, 1111.

Miner, B. G., Sultan, S. E., Morgan, S. G., Padilla D. K. and Relyea, R. A. (2005) Ecological consequences of phenotypic plasticity. *Trends in Ecology and Evolution*, **20**, 685–692.

Ohgushi, T. (2005) Indirect interaction webs: herbivore-induced effects through trait change in plants. *Annual Review of Ecology, Evolution, Systematics*, **36**, 81–105.

Orr, M. R., Selke, S. H., Benson, W. W. and Gilbert, L. E. (1995). Flies suppress fire ants. *Nature*, **373**, 292–293.

Paine, R. T. (1988) Road maps of interactions or grist for theoretical development? *Ecology*, **69**, 1648–1659.

Pangle, K. L., Peacor, S. D. and Johannsson, O. (2007) Large nonlethal effects of an invasive invertebrate predator on zooplankton population growth rate. *Ecology*, **88**, 402–412.

Peacor, S. and Werner, E. E. (1997) Trait-mediated indirect interactions in a simple aquatic food web. *Ecology*, **78**, 1146–1156.

Peacor, S. and Werner, E. E. (2001) The contribution of trait-mediated indirect effects to the net effects of a predator. *Proceedings of the National Academy of Sciences of the United States of America*, **98**, 3904–3908.

Peacor, S. and Werner, E. E. (2004) How dependent are species-pair interaction strengths on other species in the food web? *Ecology*, **85**, 2754–2763.

Pearson, D. E. (2010) Trait- and density-mediated indirect interactions initiated by an exotic invasive plant autogenic ecosystem engineer. *American Naturalist*, **176**, 394–403.

Persson, L. (1985) Asymmetrical competition: are larger animals competitively superior? *American Naturalist*, **126**, 261–266.

Piovia-Scott, J., Spiller, D. A. and Schoener, T. W. (2011) Effects of experimental seaweed deposition on lizard and ant predation in an island food web. *Science*, **331**, 461–463.

Preisser, E. L., Bolnick, D. I. and Bernard, M. F. (2005) Scared to death? The effects of intimidation and consumption in predator–prey interactions. *Ecology*, **86**, 501–509.

Raimondi, P. T., Forde, S. E., Delph, L. F. and Lively, C. M. (2000) Processes structuring communities: evidence for trait-mediated indirect effects through induced polymorphisms. *Oikos*, **91**, 353–361.

Reznick, D. N. and Ghalambor, C. K. (2001). The population ecology of contemporary adaptations: what empirical studies reveal about the conditions that promote adaptive evolution. *Genetica*, **112–113**, 183–198.

Ripple, W. J., Rooney, T. P. and Beschta, R. L. (2010) Large predators, deer, and trophic cascades in boreal and temperate ecosystems. In J. Terborgh and J. A. Estes, eds., *Trophic Cascades*. Washington DC: Island Press, pp. 141–162.

Schmitt, R. J. (1987) Indirect interactions between prey: apparent competition, predator aggregation, and habitat segregation. *Ecology*, **68**, 1887–1897.

Schmitz, O. J. (1997) Press perturbations and the predictability of ecological interactions in a food web. *Ecology*, **78**, 55–69.

Schmitz, O. J., Hawlena, D. and Trussell, G. C. (2010) Predator control of ecosystem nutrient dynamics. *Ecology Letters*, **13**, 1–11.

Schmitz, O. J., Krivan, V. and Ovadia, O. (2004) Trophic cascades: the primacy of trait-mediated indirect interactions. *Ecology Letters*, **7**, 153–163.

Schoener, T. W. (1983) Field experiments on interspecific competition. *American Naturalist*, **122**, 240–285.

Schoener, T. W. (1989) Food webs from the small to the large. *Ecology*, **70**, 1559–1589.

Schoener, T. W. (1993) On the relative importance of direct versus indirect effects in ecological communities. In H. Kawanabe, J. E. Cohen and K. Iwasaki, eds., *Mutualism and Community Organization*. Oxford: Oxford University Press, pp. 365–411.

Schoener, T. W. (2011) The newest synthesis: understanding the interplay of evolutionary and ecological dynamics. *Science*, **330**, 426–429.

Soluk, D. A. and Collins, N. C. (1988) Synergistic interactions between fish and stoneflies: facilitation and interference among stream predators. *Oikos*, **52**, 94–100.

Stelzer, R. S. and Lamberti, G. A. (1999) Independent and interactive effects of crayfish and darters on a stream benthic community. *Journal of the North American Benthological Society*, **18**, 524–532.

Werner, E. E. and Peacor, S. D. (2003) A review of trait-mediated indirect interactions in ecological communities. *Ecology*, **84**, 1083–1100.

Yodzis, P. (1988) The indeterminancy of ecological ineractions as perceived through perturbation experiments. *Ecology*, **69**, 508–515.

CHAPTER THREE

Consequences of trait changes in host–parasitoid interactions in insect communities

F. J. FRANK VAN VEEN

Centre for Ecology and Conservation, University of Exeter

and

H. CHARLES GODFRAY

Department of Zoology, University of Oxford

Introduction

Interactions between species can be direct, for example between predators and prey, or indirect where the effect of one species on another is transmitted by an intermediate species. This book is concerned with indirect effects that are mediated by changes in traits of the intermediate species. In this chapter we review evidence from research on insect host–parasitoid systems for the importance of trait-mediated indirect interactions in determining the dynamics and structure of insect food webs.

Parasitoids, like parasites, require a host organism for much of their development but unlike parasites they need to kill their host to complete development to the free-living adult stage. The parasitoid lifestyle has evolved in a number of insect groups, including the Diptera and the Coleoptera, but has reached its greatest diversity in the Hymenoptera, comprising probably over a million species (Quicke 1997). Their hosts are other arthropods, mainly insects but also spiders, for example. Parasitoid wasps have been popular subjects for biological research, in part because of their abundance and diversity which indicates their ecological significance and utility for biological control, but also because of their lifestyle which makes many aspects of their ecology and behaviour easier to study than those of predators.

Traditionally, there has been a large focus on the behavioural ecology (Godfray 1994) and population dynamics (Hassell 2000) of host–parasitoid interactions but recent decades have also seen them widely utilized for the study of quantitative food web structure (van Veen *et al.* 2006; Ings *et al.* 2009). It is within the main areas of research on insect parasitoids – population dynamics and food web structure – that we focus our review. Given the interest in trait-mediated effects and the wide utilization of host–parasitoid

Trait-Mediated Indirect Interactions: Ecological and Evolutionary Perspectives, eds. Takayuki Ohgushi, Oswald J. Schmitz and Robert D. Holt. Published by Cambridge University Press. © Cambridge University Press 2012.

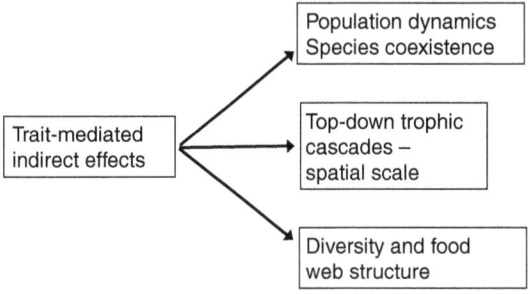

Figure 3.1 The three areas of insect parasitoid population and community ecology where trait-mediated indirect effects have been found to play a significant role. Each area is reviewed in a separate section in this chapter.

systems for research, it is striking that there are relatively few studies of their effects in these systems. A search of Web of Knowledge using the terms 'parasitoid* and trait mediated' results in a mere 25 hits of which only a small proportion actually deal with trait-mediated indirect effects (TMIEs). Nevertheless, as we hope to show, these examples provide clear evidence for the importance of TMIEs in host–parasitoid systems and reveal insights that may be applicable more widely. The main study systems represented in the literature are aphids and their hymenopteran primary and secondary parasitoids, and ants and their dipteran parasitoids, and therefore necessarily this review has a strong bias towards these systems, and especially the former. The main body of this chapter is divided into three parts that cover a broad area of population and community ecology (Fig. 3.1): (1) the stabilizing effects of modifications of the interaction between host and parasitoid by non-host species and its consequences for species coexistence; (2) trait-mediated trophic cascades, or the indirect interactions between non-adjacent trophic levels and their potential impact at the local and the meta-community scale; (3) the effects of plant–plant driven variation in herbivore traits on species diversity and network structure of insect food webs.

Foraging efficiency of parasitoids: interference by non-hosts can stabilize multispecies dynamics

The majority of hymenopteran parasitoids have rather narrow host ranges, especially when compared to the range of prey species that a predator may consume (Godfray 1994; van Veen *et al.* 2008). In coupled host–parasitoid interactions the efficiency with which parasitoids find their hosts is a major determinant of their dynamics. There is an intrinsic propensity for host–parasitoid systems to show divergent oscillations driven by the time lags inherent in the insects' life cycle. High parasitoid densities will lead to lower host populations one generation later and hence fewer opportunities for parasitoid reproduction; this allows the host population to rebound and so, after another lag, renewed opportunities for parasitoid population growth. These oscillations will increase in amplitude until either the parasitoid or both the host and the parasitoid go extinct (Nicholson and Bailey 1935), unless one or both of two broad classes of mechanisms are present to stabilize the interaction: either there must be a refuge of some sort to allow the host population to survive through periods of high parasitoid population density, or some process must exist to reduce the efficiency of the parasitoid population when it is at high density (Hassell 2000). This theory is based on host–parasitoid pairs interacting in isolation from the wider community. As we will discus in this section, some form of interference with the parasitoid's foraging behaviour by other species in the community may represent a class of interaction modification that has the potential to stabilize inherently unstable host–parasitoid interactions.

The first example of this involves the release of volatile chemicals by plants in response to herbivore feeding that can be used by parasitoids for the location of host patches. It has been shown many times in choice tests that parasitoids prefer to move towards plants damaged by their hosts compared to undamaged plants (Turlings et al. 1990; Vet and Dicke 1992). This represents an indirect effect mediated by the host plant trait (production of volatiles), affecting the strength of the host–parasitoid interaction with potentially de-stabilizing consequences. In fact, there are many ways in which plant chemistry can affect the host–parasitoid interaction, both pre- and post-oviposition (Turlings and Benrey 1998). Here, we focus on the effects of multiple herbivore species feeding on the same host plant species. The feeding of non-host herbivores may likewise trigger the release of plant volatiles and, although there are reported cases where parasitoids can distinguish between host-infested and non-host-infested plants (Du et al. 1996; DeMoraes et al. 1998), there are many cases where parasitoids are attracted to plants damaged by non-hosts (McCall et al. 1993; Turlings et al. 1993; Agelopoulos and Keller 1994; Geervliet et al. 1996; Vos et al. 2001). The presence of non-hosts therefore has the potential to affect parasitoid searching efficiency and thereby host–parasitoid dynamics.

Vos et al. (2001) showed that the parasitoid Cotesia glomerata (Braconidae) is more attracted to cabbage leaves that have feeding damage from the caterpillars of their host, the small white butterfly Pieris rapae (Pieridae), compared to undamaged leaves. However, the wasps did not appear capable of distinguishing between leaves damaged by P. rapae and those damaged by the caterpillars of the diamond back moth Plutella xylostella (Plutellidae), which is not a suitable host. This prompted Vos and co-workers to model the effect of herbivore diversity on the stability of host–parasitoid dynamics. In their model they varied the number of host–parasitoid pairs, where each parasitoid was a specialist, attacking a single host species. Each parasitoid's functional response (i.e., the per capita attack rate as a function of host density) included a non-host density dependent term in the denominator, reflecting the idea that high densities of non-hosts will reduce the parasitoid's foraging efficiency because they will be attracted to, and spend time on, plants infested with non-hosts. When this model was simulated with a single host–parasitoid pair, it showed high amplitude oscillations in the parasitoid populations with a high likelihood of parasitoid extinction. But when the number of host–parasitoid pairs was increased, the amplitude of the oscillations decreased with increasing diversity (unless the interference was so strong that parasitoids became too inefficient to persist at high diversity). This model therefore suggests that high diversity can lead to greater stability due to a modification of parasitoid searching behaviour by non-hosts via the plant.

The diversity effect works because the relative density of each host species decreases with increasing diversity, something that occurs because of the assumption that interspecific competition amongst the herbivore species is perfectly symmetrical. Of course, for herbivore diversity to be able to affect host–parasitoid dynamics in this way, the conditions for herbivore coexistence must be met. Furthermore, the stabilizing effect would be strongest if interference increases with parasitoid density, i.e., if parasitoid density has a positive effect on the non-host density, mediated by reduced resource competition with the host population. Below we explore exactly such a case where the effects of resource competition and non-host interference are explicitly included and indirectly coupled.

We (van Veen *et al.* 2005) have explored this idea in an experimental system combined with a modelling approach to untangle the effects of (1) resource competition among herbivores, (2) inherently unstable host–parasitoid dynamics and (3) the stabilizing effect of interference by non-host herbivores. The study system consisted of two aphid species, *Acyrthosiphon pisum* (Aphididae) and *Megoura viciae* (Aphididae), sharing the host plant *Vicia faba* (Fabaceae) and the parasitoid *Aphidius ervi* (Braconidae) that attacks *A. pisum* but not *M. viciae* (Fig. 3.2). We assembled replicate communities of these species in population cages in the laboratory to test for density-mediated indirect effects (DMIEs) along the chain of direct trophic interactions. The first hypothesis we tested was that *A. ervi* would have a positive indirect effect on equilibrium densities of *M. viciae* by reducing the density of *A. pisum* and thereby freeing up a greater portion of the shared resource for *M. viciae*. The second hypothesis was that *M. viciae* in turn would have a negative indirect effect on *A. ervi* by reducing the amount of resource available for its host. These two hypotheses are indicated in Fig. 3.2 by the blue and red dashed arrows, respectively. To test them we assembled further replicate communities from which either *A. ervi* or *M. viciae* was omitted. Comparing the dynamics of *M. viciae* in the presence and absence of *A. ervi* clearly demonstrated the positive DMIE of the latter on the former (Fig. 3.3a,b). Removal of the parasitoid inevitably led to competitive exclusion of *M. viciae* in all replicates. The predicted reciprocal negative effect was, however, not seen because the *A. pisum*–*A. ervi* host–parasitoid dynamics were unstable in the absence of

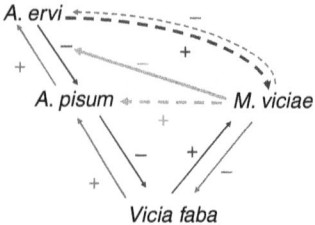

Figure 3.2 Direct and indirect interactions in the experimental community of van Veen *et al.* (2005). The two aphids *A. pisum* and *M. viciae* share the host plant *V. faba* and the parasitoid *A. ervi* attacks *A. pisum* only. Direct trophic effects and their sign are indicated by the solid red and blue arrows. Following the blue chain of direct interactions leads to the hypothesis of a positive indirect density-mediated effect of *A. ervi* on *M. viciae*. Likewise, tracing the red direct effects predicts a negative indirect effect of *M. viciae* on *A. ervi*. The solid green arrow represents the proposed interaction modification (interference with parasitoid searching behaviour) which leads to a positive trait-mediated indirect effect of *M. viciae* on *A. pisum*, as indicated by the dashed green arrow. See colour plate section.

the non-host *M. viciae* (Fig. 3.3c), leading to extinction of the host and parasitoid. In contrast, all species appeared to be able to coexist indefinitely in the full communities.

The apparent stabilizing effect of *M. viciae* on the host–parasitoid dynamics therefore requires an explanation. We used a modelling approach to examine whether this effect could be explained purely by DMIEs along the chain of trophic interactions. Perhaps the presence of the non-host would dampen the oscillations in the host–parasitoid dynamics by reducing host population growth rate. We parameterized two models, one for the competitive interaction between the two aphid species and one for the host–parasitoid interaction based on the experimental data. If trophically mediated interactions alone were responsible for the stability of the community, then combining these two models should have resulted in a match to the observed dynamics in the full community. However, we found that compared to the host–parasitoid interaction alone, adding the non-host competitor into the model actually caused the host and the parasitoid to go extinct more rapidly. This clearly indicates that there had to be some kind of non-trophic interaction that was responsible for stabilizing the community. As in the case of the infochemical interference studied by Vos *et al.* (2001) we suspected that the presence of the non-host in some way interfered with the foraging behaviour of the parasitoid. Due to the limited space in the population cages, locating host patches was unlikely to be a limiting factor for the parasitoids in this experiment. It is, however, very likely that the parasitoids would have spent time on examining non-hosts. Furthermore, the general anti-predator response of the non-host in the form of a kicking behaviour that propagates from individual to individual

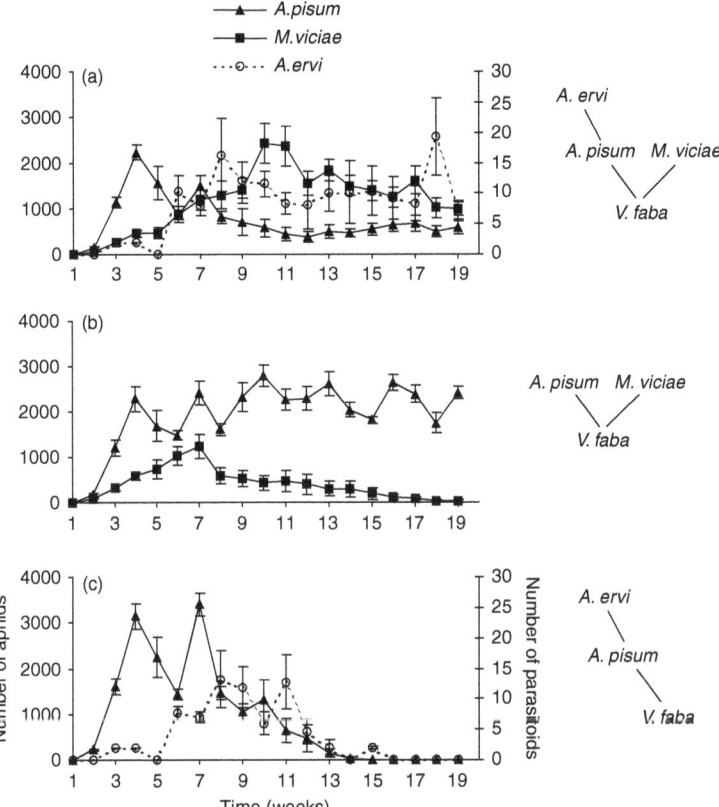

Figure 3.3 Population dynamics in replicated experimental insect communities from van Veen *et al.* (2005). (a) The full community. Note the *M. viciae* densities of 1000–2000 in the second half of the experiment and the persistence of all three insect species. (b) Aphid dynamics in the absence of the parasitoid. *M. viciae* is competitively excluded. (c) Host–parasitoid dynamics in the absence of the non-host. Both host and parasitoid go extinct in all replicates in contrast to their persistence in all replicates of the full community (a).

through a colony of aphids is likely to interfere further with the foraging behaviour of the parasitoid, providing refuges for host aphids in dense aggregations of non-hosts. When we included a non-host density-dependent interference term in the otherwise fully parameterized model of the community, we found that we could now reproduce the stable dynamics observed in the experiment. So, like Vos *et al.* (2001), we conclude that a modification of the host–parasitoid interaction by a non-host is the key process that stabilizes the community. This interaction modification is indicated in Fig. 3.2 with the solid green arrow, which results in a positive trait-mediated indirect effect of *M. viciae* on *A. pisum*, as indicated by the dashed green arrow.

This experiment provides a clear demonstration of the ability of trait-mediated effects to stabilize multispecies dynamics but the question remains whether this is a common mechanism or merely a peculiarity of the experimental system. From the experimental data we could estimate the strength of the interference effect. It appeared that in this experimental arena the parasitoid spent 1.8 times the amount of time on a non-host aphid individual as it did on a host individual. It seems unlikely that parasitoids in nature would generally spend more time on non-hosts than on hosts outside the confines of small population cages. However, further examination of the model suggests that the interference does not need to be this strong to facilitate coexistence. We explore this a bit further here. In Fig. 3.4 we plot the ratio of host to non-host at equilibrium (estimated by taking population sizes at $t = 500$ from a numerical solution of the differential equations). We varied the strength of interference, expressed as the amount of time the parasitoid spends on a non-host individual relative to the amount spent on a host individual, and also the

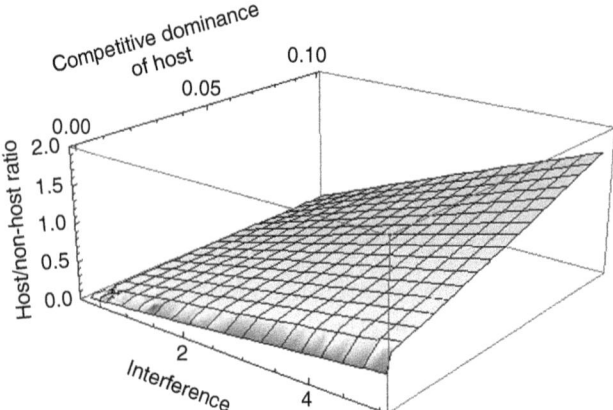

Figure 3.4 The effects of the strength of interference and competitive dominance of host on the relative abundances of host and non-host. The interference effect is expressed as the time spent on a non-host individual relative to the time spent on a host individual so that at 0 there is no interference and at 1 the time spent on non-hosts is equal to that spent on hosts, etc. The estimated interference value from the population cages was 1.8. Competitive dominance (CD) is expressed as the proportional deviation from competitive equivalence where interspecific competition equals intraspecific competition for both species. As CD increases, the interspecific competition effect on the non-host increases while that on the host decreases. The estimated CD from the population cages was 0.04. When interference = 0 the host goes extinct due to the unstable host–parasitoid dynamics. As the strength of the interference effect increases, the host increases in density at the expense of the non-host. The slope of this effect increases with increasing competitive dominance of the host. The estimated parameter values from the population cages were 1.8 for the interference effect and 0.04 for competitive dominance.

strength of interspecific competition. The latter was modelled as the proportional deviation of the interspecific competition coefficients from the intraspecific competition coefficients, with the host assumed to be the dominant competitor. Model details and parameter values are as in van Veen *et al.* (2005). As Fig. 3.4 shows, the host cannot persist in the absence of interference from the non-host irrespective of its competitive dominance (host/non-host ratio = 0). It can however persist when there is just a small degree of interference, even if it is competitively equivalent to the non-host. The host's relative density increases with increasing strength of interference and the slope of this relationship increases with increasing competitive dominance of the host. The result indicates that this trait-mediated indirect effect from non-host competitors on hosts can be an important factor stabilizing otherwise inherently unstable host–parasitoid population dynamics.

There is also evidence that this kind of interference effect is not confined to host–parasitoid systems. For example, Kratina *et al.* (2007) found in a microcosm experiment that the predation rate of protozoa by flatworms decreased with increasing non-prey density and diversity. In a field study we have found that the positive indirect effect of *M. viciae* on *A. pisum* can also be mediated by insect predators and can leave its signature on the spatial distribution of species (van Veen *et al.* 2009). Similarly, Palomo *et al.* (2003) reported a decreased impact of birds on polychaete worms in estuaries in the presence of burrowing crabs. There are also many examples of cases in which greater diversity at a particular trophic level is associated with a decreased impact of species that prey on that trophic level (Leibold 1989; Andow 1991; Strong 1992; Steiner 2001; Duffy 2002; Hillebrand and Cardinale 2004; Borer *et al.* 2005; Duffy *et al.* 2005). Perhaps some of these cases can be explained by a similar type of interaction modification to that we have described here.

If these effects are widespread then there are consequences for the extent of secondary extinctions following the loss of a species from a community. Much attention is currently given to predicting these so-called cascading extinctions from networks of trophic interactions (food webs). One simple and popular approach to this is to perform in silico experiments by removing species from food webs and recording the resulting secondary extinctions under the assumption that species will only go extinct when they have no prey species left to feed on (Sole and Montoya 2001; Montoya *et al.* 2006). Analyses of food webs based on this approach suggest that they are resilient to random extinctions but that extinctions of species with many connections can lead to a cascade of secondary extinctions. It is possible that this approach overestimates cascading extinctions because predators may switch to feeding on alternative prey species, i.e., forge new links in the food web, when the preferred prey is removed. On the other hand, if the modifications of host–parasitoid and predator–prey interactions by non-host and non-prey do indeed play a key role in stabilizing inherently unstable trophic interactions, then the removal of species could lead to further 'unexpected' cascading extinctions. Estimates based on Sole and Montoya's approach may, therefore, actually be conservative. For example, in our population cage communities, extinction of *M. viciae* leads to the extinction of the resource competitor and its parasitoid, an effect that could not be predicted from the network of trophic interactions alone.

We must add a note of caution with regard to the importance of this kind of effect in host–parasitoid systems in the field. The evidence that we have presented here is based on laboratory experiments and models. In the late 1960s–early 1970s interference among parasitoid individuals was the dominant hypothesis for how simple host–parasitoid interactions were stabilized (Hassell and Varley 1969; Hassell 1971) but it was largely abandoned as laboratory studies overestimated its importance in the field. Clear evidence of the importance of non-host interference on the dynamics of populations in the field is also still scarce. One example was provided by a study of the effect of natural enemy diversity on the control of pest insects in large field cages. Cardinale *et al.* (2003) demonstrated that the combined impact of a parasitoid and a predator on the aphid *A. pisum* was greater than the sum of their effects when in isolation. They argued that this was due to the presence of a second (non-host) aphid species that interfered with the searching behaviour of the parasitoid. The ladybird *Harmonia axyridis* (Coccinellidae) caused a decrease in the population density of this non-host, leading to reduced interference and higher attack rates of the parasitoid on *A. pisum*. The trait-mediated effect was therefore responsible for the greater than additive effect of parasitoids and predators.

Further field evidence shows that non-hosts do not always affect host–parasitoid interactions in the same way. In fact, one of the few examples of a host–parasitoid interaction modification by a non-host in the field showed the opposite effect to the one discussed here: increased attack rate in the presence of the non-host (LeBrun and Feener 2002). LeBrun and Feener studied the indirect effects of a competing ant species on the ant *Pheidole diversipilosa* (Formicidae), mediated by the parasitoid fly *Apocephalus* 'sp. 8' (Phoridae). They found that baits with foraging *P. diversipilosa* workers attracted parasitoid flies faster and in greater numbers when a heterospecific competitor was introduced. A likely explanation for this is that the ants produce alarm pheromones when they detect competitors, which are utilized as a host-finding cue by the phorids (LeBrun and Feener 2002). In this case, the trait-mediated effect has a potentially de-stabilizing effect on the host–parasitoid dynamics, although this could not be established within the timescale of the study. Interestingly, parasitoids in this system have also been shown to interact with each other via induced defensive behaviour in the host which can lead to the exclusion of one parasitoid species by another (LeBrun et al. 2009).

A related effect to the one discussed here may occur when two host species share a parasitoid that has a preference for one of the hosts, as is the case with the beetles Galerucella tenella (Chrysomelidae) and G. calmariensis in Sweden (Hambäck et al. 2006). When both beetles occur together, G. tenella experiences higher levels of parasitism than when it occurs in mono-specific populations, presumably due to classic density-mediated apparent competition (Holt 1977; Chaneton and Bonsall 2000). However, due to a strong preference of the parasitoid for G. tenella, G. calmariensis actually experiences lower levels of parasitism in mixed as opposed to mono-specific populations, despite higher densities of parasitoids. The latter species therefore benefits from a positive trait-mediated indirect effect from its apparent competitor, at least at the local scale.

Trait-mediated trophic cascades

Trophic cascades are indirect interactions between species, or species assemblages, at non-adjacent trophic levels (Strong 1992; Polis 1999; Polis et al. 2000). Top-down trophic cascades are the indirect effects of higher level consumers on lower trophic levels. The term is used widely to refer to the effects of predator removal on the diversity and total biomass of lower trophic levels, as well as on subsets of communities and single species populations (Strong 1992; Schmitz et al. 2000; Terborg and Estes 2010). Trophic cascades are perhaps most obvious in aquatic systems (Strong 1992; Shurin et al. 2002); however, there is ample evidence for trophic cascades in terrestrial systems as well (Schmitz et al. 2000; Halaj and Wise 2001; Mooney et al. 2010). In terrestrial systems the higher trophic levels often affect not so much overall producer biomass, but rather herbivory damage and reproductive output of individual plants. Terrestrial trophic cascades, therefore, appear to be mainly species-level phenomena (Halaj and Wise 2001).

As with other indirect interactions, trophic cascades too can be mediated not just by purely trophic effects but also by changes in behaviour or other traits in intermediate species, affecting the strength of trophic interactions (Schmitz et al. 1997). This was very clearly demonstrated by Beckerman et al. (1997) who reported a positive effect of spiders on plants even when the spiders' mouthparts were glued together and they could therefore not consume prey. The effect of the presence of the spiders was that crickets spent less time foraging thus reducing their per capita impact on the plants. These kinds of effects are not limited to arthropods but appear to be widespread and have, for example, been demonstrated to be in part responsible for the positive effect of wolf reintroduction on the re-growth of aspen in Yellowstone Park (Ripple and Beschta 2007) through a shift in habitat use by elk, while in another recent example the presence of artificial perches for

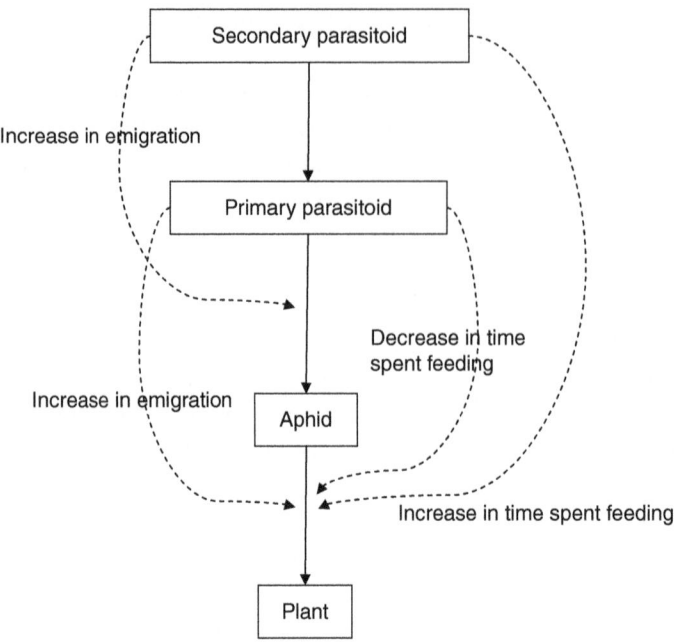

Figure 3.5 Interaction modifications in a four-level trophic cascade. Empirical evidence for each of the effects indicated by the dashed arrows is discussed in the text, however it is not yet clear how important these effects are for community dynamics at the local or meta-community scales.

predatory birds caused a shift in the diet of insectivorous lizards (Hawlena and Perez-Mellado 2009). Trait-mediated trophic cascades have not been reported in host–parasitoid systems, as far as we are aware, but as we will show, there is evidence for significant trait changes in insects in response to higher trophic levels that can be expected to alter their interactions with lower trophic levels. We focus in particular on an aphid–parasitoid system with four trophic levels (Fig. 3.5) and we discuss how observed trait changes can lead to trophic cascades with different effects at different spatial scales.

First, we look at the indirect effects of the third trophic level on the first. To feed, aphids need to have their mouthparts inserted into the phloem of the host plant and establishing such a feeding position takes time. When aphids are disturbed by a predator or parasitoid they will, however, often leave their feeding site and move to a different area of the plant or may even throw themselves off the plant altogether, a reaction that can also be triggered by alarm pheromones (Montgomery and Nault 1977). This behaviour, in response to a natural enemy, can therefore result in a decrease in the proportion of time aphids spend feeding and hence reduced damage to the host plant, exactly as in Beckerman *et al.* 's study of spiders and crickets.

This positive indirect effect of the third trophic level on the first can be further enhanced by a well-documented effect of predators and parasitoids on the dispersal of aphids: most species reproduce parthenogenetically for part or all of the year with females giving birth to live clonal offspring. Despite the genetic homogeneity of the offspring, in most species nymphs can grow into either wingless (apterae) or winged (alatae) adults, depending on environmental cues. In general, winged morphs pay for their dispersal ability with decreased fecundity (Dixon 1998). It has long been known that crowding on a host plant triggers the development of alatae, but in the last decade it has also been demonstrated that the presence of predators and parasitoids can lead to an increase in the proportion of nymphs that develop into alatae (Weisser et al. 1999; Sloggett and Weisser 2002; Kunert et al. 2007) and that this too may be triggered by alarm pheromone (Kunert et al. 2005). This can be viewed as an interaction modification whereby the higher trophic level changes the traits of individuals at the intermediate trophic level, leading to higher emigration and reduced impact of local herbivory. Intriguingly, it appears that the strength of the aphid dispersal response to predators and parasitoids increases under elevated CO_2 and O_3 levels respectively, suggesting that the importance of this trait-mediated effect may increase as a result of human-induced atmospheric changes (Mondor et al. 2004).

These trait-mediated effects of the third trophic level on the first are positive at the local level. At a larger spatial scale, however, the effects may be different. First, increased dispersal rates may lead to a larger proportion of the plant population experiencing herbivory. Second, aphids, like other sap-feeding insects, are important vectors for many plant diseases (Gray and Gildow 2003; Ng and Falk 2006; Hogenhout et al. 2008) so an increase in aphid dispersal can be expected to lead to higher disease transmission rates. Finally, at the metapopulation scale, the impact of increased dispersal would be an increased colonization rate which would have a positive effect on population persistence when the risk of local extinction is high (Levins 1969; Hanski 1999).

The parasitoids of aphids are themselves vulnerable to attack by so-called secondary parasitoids (Fig. 3.5). Levels of secondary parasitism can be very high with the potential to limit primary parasitoid populations (Höller et al. 1993; Müller et al. 1999). However, Höller et al. (1993) argued that high levels of secondary parasitism alone could not explain the relatively low levels of primary parasitism they observed in cereal aphids. They hypothesized that this is due to increased dispersal of primary parasitoid females in response to high densities of secondary parasitoids and their analysis showed the predicted negative correlation between parasitism rates of aphids and secondary parasitoid density. This behavioural response of the primary parasitoids may lead to reduced mortality of aphids at a local scale and again has the potential to affect metapopulation dynamics. Thus, there is the potential for a trophic cascade over four trophic levels in this system in which, in addition to the trophic effects, there is the scope for two types of interaction modification that locally strengthen the net negative effect of the secondary parasitoids on the plants (Fig. 3.5), while at a larger scale they have the potential to increase stability of the meta-community through increased dispersal.

Trait changes do not just occur in response to species at adjacent trophic levels. For example, in laboratory experiments it was shown that aphids exposed to volatiles from secondary parasitoids produced more offspring than a control group (Boenisch *et al.* 1997). In order to test the significance of this result under more realistic conditions, we conducted an experiment in the field in which we placed secondary parasitoids in mesh bags near experimental aphid colonies. We too found that aphids in the experimental colonies produced more offspring than those in a control group that had been exposed to empty mesh bags (van Veen *et al.* 2001). We suspect this is due to a modification of the aphid's response to disturbance. As we discussed above, leaving a feeding site is costly for an aphid but they may do so when they detect the presence of a natural enemy. Perhaps the aphids respond to volatiles from the secondary parasitoids by increasing their disturbance threshold for leaving the feeding site. Such a behavioural response may have evolved if these volatiles give an honest indication of the risk of parasitism (we tried, but failed, to measure this in our experiment) because primary parasitoids may disperse away from areas with high densities of secondary parasitoids. If this is true, we would also hypothesize that alatae production in the aphids would be reduced in the presence of secondary parasitoid volatiles. Whatever the ultimate cause of the effect, the increased amount of time spent feeding would explain the higher fecundity of aphids in the presence of secondary parasitoids. This again increases the negative effect of the secondary parasitoids on the plant via an interaction modification.

This complex trophic cascade is summarized in Fig. 3.5. It should be noted that although there is empirical evidence for each of the interaction modifications in this figure, we still have little idea of their importance in the dynamics of this system. To determine the true significance of trait-mediated trophic cascades in these and other systems the effects will need to be studied in the field at a range of spatial scales.

Foraging behaviour, host traits and the structure of parasitoid communities

So far, we have focused on how changes in traits or behaviour of species in response to other species can affect the dynamics and persistence of multispecies communities. In this section we briefly review how changes in traits may affect the structure and composition of communities. This is an aspect of trait-mediated interactions that is not often highlighted but recent evidence points towards a potentially important role for trait changes in explaining observed differences in food web structure.

Two important food web metrics are species diversity and connectance (the proportion of all possible links that is realized). Both of these are known to affect the stability and resilience of food webs. However, until recently, most food web assembly models treated these as exogenous parameters (Cohen *et al.* 1990; Williams and Martinez 2000) rather than emergent properties. Beckerman *et al.* (2006) used optimal foraging theory to predict aspects of connectance by assuming that the diet breadth of species is determined by energy-maximizing behaviour of the individuals. A more general version of this model assumes that handling time and profit depend on the body size of predator and prey. This model not only predicts the level of connectance well, i.e., the number of links for a given diversity, but also which actual links are realized with up to 60% accuracy (Petchey *et al.* 2008). This suggests that variation in food web structure can be determined by individual behaviour in response to traits such as body size. It may be expected that theory based on the single niche dimension of body size will be most applicable to aquatic pelagic systems where trophic interactions are largely determined by body size, rather than species identity (Law *et al.* 2009). However, even in terrestrial host–parasitoid systems, where we know that host ranges are taxonomically constrained, plasticity in traits of one species in response to another species can also have quite dramatic effects on the trophic structure of communities (Bukovinszky *et al.* 2008).

Bukovinszky *et al.* observed that the aphid–parasitoid–secondary parasitoid food webs on two different strains of *Brassica oleracea* (1st trophic level) were markedly different in their diversity and connectance and that this effect was largely due to a greater diversity and density of a generalist guild of secondary parasitoids (4th trophic level) on one of the *Brassica* strains, while there were little differences in the host–specialist guild. This was unexpected because species composition at the herbivore level was not affected. It was observed, however, that the aphids on one *Brassica* strain grew to a larger size than those on the other and as a result the parasitized aphids, so-called mummies, were also larger. Furthermore, the generalist secondary parasitoids were more likely to emerge from larger hosts. This indicated that the strong influence of plant genotype on community structure at the highest trophic level was caused by an effect of the plant on herbivore body size, which translated into primary parasitoid (mummy) body size which, in turn, affected the foraging behaviour of the guild of secondary parastoids that have wide host ranges. This probably arose because foraging females of these species are more likely to leave areas where host quality is low, especially if there are alternative host species in the vicinity. The specialist guild of hyperparasitoids showed no such response, presumably because their options are more limited. Further

evidence that the secondary parasitoids were making active foraging decisions based on host size comes from sex-ratio data. Because parasitoid wasps
have haplo-diploid sex determination, mothers are able to affect the sex of
their offspring by allowing eggs to be fertilized or not. Fitness is generally
more strongly related to body size in female parasitoids than it is in males.
As expected, Bukovinszky *et al.* found that females were more likely to emerge
from larger hosts and that population sex ratio was more female-biased on the
plots where average host size was greater. So plasticity in body size in the
aphids in response to genetically determined traits of their host plants inter-
acted with the foraging behaviour of parasitoids at the 4th trophic level with
strong effects on food web structure. Mean aphid body size explained 73% of
the variation in secondary parasitoid diversity among experimental plots and
76% of the variation in the proportion of generalists of the total secondary
parasitoid community, hence influencing food web connectance.

If genetically determined plant traits can have such dramatic effects on
insect communities it is to be expected that the degree of genetic variation in
plant populations will affect insect communities. Indeed it has been demonstrated several times that increased genetic diversity leads to higher arthropod species diversity at different trophic levels (Underwood and Rausher
2000; Johnson *et al.* 2006). One explanation for these diversity effects is that
greater genetic diversity in the plant leads to a broader niche space for
herbivores to exploit, allowing more species to coexist because each species
specializes on only part of the niche space. It has, for example, been demonstrated that different herbivorous insects on *Brassica* are specialized on genotypes with different chemical profiles (Poelman *et al.* 2009). A greater diversity
of herbivores in turn represents a wider resource base for higher trophic level
consumers, leading to cascading diversity effects at higher trophic levels.
A recent experiment in which the genetic diversity of a pasture grass was
manipulated suggests that responses in species diversity at higher trophic
levels may also be due to trait-mediated effects rather than cascading diversity
effects (Jones *et al.* 2011). In this case, herbivore diversity did not respond to
host-plant genetic diversity, while primary (3rd trophic level) and especially
secondary (4th trophic level) parasitoid diversity did. Clearly the response at
higher trophic levels cannot be explained by a cascading species diversity
effect because herbivore diversity did not vary and it is therefore likely that
the effects were trait-mediated. Unfortunately, in this case the traits responsible could not be identified.

Concluding remarks

The examples we have described above make it clear that trait-mediated
interactions can play an important role in determining the dynamics and
structure of insect communities. They can help explain the stable coexistence

of species and provide insight into the mechanisms behind unexpected cascading extinctions. The potential for interaction modifications in a cascade involving four trophic levels revealed that these tend to strengthen the top-down effects at a local scale. However, there may be different effects at larger spatial scales mediated by changes in the probability of dispersal. It would be interesting to explore the consequences of enemy-induced dispersal for multi-trophic meta-community dynamics. Finally, we have seen how behavioural responses of parasitoids to the traits of species at lower trophic levels can have significant effects on the diversity and connectance of insect food webs. The common thread in all these examples is behavioural flexibility. Parasitoids make oviposition and foraging decisions in response to the spectrum of suitable and unsuitable hosts they encounter, as well as how they are distributed in the environment. These decisions have long been studied in the context of evolutionary ecology and biological control, and to a lesser degree population dynamics. An exciting challenge is to explore the degree to which an explicit consideration of behavioural ecology can provide insight into food web structure and help explain the persistence and resilience of these complex multitrophic communities.

References

Agelopoulos, N. G. and Keller, M. A. (1994) Plant–natural enemy association in the tritrophic system *Cotesia rubecula–Pieris rapae*–Brassicaceae (Cruciferae). II. Preference of *C. rubecula* for landing and searching. *Journal of Chemical Ecology*, **20**, 1735–1748.

Andow, D. A. (1991) Vegetational diversity and arthropod population response. *Annual Review of Entomology*, **36**, 561–586.

Beckerman, A. P., Petchey, O. L. and Warren, P. H. (2006) Foraging biology predicts food web complexity. *Proceedings of the National Academy of Sciences of the United States of America*, **103**, 13745–13749.

Beckerman, A. P., Uriarte, M. and Schmitz, O. J. (1997) Experimental evidence for a behavior-mediated trophic cascade in a terrestrial food chain. *Proceedings of the National Academy of Sciences of the United States of America*, **94**, 10735–10738.

Boenisch, A., Petersen, G. and Wyss, U. (1997) Influence of the hyperparasitoid *Dendrocerus carpenteri* on the reproduction of the grain aphid *Sitobion avenae*. *Ecological Entomology*, **22**, 1–6.

Borer, E. T., Seabloom, E. W., Shurin, J. B. *et al.* (2005) What determines the strength of a trophic cascade? *Ecology*, **86**, 528–537.

Bukovinszky, T., van Veen, F. J. F., Jongema, Y. and Dicke, M. (2008) Direct and indirect effects of resource quality on food web structure. *Science*, **319**, 804–807.

Cardinale, B. J., Harvey, C. T., Gross, K. and Ives, A. R. (2003) Biodiversity and biocontrol: emergent impacts of a multi-enemy assemblage on pest suppression and crop yield in an agroecosystem. *Ecology Letters*, **6**, 857–865.

Chaneton, E. J. and Bonsall, M. B. (2000) Enemy-mediated apparent competition: empirical patterns and the evidence. *Oikos*, **88**, 380–394.

Cohen, J. E., Briand, F. and Newman, C. M. (1990) *Community Food Webs: Data and Theory*. Berlin, Germany: Springer-Verlag.

De Moraes, C. M., Lewis, W. J., Pare, P. W., Alborn, H. T. and Tumlinson, J. H. (1998)

Herbivore-infested plants selectively attract parasitoids. *Nature*, **393**, 570–573.

Dixon, A. F. G. (1998) *Aphid Ecology*. London: Chapman and Hall.

Du, Y. J., Poppy, G. M. and Powell, W. (1996) Relative importance of semiochemicals from first and second trophic levels in host foraging behavior of *Aphidius ervi*. *Journal of Chemical Ecology*, **22**, 1591–1605.

Duffy, J. E. (2002) Biodiversity and ecosystem function: the consumer connection. *Oikos*, **99**, 201–219.

Duffy, J. E., Richardson, J. P. and France, K. E. (2005) Ecosystem consequences of diversity depend on food chain length in estuarine vegetation. *Ecology Letters*, **8**, 301–309.

Geervliet, J. B. F., Vet, L. E. M. and Dicke, M. (1996) Innate responses of the parasitoids *Cotesia glomerata* and *C. rubecula* (Hymenoptera: Braconidae) to volatiles from different plant–herbivore complexes. *Journal of Insect Behavior*, **9**, 525–538.

Godfray, H. C. J. (1994) *Parasitoids: Behavioral and Evolutionary Ecology*. Princeton, NJ: Princeton University Press.

Gray, S. and Gildow, F. E. (2003) Luteovirus–aphid interactions. *Annual Review of Phytopathology*, **41**, 539–566.

Halaj, J. and Wise, D. H. (2001) Terrestrial trophic cascades: how much do they trickle? *American Naturalist*, **157**, 262–281.

Hambäck, P. A., Stenberg, J. A. and Ericson, L. (2006) Asymmetric indirect interactions mediated by a shared parasitoid: connecting species traits and local distribution patterns for two chrysomelid beetles. *Oecologia*, **148**, 475–481.

Hanski, I. (1999) *Metapopulation Ecology*. Oxford: Oxford University Press.

Hassell, M. P. (1971) Mutual interference between searching insect parasites. *Journal of Animal Ecology*, **40**, 473.

Hassell, M. P. (2000) Host–parasitoid population dynamics. *Journal of Animal Ecology*, **69**, 543–566.

Hassell, M. P. and Varley, G. C. (1969) New inductive population model for insect

parasites and its bearing on biological control. *Nature*, **223**, 1133–1137.

Hawlena, D. and Perez-Mellado, V. (2009) Change your diet or die: predator-induced shifts in insectivorous lizard feeding ecology. *Oecologia*, **161**, 411–419.

Hillebrand, H. and Cardinale, B. J. (2004) Consumer effects decline with prey diversity. *Ecology Letters*, **7**, 192–201.

Hogenhout, S. A., Ammar, E. D., Whitfield, A. E. and Redinbaugh, M. G. (2008) Insect vector interactions with persistently transmitted viruses. *Annual Review of Phytopathology*, **46**, 327–359.

Höller, C., Borgemeister, C., Haardt, H. and Powell, W. (1993) The relationship between primary parasitoids and hyperparasitoids of cereal aphids: an analysis of field data. *Journal of Animal Ecology*, **62**, 12–21.

Holt, R. D. (1977) Predation, apparent competition, and structure of prey communities. *Theoretical Population Biology*, **12**, 197–229.

Ings, T. C., Montoya, J. M., Bascompte, J. *et al.* (2009) Ecological networks: beyond food webs. *Journal of Animal Ecology*, **78**, 253–269.

Johnson, M. T. J., Lajeunesse, M. J. and Agrawal, A. A. (2006) Additive and interactive effects of plant genotypic diversity on arthropod communities and plant fitness. *Ecology Letters*, **9**, 24–34.

Jones, T. S., Allen, E., Härri, S. A., Krauss, J., Müller, C. B. and van Veen, F. J. F. (2011) Effects of genetic diversity of grass on insect species diversity at higher trophic levels are not due to cascading diversity effects. *Oikos*, **120**, 1031–1036.

Kratina, P., Vos, M. and Anholt, B. R. (2007) Species diversity modulates predation. *Ecology*, **88**, 1917–1923.

Kunert, G., Otto, S., Rose, U. S. R., Gershenzon, J. and Weisser, W. W. (2005) Alarm pheromone mediates production of winged dispersal morphs in aphids. *Ecology Letters*, **8**, 596–603.

Kunert, G., Trautsch, J. and Weisser, W. W. (2007) Density dependence of the alarm pheromone effect in pea aphids, *Acyrthosiphon pisum* (Sternorrhyncha: Aphididae). *European Journal of Entomology*, **104**, 47–50.

Law, R., Plank, M. J., James, A. and Blanchard, J. L. (2009) Size-spectra dynamics from stochastic predation and growth of individuals. *Ecology*, **90**, 802–811.

LeBrun, E. G. and Feener, D. H. (2002) Linked indirect effects in ant–phorid interactions: impacts on ant assemblage structure. *Oecologia*, **133**, 599–607.

LeBrun, E. G., Plowes, R. M. and Gilbert, L. E. (2009) Indirect competition facilitates widespread displacement of one naturalized parasitoid of imported fire ants by another. *Ecology*, **90**, 1184–1194.

Leibold, M. A. (1989) Resource edibility and the effects of predators and productivity on the outcome of trophic interactions. *American Naturalist*, **134**, 922–949.

Levins, R. (1969) Some demographic and genetic consequences of environmental heterogeneity for biological control. *Bulletin of the Entomological Society of America*, **15**, 237–240.

McCall, P. J., Turlings, T. C. J., Lewis, W. J. and Tumlinson, J. H. (1993) Role of plant volatiles in host location by the specialist parasitoid *Microplitis croceipes cresson* (Braconidae, Hymenoptera). *Journal of Insect Behavior*, **6**, 625–639.

Mondor, E. B., Tremblay, M. N. and Lindroth, R. L. (2004) Transgenerational phenotypic plasticity under future atmospheric conditions. *Ecology Letters*, **7**, 941–946.

Montgomery, M. E. and Nault, L. R. (1977) Comparative response of aphids to alarm pheromone, (e)-β-farnesene. *Entomologia Experimentalis et Applicata*, **22**, 236–242.

Montoya, J. M., Pimm, S. L. and Sole, R. V. (2006) Ecological networks and their fragility. *Nature*, **442**, 259–264.

Mooney, K. A., Gruner, D. S., Barber, N. A. *et al.* (2010) Interactions among predators and the cascading effects of vertebrate insectivores on arthropod communities and

plants. *Proceedings of the National Academy of Sciences of the United States of America*, **107**, 7335–7340.

Müller, C. B., Adriaanse, I. C. T., Belshaw, R. and Godfray, H. C. J. (1999) The structure of an aphid–parasitoid community. *Journal of Animal Ecology*, **68**, 346–370.

Ng, J. C. K. and Falk, B. W. (2006) Virus–vector interactions mediating nonpersistent and semipersistent transmission of plant viruses. *Annual Review of Phytopathology*, **44**, 183–212.

Nicholson, A. J. and Bailey, V. A. (1935) The balance of animal populations. *Proceedings of the Zoological Society of London*, **1**, 551–598.

Palomo, G., Botto, F., Navarro, D., Escapa, M. and Iribarne, O. (2003) Does the presence of the SW Atlantic burrowing crab *Chasmagnathus granulatus* Dana affect predator–prey interactions between shorebirds and polychaetes? *Journal of Experimental Marine Biology and Ecology*, **290**, 211–228.

Petchey, O. L., Beckerman, A. P., Riede, J. O. and Warren, P. H. (2008) Size, foraging, and food web structure. *Proceedings of the National Academy of Sciences of the United States of America*, **105**, 4191–4196.

Poelman, E. H., van Dam, N. M., van Loon, J. J. A., Vet, L. E. M. and Dicke, M. (2009) Chemical diversity in *Brassica oleracea* affects biodiversity of insect herbivores. *Ecology*, **90**, 1863–1877.

Polis, G. A. (1999) Why are parts of the world green? Multiple factors control productivity and the distribution of biomass. *Oikos*, **86**, 3–15.

Polis, G. A., Sears, A. L. W., Huxel, G. R., Strong, D. R. and Maron, J. (2000) When is a trophic cascade a trophic cascade? *Trends in Ecology and Evolution*, **15**, 473–475.

Quicke, D. L. J. (1997) *Parasitic Wasps*. London: Chapman and Hall.

Ripple, W. J. and Beschta, R. L. (2007) Restoring Yellowstone's aspen with wolves. *Biological Conservation*, **138**, 514–519.

Schmitz, O. J., Beckerman, A. P. and O'Brien, K. M. (1997) Behaviorally mediated trophic cascades: effects of predation risk on food web interactions. *Ecology*, **78**, 1388–1399.

Schmitz, O. J., Hambäck, P. A. and Beckerman, A. P. (2000) Trophic cascades in terrestrial systems: a review of the effects of carnivore removals on plants. *American Naturalist*, **155**, 141–153.

Shurin, J. B., Borer, E. T., Seabloom, E. W. *et al.* (2002) A cross-ecosystem comparison of the strength of trophic cascades. *Ecology Letters*, **5**, 785–791.

Sloggett, J. J. and Weisser, W. W. (2002) Parasitoids induce production of the dispersal morph of the pea aphid, *Acyrthosiphon pisum*. *Oikos*, **98**, 323–333.

Sole, R. V. and Montoya, J. M. (2001) Complexity and fragility in ecological networks. *Proceedings of the Royal Society of London, Series B*, **268**, 2039–2045.

Steiner, C. F. (2001) The effects of prey heterogeneity and consumer identity on the limitation of trophic-level biomass. *Ecology*, **82**, 2495–2506.

Strong, D. R. (1992) Are trophic cascades all wet? Differentiation and donor-control in speciose ecosystems. *Ecology*, **73**, 747–754.

Terborg, J. and Estes, J. A. (eds.) (2010) *Trophic Cascades: Predators, Prey and the Changing Dynamics of Nature*. Washington DC: Island Press.

Turlings, T. C. J. and Benrey, B. (1998) Effects of plant metabolites on the behavior and development of parasitic wasps. *Ecoscience*, **5**, 321–333.

Turlings, T. C. J., McCall, P. J., Alborn, H. T. and Tumlinson, J. H. (1993) An elicitor in caterpillar oral secretions that induces corn seedlings to emit chemical signals attractive to parasitic wasps. *Journal of Chemical Ecology*, **19**, 411–425.

Turlings, T. C. J., Tumlinson, J. H. and Lewis, W. J. (1990) Exploitation of herbivore-induced plant odors by host-seeking parasitic wasps. *Science*, **250**, 1251–1253.

Underwood, N. and Rausher, M. D. (2000) The effects of host-plant genotype on herbivore population dynamics. *Ecology*, **81**, 1565–1576.

van Veen, F. J. F., Brandon, C. E. and Godfray, H. C. J. (2009) A positive trait-mediated indirect effect involving the natural enemies of competing herbivores. *Oecologia*, **160**, 195–205.

van Veen, F. J. F., Morris, R. J. and Godfray, H. C. J. (2006) Apparent competition, quantitative food webs, and the structure of phytophagous insect communities. *Annual Review of Entomology*, **51**, 187–208.

van Veen, F. J. F., Mueller, C. B., Pell, J. K. and Godfray, H. C. J. (2008) Food web structure of three guilds of natural enemies: predators, parasitoids and pathogens of aphids. *Journal of Animal Ecology*, **77**, 191–200.

van Veen, F. J. F., Rajkumar, A., Müller, C. B. and Godfray, H. C. J. (2001) Increased reproduction by pea aphids in the presence of secondary parasitoids. *Ecological Entomology*, **26**, 425–429.

van Veen, F. J. F., van Holland, P. D. and Godfray, H. C. J. (2005) Stable coexistence in insect communities due to density- and trait-mediated indirect effects. *Ecology*, **86**, 3182–3189.

Vet, L. E. M. and Dicke, M. (1992) Ecology of infochemical use by natural enemies in a tritrophic context. *Annual Review of Entomology*, **37**, 141–172.

Vos, M., Berrocal, S. M., Karamaouna, F., Hemerik, L. and Vet, L. E. M. (2001) Plant-mediated indirect effects and the persistence of parasitoid-herbivore communities. *Ecology Letters*, **4**, 38–45.

Weisser, W. W., Braendle, C. and Minoretti, N. (1999) Predator-induced morphological shift in the pea aphid. *Proceedings of the Royal Society of London, Series B*, **266**, 1175–1181.

Williams, R. J. and Martinez, N. D. (2000) Simple rules yield complex food webs. *Nature*, **404**, 180–183.

CHAPTER FOUR

The impact of trait-mediated indirect interactions in marine communities

JEREMY D. LONG

*Department of Biology and Coastal Marine Institute Laboratory,
San Diego State University*

and

MARK E. HAY

School of Biology, Georgia Institute of Technology

Introduction

Marine communities have served as productive laboratories for the discovery of fundamental processes and mechanisms driving community structure and function (e.g., Paine 1966; Connell 1961; Dayton 1975). Within these communities, inducible responses are ubiquitous; extending from the seafloor to the sea-surface, and from microscopic plankton to charismatic marine megafauna (e.g., Harvell 1990; Toth and Pavia 2007; Hay 2009; Vaughn and Allen 2010). These changes may influence energy and nutrient cycling within and among ecosystems (Hay and Kubanek 2002; Long *et al.* 2007b). Over the past decade, marine ecologists have begun to demonstrate the often dramatic, cascading consequences of these trait modifications on co-occurring species and thus community structure and function. These trait-mediated indirect interactions (TMIIs) influence trophic cascades (Trussell *et al.* 2002; Grabowski and Kimbro 2005), competition between herbivores (Denno *et al.* 2000; Long *et al.* 2007a, 2011), apparent competition (Schmitt 1987), herbivore population dynamics (Denno *et al.* 2000), linkage across ecosystems (Nevitt *et al.* 1995) and energy flow and food chain length (Trussell *et al.* 2006b). In this chapter we discuss how TMIIs affect marine populations, communities and sometimes ecosystems, and consider the insights that can be gained from understanding and contrasting marine patterns and processes with those occurring in terrestrial or freshwater systems.

Before the role of TMIIs was broadly recognized in marine systems, marine ecologists recognized and documented the importance of indirect interactions in structuring communities such as rocky shores (Paine 1966; Lubchenco 1978; Menge 1978) and kelp forests (Estes *et al.* 1998). Recent evidence highlights that many of these indirect interactions may be trait-mediated. In a recent field manipulation, predation risk alone was enough

Trait-Mediated Indirect Interactions: Ecological and Evolutionary Perspectives, eds. Takayuki Ohgushi, Oswald J. Schmitz and Robert D. Holt. Published by Cambridge University Press. © Cambridge University Press 2012.

to drive the cascade from predatory crabs, through grazing snails, to seaweeds on New England rocky shores (Trussell *et al.* 2004). It appears that considerable portions of the total indirect effects seen in previous marine experiments may be trait-mediated, rather than density-mediated (density-mediated indirect interactions, DMII). This hypothesis seems possible because most studies of marine trophic cascades either (1) assumed that changes in the abundance of intermediate consumers (e.g., snails) were the result of predators eating prey or (2) restricted herbivore movements and predator avoidance strategies in ways that prevented expression of some important TMIIs. Thus, some influential field studies could have confounded density-mediated effects with trait-mediated effects.

Two properties of marine systems suggest that studies of TMIIs may be especially productive in these systems. First, marine systems harbour greater phyletic diversity than terrestrial or freshwater systems, allowing an assessment of the generality of TMIIs across a greater range of organism types. Second, a greater proportion of marine consumers are generalists as opposed to specialists (insects dominate species numbers in terrestrial systems and many are relatively specialized feeders; Hay 1992). If TMIIs are stronger, or more frequent, in 'interaction webs with several interactions pathways' (e.g., those containing generalist herbivores; Schmitz 1998), then TMIIs should be of considerable importance in marine communities. One of the only tests comparing the relative strength of TMIIs and DMIIs in the sea revealed that TMIIs accounted for 60–80% of the total indirect effect (TMII + DMII) of predator cues on basal resources (Trussell *et al.* 2008).

For this chapter, we focused on indirect interactions mediated by trait changes that involve at least three species. Such trait changes are examples of inducible responses whereby one species elicits a behavioural, morphological or chemical change in a second species. This focus is consistent with the TMII definition provided by Werner and Peacor (2003), 'a species reacts to the presence of a 2nd species by altering its phenotype, then the trait changes in the reacting species alter the per capita effect of the reacting species on other species'. Thus, studies of inducible defences demonstrating effects on other individuals of the same inducing consumer species (e.g., inducible herbivore deterrents in seaweeds), though growing in number (Toth and Pavia 2007), are not considered here.

This review is not exhaustive but aims to demonstrate that TMIIs are common, pervasive and ecologically important in the sea. Our goals in this chapter are to: (1) review the diverse types of species, habitats and experiments that have documented TMIIs, (2) highlight exciting examples of predator-induced cascades, especially those that scale-up to influence broader patterns of energy transfer, (3) describe how human activity may be altering

TMIIs, (4) identify TMIIs that extend beyond classic trophic cascades, (5) identify areas of future research that are likely to be productive or novel and (6) discuss how understanding TMIIs will change the way we address ecological questions.

Types of experimental design

Most studies of TMIIs in marine systems have focused on shallow, coastal habitats and examined indirect interactions among macroscopic species. This mirrors our better understanding of coastal benthic, as opposed to pelagic, communities due to their access, ease of manipulation and the history of greater experimentation in benthic systems. Demonstrations of TMIIs in non-coastal habitats or for microscopic species appear less common, but may simply result from fewer attempts to assess TMIIs in these systems. However, several recent works demonstrate that (1) inducible responses do occur among marine plankton (Selander *et al.* 2006; Long *et al.* 2007b) and (2) microscopic species may contribute to TMIIs (Mouritsen and Poulin 2005; Wood *et al.* 2007). These findings suggest that TMIIs may be common among small species, also occur in planktonic systems and thus may be common throughout marine species and communities.

As with terrestrial and freshwater systems, the dominant type of TMII documented from marine systems is when a predator produces cascading indirect effects on basal resources via trait changes (especially predator avoidance) of an intermediate consumer (a three species food-chain per Werner and Peacor 2003). Predator avoidance behaviours are strong and rapidly expressed so they can produce rapid and spatially large effects (i.e., consumers kill one prey but frighten many and alter their behaviour) that can be documented in short-term studies. TMIIs involving other trait changes in prey (e.g., morphological or chemical changes) or those where prey elicit trait changes in consumers are less well-known. Lack of data showing prey traits cascading up to affect consumer traits could be due to: (1) top-down selection for consumer avoidance being stronger than the bottom-up selection for acquiring food (i.e., the life versus dinner principle; Dawkins and Krebs 1979), (2) the generalized feeding habits of most marine consumers allowing them to shift from induced prey to alternative, less defended prey rather than altering their own traits when encountering induced prey, or (3) it possibly taking longer for them to be expressed so they are less easily demonstrated in short-term experiments.

Because experiments assessing TMIIs require assessing multiple species and how they interact with each other, marine experiments are often conducted in laboratory mesocosms with flow-through seawater (but see Raimondi *et al.* 2000; Trussell *et al.* 2004). Although such experiments have been productive, they commonly fail to mimic parameters such as flow rate and consumer density that could critically influence TMIIs. The slower and less turbulent flow characteristic of mesocosm experiments compared to field conditions may present prey with elevated concentrations of predator cues compared to the field. Additionally, as with most terrestrial studies, TMII experiments in marine systems have been short term, lasting no more than a few months (Schmitz et al. 2004). Longer-term effects are uncertain; it is unclear whether responding species would acclimate to inducing species over the longer term (e.g., TMIIs may weaken over long-term experiments as starved prey begin to take more risks).

Cascading effects of predator avoidance

Influential predators that elicit defensive responses in multiple prey species may produce community-wide cascades and could therefore be considered keystone inducers (producing large trait changes in many prey species despite low densities of the consumers themselves). Chemical cues from the green crab, *Carcinus maenus*, induce responses in mussels (Leonard et al. 1999), clams (Griffiths and Richardson 2006), two species of herbivorous snails (Trussell 1996; Trussell et al. 2002) and a carnivorous snail (Trussell et al. 2008). They even elicit responses in non-native prey with which they have co-occurred for only short periods, such as the predatory snail *Acanthinucella spirata* that has co-occurred with green crabs for only ~10–15 years (Kimbro et al. 2009). Through these trait modifications, green crabs have cascading, positive, indirect effects on fucoid algae (Trussell et al. 2002), ephemeral algae (Trussell et al. 2004), barnacles (Trussell et al. 2008) and mussels (Trussell et al. 2008); and negative indirect effects on non-responding bivalve prey (Griffiths and Richardson 2006).

Similarly, tiger sharks, *Galeocerdo cuvier*, induce avoidance responses in bottlenose dolphin (Heithaus and Dill 2002), sea turtles (Heithaus et al. 2007), cormorants (Heithaus et al. 2009), dugongs (Wirsing et al. 2007) and harbour seals (Wirsing et al. 2008). Although the risk of tiger shark attack and resource availability interact to influence habitat use by these prey, the cascading effects of tiger sharks on other species via induced responses in shark prey have not been examined experimentally. However, the high metabolic needs of endothermic prey in cold marine waters suggests that the cascading effects of changed behaviour in animals such as dolphins, seals, dugongs and birds could be considerable. Both green crabs and tiger sharks are mobile, generalist predators, so strong responses to these consumers by multiple prey species might be predicted. Identifying other keystone inducers and shared traits between these consumers will be an important area of future research. Seastars, for example, that are known to elicit rapid behavioural avoidance in several mollusc prey (Phillips 1975), are good candidates. Interestingly, even the classic keystone ochre sea star, *Pisaster ochraceus*, that is believed to exert community control largely through consumption of sessile invertebrates may also affect communities of mobile invertebrates via induced avoidance responses (Phillips 1975; Miller 1986; Peckarsky *et al.* 2008). As with tiger sharks, however, the cascading effects of predator avoidance by ochre sea star prey has not been studied.

While most studies suggest that predators can indirectly influence the *abundance* of basal prey via trait changes within intermediate consumers, predators may also indirectly influence the *traits* of basal prey via trait changes within intermediate consumers. In such a 'trait cascade', predators induce avoidance strategies in herbivores that, in turn, reduce inducible defences of seaweeds. In a terrestrial example of this concept, behavioural responses of tobacco hornworm, *Manduca sexta*, larvae to predator cues resulted in lower induced resistance in West Indian nightshade, *Solanum ptychanthum* (Griffin and Thaler 2006). In the only marine test of this hypothesis, Reynolds and Sotka (2011) found that amphipods reduce feeding on the seaweed *Sargassum filipendula* in the presence of predatory pinfish cues and these predator-mediated changes in herbivory had a cascading effect on *Sargassum* palatability – amphipods preferred *Sargassum* thalli that had been exposed to amphipods and cues from pinfish compared to thalli exposed to only amphipods. Trait cascades might be effectively studied in intertidal communities compared to other marine communities given the background of well-described inducible responses of snails to predators (Trussell *et al.* 2002) and the inducible responses of basal intertidal prey (barnacles and seaweeds) to these same snail species (Lively 1986; Toth and Pavia 2007). However, there is no compelling logic to suggest that such TMIIs are not also common in the many other marine systems in which such background information is less available.

The common discovery of TMIIs in the limited number of marine communities where it has been investigated suggests TMIIs could be widespread – as might be assumed from the many demonstrations of TMIIs in a host of terrestrial, marine and freshwater systems. If TMIIs are common, then understanding the mechanisms driving community structure and function will mandate additional efforts in how ecologists design experiments. Experiments commonly measure changes in organism distribution or abundance. The effects of TMIIs may be captured by such measurements if the trait changes involve migration and if the full range of species being affected is sampled. However, if trait changes are expressed as changes in per capita behaviour (e.g., feeding rates) without alterations in density or distribution, then the effects of TMIIs may be missed. As examples, densities of salt marsh herbivores do not change between higher versus lower latitude marshes, but per capita grazing is higher for lower latitude grazers causing them to have impacts on marsh vegetation that higher latitude grazers do not produce despite their similar densities (Pennings and Silliman 2005). Similarly, when snails sense feeding predators, they may lower their feeding rate or change use of protected versus exposed microhabitats, thus producing significant changes in their effects on their prey without alterations in local snail density (e.g., Trussell *et al.* 2006a, 2008). In such cases, measures of processes (e.g., feeding rates) will detect alterations in critical processes that measurements of consumer density alone may miss.

Context dependency

As with most ecological interactions, the trait-mediated cascading effects of predators on basal prey in marine systems are context dependent, with variation arising from (1) occupancy of risky versus refuge habitats and (2) small and large individual prey or predators that produce variance in predation risk. From non-marine systems, it is well known that organisms alter habitat use based on predation risk (Lima and Dill 1990; Schmitz *et al.* 1997). In the presence of predators, organisms may switch from foraging in rewarding, but risky, habitats to less rewarding, refuge habitats. Using flow-through mesocosms, Trussell *et al.* (2006a) examined the indirect effects of green crabs on barnacles mediated by changes in the behaviour of barnacle predators (dogwhelks, *Nucella lapillus*). They compared the strength of these TMIIs in habitats differing in their perceived risk of predation. Tiles placed inside the mesocosms with foraging whelks contained high reward, risky habitats (high barnacle density on the tops of tiles) and low reward, refuge habitats (low barnacle density on the shaded undersides of tiles); half of the replicate mesocosms received green crab cues, the other half did not. Consistent with findings from terrestrial systems, the magnitude and direction of the indirect effect of green crab cues on barnacles via altered dogwhelk predation contrasted strikingly between risky and refuge habitats (Trussell *et al.* 2006a). Because dogwhelks spent more time in refuges after detecting predators, the presence of crab cues resulted in communities with 580% more barnacles in risky habitats but 85% fewer barnacles in the refuge habitat, suggesting that dogwhelks were switching foraging from risky to refuge habitats in the presence of predator cues (Fig. 4.1). In a subsequent experiment, the presence of crab cues again reduced whelk foraging in risky barnacle habitats but had a much weaker effect in risky mussel habitats leading to a six-times stronger positive indirect effect of green crab cues in the barnacle compared to the mussel food chain (Trussell *et al.* 2008). Additionally, Trussell *et al.* (2008) compared the effect sizes of TMIIs and DMIIs (using periodic whelk removal to simulate lethal effects) and found that the trait-mediated indirect effect was larger than the DMII with barnacles but they were equivalent with mussels. This difference may arise because mussels are more topographically complex and may be perceived as a safer habitat for whelks (Trussell *et al.* 2008).

In contrast to the variance across habitats shown above, Grabowski and Kimbro (2005) found that the positive trait-mediated effect of predatory

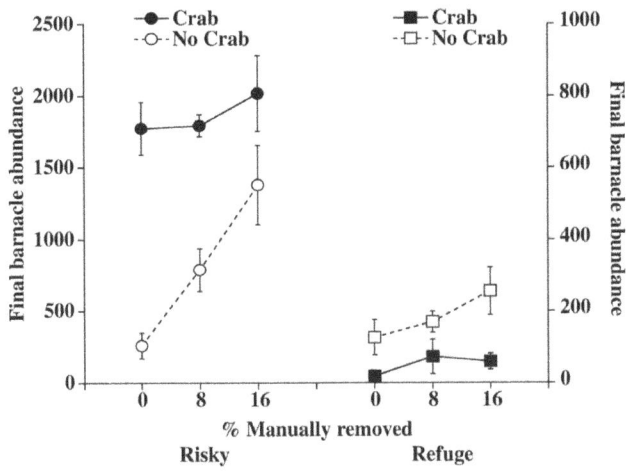

Figure 4.1 The indirect effect of crab cues on barnacle abundance via behavioural changes in intermediate consumers, dogwhelks, varies between risky and refuge habitats. Mean (± SE) barnacle abundance in risky and refuge habitats after exposure to the presence and absence of predation risk (crab and no crab) and different levels of snail removal (0, 8 and 16%) for 36 days. The TMII was positive in risky habitats but negative in refuge habitats (reprinted from Trussell *et al.* 2006a, Fig. 2B).

toadfish, *Opsanus tau*, on juvenile hard clams, *Mercenaria mercenaria*, extended to refuge habitats. Similar to the studies of New England intertidal communities (Trussell *et al.* 2006a, 2008), intermediate consumers (the mud crab *Panopeus herbstii*) hid in refuges in the presence of predator risk cues. Although toadfish shifted the foraging habitats of mud crabs to refuges between oyster shells, this did not lead to enhanced predation on clams residing within these refuges. The apparent discrepancy between these studies may result from an important experimental difference – the availability of basal prey in risky habitats. Dogwhelk prey (barnacles) were present in both risky and refuge habitats (Trussell *et al.* 2006a), whereas mud crab prey (hard clams) only resided in refuge habitats (Grabowski and Kimbro 2005). Thus although all authors found an overall reduction in feeding in the presence of crab cues, the opposing direction in the effects of predators on basal prey in risky and refuge habitats was only evident when prey were available in risky habitats.

The ratio of predator size to prey size influences prey susceptibility (Schmitz *et al.* 2004) and thus, may determine how prey respond to predators. Two marine tests of this hypothesis found conflicting results. In support of this hypothesis, small juvenile sea urchins (*Strongylocentrotus franciscanus*), but not large urchins, reduced kelp consumption after encountering waterborne cues from a predatory seastar, *Pycnopodia helianthoides* (Freeman 2005). This ontogenetic behavioural switch appeared adaptive because seastars selectively consumed small urchins. In contrast, all sizes of a congeneric urchin, *Stronglyocentrotus purpuratus*, responded to cues from predatory lobsters that had been fed urchins, and there was no interaction between urchin size and the influence of lobster cues on kelp consumption (Matassa 2010). Surprisingly, when lobster diet was switched to fish, medium to large urchins reduced kelp consumption by 57% but small urchins did not respond (Matassa 2010). The inconsistent patterns in these experiments suggest that additional factors influence the role of predator to prey size ratios in affecting TMIIs.

Additionally, differential responses between prey size classes may result from factors unrelated to differential predation. As an example, juvenile male dolphins tend to enter profitable but riskier foraging habitats containing tiger sharks more frequently than do other size and sex classes of dolphins (Heithaus and Dill 2002). Although these habitats are more risky, they are especially valuable to juvenile males because foraging in the riskier, but more profitable, habitats may (1) increase growth rates of young males, (2) facilitate earlier alliance formation with other males, and thus (3) produce an earlier competitive advantage for access to females (Heithaus and Dill 2002).

Just as the size of intermediate consumers may determine the strength of these indirect interactions, predator size may influence perceived risk by prey. In seagrass beds of the southeastern United States, pinfish are important predators of blue crabs. Adult pinfish, but not juveniles, elicit a behavioural change in blue crab prey that increases scallop survivorship (Bishop and Wear 2005) presumably because adult pinfish present a more serious risk to blue crabs (even though blue crab mortality was low and did not differ between treatments in this experiment). This result is consistent with the observed failure of scallop recruitment in the spring when most pinfish are juveniles and not large enough to threaten blue crabs foraging on juvenile scallops (Bishop and Wear 2005). In the autumn, when pinfish are large, they suppress blue crab foraging and shift it to primarily nocturnal periods when the pinfish do not forage; successful scallop recruitment occurs during this period (Bishop and Wear 2005).

The cascading effects of predator avoidance extend beyond three species

Trait-mediated interactions not only influence the traits and abundance of a third species, they may have cascading effects that extend beyond the three interacting species. In marine systems, TMIIs influence energy transfer within and between communities, and could potentially alter food chain length, vertical movement of energy through the water column and sediments, and the coupling of benthic–pelagic systems (see discussions in Hay and Kubanek 2002, Trussell *et al.* 2006a, Hay 2009). TMIIs may also affect marine–terrestrial subsidies, fish and mammal kills from harmful algal blooms, the clearance rates of filter feeders and thus light transmission through the water column; however, some of these connections are based on reasonable arguments more than hard data and need experimental confirmation (Hay and Kubanek 2002; Hay 2009).

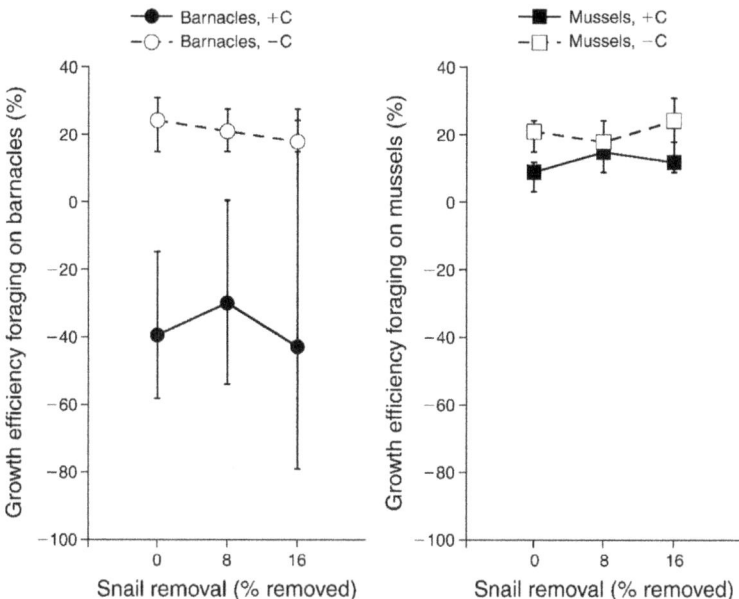

Figure 4.2 Predator risk suppression of growth efficiency was context dependent. Mean (± 95% CI) growth efficiency of dogwhelks foraging on barnacles and mussels in the presence (+C) and absence (−C) of predation risk and at different levels of snail removal (0, 8, 16%). Predator risk more strongly reduced growth efficiency in barnacle versus mussel habitats (reprinted from Trussell *et al.* 2008, Fig. 5).

The stress of nearby predators also alters consumer conversion efficiency and energy movement through food chains. In the presence of green crab cues *N. lapillus* not only reduces feeding rates on barnacles, but the stress of predator risk also suppresses growth efficiency (Fig. 4.2). This effect on metabolic efficiency was context dependent; growth efficiency was reduced most when whelks were at low densities and feeding on barnacles rather than mussels (Trussell *et al.* 2006b, 2008) perhaps because mussels provide some refuge from crab predation. The weaker effect of crab cues on growth efficiency when whelks were at high density suggests that either increasing prey density decreased the perceived risk of predation for each individual or that the increased competition from conspecifics increased the cost of responding and that whelks were trading off responding to predators versus competitors. Fear-induced reductions in growth efficiency may lead to communities with shortened food chains (Trussell *et al.* 2006b).

Predator cues can also influence the vertical migration of prey and, in turn, the vertical transfer of energy via predation and waste excretion (Loose and Dawidowicz 1994; Hays 2003). A classic example of vertical migration is zooplankton migration to deeper waters during the day, presumably to avoid visually hunting predators in well-lit surface waters. While physical cues can affect planktonic migrations, predator presence also influences migrations. For example, copepods migrated vertically in large mesocosms holding free-ranging fish, but not in similar fish-free mesocosms (Bollens and Frost 1989). This effect was not driven by water-borne chemical cues as caged, feeding fish did not elicit similar responses (Bollens and Frost 1989). Predator-induced planktonic migrations could indirectly influence other species, but no study has specifically examined the indirect effect of predators on other species via induced trait changes in zooplankton.

Within the plankton, inducible responses could influence how energy moves through food webs. For example, microzooplanktonic ciliates and flagellates represent important grazers whose feeding rates can exceed those of meso-zooplanktonic copepods in epipelagic communities. Unfortunately, previous assessments of microzooplankton grazing rates were predominantly con-ducted in the absence of copepod predators, i.e., without possibility of TMIIs, even though ciliates display avoidance responses towards copepods (Verity 2008). Thus, ciliate grazing rates measured in the absence of predator cues likely have overestimated natural grazing rates. In a second example, the heteromorphic phytoplankton *Phaeocystis globosa* alternates between solitary cells of only a few microns in diameter and colonies that can be several centimetres, in part as a response to chemical cues from grazers. Ciliates that feed best on solitary cells produce chemical cues that enhance colony formation. In contrast, large copepods that feed best on colonies produce chemical cues that suppress colony formation (Long et al. 2007b). Thus, this simple phytoplankton not only responds to grazer cues but responds in grazer-specific ways (Long et al. 2007b). Because the ultimate fate of primary production partially depends on the identity of phytoplankton, microzoo-plankton and mesozooplankton, inducible responses could alter the flow of energy between surface and deeper waters (e.g., the biological pump) and affect whether energy and nutrients are delivered to deep waters below the thermocline and made unavailable for primary production or retained in shallow waters and recycled.

In addition to shapeshifting, phytoplankton can respond to the threat of attack by changing their chemical composition. The microalga *Alexandrium minutum* increases its concentration of paralytic shellfish toxins when exposed to chemical cues from copepod grazers (Selander et al. 2006). This response appears adaptive because induced algae are grazed less than non-induced conspecifics. Interestingly, inducible responses in *A. minutum* are grazer-specific; toxin production increased in response to cues from the copepods *Centropages* and *Acartia* but not *Pseudocalanus* (Bergkvist et al. 2008). Currently, there is no evidence that inducible responses in microalgae play a role in marine TMIIs. However, given the ability of harmful algal toxins to alter entire ecosystems, especially via fish and mammal kills, copepods may indirectly impact large vertebrates via induced chemical changes in phyto-plankton (Hay and Kubanek 2002).

The evidence that predator-induced vertical migrations could produce cas-cading effects is suggestive for pelagic communities, but well-documented for some benthic systems. Within experimental mesocosms, free-roaming killi-fish, *Fundulus heteroclitus*, restricted grass shrimp, *Palaemonetes pugio*, to shallow habitats, thereby increasing survivorship of shrimp prey (the clam *Mulinia lateralis*) in deeper habitats (Posey and Hines 1991). These changes were assumed to be trait-mediated because overall shrimp densities were equiva-lent with and without fish. In a second example, the clam *Macoma balthica* moved deeper within sediments in the presence of green crab cues but the co-occurring cockle, *Cerastoderma edule*, did not (Griffiths and Richardson 2006). These differential responses have large consequences. When green crabs were quickly placed into feeding arenas with both prey species, they attacked *Cerastoderma* 1.5 times more frequently than *Macoma*. However, if crab cues were introduced before the crab was allowed to feed, then *Macoma* moved deeper and when the crab was finally introduced, *Cerastoderma* was attacked 15 times more frequently than *Macoma*. Thus, the differential response of these bivalves dramatically redirected green crab foraging to the non-responding prey species.

While TMIIs may influence the vertical movement of energy within the water column and sediments, they also may influence the 'horizontal' trans-fer of energy between marine and terrestrial ecosystems. When Antarctic zooplankton graze phytoplankton rich in dimethyl sulfioproponate (DMSP), the volatile compound dimethyl sulfide (DMS) is released into the water column and atmosphere (Dacey and Wakeham 1986). This metabolite appears to be used as an olfactory signpost by seabirds foraging for zooplankton prey over an otherwise featureless ocean surface (Nevitt 2008). To examine whether seabirds such as petrels might use this compound to find zooplank-ton, researchers created surface slicks of DMS and examined the number of seabirds sited on these slicks compared to controls. Several seabirds, espe-cially petrels, were sited 1.5–2.5 times more frequently at DMS compared to control slicks (Nevitt et al. 1995; Nevitt 2008). Some of these seabirds carry marine prey to their young in land-based nests hundreds to thousands of kilometres away where the nutrients flow to terrestrial communities via bird defecation, death of juveniles and predation on eggs, nestlings or adult birds by terrestrial consumers. This DMS-mediated transfer of energy and nutrients can drive productivity and diversity patterns in the terrestrial sys-tems to which the adult birds return, affecting organisms as diverse as cacti, coyotes, beetles and spiders (see review by Hay 2009).

Two final examples suggest that inducible responses in marine settings could have cascading effects beyond the interacting species. First, chemical cues from predatory blue crabs, *Callinectes sapidus*, and knobbed whelks, *Busycon carica*, reduce the pumping, or filter feeding activity, of hard clams, *Mercenaria mercenaria* (Smee and Weissburg 2006; Ferner *et al.* 2009); this appears to reduce the chemical cues leaking from these prey and makes them less obvious to predators. Hence, predators that elicit avoidance strategies in filter-feeding bivalves could have a positive indirect effect on plankton and water-column standing stock, especially in areas with dense coverage of bivalves (e.g., mussel beds). Second, in marine systems, there have been few examinations of tritrophic signalling whereby consumer attack of a basal prey releases a signal that attracts a top predator. In terrestrial systems, caterpillar-grazed plants can release volatile compounds that attract parasitoid predators (Dicke 1999). In the only direct marine test of this concept, cues from the seaweed *Ascophyllum nodosum* being attacked by the smooth periwinkle *Littorina obtusata* were more attractive to fish and green crab predators in a Y-maze, laboratory arena compared to cues from unattacked plants or periwinkles alone (Coleman *et al.* 2007). Unlike the numerous studies documenting inducible responses of prey towards green crabs, this study appears to be an exception in that green crabs exploit inducible responses to their advantage. In marine systems, there appears to be considerable scope for increased study of the frequency of tritrophic interactions, their cascading effects, and the mechanisms involved.

Humans alter indirect interactions

Through introduction of non-native species, overfishing, ocean acidification and other separate and synergistic stresses, humans may be influencing the occurrence, strength and effects of TMIIs. Introduced species may cause failure of trait-mediated cascades because species without a shared evolutionary history may not recognize and appropriately respond to the presence of novel predators or prey (Aschaffenburg 2008; Edgell and Neufeld 2008; Kimbro *et al.* 2009). For example, native prey along the Northeast Pacific Ocean may not recognize the threat of green crab predation because green crabs invaded this region only 10–15 years ago. Indeed, this region's native whelks displayed induced behavioural (Kimbro *et al.* 2009) and morphological (Edgell and Neufeld 2008) responses to native *Cancer* crabs but failed to respond to introduced green crabs; thus the positive cascade affecting basal oyster resources that occurred in the presence of native crabs was lost with invasion of the green crab (Kimbro *et al.* 2009). Similarly, native *Cancer* crabs produce no trait-mediated cascade of effects on oysters in the presence of invasive whelks that do not avoid *Cancer* cues (Kimbro *et al.* 2009). In contrast, invasive green crabs in New England have strong, positive indirect effects on basal resources via avoidance behaviours in several native prey (Leonard *et al.* 1999; Griffiths and Richardson 2006). The disparity in the responses of native prey to green crabs in the Northeast Pacific versus New England may result from different histories (both recent and evolutionary) of prey with invasive green crab predators. Over evolutionary time, Northeast Pacific prey species and green crabs have co-occurred only recently, whereas founding populations of New England prey species from Europe likely co-occurred with green crabs prior to the last glacial maximum (Wares and Cunningham 2001). Over more recent history, green crabs have been present in New England for about 100 years compared to the Northeast Pacific, where they have occurred for only the last 10–15 years. Thus, the strong responses of New England, but not Northeast Pacific, intermediate consumers is consistent with the hypothesis that the effect of invasive species on trait-mediated cascades will vary due to the duration of historical overlap (Kimbro et al. 2009); however, with only one instance of each to compare, it is premature to consider this a test of the hypothesis.

Another factor that may lead to failure of TMIIs is human-induced decline in the abundance and diversity of large predatory fishes, including sharks, due to overfishing and global climate change (Worm et al. 2005). Recently, these alarming decreases in top predators were recognized to enhance intermediate prey and to produce cascading indirect effects on the bivalves that these intermediate prey consumed (Myers et al. 2007; Prugh et al. 2009). Because predator avoidance plays a dominant role in structuring marine communities (Bollens and Frost 1989; Trussell et al. 2002; Dill et al. 2003; Smee and Weissburg 2006), these declines could lead to 'fear-released systems' with weaker TMIIs (Frid et al. 2008; Madin et al. 2010). In a field manipulation on Bahamian coral reefs, the presence of a threatened top predator (Nassau grouper, *Epinephelus striatus*) positively impacted recruitment of basal fish species via altered behaviour of intermediate predatory fishes (Fig. 4.3). In the presence of Nassau grouper, smaller grouper species tended to hide more and grew at slower rates (Stallings 2008). Similarly, a model predicted that the removal of tiger sharks would indirectly affect predation on fish via behavioural changes in seals that feed on fishes (Frid et al. 2008). In the presence of sharks, seals tend to forage on shallow, low-value fish species such as herring because deeper habitats are more risky for seals. However, under shark removal scenarios, seals shifted to deeper strata to forage on more profitable prey such as haddock (Frid et al. 2008). Given the tendency of fisheries to target species from high trophic levels (Pauly et al. 1998), it is likely that fishing will release mesopredator densities, possibly degrading trait-mediated cascades and de-stabilizing communities (Prugh et al. 2009).

Finally, there is evidence that ocean acidification can weaken inducible defences and, therefore, has the potential to affect indirect interactions between species – though these indirect effects have yet to be demonstrated. *Amphiprion percula* fish larvae avoid settling in habitats where they detect predators' odours. However, larvae raised under predicted future ocean pH

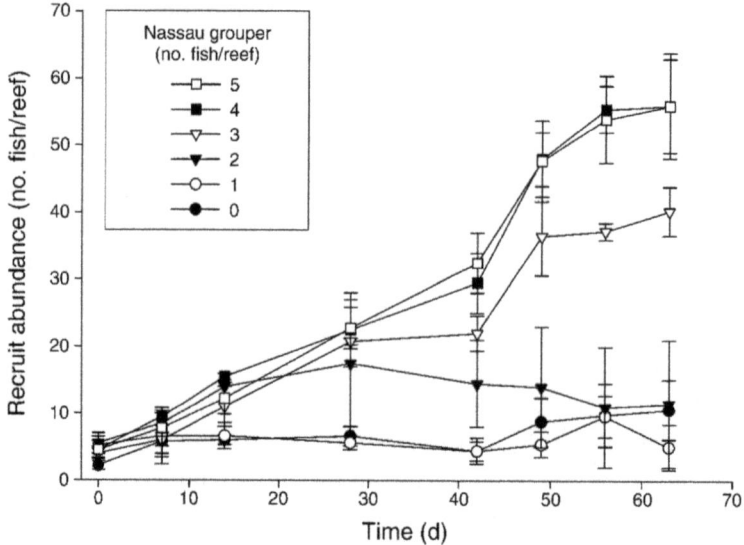

Figure 4.3 A threatened top predator (Nassau grouper) increases abundance of fish recruits via behavioural modification of intermediate piscivores. Temporal changes in abundance (mean ± SE) of fish recruits on reefs occupied by different numbers of Nassau grouper (six levels, 0–5 fish/reef). These differences were unrelated to changes in the abundance of intermediate piscivores (reprinted from Stallings 2008, Fig. 3).

levels became strongly attracted to predator odours (Dixson et al. 2010) and field experiments demonstrate that larvae raised under acidified conditions are bolder, move further from safe structures and suffer dramatically higher rates of predation on natural coral reefs (Munday et al. 2010). Under these scenarios, weakened inducible responses of prey to their predators may result in weaker TMIIs but stronger DMIIs as ineffectively responding prey are consumed by predators.

Moving beyond indirect interactions between predators and the resources of their prey

Several studies have documented predator avoidance responses and their cascading influences on the resources of their prey in marine settings, but few have investigated other indirect interactions mediated by trait changes in marine systems – such as the cascading effects of prey-induced predator traits or TMIIs between species at a given trophic level. This is surprising given that these other indirect interactions appear common and important in terrestrial systems (Denno and Kaplan 2007). Studies that examine these other indirect interactions in the sea typically examine induced chemical and morphological changes, rather than behavioural changes, perhaps because the transmitting species are often sessile or sedentary and are limited in their repertoire of observable behavioural responses.

Trait-mediated grazer–grazer interactions

Herbivores may interact indirectly with each other via grazer-induced changes in plant traits (i.e., a three species shared-resource web, Werner and Peacor 2003). In plant–insect interactions, over half of all the examples of competition between herbivorous insects are mediated by inducible plant traits (Denno and Kaplan 2007). Although this topic has received less attention in the marine literature, three studies demonstrate that grazing-induced plant responses may negatively affect other grazer species. The first two examples involve grazer–grazer interactions mediated by grazer-induced changes in the marsh cordgrass *Spartina alterniflora*. First, both laboratory and field manipulations revealed that prior feeding by sap-feeding planthoppers on *Spartina* influenced a congeneric planthopper by delaying development time by several days, reducing survivorship by >30% and reducing adult size (Denno *et al*. 2000). Such interactions were asymmetric with one species of planthopper having a greater negative effect on the other than vice versa. Because of this asymmetry, these interactions may influence the large-scale migration of the inferior competitor away from marsh habitats containing the superior competitor (Denno and Roderick 1992; Denno *et al*. 2000).

Second, *Spartina* that was previously attacked by a natural suite of consumers was a less attractive habitat to the marsh periwinkle, *Littoraria irrorata*, and to the planthopper *Prokelisia* sp.; *Littoraria* also consumed previously attacked *Spartina* at lower rates than non-attacked *Spartina* (Long *et al*. 2011). Water-soluble extracts from grazed plants reduced feeding by *Littoraria*, suggesting an induced chemical trait; other plant traits did not differ significantly. Interestingly, these indirect interactions were evident in South Carolina, but not Maine. Currently, it is not known if this variation represents genetic differences in plants or responses to different environmental factors.

In a third example, previous attack by the specialist smooth periwinkle, *Littorina obtusata*, but not the generalist common periwinkle, *L. littorea*, lowered the palatability of *Fucus* towards three consumer species (two snails and an isopod) by at least 52% (Long *et al*. 2007a). Transplanted *Fucus* that had previously been attacked by smooth periwinkles was colonized by half as many common periwinkles in the field compared to ungrazed control *Fucus* (Fig. 4.4). Thus, the smooth periwinkle had a negative indirect effect on a congeneric snail via induced changes in a shared host.

These TMIIs among herbivores have been documented in terrestrial (Denno and Kaplan 2007), freshwater (Bolser and Hay 1998) and marine systems (Long *et al*. 2007a, 2011); they are probably common but have been too

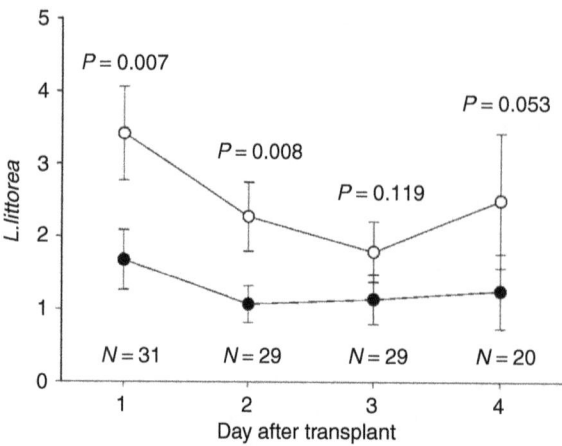

Figure 4.4 Previous grazing by a specialist snail herbivore, *Littorina obtusata*, decreases abundance of a generalist snail herbivore, *Littorina littorea*, via trait changes in the seaweed *Fucus vesiculosus*. Mean (± SE) *L. littorea* density (number per *Fucus* transplant) on *Fucus* transplants in the field. Prior to transplantation, *Fucus* was grazed by *L. obtusata* (solid circles) or was ungrazed in controls (open circles; reprinted from Long *et al.* 2007a, Fig. 4C).

infrequently studied in marine or freshwater systems to assess rigorously either their frequency or importance. However, the few available studies suggest that herbivore effects via a shared host are commonly asymmetric, providing the opportunity for one herbivore to suppress another via induced changes in a shared host. More studies addressing these interactions in marine systems would be useful.

Trait-mediated interactions between basal species induced by consumers

The carnivorous whelk *Acanthina angelica* induces a bent, whelk-resistant barnacle morph when whelks crawl over barnacles (Lively 1986; Raimondi *et al.* 2000). The timing of barnacle encounters with whelks determines the cascading consequences of these interactions on the surrounding community. In a field study, two contrasting communities developed depending on whether barnacles encountered whelks early in the year as juveniles or later in the year as adults (Fig. 4.5). When older barnacles, with a lower capacity to respond to predators, encountered whelks late in the year, then whelks fed upon barnacles leading to a positive indirect effect on mussels because mussels can recruit to the empty tests of dead barnacles. However, when juvenile barnacles encountered whelks early in the year, then a proportion of barnacles developed into bent morphs leading to a positive indirect effect on an encrusting alga, perhaps because mussels had fewer suitable sites for recruitment. In this example, consumers had an indirect positive effect on a basal species via trait changes in another basal species. However, species also may have indirect negative effects on others (see Griffiths and Richardson 2006). At present, there are too few studies of these interactions between basal species to justify even speculative statements about general trends and their potential impacts. More studies are needed to help establish an understanding of the frequency, importance and variance in effect of these interactions.

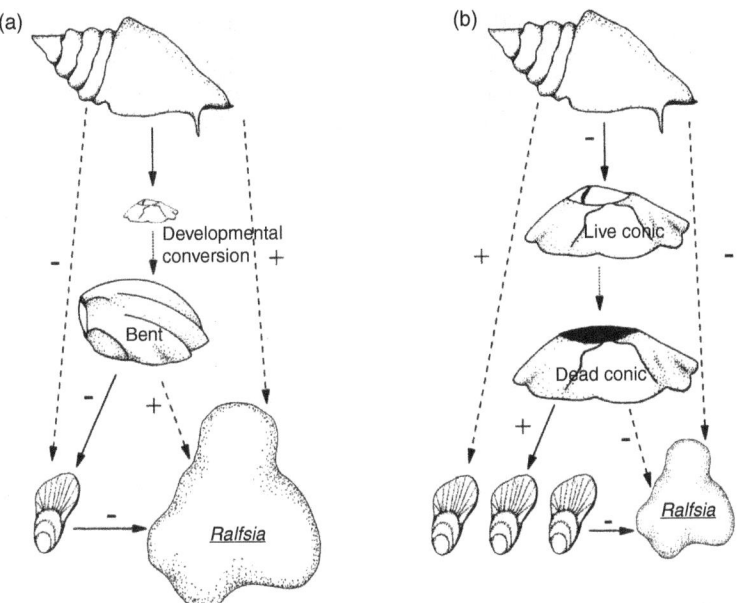

Figure 4.5 Predatory whelks (*Acanthina*) influence the structure of basal species via induced morphological changes in barnacle prey (*Chthamalus*). Solid lines represent direct interactions and dashed lines represent indirect interactions. (a) Whelks have a positive indirect effect on encrusting algae, *Ralfsia*, when juvenile, inducible barnacles encounter whelks. (b) However, whelks have a negative indirect effect on *Ralfsia* and a positive indirect effect on mussels when adult, non-inducible barnacles encounter whelks (reprinted from Raimondi *et al.* 2000, Fig. 2).

Trait-mediated interactions between prey species induced by prey

While most examples of marine TMIIs centre around consumer-induced trait changes, there are several examples that include prey-induced changes in consumers that influence other prey indirectly. Prey may interfere with predators attempting to feed on other prey by increasing prey crypsis or by physically interfering with prey capture. As an example, chemically defended seaweeds can provide an associational refuge for more palatable seaweed species. In some heavily grazed South American kelp beds, the giant kelp *Macrocystis pyrifera* can only recruit to areas encircled by the seaweed *Desmarestia ligulata* – whose acidic surfaces act as 'acid brooms' that deter urchin foraging (Dayton 1985). Likewise, the unpalatable brown alga *Sargassum filipendula* provides a safe habitat for the palatable red seaweed *Hypnea*; *Sargassum* competitively suppresses *Hypnea* in the absence of herbivores, but when herbivores are common, *Hypnea* is driven locally extinct when not associated with *Sargassum* (Hay 1986). Because less palatable species such as *Sargassum* may induce greater defences when attacked by herbivores

(Taylor *et al.* 2002), there could be complex interactions among herbivores, their inducible prey and the alternative prey relying on their inducible, defended competitors. Such interactions have not been investigated, but seem possible, if not likely.

Conclusions

Consistent with studies from terrestrial and freshwater systems, the limited data for marine systems suggest that TMIIs are omnipresent and produce patterns of ecological consequence at the levels of individuals, populations, communities and even ecosystems. TMIIs affect vertical migrations of marine consumers in ways that will impact the productivity, biogeochemistry and patterns of nutrient dynamics in shallow, productive waters above the thermocline versus deeper waters, and TMIIs even produce impacts that sometimes cascade across the borders of marine and terrestrial systems. More studies of TMIIs and their consequences are needed, especially in planktonic communities where inducible responses occur, but are poorly investigated. Complex interactions within these communities could be especially important given the critical role these ecosystems play in Earth's biogeochemical cycles. Other especially productive areas are likely to include (1) increasing our understanding of how prey induced by one consumer will have cascading effects on both other consumers and the other prey with which these consumers and prey interact and (2) understanding how anthropogenic impacts on the sea will affect risk perception and other chemically mediated interactions (some of which are reversed in maladaptive ways by changes such as ocean acidification; Munday *et al.* 2010). The demonstrated importance of TMIIs in many communities and ecosystems also mandates that we add assessments of critical behaviours and biotic processes (herbivory, etc.) to our standard studies of organism distribution and abundance.

Acknowledgements

Work on this manuscript was supported by funding from the National Science Foundation (OCE 0825846 to JL; OCE 0929119 to MEH). This is Contribution No. 11 of the Coastal and Marine Institute Laboratory, San Diego State University.

References

Aschaffenburg, M. D. (2008) Different crab species influence feeding of the snail *Nucella lapillus* through trait-mediated indirect interactions. *Marine Ecology: An Evolutionary Perspective*, **29**, 348–353.

Bergkvist, J., Selander, E. and Pavia, H. (2008) Induction of toxin production in dinoflagellates: the grazer makes a difference. *Oecologia*, **156**, 147–154.

Bishop, M. and Wear, S. (2005) Ecological consequences of ontogenetic shifts in predator diet: seasonal constraint of a behaviorally mediated indirect interaction. *Journal of Experimental Marine Biology and Ecology*, **326**, 199–206.

Bollens, S. M. and Frost, B. W. (1989) Predator-induced diel vertical migration in a planktonic copepod. *Journal of Plankton Research*, **11**, 1047–1065.

Bolser, R. C. and Hay, M. E. (1998) A field test of inducible resistance to specialist and generalist herbivores using the water lily *Nuphar luteum*. *Oecologia*, **116**, 143–153.

Coleman, R. A., Ramchunder, S. J., Davies, K. M., Moody, A. J. and Foggo, A. (2007) Herbivore-induced infochemicals influence foraging behaviour in two intertidal predators. *Oecologia*, **151**, 454–463.

Connell (1961) Influence of interspecific competition and other factors on distribution of barnacle *Chthamalus stellatus*. *Ecology*, **42**, 710–723.

Dacey, J. W. H. and Wakeham, S. G. (1986) Oceanic dimethylsulfide: production during zooplankton grazing on phytoplankton. *Science*, **233**, 1314–1316.

Dawkins, R. and Krebs, J. R. (1979) Arms races between and within species. *Proceedings of the Royal Society of London Series B*, **205**, 489–511.

Dayton, P. K. (1975) Experimental evaluations of ecological dominance in a rocky intertidal community. *Ecological Monographs*, **45**, 137–159.

Dayton, P. K. (1985) The structure and regulation of some South American kelp communities. *Ecological Monographs*, **55**, 447–468.

Denno, R. F. and Kaplan, I. (2007) Plant-mediated interactions in herbivorous insects: mechanisms, symmetry, and challenging the paradigms of competition past. In T. Ohgushi, T. P. Craig and P. W. Price, eds., *Ecological Communities: Plant Mediation in Indirect Interaction Webs*. Cambridge: Cambridge University Press, pp. 19–50.

Denno, R. F. and Roderick, G. K. (1992) Density-related dispersal in planthoppers: effects of interspecific crowding. *Ecology*, **73**, 1323–1334.

Denno, R. F., Peterson, M. A., Gratton, C. *et al.* (2000) Feeding-induced changes in plant quality mediate interspecific competition between sap-feeding herbivores. *Ecology*, **81**, 1814–1827.

Dicke, M. (1999) The evolution of induced indirect defense of plants. In R. Tollrian and C. D. Harvell, eds., *The Ecology and Evolution of Inducible Defenses*. Princeton, NJ: Princeton University Press, pp. 62–88.

Dill, L. M., Heithaus, M. R. and Walters, C. J. (2003) Behaviorally mediated indirect interactions in marine communities and their conservation implications. *Ecology*, **84**, 1151–1157.

Dixson, D. L., Munday, P. L. and Jones, G. P. (2010) Ocean acidification disrupts the innate ability of fish to detect predator olfactory cues. *Ecology Letters*, **13**, 68–75.

Edgell, T. C. and Neufeld, C. J. (2008) Experimental evidence for latent developmental plasticity: intertidal whelks respond to a native but not an introduced predator. *Biology Letters*, **4**, 385–387.

Estes, J. A., Tinker, M. T., Williams, T. M. and Doak, D. F. (1998) Killer whale predation on sea otters linking oceanic and nearshore ecosystems. *Science*, **282**, 473–476.

Ferner, M. C., Smee, D. L. and Weissburg, M. J. (2009) Habitat complexity alters lethal and non-lethal olfactory interactions between predators and prey. *Marine Ecology: Progress Series*, **374**, 13–22.

Freeman, A. S. (2005) Size-dependent trait-mediated indirect interactions among sea urchin herbivores. *Behavioral Ecology*, **17**, 182–187.

Frid, A., Baker, G. G. and Dill, M. L. (2008) Do shark declines create fear-released systems? *Oikos*, **117**, 191–201.

Grabowski, J. H. and Kimbro, D. L. (2005) Predator-avoidance behavior extends trophic cascades to refuge habitats. *Ecology*, **86**, 1312–1319.

Griffin, C. A. M. and Thaler, J. S. (2006) Insect predators affect plant resistance via density- and trait-mediated indirect interactions. *Ecology Letters* **9**, 338–346.

Griffiths, C. L. and Richardson, C. A. (2006) Chemically induced predator avoidance behaviour in the burrowing bivalve *Macoma balthica. Journal of Experimental Marine Biology and Ecology*, **331**, 91–98.

Harvell, C. D. (1990) The ecology and evolution of inducible defenses. *Quarterly Review of Biology* **65**, 323–340.

Hay, M. E. (1986) Associational plant defenses and the maintenance of species diversity: turning competitors into accomplices. *American Naturalist*, **128**, 617–641.

Hay, M. E. (1992) The role of seaweed chemical defenses in the evolution of feeding specialization and in the mediation of complex interactions. In V. J. Paul, ed., *Ecological Roles for Marine Natural Products.* Ithaca, NY: Comstock Press, pp. 93–118.

Hay M. E. (2009) Marine chemical ecology: chemical signals and cues structure marine populations, communities, and ecosystems. *Annual Review of Marine Sciences*, **1**, 193–212.

Hay, M. E. and Kubanek, J. (2002) Community and ecosystem level consequences of chemical cues in the plankton. *Journal of Chemical Ecology*, **28**, 2001–2016.

Hays, G. C. (2003) A review of the adaptive significance and ecosystem consequences of zooplankton diel vertical migrations. *Hydrobiologia*, **503**, 163–170.

Heithaus, M. R. and Dill, L. M. (2002) Food availability and tiger shark predation risk influence bottlenose dolphin habitat use. *Ecology*, **83**, 480–491.

Heithaus, M. R., Frid, A., Wirsing, A. J. et al. (2007) State-dependent risk-taking by green sea turtles mediates top-down effects of tiger shark intimidation in a marine ecosystem. *Journal of Animal Ecology*, **76**, 837–844.

Heithaus, M. R., Wirsing, A. J., Burkholder, D., Thomson, J. and Dill, L. M. (2009) Towards a predictive framework for predator risk effects: the interaction of landscape features and prey escape tactics. *Journal of Animal Ecology*, **78**, 556–562.

Kimbro, D. L., Grosholz, E. D., Baukus, A. J. et al. (2009) Invasive species cause large-scale loss of native California oyster habitat by disrupting trophic cascades. *Oecologia*, **160**, 563–575.

Leonard, G. H., Bertness, M. D. and Yund, P. O. (1999) Crab predation, waterborne cues, and inducible defenses in the blue mussel, *Mytilus edulis. Ecology*, **80**, 1–14.

Lima, S. L. and Dill, L. M. (1990) Behavioral decisions made under the risk of predation: a review and prospectus. *Canadian Journal of Zoology/Revue Canadienne De Zoologie*, **68**, 619–640.

Lively, C. M. (1986) Predator-induced shell dimorphism in the acorn barnacle *Chthamalus anisopoma. Evolution*, **40**, 232–242.

Long, J. D., Hamilton, R. S. and Mitchell, J. L. (2007a) Asymmetric competition via induced resistance: specialist herbivores indirectly suppress generalist preference and populations. *Ecology*, **88**, 1232–1240.

Long, J. D., Mitchell, J. L. and Sotka, E. E. (2011) Local consumers induce resistance differentially between *Spartina alterniflora* populations in the field. *Ecology*, **92**, 180–188.

Long, J. D., Smalley, G. W., Barsby, T. A., Anderson, J. T. and Hay, M. E. (2007b) Chemical cues induce consumer-specific defenses in a bloom-forming marine phytoplankton. *Proceedings of the National Academy of Sciences of the United States of America*, **104**, 10512–10517.

Loose, C. J. and Dawidowicz, P. (1994) Trade-offs in diel vertical migration by zooplankton: the costs of predator avoidance. *Ecology*, **75**, 2255–2263.

Lubchenco, J. (1978) Plant species diversity in a marine intertidal community: importance of herbivore food preference and algal competitive abilities. *American Naturalist*, **112**, 23–39.

Madin, E. M. P., Gaines, S. D., Madin, J. S. and Warner, R. R. (2010) Fishing indirectly structures macroalgal assemblages by altering herbivore behavior. *American Naturalist*, **176**, 785–801.

Matassa, C. M. (2010) Purple sea urchins *Strongylocentrotus purpuratus* reduce grazing rates in response to risk cues from the spiny lobster *Panulirus interruptus*. *Marine Ecology Progress Series*, **400**, 283–288.

Menge, B. A. (1978) Predation intensity in a rocky intertidal community: relation between predator foraging activity and environmental harshness. *Oecologia*, **34**, 1–16.

Miller, M. L. (1986) Avoidance and escape responses of the gastropod *Nucella emarginata* (Deshayes, 1839) to the predatory seastar *Pisaster ochraceus* (Brandt, 1835). *Veliger*, **28**, 394–396.

Mouritsen, K. N. and Poulin, R. (2005) Parasites boost biodiversity and change animal community structure by trait-mediated indirect effects. *Oikos*, **108**, 344–350.

Munday, P. L., Dixson, D. L., McCormick, M. I. *et al.* (2010) Replenishment of fish populations is threatened by ocean acidification. *Proceedings of the National Academy of Sciences of the United States of America*, **107**, 12930–12934.

Myers, R. A., Baum, J. K., Shepherd, T. D., Powers, S. P. and Peterson, C. H. (2007) Cascading effects of the loss of apex predatory sharks from a coastal ocean. *Science*, **315**, 1846–1850.

Nevitt G. A. (2008) Sensory ecology on the high seas: the odor world of procellariiform sea birds. *Journal of Experimental Biology*, **211**, 1706–1713.

Nevitt, G. A., Veit, R. R. and Kareiva, P. (1995) Dimethyl sulphide as a foraging cue for Antarctic Procellariiform seabirds. *Nature*, **376**, 680–682.

Paine, R. T. (1966) Food web complexity and species diversity. *American Naturalist*, **100**, 65–75.

Pauly, D., Christensen, V., Dalsgaard, J., Froese, R. and Torres, F. (1998) Fishing down marine food webs. *Science*, **279**, 860–863.

Peckarsky, B. L., Abrams, P. A., Bolnick, D. I. *et al.* (2008) Revisiting the classics: considering nonconsumptive effects in textbook examples of predator–prey interactions. *Ecology*, **89**, 2416–2425.

Pennings, S. C. and Silliman, B. R. (2005) Linking biogeography and community ecology: latitudinal variation in plant–herbivore interaction strength. *Ecology*, **86**, 2310–2319.

Phillips, D. W. (1975) Distance chemoreception-triggered avoidance-behavior of limpets *Acmaea* (Collisella) *limulata* and *Acmaea* (Notoacmea) *scutum* to predatory starfish *Pistaster ochraceus*. *Journal of Experimental Zoology*, **191**, 199–209.

Posey, M. H. and Hines, A. H. (1991) Complex predator–prey interactions within an estuarine benthic community. *Ecology*, **72**, 2155–2169.

Prugh, L. R., Stoner, C. J., Epps, C. W. *et al.* (2009) The rise of the mesopredator. *BioScience*, **59**, 779–791.

Raimondi, P. T., Forde, S. E., Delph, L. F. and Lively, C. M. (2000) Processes structuring communities: evidence for trait-mediated indirect effects through induced polymorphisms. *Oikos*, **91**, 353–361.

Reynolds, P. L. and Sotka, E. E. (2011) Non-consumptive predator effects indirectly influence marine plant biomass and palatability. *Journal of Ecology*, **99**, 1272–1281.

Schmitt, R. J. (1987) Indirect interactions between prey: apparent competition, predator aggregation, and habitat segregation. *Ecology*, **68**, 1887–1897.

Schmitz, O. J. (1998) Direct and indirect effects of predation and predation risk in old-field interaction webs. *American Naturalist*, **151**, 327–342.

Schmitz, O. J., Beckerman, A. and O'Brien, K. M. (1997) Behaviorally mediated trophic cascades: effects of predation risk on food web interactions. *Ecology*, **78**, 1388–1399.

Schmitz, O. J., Krivan, V. and Ovadia, O. (2004) Trophic cascades: the primacy of trait-mediated indirect interactions. *Ecology Letters*, **7**, 153–163.

Selander, E., Thor, P., Toth, G. and Pavia, H. (2006) Copepods induce paralytic shellfish toxin production in marine dinoflagellates. *Proceedings of the Royal Society of London, Series B*, **273**, 1673–1680.

Smee, D. L. and Weissburg, M. J. (2006) Clamming up: environmental forces diminish perceptive ability of bivalve prey. *Ecology*, **87**, 1587–1598.

Stallings, C. D. (2008) Indirect effects of an exploited predator on recruitment of coral-reef fishes. *Ecology*, **89**, 2090–2095.

Taylor R. B., Sotka, E. and Hay, M. E. (2002) Tissue-specific induction of herbivore resistance: seaweed response to amphipod grazing. *Oecologia*, **132**, 68–76

Toth, G. B. and Pavia, H. (2007) Induced herbivore resistance in seaweeds: a meta-analysis. *Journal of Ecology*, **95**, 425–434.

Trussell, G. C. (1996) Phenotypic plasticity in an intertidal snail: the role of a common crab predator. *Evolution*, **50**, 448–454.

Trussell, G. C., Ewanchuk, P. J. and Bertness, M. D. (2002) Field evidence of trait-mediated indirect interactions in a rocky intertidal food web. *Ecology Letters*, **5**, 241–245.

Trussell, G. C., Ewanchuk, P. J., Bertness, M. D. and Silliman, B. R. (2004) Trophic cascades in rocky shore tide pools: distinguishing lethal and nonlethal effects. *Oecologia*, **139**, 427–432.

Trussell, G. C., Ewanchuk, P. J. and Matassa, C. M. (2006a) Habitat effects on the relative importance of trait- and density-mediated indirect interactions. *Ecology Letters*, **9**, 1245–1252.

Trussell, G. C., Ewanchuk, P. J. and Matassa, C. M. (2006b) The fear of being eaten reduces energy transfer in a simple food chain. *Ecology*, **87**, 2979–2984.

Trussell, G. C., Ewanchuk, P. J. and Matassa, C. M. (2008) Resource identity modifies the influence of predation risk on ecosystem function. *Ecology*, **89**, 2798–2807.

Vaughn, D. and Allen, J. D. (2010) The peril of the plankton. *Integrative and Comparative Biology*, **50**, 552–570.

Verity, P. G. (2008) Are feeding and growth rates of planktonic ciliates overestimated from experiments in the absence of copepod predators? National Science Foundation Biological Oceanography Award 451347.

Wares, J. P. and Cunningham, C. W. (2001) Phylogeography and historical ecology of the North Atlantic intertidal. *Evolution*, **55**, 2455–2469.

Werner, E. E. and Peacor, S. D. (2003) A review of trait-mediated indirect interactions in ecological communities. *Ecology*, **84**, 1083–1100.

Wirsing, A. J., Heithaus, M. R. and Dill, L. M. (2007) Fear factor: do dugongs (*Dugong dugon*) trade food for safety from tiger sharks (*Galeocerdo cuvier*)? *Oecologia*, **153**, 1031–1040.

Wirsing, A. J., Heithaus, M. R., Frid, A. and Dill, L. M. (2008) Seascapes of fear: evaluating sublethal predator effects experienced and generated by marine mammals. *Marine Mammal Science*, **24**, 1–15.

Wood, C. L., Byers, J. E., Cottingham, K. L. *et al.* (2007) Parasites alter community structure. *Proceedings of the National Academy of Sciences of the United States of America*, **104**, 9335–9339.

Worm, B., Sandow, M., Oschlies, A., Lotze, H. K. and Myers, R. A. (2005) Global patterns of predator diversity in the open oceans. *Science*, **309**, 1365–1369.

CHAPTER FIVE

Trait-mediated indirect interactions in size-structured populations: causes and consequences for species interactions and community dynamics

VOLKER H. W. RUDOLF

Department of Ecology and Evolutionary Biology, Rice University

Introduction

Ecological communities are complex networks of interacting species, and it has been a central challenge in community ecology to understand and predict their dynamics. To deal with this daunting complexity, scientists typically abstract these communities into more tractable subcomponents, such as food web modules (e.g., food chains, predator–prey interactions, competitive interactions) (Holt 1997), to elucidate the causal mechanisms that determine the dynamics of species interactions. However, even at this reduced level of complexity, researchers face the challenge of how much detail should be included to capture the full dynamics of natural communities without getting tangled up in details or losing generality.

Much of our conceptual foundation for species interactions is derived from basic models such as the Lotka–Volterra equations and extensions, which assume that per capita interaction strengths between species are on average the same across individuals within a population and that community dynamics are solely governed by changes in population densities. Similarly, food web theory traditionally treats a species as a single node in which all individuals within a species are expected to experience the same type and strength of species interactions (reviewed in Pascual and Dunne 2005; Montoya *et al.* 2006). While these assumptions make ecological systems much more tractable, they also sacrifice important biological details below the species level that may influence the dynamics of communities. In particular, by focusing on the species level we inherently assume that either all individuals within a population are identical, or at least, that they are on average the same and any variation around this mean does not alter the dynamics of the system and can safely be ignored. However, no population is truly homogenous and individuals within populations often vary considerably in their ecology. The question is: does this intraspecific variation matter?

Trait-Mediated Indirect Interactions: Ecological and Evolutionary Perspectives, eds. Takayuki Ohgushi, Oswald J. Schmitz and Robert D. Holt. Published by Cambridge University Press. © Cambridge University Press 2012.

By far the largest source of this intraspecific variation stems from differ-ences in developmental stage and size (Wilbur 1980; Werner and Gilliam 1984; Ebenman and Persson 1988; Benton et al. 2006). Body size is arguably one of the most important traits of an individual. It determines its growth and foraging rates, mortality risk, behaviour, diet choice, habitat use and to a large extent, species interactions. For example, large individuals are able to consume larger prey and often face a smaller risk of predation. During their ontogeny, most organisms undergo substantial changes in size, often cover-ing several orders of magnitude (Werner and Gilliam 1984), resulting in considerable ecological variation between individuals within populations that is often larger than differences between similar sized species (Polis 1984; Munoz and Ojeda 1998; Woodward and Hildrew 2002). As a conse-quence, the type and strength of ecological interactions often vary among different ontogenetic stages within a species. Although ecologists have long recognized the importance of size-structured interactions for population and community dynamics (reviewed in Ebenman and Persson 1988; De Roos et al. 2003; Claessen et al. 2004; Woodward et al. 2005), they have largely overlooked that different sized cohorts often represent distinct functional groups and could thus create trait-mediated indirect interactions (TMIIs) even in two-species systems (Rudolf 2006, 2008b, c).

Ecologists have become increasingly aware that TMIIs are replete in natural communities and can have important consequences for species interactions and the dynamics of natural communities (for reviews see Werner and Peacor 2003; Schmitz et al. 2004; and various chapters in this book). TMIIs arise when the interaction between individuals A and B is altered in the presence of individual C through plastic changes in the behaviour, physiology or other key life-history traits of either A or B (Werner and Peacor 2003). Thus TMIIs can alter the strength or even the direction in the interaction between species and their effects can often be larger than density-mediated indirect interac-tions (DMIIs; e.g., Schmitz et al. 1997; Werner and Peacor 2003; Krivan and Schmitz 2004; Schmitz et al. 2004; Preisser et al. 2005). Traditional studies on TMII have focused on three or more species systems. However, it has largely been neglected that because ontogenetic stages and size classes within populations often represent distinct functional groups, such TMIIs are also possible in two or even one species systems (Rudolf 2006). If TMIIs between individuals within species are present, this will result in non-linear interac-tions and the classical approach of averaging across individuals within pop-ulations will lead to erroneous predictions.

Several steps are required to identify when TMIIs arise from ecological variation among stages within populations and how they influence the struc-ture and dynamics of natural communities. First, we need to establish that size-classes indeed represent distinct functional groups and how they differ.

Given the difference, the next step is to ask whether this commonly leads to TMII in nature and what the underlying mechanisms are. Finally, the question is whether and how this influences the functional relationship between species and the long-term dynamics of communities. Here I review empirical and theoretical studies documenting size-structured TMIIs and identify the underlying mechanisms and their consequences for species interactions and community dynamics. Recent studies indicate that ecological communities are replete with such size-structured TMIIs and that their effects can have measurable effects in natural communities.

Size classes as distinct functional groups

Differences in body size among developmental stages often result in ontogenetic shifts in the ecological interactions of individuals (Werner and Gilliam 1984; Yang and Rudolf 2010). For example in predator–prey systems, early stages of predators often compete with their future prey resulting in a shift from competition to predation during ontogeny (Maly 1976; Werner and Gilliam 1984; Persson and Greenberg 1990; Persson et al. 1999). Similarly, when the prey outgrows the predator, this can shift predation to competition (Boone et al. 2002; Rudolf and Armstrong 2008), or even reverse predator–prey interactions (Polis et al. 1989; Wissinger 1992; Magalhaes et al. 2005). For example, young of the predatory mite are attacked and killed by young thrips, which in turn are consumed by adult mites (Magalhaes et al. 2005). Such shifts in species interactions are ubiquitous in terrestrial and aquatic systems in a diverse range of taxa, ranging from small invertebrates to predatory mammals (reviewed in Werner and Gilliam 1984; Polis et al. 1989; Yang and Rudolf 2010).

Even without such changes in interaction type, there is often a continuous variation in the per capita interaction strength across ontogenetic stages. For example in predator–prey interactions, predation rates generally depend on the relative size differences between predator and prey (Streams 1994; Persson et al. 1998; Wahlstrom et al. 2000; Aljetlawi et al. 2004; Rudolf 2008a). With gape-limited predation, prey individuals may face higher predation risk by large stages of their predator than by small stages (Taylor et al. 2001; Rudolf 2006; Urban 2007). If predators are size selective, predation risk will vary among different sized prey stages (Rudolf 2008a, b). Similarly, competitive dominance often scales with relative size difference (Werner 1994; Persson et al. 2000; Bystrom and Andersson 2005; Rudolf 2006).

In general, these studies clearly demonstrate that species interactions are typically not constant across different sized individuals within populations. Instead, different size classes (i.e., groups of individuals within a given size range) often represent distinct functional groups that engage in different ecological interactions. As most populations are size-structured, this also implies that there can be at least three distinct functional groups even in two-species systems, thus creating the potential for indirect interactions that could influence the dynamics of species interactions. How strong the indirect interactions are is likely to depend on the functional difference and thus the relative size difference between different size classes.

Stage-structured indirect interactions

A central challenge in community ecology is to understand and predict how the consumption rate of a predator is impacted by changes in the predator and prey density, because it links the dynamics of predator and prey populations. Thus, understanding how size-structured TMIIs influence the consumption rate of a predator population is key to predicting short- and long-term dynamics of natural communities. What types of indirect interactions are possible and what the underlying mechanisms are depends on whether stage-structure occurs in the consumer or in the resource. Here I will first review indirect interactions that arise due to stage-structure in the consumer followed by indirect interactions arising from stage-structure in the resource.

Size-structured predators

As different stages vary in their impact on other species, it is no surprise that they also have different effects on the behaviour of other individuals. For example, small larvae of the salamander *Eurycea cirrigera* differentially respond to the changes in the abundance of large versus small larvae of its predator *Gyrinophilus porphyriticus*. In the presence of large stages of the predator, *E. cirrigera* reduces activity and increases refuge use during the night but increases day activity (Rudolf 2006). As the predator is mostly night-active, this changed activity schedule likely reflects a classical antipredator behaviour. Interestingly, changes in the density of small predatory stages did not alter prey behaviour, indicating that the prey discriminated between predator stages. Laboratory studies indicate that predation rates increase with differences in size between both species (Rudolf, unpubl. data) suggesting that this size-specific behaviour could be based on differences in size-specific predation risk.

If prey behaviour changes based on what predator stages are present, this also means that per capita consumption rates of predator stages are not independent of each other. We can explicitly test and quantify such TMII between stages by independently manipulating the densities or presence/absence of different size classes. This has been done in the salamander system. Here, prey survival was 14% higher in treatments with both small and large predator stages than expected based on independent effects of each stage (Rudolf 2006) (Fig. 5.1). This significant interaction effect indicated that the per capita consumption rates of both stages were not independent of each other. As no mortality of small predators occurred, this suggested a non-

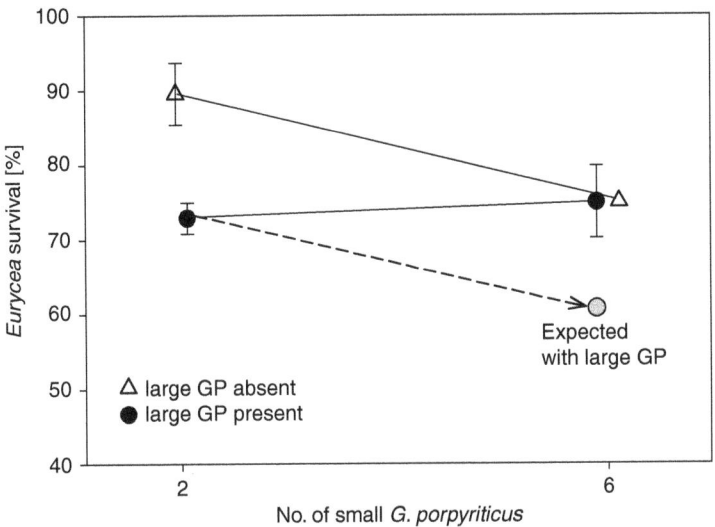

Figure 5.1 Example of non-independent effects of two predator stages of the salamander *G. porphyriticus* (GP). The dashed arrow indicates the expected value for the treatment with high density of small predators and large predators present. The expected value was calculated using a multiplicative risk model that assumes that both treatment effects are independent (= additive). The effect of both treatments are not independent; interaction term $F_{1,7.97} = 9.35$, $P = 0.0157$. With large conspecifics present, increasing the density of small GP resulted in 14.3% higher survival then expected when assuming independent effects of both predator size classes (dashed arrow). Values are ± 1 SE. Modified from Rudolf (2006). Copyright 2006 by Ecological Society of America. Reproduced with permission of Ecological Society of America in the format Tradebook via Copyright Clearance Center.

lethal, behaviourally mediated indirect effect of large predators on prey survival (Fig. 5.2a). While few studies have explicitly tested for such indirect interactions between predator stages, there is ample evidence that prey individuals alter their behaviour based on the perceived predation risk (e.g., Werner and Hall 1988; Tejedo 1993; Eklöv and Werner 2000; Ziemba *et al.* 2000), suggesting that such behavioural indirect interactions between size-classes are likely to be present in many systems.

Changes in morphological traits of the prey may also influence the interaction between predator stages. For instance, tadpoles of the Japanese brown frog (*Rana pirica*) develop a bulgier body in response to risk cues from predator larvae of the salamander *Hynobios retardatus*. This changed phenotype effectively reduces predation rates by the salamander, but increases mortality rates by 12% in size-structured salamander populations, most likely due to increased cannibalism rates (Kishida *et al.* 2009). This study illustrates how changes in the morphology of a species alter the interaction between stages within populations.

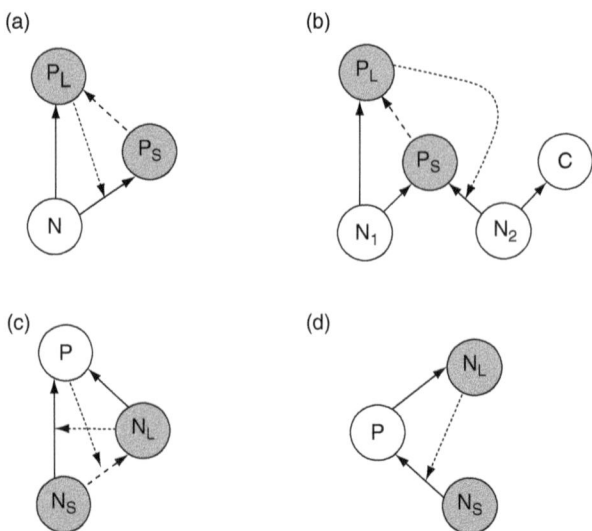

Figure 5.2 Examples of observed behavioural-mediated indirect interactions in size-structured predator (P)–prey (N) systems. (a) Presence of a large predator stage (P_L) can alter prey survival through changing the consumption rates of small conspecifics (P_S) by (i) altering behaviour of the prey or (ii) the behaviour of small conspecifics in cannibalistic predators. (b) Cannibalism or aggressive interactions between large and small predator stages can result in a resources/habitat shift in small predators (from N_1 to N_2), which can indirectly alter competitive interactions with other another species (C). (c) The net effect of the predator on prey survival can be reduced because (i) the risk of cannibalism can alter predation rates, or ii) the risk of predation can reduce cannibalism rates by changing the behaviour of small (N_S) or large (N_L) prey stages. (d) When large prey stages outgrow their former predator, the resulting mutual predation loop can alter the behaviour of predators thereby reducing their predation rates on small prey stages. Grey circles represent two stages (small S and large L) of the same species. Solid arrows indicate feeding relationships from resource to consumer, dashed arrows indicate potential cannibalistic or aggressive interactions and dotted arrows indicate behaviourally mediated indirect changes in interactions.

Another type of TMII can arise due to interactions within the population of the predator: interactions between predator stages can alter the behaviour of the predator itself, thus indirectly influencing survival of its prey. Size differences between predator stages often lead to asymmetric interactions, including interference competition or cannibalism (Polis 1981; Wissinger *et al.* 2010). To avoid injury or lethal interactions, small individuals often alter their behaviour in response to the presence of large conspecifics. For example, small chameleons alter their habitat use to avoid areas with large cannibalistic conspecifics (Keren-Rotem *et al.* 2006). Similar habitat shifts have been documented in a variety of cannibalistic fish (Persson and Eklov 1995; Greenberg *et al.* 1997; Biro *et al.* 2003; Nilsson 2006), amphibians (Rudolf 2006, 2008c; Wissinger *et al.* 2010), insects (Sih 1982; Claus-Walker *et al.* 1997; Rudolf and Armstrong 2008) and isopods (Leonardsson 1991). Interestingly, individuals may show a stronger behavioural response to conspecific than to heterospecific predators. For example small larvae of the dragonfly *Aeshna umbrosa* keep a larger distance from large cannibalistic conspecifics than from heterospecific predators. When small predators are unable to move to habitats without large conspecifics, they also commonly change their activity and foraging behaviour in the presence of large conspecifics. This behaviour has been observed in a large range of species and seems to be ubiquitous in cannibalistic systems (Johansson 1992; Van Buskirk 1992; Claus-Walker *et al.* 1997; Ziemba *et al.* 2000; Biro *et al.* 2003; Rudolf 2006, 2008b, c; Rudolf and Armstrong 2008; Wissinger *et al.* 2010).

In general, these changes in behaviour also translate to changes in the per capita consumption rates of small predator stages, indicating a behaviourally mediated indirect interaction where large predator stages indirectly influence the survival of prey consumed by their smaller conspecifics (Fig. 5.2a, b). Depending on the location of the prey, this change in behaviour can increase or decrease consumption rates. When small predators shift habitats or reduce consumption rates in the presence of large conspecifics this should also reduce the consumption rates of small stages (Fig. 5.2a). Indeed, most studies that have examined the combined effects of different predator stages found an interaction effect between size-classes that resulted in 10–45% higher prey survival than expected based on independent effects of predator stages (Sih 1982; Crumrine 2005; Griffen and Byers 2006; Rudolf 2006; Rudolf and Armstrong 2008; Crumrine 2010). These studies indicate that the magnitude of size-structured TMIIs can be very large in natural systems. The corresponding increase in prey survival due to the interference between predator stages is similar to 'risk reduction' that arises between predator species in three-species systems (Sih *et al.* 1998; Crumrine and Crowley 2003).

However, if small predator stages shift habitats, this shift often corresponds with a concurrent change in the resources consumed by small predators (Fig. 5.2b). For example, in the presence of large cannibalistic stages, small perch move to refuges in the vegetation of the littoral zone and change from feeding largely on zooplankton (cyclopoid copepods) to feeding on chironomids, benthic cladocerans and other macro-invertebrates (Persson and Eklov 1995). Interestingly, this diet change also increased resource overlap between small predator (perch) stages and their competitors (roach). This suggests that such behavioural interactions between predator stages may also indirectly alter competitive interactions between species.

Size-structured prey

If the prey is size-structured, several types of non-lethal interactions are possible that can change interactions between species by either altering interspecific predation rates or intraspecific predation rates (Fig. 5.2c). Predators are often size selective. For example, larvae of the dragonfly *A. junius* preferentially consume large stages of their prey *Plathemis lydia* (Rudolf 2008b). When both stages are present, predators spend more time pursuing large prey stages and less consuming small prey. Consequently, the presence of large stages of *P. lydia* decreases the predation rate of small *P. lydia* by altering the behaviour of *A. junius* (Rudolf 2008b).

In cannibalistic populations, additional nonlethal indirect interaction can alter predator–prey dynamics. As described above, cannibals often stimulate antipredator behaviour in their conspecific prey, including changes in habitat use and activity. Such general antipredator behaviours are also likely to be effective against heterospecific predators, thereby reducing predation rates (Fig. 5.2c). In turn, heterospecific predators can induce behavioural changes in the small prey stage or the cannibalistic stage that can then reduce cannibalism rates (Fig. 5.2c). Support for all these indirect interactions comes from experiments in a dragonfly larvae system (Rudolf 2008b). In field and laboratory experiments, cannibalistic prey showed strongly reduced activity, foraging rates and cannibalistic attacks in the presence of a predator. In addition, small prey reduced activity in the presence of both predators and cannibals. In a field experiment, the interaction between cannibals and predators resulted in 89% lower mortality of small prey than expected without indirect interactions (Fig. 5.3). A model that separated all density-mediated and behaviourally mediated indirect interactions revealed that 59% of this reduction could be explained by behaviourally mediated indirect interactions that reduced both predation rates and cannibalism rates (Fig. 5.3). Interestingly, the behaviourally mediated indirect effects were almost twice as strong as DMIIs (i.e., consumption of cannibals) (Rudolf 2008b).

Size-structured mutual predation

In systems where the prey reaches a similar or larger size than its predator, large prey stages can also trigger antipredator behaviour in their former predators (Fig. 5.2d). One such example comes from a study on the dragonfly larvae *A. junius* and *A. umbrosa*, which commonly experience reversal in predator–prey roles during their ontogeny. In the presence of a large prey stage that was identical in size to the predator, predators shifted their microhabitat use towards higher vegetation, which effectively reduced the overlap in microhabitat use with small prey stages, which stayed at lower vegetation heights. In addition, predators actively moved away from large prey stages, and reduced activity in their presence. As a result, when large prey stages were

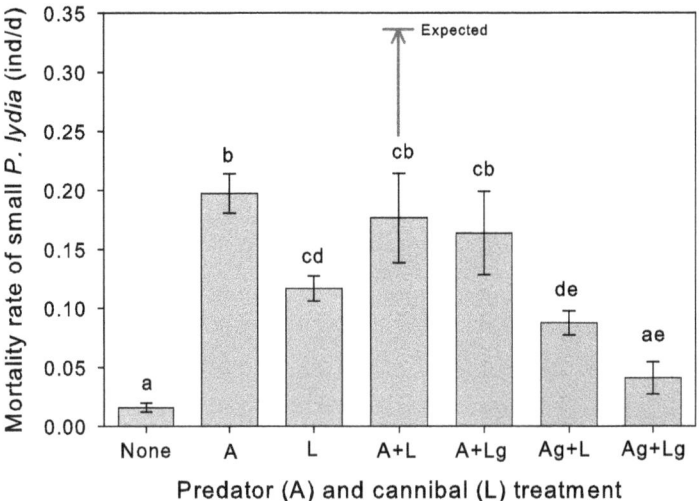

Figure 5.3 Mean (± 1 SE) mortality rate observed in all treatments manipulating the presence/absence of the top predator *A. junius* (A) and the large cannibalistic prey stage of *P. lydia* (L) and their feeding ability (g = glued mouth parts). The arrow indicates the *expected* mortality based on a multiplicative risk model that assumes independent effects of (A) and (L). Treatments with different letters are significantly different at P = 0.05 corrected for multiple comparisons. Additional randomization tests that corrected for mortality in the appropriate controls (None, Ag + Lg) indicate that cannibalism and predation rates were reduced by 59% (P < 0.02) and 33% respectively (P > 0.05) in the presence of nonlethal predators or cannibals. Modified from Rudolf (2008b). Copyright 2008 by Ecological Society of America. Reproduced with permission of Ecological Society of America in the format Tradebook via Copyright Clearance Center.

present together with predators, survival of small prey stages was 1.4 times higher than expected from their independent effects, indicating a behaviourally mediated indirect effect (Rudolf and Armstrong 2008). However, because small prey stages also altered their behaviour in response to large stages, it was unclear what percentage of this increase in survival was due to the behavioural interference between large prey stages and predators.

Structural versus numerical changes
In general, these studies clearly indicate that the effects of different ontogenetic stages are not independent of each other. This has several important implications. First, the presence of size-structured TMIIs indicate that the classical approach of averaging across individuals will often result in erroneous predictions. This argues for resolving species interactions below the species level. Second, this also implies that the functional relationship

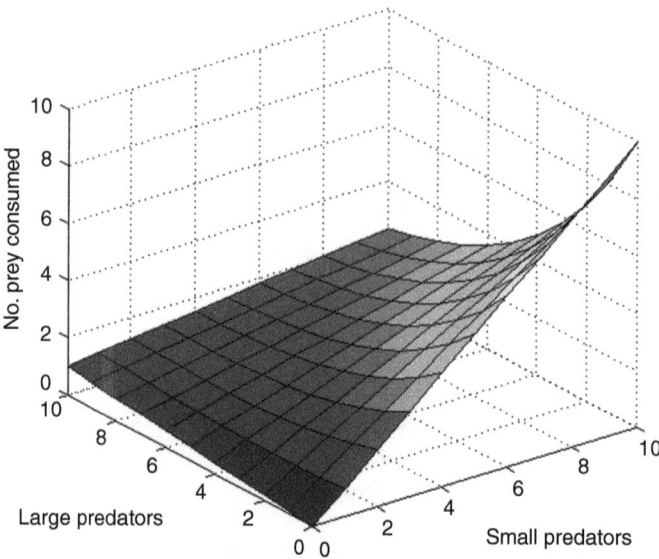

Figure 5.4 Hypothetical examples of how behavioural interference between large and small predator stages affects predator consumption rate (total number of prey consumed by small + large predators) as a function of changes in the abundance of large and small predator stages. Consumption rates were modelled after Rudolf (2007a) with the total number of consumed prey = $N(-a_L P_L - a_S e^{(-\gamma PL)} P_S)$, with the realized attack rate of small predators = $a_S e^{(-\gamma PL)}$ and a_L (= attack rate of large predators) = 0.01, a_S = 0.1, N = 10, and γ = 0.2. See colour plate section.

between species often depends on the population size-structure. For example, in the presence of size-structured indirect interactions, changes in the prey's or predator's stage-structure will alter the net effect of a predator on its prey. In this case, the per capita interaction strength between species is a function of the relative abundance of the respective stages (Fig. 5.4). This has important implications for the dynamics of species interactions.

Population dynamics can be characterized in at least two ways that are inherently connected: (1) *numerical* changes, such that the total number of individuals within a population increases or decreases. This represents the classical unstructured approach that does not differentiate between stages; (2) *structural* changes, where the relative abundance of different ontogenetic stages changes within a population (sensu Rudolf 2006). Much of the conceptual basis of community ecology assumes that dynamics are solely governed by changes in the densities of populations (i.e., numerical changes). However, intrinsic numerical changes (e.g., population growth) are inherently coupled with structural changes. For example, when an iteroparous species reproduces, this will increase the total number of individuals, but only early ontogenetic stages increase while later stages will either remain constant or decrease.

Similarly, natural or anthropogenic factors often have differential effects on the mortality of different stages (e.g., size-selected harvesting).

As outlined above, in the presence of size-structured indirect interactions the per capita interaction strength between species depends on the relative frequency of different stages within a species. Consequently, the effect of a numerical change (i.e., change in *total* number of individuals within a population) is determined by the concurrent structural change within a population. This has been empirically shown in predator–prey interactions in a salamander system, where, in the absence of large conspecifics, increasing the density of small predators increased mortality in its prey. However, when large conspecifics were present, increasing the density of small predators had no significant effect on prey mortality (Rudolf 2006) (Fig. 5.1). This study shows that numerical changes are clearly important, but their effect is often context dependent, i.e., depends on the size-distribution (structural changes) within populations. Thus, predicting the dynamics of species interactions solely based on numerical changes without accounting for the population structure will often lead to erroneous predictions. Instead we need to account for stage-structured indirect interactions within and between populations to predict correctly the dynamics of species interactions.

Effects on short- and long-term dynamics

As reviewed above, there is increasing evidence that stage-structured indirect interactions can alter the functional relationship (i.e., per capita interaction strength) between species, thereby indirectly altering predator–prey or competitive interactions. The question is, however, how does this influence the short-term (i.e., in less than a generation) and long-term (i.e., multiple generation) dynamics of natural communities?

Short-term dynamics

Functional responses determine how the consumption rate of a predator scales with the prey and its own density. Because functional responses link predator and prey densities, understanding the factors that determine the functional response is imperative to predict population and community dynamics. Current formulations of functional responses are dominated by Holling's (1959) three types of functional responses and their extensions, which generally assume that populations are homogenous. As reviewed above, this assumption is often not met, and stages within populations influence each other's per capita consumption rates. To obtain a more general understanding of how stage-structured TMIIs influence the consumption rate of a predator population, we can integrate the empirical observed indirect interactions into the functional response of a species. This has been done in a recent study for a stage-structured predator–prey model (Rudolf 2007a). In this model, the predation rate of small, juvenile predator stages decreased as a function of the density of large, adult predator stages (Fig. 5.4). Analysis of this model indicates that the short-term (i.e. within a generation) effect of the predator on its prey depends on the relative frequency of both stages and specific parameter combination (Fig. 5.4). Interestingly, increasing the abundance of large predators in this scenario can either have a short-term negative effect, or a counterintuitive positive effect depending on their relative abundance and strength of interference between stages (Fig. 5.4). While the conditions for the positive effect depend on the specific parameter combination (Rudolf 2007a) the effect is generally stronger at relatively high densities of small predator stages (Fig. 5.4), and when the per capita attack rate of small predators is larger than that of small predators. This example suggests that behaviourally mediated indirect interactions between stage-structured predators can not only alter the short-term effect of numerical changes in a predator on its prey (i.e., the functional response), but fundamentally alter and even reverse the net effect of such a numerical effect (i.e., change it from a negative effect to a positive effect). While we currently lack experiments that test this explicitly, this could be easily done using response-surface experiments that manipulate the relative frequency of different predator stages.

Long-term dynamics

One of the immediate questions that always arise with behavioural interactions is whether the observed short-term effects impact long-term dynamics. This is particularly challenging to study empirically and we generally lack long-term studies that explicitly test this. However, following the approach in studies of three-species systems (e.g., Krivan and Schmitz 2004), we can use empirical observation on how non-lethal indirect interactions influence the functional relationship between stage-structured species to expand classical predator–prey models to make some long-term predictions.

Given that stage-structured indirect interactions alter the per capita consumption rate of a predator we would expect that they have at least some long-term effect on population densities and stability of the system. Indeed, recent analyses suggest that behavioural interference between predator stages can alter long-term dynamics of predator–prey system in various ways. Model analysis suggest that if small predators reduce their foraging rates in response to large conspecifics, this behaviourally mediated indirect interaction can stabilize otherwise unstable predator–prey dynamics (Rudolf 2007a). This intriguing result suggests that the simple behavioural interference between predator stages could provide an intrinsic mechanism to stabilize the dynamics of natural populations and counteract potential de-stabilizing effects such as enrichment of the resource (Rosenzweig and MacArthur 1963; Rosenzweig 1971) or intraspecific competition between predator cohorts.

Indirect behavioural interactions between predator stages can also alter the dynamics of trophic cascades. In classical unstructured predator–prey models, increasing the productivity of the prey will only result in an increase in the predator density but not alter the prey density (Oksanen et al. 1981; Leibold et al. 1997). As a consequence, predator and prey abundances are expected to be uncorrelated across a resource productivity gradient. This, however, is inconsistent with the positive correlation between predator and prey abundances observed in most natural systems (McQueen et al. 1986; Ginzburg and Akcakaya 1992; Leibold et al. 1997). When large stages of the predator reduce foraging rates of smaller conspecifics, this can completely change the classical predictions: now both predator and prey abundances can be positively correlated across a productivity gradient, consistent with empirical observations. Increasing the behavioural interference between stages generally increases the positive effect on prey density. In addition, the behavioural interference can also decrease the strength of top-down cascades (e.g., due to increased predator mortality). However, this long-term effect on the strength of trophic cascades is only expected in systems where the reduced foraging rate of the small predator stage also reduces their own developmental and maturation rates. This makes intuitive sense; if reducing the attack rate of small predators has no impact on their development or growth rate, the behavioural change only alters the interaction strength but it is disconnected from the numerical response of the predator (i.e., it won't affect the growth rate or equilibrium density of the predator). If the maturation rate is affected, this will result in a density-dependent feedback that regulates the predator population (i.e. the growth rate of the predator will depend on its own density). The strength of the behavioural interference then determines how strongly the predator density is regulated. This indicates that TMII can have important delayed effects through changes in growth rates that need to be considered when examining the importance of TMII in size-structured systems.

The effect on the maturation rate of the predator is commonly observed in systems where small predators cannot shift to alternative resources or another habitat. For example, in several salamander species, small individuals not only exhibit reduced foraging rates, but also strongly reduced growth rates in the presence of large conspecifics (Rudolf 2006; Wissinger et al. 2010). If, however, small individuals can shift to an alternative resource this is likely to have little effect on their maturation rates. For example, in the presence of large cannibalistic conspecifics, small perch shift from the pelagic, open water to the littoral, vegetation zone of lakes. While this reduces feeding rates on zooplankton in the presence of large conspecifics, this reduction has little effect on the maturation rate as individuals simply shift their diet to feeding on other invertebrates (Persson and Eklov 1995). In such systems, indirect behavioural interactions between stages are unlikely to change qualitatively the general relationship of trophic cascades. However, the indirect interactions will still lead to quantitative changes in the dynamic of the system (e.g., increase equilibrium densities of the prey).

In general, these theoretical studies suggest that non-lethal indirect interactions can have profound long-term consequences for the dynamics and structure of natural communities. However, how the long-term dynamics are changed depends on whether behavioural interactions also change developmental rates or reproductive output of populations. Thus, long-term effects of stage-structured indirect interactions are likely to differ among systems, but using general knowledge about the natural history of species can provide important insight into the long-term consequences for community dynamics and structure. In general these studies suggest that we need to account for non-lethal stage-structured interactions within and between populations to make reliable predictions about the short- and long-term dynamics of natural communities and how they respond to natural or anthropogenic changes. They also emphasize that we need to account for the effects of TMII on the developmental and growth rates of the predator to predict the long-term dynamics of natural communities.

Expanding the TMII concept

In the current literature, non-lethal indirect interactions are commonly termed TMIIs (Abrams 1995). Traditionally this term has been restricted to three or more species systems. However, as reviewed above, different size classes within populations often represent distinct functional groups that exhibit similar non-lethal indirect interactions in two-species systems. Indeed, many of the mechanisms underlying TMIIs can be observed in such size-structured two-species systems. For example, individuals often show similar changes in foraging behaviour when exposed to conspecifics or heterospecific predators if both pose similar predation threats (Rudolf and Armstrong 2008). Interactions between different predator stages typically result in higher than expected prey survival, similar to 'risk-reduction' observed in multi-predator species systems (Sih et al. 1998; Crumrine and Crowley 2003). Increasing evidence suggests that it is typically not the species identity that matters, but rather the type and strength of interaction between individuals. Given that the ecological variation among different ontogenetic stages within populations often exceeds the variation among similar sized species (Polis 1984; Werner and Gilliam 1984; Munoz and Ojeda 1998; Woodward and Hildrew 2002; Griffen and Byers 2006; Rudolf 2008c; Rudolf and Armstrong 2008), the term TMII should refer to the mechanism and be extended to include indirect interaction between distinct functional groups, irrespective of the number of species involved. Indeed, TMII may even occur within a single species if at least three different functional groups are present. For example, in dragonfly larvae, cannibalism occurs with only one or two instar differences between individuals (Wissinger 1988), effectively creating a 'within-species food chain'. Recent experiments suggest that this can lead to TMII between different sized instars that alter cannibalism rates and population dynamics (Rudolf unpublished.).

While many mechanisms (e.g., changes in behaviour) of size-structured TMII often resemble those of classical TMII in unstructured systems between species, it is important to keep the differences in mind. In size-structured systems, different functional groups are actually members of the same species and thus their dynamics are inherently linked. This link has important implications for the long-term consequences of TMII for population dynamics. As reviewed above, changes in the growth rates or developmental rates can fundamentally alter the long-term dynamics of predator–prey systems because they alter transition rates between stages. These changes in transition rates lead to density-mediated effects that alter the size-structure of a population, which determines species interactions and population dynamics. This emphasizes the importance of accounting for changes in growth rates and transition rates within populations when studying size-structured TMII.

Future directions

Although the importance of shifts in ecological interactions during the ontogeny of individuals for the dynamics of populations has long been recognized by ecologists (Werner and Gilliam 1984; Ebenman and Persson 1988; Persson 1999; De Roos et al. 2003), we still know relatively little about the consequences of these shifts for the dynamics and structure of natural communities. Increasing evidence indicates that because different ontogenetic stages or size classes often represent distinct functional groups, this allows for complex indirect interactions that can result in dynamics that cannot be predicted by classical unstructured models (Rudolf 2006; Persson et al. 2007; Rudolf 2007b; De Roos et al. 2008). However, most of this work has focused on density-mediated interactions, and little attention has been paid to the importance of nonlethal indirect interactions. There is now a large body of work demonstrating the importance of TMII for the dynamics and structure of three and more species systems (reviewed in Werner and Peacor 2003; Schmitz et al. 2004). Here I have shown that TMIIs often arise in stage-structured populations, and their short- and long-term effects can be similar to or even larger than DMIIs (Rudolf 2008b, c). In addition, DMIIs and TMIIs are inherently coupled. For example, the act of cannibalism also generally triggers antipredator behaviour in potential victims, and recent studies indicate that the behavioural effects will enhance the effect of cannibalism on trophic cascades (Rudolf 2007a). This suggests that we often need to account for both DMIIs and

TMIIs in stage-structured systems to predict the net effect of species on other members in the community and the resulting community dynamics.

Much more work is needed, however, to understand fully the short- and long-term consequences of these indirect interactions in stage-structured systems. In particular we need to understand (1) how they alter the functional relationship between interacting species and (2) what the underlying mechanisms are. Although several studies indicate that the interactions between different predator stages are often non-linear, the underlying mechanisms have rarely been identified. Are they consistent across different types of systems? Similarly, there are almost no studies that have looked at how stage structure in the prey alters the functional relationship between species and what indirect interactions are involved. In this context, it is also important to realize that although current studies largely focus on a few distinct size classes, many natural populations consist of a range of different sized individuals. Given that size typically determines interaction strength, this also suggests that TMII should change with changes in relative size, but this has not been tested empirically. Increasing the size range within populations increases the number of possible TMII, and it is unclear how they will add up (i.e., whether they are independent of each other) to determine the net effect on species interactions. However, this knowledge is especially important given that the size-distribution of populations often changes within and between seasons.

There is also an apparent bias in what types of interactions are studied. Almost all studies on size-structured indirect interactions focus on predation, and little is known about their importance for other species interactions, such as competition, parasitism or mutualism. Consequently, it is unclear whether the observed patterns hold true across various systems and interactions, and in which systems we expect stage-structured indirect interactions to be more important. For example most studies on stage-structured interactions have been conducted in aquatic systems. Yet, there is ample evidence that ontogenetic niche shifts in the ecology of individuals are also occurring in terrestrial systems, including plants (Polis 1991; Boege and Marquis 2005). In addition, current studies on size-structured TMII are largely focused on behaviour, while other traits are largely neglected despite their potential importance for species interactions. For example, in the red-eyed treefrog, predators that feed on arboreal eggs induce early hatching of tadpoles, which are consequently smaller and potentially more vulnerable to predation in their aquatic stage. However, because early hatchlings grow faster (Vonesh and Osenberg 2003), this reduced predation by larval predators by 44%, indicating a TMII between egg and larval predators that would otherwise not interact with each other. In this scenario, interactions in one stage actually changed multiple physiological traits of the subsequent stage. Phenotypic plasticity in growth and developmental rates are common in a variety of taxa, suggesting that similar indirect effects are likely to be present in a variety of other systems, but little is known about their consequences. Thus, much work remains to be done to understand fully the extent to which size-structured TMII influence the dynamics and structure of natural communities.

Acknowledgements

I would like to thank the editors for giving me the opportunity to write this chapter, and A. E. Dunham and two anonymous reviewers for their valuable comments on an earlier version of the manuscript. This work was partially supported by NSF DEB- 0841686 to VHWR.

References

Abrams, P. A. (1995) Implications of dynamically variable traits for identifying, classifying, and measuring direct and indirect effects in ecological communities. *American Naturalist*, **146**, 112–134.

Aljetlawi, A. A., Sparrevik, E. and Leonardsson, K. (2004) Prey–predator size-dependent functional response: derivation and rescaling to the real world. *Journal of Animal Ecology*, **73**, 239–252.

Benton, T. G., Plaistow, S. J. and Coulson, T. N. (2006) Complex population dynamics and complex causation: devils, details and demography. *Proceedings of the Royal Society of London, Series B*, **273**, 1173–1181.

Biro, P. A., Post, J. R. and Parkinson, E. A. (2003) From individuals to populations: prey fish risk-taking mediates mortality in whole-system experiments. *Ecology*, **84**, 2419–2431.

Boege, K. and Marquis, R. J. (2005) Facing herbivory as you grow up: the ontogeny of resistance in plants. *Trends in Ecology and Evolution*, **20**, 441–448.

Boone, M. D., Scott, D. E. and Niewiarowski, P. H. (2002) Effects of hatching time for larval ambystomatid salamanders. *Copeia*, **2**, 511–517.

Bystrom, P. and Andersson, J. (2005) Size-dependent foraging capacities and intercohort competition in an ontogenetic omnivore (Arctic char). *Oikos*, **110**, 523–536.

Claessen, D., De Roos, A. M. and Persson, L. (2004) Population dynamic theory of size-dependent cannibalism. *Proceedings of the Royal Society of London, Series B*, **271**, 333–340.

Claus-Walker, D. B., Crowley, P. H. and Johansson, F. (1997) Fish predation, cannibalism, and larval development in the dragonfly *Epitheca cynosura*. *Canadian Journal of Zoology*, **75**, 687–696.

Crumrine, P. (2005) Size structure and substitutability in an odonate intraguild predation system. *Oecologia*, **145**, 132–139.

Crumrine, P. W. (2010) Size-structured cannibalism between top predators promotes the survival of intermediate predators in an intraguild predation system. *Journal of the North American Benthological Society*, **29**, 636–646.

Crumrine, P. W. and Crowley, P. H. (2003) Partitioning components of risk reduction in a dragonfly-fish intraguild predation system. *Ecology*, **84**, 1588–1597.

De Roos, A. M., Persson, L. and McCauley, E. (2003) The influence of size-dependent life-history traits on the structure and dynamics of populations and communities. *Ecology Letters*, **6**, 473–487.

De Roos, A. M., Schellekens, T., Van Kooten, T. and Persson, L. (2008) Stage-specific predator species help each other to persist while competing for a single prey. *Proceedings of the*

National Academy of Sciences of the United States of America, **105**, 13930–13935.

Ebenman, B. and Persson, L. (1988) *Size-Structured Populations: Ecology and Evolution*. Berlin: Springer-Verlag.

Eklöv, P. and Werner, E. E. (2000) Multiple predator effects on size-dependent behavior and mortality of two species of anuran larvae. *Oikos*, **88**, 250–258.

Ginzburg, L. R. and Akcakaya, H. R. (1992) Consequences of ratio-dependent predation for steady-state properties of ecosystems. *Ecology*, **73**, 1536–1543.

Greenberg, L. A., Bergman, E. and Eklov, A. G. (1997) Effects of predation and intraspecific interactions on habitat use and foraging by brown trout in artificial streams. *Ecology of Freshwater Fish*, **6**, 16–26.

Griffen, B. D. and Byers, J. E. (2006) Intraguild predation reduces redundancy of predator species in multiple predator assemblage. *Journal of Animal Ecology*, **75**, 959–966.

Holling, C. S. (1959) Some characteristics of simple types of predation and parasitism. *Canadian Entomologist*, **91**, 385–398.

Holt, R. D. (1997) Community modules. In A. C. Gange and V. K. Brown, eds., *Multitrophic Interactions in Terrestrial Systems*. Oxford: Blackwell Science, pp. 333–349.

Johansson, F. (1992) Effects of zooplankton availability and foraging mode on cannibalism in 3 dragonfly larvae. *Oecologia*, **91**, 179–183.

Keren-Rotem, T., Bouskila, A. and Geffen, E. (2006) Ontogenetic habitat shift and risk of cannibalism in the common chameleon (*Chamaeleo chamaeleon*). *Behavioral Ecology and Sociobiology*, **59**, 723.

Kishida, O., Trussell, G. C., Nishimura, K. and Ohgushi, T. (2009) Inducible defenses in prey intensify predator cannibalism. *Ecology*, **90**, 3150–3158.

Krivan, V. and Schmitz, O. J. (2004) Trait and density mediated indirect interactions in simple food webs. *Oikos*, **107**, 239–250.

Leibold, M. A., Chase, J. M., Shurin, J. B. and Downing, A. L. (1997) Species turnover and the regulation of trophic structure. *Annual Review of Ecology and Systematics*, **28**, 467–494.

Leonardsson, K. (1991) Effects of cannibalism and alternative prey on population-dynamics of *Saduria entomon* (Isopoda). *Ecology*, **72**, 1273–1285.

McQueen, D. J., Post, J. R. and Mills, E. L. (1986) Trophic relationships in fresh-water pelagic ecosystems. *Canadian Journal of Fisheries and Aquatic Sciences*, **43**, 1571–1581.

Magalhaes, S., Janssen, A., Montserrat, M. and Sabelis, M. W. (2005) Prey attack and predators defend: counterattacking prey trigger parental care in predators. *Proceedings of the Royal Society of London, Series B*, **272**, 1929–1933.

Maly, E. J. (1976) Resource overlap between co-occurring copepods: Effects of predation and environmental fluctuation. *Canadian Journal of Zoology*, **54**, 933–940.

Montoya, J. M., Pimm, S. L. and Sole, R. V. (2006) Ecological networks and their fragility. *Nature*, **442**, 259–264.

Munoz, A. A. and Ojeda, F. P. (1998) Guild structure of carnivorous intertidal fishes of the Chilean coast: implications of ontogenetic dietary shifts. *Oecologia*, **114**, 563–573.

Nilsson, A. P. (2006) Avoid your neighbours: size-determined spatial distribution patterns among northern pike individuals. *Oikos*, **113**, 251–258.

Oksanen, L., Fretwell, S. D., Arruda, J. and Niemela, P. (1981) Exploitation ecosystems in gradients of primary productivity. *American Naturalist*, **118**, 240–261.

Pascual, M. and Dunne, J. A. (2005) *Ecological Networks: Linking Structure to Dynamics in Food Webs*. New York: Oxford University Press.

Persson, L. (1999) Trophic cascades: abiding heterogeneity and the trophic level concept at the end of the road. *Oikos*, **85**, 385–397.

Persson, L. and Eklov, P. (1995) Prey refuges affecting interactions between piscivorous perch and juvenile perch and roach. *Ecology*, **76**, 70–81.

Persson, L. and Greenberg, L. A. (1990) Juvenile competitive bottlenecks: the perch (*Perca fluviatilis*)–roach (*Rutilus rutilus*) interaction. *Ecology*, **71**, 44.

Persson, L., Amundsen, P.-A., De Roos, A. M. *et al.* (2007) Culling prey promotes predator recovery: alternative states in a whole-lake experiment. *Science*, **316**, 1743–1746.

Persson, L., Bystrom, P. and Wahlstrom, E. (2000) Cannibalism and competition in Eurasian perch: population dynamics of an ontogenetic omnivore. *Ecology*, **81**, 1058–1071.

Persson, L., Bystrom, P., Wahlstrom, E., Andersson, J. and Hjelm, J. (1999) Interactions among size-structured populations in a whole-lake experiment: size- and scale-dependent processes. *Oikos*, **87**, 139–156.

Persson, L., Leonardsson, K., de Roos, A. M., Gyllenberg, M. and Christensen, B. (1998) Ontogenetic scaling of foraging rates and the dynamics of a size-structured consumer–resource model. *Theoretical Population Biology*, **54**, 270–293.

Polis, G. A. (1981) The evolution and dynamics of intraspecific predation. *Annual Review of Ecology and Systematics*, **12**, 225–251.

Polis, G. A. (1984) Age structure component of niche width and intraspecific resource partitioning: can age groups function as ecological species? *American Naturalist*, **123**, 541–564.

Polis, G. A. (1991) Complex trophic interactions in deserts: an empirical critique of food-web theory. *American Naturalist*, **138**, 123–155.

Polis, G. A., Myers, C. A. and Holt, R. D. (1989) The ecology and evolution of intraguild predation: potential competitors that eat each other. *Annual Review of Ecology and Systematics*, **20**, 297–330.

Preisser, E. L., Bolnick, D. I. and Benard, M. F. (2005) Scared to death? The effects of intimidation and consumption in predator–prey interactions. *Ecology*, **86**, 501–509.

Rosenzweig, M. L. (1971) Paradox of enrichment: destabilization of exploitation ecosystems in ecological time. *Science*, **171**, 385–387.

Rosenzweig, M. L. and MacArthur, R. H. (1963) Graphical representation and stability conditions of predator–prey interactions. *American Naturalist*, **97**, 209–223.

Rudolf, V. H. W. (2006) The influence of size-specific indirect interactions in predator–prey systems. *Ecology*, **87**, 362–371.

Rudolf, V. H. W. (2007a) Consequences of stage-structured predators: cannibalism, behavioral effects and trophic cascades. *Ecology*, **88**, 2991–3003.

Rudolf, V. H. W. (2007b) The interaction of cannibalism and omnivory: consequences for community dynamics. *Ecology*, **88**, 2697–2705.

Rudolf, V. H. W. (2008a) Consequences of size structure in the prey for predator–prey dynamics: the composite functional response. *Journal of Animal Ecology*, **77**, 520–528.

Rudolf, V. H. W. (2008b) The impact of cannibalism in the prey on predator–prey dynamics. *Ecology*, **89**, 3116–3127.

Rudolf, V. H. W. (2008c) Impact of cannibalism on predator–prey dynamics: size-structured interactions and apparent mutualism. *Ecology*, **89**, 1650–1660.

Rudolf, V. H. W. and Armstrong, J. (2008) Emergent impacts of cannibalism and size refuges in the prey on intraguild predation systems. *Oecologia*, **157**, 675–686.

Schmitz, O. J., Beckerman, A. P. and Obrien, K. M. (1997) Behaviorally mediated trophic cascades: effects of predation risk on food web interactions. *Ecology*, **78**, 1388–1399.

Schmitz, O. J., Krivan, V. and Ovadia, O. (2004) Trophic cascades: the primacy of trait-mediated indirect interactions. *Ecology Letters*, **7**, 153–163.

Sih, A. (1982) Foraging strategies and the avoidance of predation by an aquatic insect, *Notonecta hoffmanni*. *Ecology*, **63**, 786–796.

Sih, A., Englund, G. and Wooster, D. (1998) Emergent impacts of multiple predators on prey. *Trends in Ecology and Evolution*, **13**, 350–355.

Streams, F. A. (1994) Effect of prey size on attack components of the functional-response by *Notonecta undulata*. *Oecologia*, **98**, 57–63.

Taylor, R. C., Trexler, J. C. and Loftus, W. F. (2001) Separating the effects of intra- and interspecific age-structured interactions in an experimental fish assemblage. *Oecologia*, **127**, 143–152.

Tejedo, M. (1993) Size-dependent vulnerability and behavioral responses of tadpoles of two anuran species to beetle larvae predators. *Herpetologica*, **49**, 287–294.

Urban, M. (2007) Predator size and phenology shape prey survival in temporary ponds. *Oecologia*, **154**, 571–580.

Van Buskirk, J. (1992) Competition, cannibalism, and size class dominance in a dragonfly. *Oikos*, **65**, 455–464.

Vonesh, J. R. and Osenberg, C. W. (2003) Multi-predator effects across life-history stages: non-additivity of egg- and larval-stage predation in an African treefrog. *Ecology Letters*, **6**, 503–508.

Wahlstrom, E., Persson, L., Diehl, S. and Bystrom, P. (2000) Size-dependent foraging efficiency, cannibalism and zooplankton community structure. *Oecologia*, **123**, 138–148.

Werner, E. E. (1994) Ontogenic scaling of competitive relations: size-dependent effects and responses in two anuran larvae. *Ecology*, **75**, 197–213.

Werner, E. E. and Gilliam, J. F. (1984) The ontogenetic niche and species interactions in size structured populations. *Annual Review of Ecology and Systematics*, **15**, 393–425.

Werner, E. E. and Hall, D. J. (1988) Ontogenetic habitat shifts in bluegill: the foraging rate-predation risk trade-off. *Ecology*, **69**, 1352–1366.

Werner, E. E. and Peacor, S. D. (2003) A review of trait-mediated indirect interactions in ecological communities. *Ecology*, **84**, 1083–1100.

Wilbur, H. M. (1980) Complex life-cycles. *Annual Review of Ecology and Systematics*, **11**, 67–93.

Wissinger, S. A. (1988) Effects of food availability on larval development and inter-instar predation among larvae of *Libellula lydia* and *Libellula luctuosa* (Odonata, Anisoptera). *Canadian Journal of Zoology*, **66**, 543–549.

Wissinger, S. A. (1992) Niche overlap and the potential for competition and intraguild predation between size-structured populations. *Ecology*, **73**, 1431–1444.

Wissinger, S. A., Whiteman, H. H., Denoel, M., Mumford, M. L. and Aubee, C. B. (2010) Consumptive and nonconsumptive effects of cannibalism in fluctuating age-structured populations. *Ecology*, **91**, 549–559.

Woodward, G., Ebenman, B., Emmerson, M. *et al.* (2005) Body size in ecological networks. *Trends in Ecology and Evolution*, **20**, 402–409.

Woodward, G. and Hildrew, A. G. (2002) Body-size determinants of niche overlap and intraguild predation within a complex food web. *Journal of Animal Ecology*, **71**, 1063–1074.

Yang, L. H. and Rudolf, V. H. W. (2010) Phenology, ontogeny, and the effects of climate change on the timing of species interactions. *Ecology Letters*, **13**, 1–10.

Ziemba, R. E., Myers, M. T. and Collins, J. P. (2000) Foraging under the risk of cannibalism leads to divergence in body size among tiger salamander larvae. *Oecologia*, **124**, 225–231.

CHAPTER SIX

Trait-mediated effects, density dependence and the dynamic stability of ecological systems

ROBERT D. HOLT and MICHAEL BARFIELD

Department of Biology, University of Florida

Introduction

One of the most basic things that can be said about the ecology of a species is that the traits of its members (in a given environmental setting) influence that species' demographic rates – births, deaths and movements – as well as transitions such as changes in body size or condition. When species interact, their demographic rates interlock, so changes in the abundance or traits of one species affect the vital rates of the others (and vice versa, sometimes). A message elaborated throughout this volume is that due to the fundamental connection between traits and demography, the traits of interacting individuals, and in particular plasticity in those traits, can have ramifying influences for many different issues in ecology and evolutionary biology, including in particular the percolation of indirect interactions through complex communities (e.g., Krivan and Schmitz 2004; Ohgushi 2005; Abrams 2010; Utsumi *et al.* 2010).

A perennial issue in population ecology is understanding how density dependence influences the abundance and distribution of species. Here we argue that an explicit concern with density dependence is – or should be – of central interest as well in articulating the ecological impact of trait-mediated indirect effects (TMIEs). We start with general comments on how the traits involved in indirect interactions can also influence density dependence. We provide a brief overview of prior literature that touches on density dependence in the context of trait-mediated interactions and present a simple theoretical model to illustrate how trait-mediated effects can influence density dependence. The term 'stability' in ecology encompasses several distinct concepts, such as the return of a population or community to a reference state following disturbance, or the persistence of ecological systems over time (Grimm and Wissel 1997). Understanding the causes of stability for essentially all these meanings requires one to gauge how density dependence operates over both short and long timescales, via both direct and indirect mechanisms.

Trait-Mediated Indirect Interactions: Ecological and Evolutionary Perspectives, eds. Takayuki Ohgushi, Oswald J. Schmitz and Robert D. Holt. Published by Cambridge University Press. © Cambridge University Press 2012.

Density dependence and trait-mediated interactions

Why is the topic of density dependence relevant in a volume on trait-mediated indirect interactions (TMIIs)? A full understanding of trait-mediated effects in community ecology, we suggest, requires one to consider carefully intraspecific density dependence, and shifts in density dependence that accompany changes in traits. Consider a chain of interacting species, A → B → C. The total effect of B on C (measured say in an altered growth rate) is given by $a_{CB}N_B$, a per capita effect of B on C, times the abundance of B. The per capita effect should reflect various traits in species B. A change in the abundance or activity of species A can indirectly influence species C via either of these terms, which we might represent as $\Delta a_{CB}N_B$ and $a_{CB}\Delta N_B$. The second of these denotes a 'density-mediated effect', and the first, a 'trait-mediated effect'. Density dependence can enter this formulation in two distinct ways. First, when traits are fixed, the magnitude of direct density dependence can influence the magnitude of ΔN_B. For instance, a change in herbivore mortality due to predation may not greatly affect the impact of a herbivore upon its plant resources via the density-mediated effect, if predation acts as a compensatory mortality source (Rosenzweig 1974). Second, if the trait that is affected influences the strength of density dependence, a change in the trait will alter ΔN_B, too.

Direct density dependence is not always important. Some taxa experience only weak or intermittent density dependence (Ziebarth et al. 2010), and density dependence may arise entirely indirectly and over long timescales via interactions with other species. But in others, it is strong, within-generation, and evident in even short time series (e.g., elephants, Sinclair 2003) or experiments (e.g., *Maculinea* butterflies, Nowicki et al. 2009). Density dependence should be influenced by species' traits, so shifts in traits due to interspecific interactions can alter density dependence, as well. Few empirical studies have as yet focused on this issue, but consider the following hypothetical – but biologically reasonable – scenarios, involving basic mechanisms of density dependence:

- Within-generation outbreaks of specialized pathogens (e.g., Bagchi et al. 2010), with density-dependent transmission, can be expressed in population models as forms of direct density dependence. One expression of physiological stress due to, say, predation (Hawlena and Schmitz 2010) may be reduced resources, suppressing immune functionality (Lochmiller and Deerenberg 2000). Horak et al. (2006), for instance, showed that exposure to the odour of a predator (a cat) reduced the immunocompetence of prey (rats). If infected rats stay infected longer in the presence of a predator, this enhances disease transmission and could strengthen this mechanism of density dependence. Alternatively, if stressed, infected rats more rapidly die, disease prevalence declines and infectious disease could be less important as a source of density dependence.

- Given direct interference among predators while foraging, the functional response can depend on predator as well as prey density (Beddington 1975; DeAngelis *et al.* 1975; Skalski and Gilliam 2001; Kratina *et al.* 2009) implying direct density dependence.
- Consumers which rapidly consume depletable resources experience strong direct density dependence (Schoener 1973; Schmitz 1992, 2010). Trait changes that alter consumption rates, or behavioural shifts among habitats differing in resource renewal, can alter the strength of density dependence via resource competition. In many large vertebrate herbivores, density dependence is focused on reproduction (Gaillard *et al.* 2000). Predation risk can impact reproductive physiology (e.g., Creel *et al.* 2007) suppressing reproduction (Sheriff *et al.* 2009). As Creel *et al.* (2007) suggest, this could easily be mistaken for bottom-up, density-dependent limitation by resources.
- Although less frequently examined, positive density dependence also arises because of Allee effects (Courchamp *et al.* 2009; Brashares *et al.* 2010), as well as other mechanisms such as saturating predatory responses (Sinclair 2003), all of which could be involved in trait-mediated pathways of interspecific interactions. For instance, if predation restricts prey movement, this may make it harder for mates to encounter each other, contributing to an Allee effect.

A consideration of density dependence is of obvious importance in population ecology in explaining persistence, average abundance and bounds on temporal instability, but density dependence also lies at the heart of many issues in community ecology. To gauge the potential for competitive coexistence at a stable equilibrium, for instance, one compares the strength of interspecific competition to the strength of intraspecific competition (i.e., density dependence; for a formal proof for models of direct competition, see Holt 1985, pp. 489–490). If the latter is weak, relative to the former, coexistence at a stable equilibrium is precluded. As another example, if a predator depends on a single prey species, for the system to be stable rather than exhibit sustained limit cycles requires at least one species to experience direct density dependence (which may arise independently of the interaction, e.g., due to territoriality in the predator, or instead be an integral part of the interaction, e.g., a Type III functional response when prey are scarce, such that an increase in prey density leads to an immediate increase in per prey predation rates). Conversely, positive density dependence tends to be de-stabilizing, for instance leading to population outbreaks and population cycles in trophic interactions. Positive density dependence can lead to alternative community states, where species composition is strongly influenced by initial conditions (see Holt 1977, pp. 211–212 for an example).

The topic of density dependence has been addressed in a somewhat scattered fashion in the literature on TMIEs. In the predator–prey literature, there has been a long-standing concern with how nonlinear functional responses due to behavioural plasticity could lead to negative and positive density dependence, with consequences for stability (Fryxell and Lundberg 1998). Although the term 'trait-mediated' was not then in vogue, early treatments of Type III functional responses and prey switching and their impacts on competitive coexistence (e.g., Holt 1977; Roughgarden and Feldman 1975; Murdoch and Oaten 1975) all implicitly involved TMIEs. Baldly stated: prey A increases in abundance, then a predator shifts a trait – its foraging activity, microhabitat use, foraging speed or cognitive awareness of particular sensory patterns (search image) – and this altered trait affects its predation of prey B. Often (not always), flexible foraging tactics can stabilize otherwise unstable dynamics (van Baalen *et al.* 2001; Kimbrell and Holt 2004). Antipredator behaviour by prey can sometimes moderate the magnitude of unstable predator–prey dynamics (Krivan 1998), and adaptive foraging by parasitoids can likewise stabilize otherwise unstable host–parasitoid dynamics (Krivan 1997).

Peter Abrams in particular has touched on the issue of density dependence in a number of publications, and he has recently provided a concise, excellent review of his own and other contributions in this area (Abrams 2010). In Abrams (1984), for instance, he showed in a tritrophic model that adaptive foraging in a prey species facing a trade-off between benefits and risks of foraging leads to emergent direct density dependence. When predator numbers increase, prey respond by making themselves harder to catch, reducing the predator's growth rate – i.e., there is induced density dependence in the predator. This can stabilize food chain interactions (Rinaldi *et al.* 2004). More recently, Abrams (2007) briefly notes that predators which indirectly increase the food supply of their prey will thus shift the optimal foraging tactics of those prey that escape predation, in turn altering how density dependence operates in the prey itself. Abrams and Vos (2003) consider models of the effect of mortality on abundance across a food chain, assuming an adaptive trade-off between foraging and predation risk at the middle trophic level, and that each level experiences direct density dependence. They demonstrate that changes in the strength of density dependence can qualitatively influence how mortality, via alteration in a trait, changes abundance across trophic levels. The particular models they explore assume the form of density dependence itself is fixed, but the message holds even if density dependence itself is also labile and trait-dependent.

Other authors have explored how labile traits can influence the dynamics of interacting species (e.g., Denno and Kaplan 2007). We provide a representative sample, rather than attempt a comprehensive review. Ives and Dobson (1987) show for an interacting predator–prey pair that if the interaction is stable with fixed prey behaviours, labile prey behaviours that trade off predation risk with growth tend only to strengthen this stability (increasing the rate of return to the equilibrium, following a perturbation). Similar messages are delivered by Krivan (2007) for a Lotka–Volterra model in which both predator and prey behave adaptively; these interlocking adaptive behaviours stabilize the otherwise neutrally stable interaction (see also Mougi and Kishida 2009). Brown and Kotler (2007) examine the isoclines of a food chain to argue that adaptive shifts between foraging and defence against predation tend to stabilize food chain dynamics. Urbani and Ramos-Jiliberto (2010) likewise conclude that antipredator behaviours stabilize a three-species intraguild predation system, and Kondoh (2003) and Loeuille (2010) state that quite broadly, flexible behaviours tend to stabilize complex multispecies food webs.

But in some circumstances, labile, adaptive traits can destabilize. In a remarkably prescient paper, Gause *et al.* (1936) developed a predator–prey model in which predator body size varied along with predator and prey numbers, with de-stabilizing effects on the interaction (we thank Vlastimil Krivan for this historical reference, which surely is the earliest ecological model acknowledging phenotypic plasticity). Allowing traits to vary alongside abundances adds multiple degrees of freedom along which systems can move, and time-lagged feedbacks can occur, permitting novel sources of instability. Abrams (2010, p. 13) outlines a number of scenarios in which flexible behaviours are de-stabilizing. For instance, Abrams and Kawecki (1999) show that if predators take time in shifting between habitats, this can destabilize predator–prey dynamics, compared to systems in which predators have fixed patterns of habitat use. Kopp and Gabriel (2006) show that adding inducible prey defence to a Nicholson–Bailey model (which tends to be unstable because of the inherent time-lag in discrete generations) can facilitate persistence, but at the same time destroy stability, if the prey response is very strong. These counterexamples involve pairwise interactions, but there is no reason to think that the situation would be different with more species. Indeed, Berec *et al.* (2010) found it difficult to show that plasticity led to any clear trends in the stability of complex food webs.

In a multispecies community at an equilibrium, one evaluates local stability by analyzing eigenvalues of the Jacobian (the matrix expressing the net effect on each species' growth rate of a small perturbation in the abundance of each species). If population dynamics are described by sets of differential or difference equations, a necessary condition for stability (in the sense of tending to return to a point equilibrium following a small perturbation) is that at least one diagonal entry in the Jacobian matrix describing intraspecific density dependence be negative (May 1974; Logofet 1993). Conversely, strong positive density dependence can be de-stabilizing and imply phenomena ranging from cycles and chaos to transitions between alternative states (Scheffer 2009).

Trait lability can influence the strength and even sign of direct intraspecific density dependence, and thus community stability. Given time lags, strong density dependence can itself be de-stabilizing (e.g., Wissinger *et al.* 2010).

In general, weak or time-lagged density dependence tends to lead to unstable dynamics, with a wide range of potential consequences for community composition. On the one hand, fluctuations in abundance may imply that species spend time near zero density, and so are likely to go extinct. This drives local extinctions in strong predator–prey interactions, and provides one of the reasons that spatial processes appear to be of great importance in permitting persistence of strong predator–prey interaction (Briggs and Hoopes 2004). Conversely, instability can at times create novel mechanisms of coexistence (e.g., Huisman and Weissing 1999). For instance, if two consumer species with saturating functional responses consume a shared biotic resource, unstable dynamics can permit temporal niche partitioning of the consumers in terms of their relative effectiveness at competing for that resource, permitting coexistence that is impossible when the dynamics are stable (Armstrong and McGehee 1976; Abrams and Holt 2002). Adding direct density dependence can destroy such coexistence.

Thus, a key concern of population ecology – intraspecific density dependence – is inescapably at the heart of many classical issues in community ecology. We suggest that characterizing how trait-mediated effects alter density dependence in like manner is germane to understanding the broader community implications of trait-mediated effects.

To illustrate these points, we now consider a simple model, in which trait plasticity alters the strength of direct density dependence in a focal species, which then alters the magnitude of indirect effects mediated through this species.

Timescales and trait effects

As reviewed in Abrams (2010) and discussed in other chapters in this volume (e.g., Peacor and Cressler, this volume, Chapter 8), the issue of timescale in articulating trait effects in community models can be particularly important. Consider the following schematic model, where $\mathbf{N}$ and $\mathbf{E}$ are respectively vectors of species' local abundances and environmental conditions, and $\mathbf{T}$ is a matrix of trait values (i indexes species, and j traits):

$$\frac{dN_i}{dt} = f_i(\mathbf{N}, \mathbf{T}, \mathbf{E}) \tag{6.1}$$

$$\frac{dT_{ji}}{dt} = g_{ji}(\mathbf{N}, \mathbf{T}, \mathbf{E}); \tag{6.2}$$

f_i is the growth rate of species i, and g_{ji} is the rate of change in its trait j. The timescale issue may be resolved in three ways (expanding a bit on points made by Abrams 1995): (1) *Fixed species traits.* If all traits are species-specific attributes, with no plasticity or evolution, Equation (6.2) is simply zero. It is fair to say this is the unstated assumption of classical (e.g., Lotka–Volterra) models in ecology; (2) *Determinate but labile traits.* If behaviour or other traits equilibrate rapidly relative to changes in abundance or the environment, in some cases one can find a one-to-one correspondence between trait values and other variables in a model. In this case, trait variation can be subsumed in more complex forms of the functions (f_i) in Equation (6.1). For instance, an idealized optimal forager might alter its diet as resource abundances change in accord with optimal diet rules, with no lag in its response. Such responses can lead to complex expressions for functional responses (e.g., Beckerman 2005); (3) *Time-varying traits.* If traits change in response to shifts in abundance or the environment, but slowly, one needs to increase the dimensionality of the system. Adding more degrees of freedom can permit novel dynamical behaviours to arise. Distinctions among these are not hard-and-fast, but depend on the timescales and mechanisms operating in the system at hand.

Trait-mediated effects and direct density dependence

Note that our chapter title refers to 'trait-mediated effects', not just 'trait-mediated indirect effects'. The reason for this broadening of scope is that a key interaction that must always be considered in ecological models is the interaction of a species with itself. Imagine effects that propagate across three species: $M \rightarrow N \rightarrow C$. The effects on C that are caused by N could involve changes in the density of N, or in the traits of N – or both. The changes in traits of N could be within generation (e.g., behavioural responses, as is the focus of much of this volume), or among generations (including evolution, see e.g., Lau, this volume, Chapter 12). The issue of TMIIs is how changes in traits of N, caused by M, influence C. One way this can happen is for species M to influence traits in species N which affect direct density dependence, leading to changes in the abundance of species N. The third species, C, could then be impacted by this change in abundance.

As a simple theoretical example, consider a species which grows logistically when it has a fixed value of trait B (e.g., body size). Two parameters are needed to describe logistic growth: an intrinsic (low density) growth rate r, and a strength of density dependence c (which together with r determines carrying capacity), where both can be functions of trait B. Another species is present, with density M, which influences the intrinsic growth rate, and the strength of direct density dependence of N. If B is constant, and by assumption the population has logistic growth, the rate of change of the focal population density (N) is given by:

$$\frac{dN}{dt} = N[r(B,M)-c(B,M)N] = N\omega(N,B,M), \tag{6.3}$$

where $\omega(N,B,M) = r-cN$ is the instantaneous per capita growth rate of the population, a measure of fitness. For now we assume M is fixed. If the trait is labile, it is reasonable to assume that it will change in a direction that increases fitness. The rate at which the trait changes might be described by

$$\frac{dB}{dt} = G\frac{d\omega}{dB} = G\left(\frac{dr}{dB}-N\frac{dc}{dB}\right), \tag{6.4}$$

where G is a scalar determining how rapidly the trait can adaptively shift. Equation (6.4) could describe evolution of a quantitative trait (in which case G is a measure of genetic variation), or alternatively a plastic response (here G governs the rapidity of learning or other mechanisms involved in trait lability). We assume here that G is constant, and we impose no limits on values B can take. In our examples, there is an intermediate optimal B toward which the actual B will evolve. In some cases, it might be necessary to impose limits on B (for example a minimum of 0, or some maximum if B otherwise could evolve towards very high values). This can be done by making G a function of B that goes to 0 at the limits (e.g., $G = G_0(B-B_{min})(B_{max}-B)$; Cortez and Ellner 2010).

The above treatment of evolution and adaptive plasticity pertains to some mechanisms of density dependence, but not to others. We assume above that c for each individual depends on the trait value of that individual, but not the mean trait value of the population as a whole. For instance, a species may experience costs because of metabolic wastes exuded by its population. An individual's tolerance to such wastes may come at a cost, as measured in its intrinsic growth rate. As observed by V. Krivan (pers. comm.), if density dependence instead emerges from say aggressive interactions between individuals differing in size, and larger individuals win, one has to account explicitly for aspects of frequency-dependent selection that can arise in the evolution of density dependence.

Assume trait evolution is much more rapid than population dynamics, so the trait always tracks its evolutionary or behavioural equilibrium (assuming one exists), defined by $dB/dt = 0$, or equivalently, from Equation (6.4), $d\omega/dB = 0$. Since we have assumed that $\omega = r-cN$, $d\omega/dB = 0$ implies

$$\frac{dr}{dB} = N\frac{dc}{dB}. \tag{6.5}$$

Trait B should evolve so as to increase the realized growth rate, but this involves a trade-off between the intrinsic density-independent growth rate and the sensitivity of the species to density dependence. If the population is also at demographic equilibrium, then

$$N^* = \frac{r(B,M)}{c(B,M)}. \tag{6.6}$$

If we assume r and c are monotonic functions of B, Equations (6.5) and (6.6) imply that

$$N^* = \frac{dr/dB}{dc/dB} = \frac{dr}{dc} = \frac{r(B,M)}{c(B,M)}. \tag{6.7}$$

For an optimum to exist, it also must represent a local maximum, which requires that

$$\frac{d^2\omega}{dB^2} = \frac{d^2r}{dB^2} - N^*\frac{d^2c}{dB^2} < 0 \tag{6.8}$$

for B and N satisfying Equation (6.5). Inequality (6.8) can be satisfied, regardless of N, if $r(B,M)$ is concave down (with respect to B) and $c(B,M)$ is concave up, or if one is linear (or constant) and the other is concave (down for r, or up for c).

We use as an illustration a simple case:

$$r = r_0(M) - q(M)B^2 \text{ and } c = c_0(M) - q'(M)B. \tag{6.9}$$

This expression assumes the peak r occurs at $B = 0$, but we can always measure B relative to the value giving the maximum r, so this scaling is for convenience. Moreover, we implicitly assume that B varies only over a range that permits negative density dependence (e.g., c never becomes negative). Inserting Equations (6.9) into Equation (6.5) and solving yields

$$B_{opt} = \frac{q'(M)}{2q(M)} N. \tag{6.10}$$

Substituting back into the population growth Equation (6.3) leads to

$$\frac{dN}{dt} = N[(r_0 - c_0 N) + \gamma N^2], \tag{6.11}$$

where $\gamma = q'^2/(4q)$ (for simplicity we drop the dependence on M). The per capita growth rate assuming B tracks the optimum is the bracketed term. The population equilibrates at

$$N^* = c_0/(2\gamma) - \sqrt{(c_0/2\gamma)^2 - r_0/\gamma}. \tag{6.12}$$

This expression is biologically meaningful only if $\sqrt{r_0/q} \le c_0/q'$, which we assume true.

Changes in the abundance of an interacting species can shift the optimal trait B_{opt} and hence indirectly alter density dependence. This in turn changes the equilibrial abundance of the focal species. If this species interacts with a third species, C, we can have a form of trait-mediated indirect interaction as follows: a changed abundance of M influences growth parameters in N, which leads to a shift in a trait so as to increase fitness. This alters density

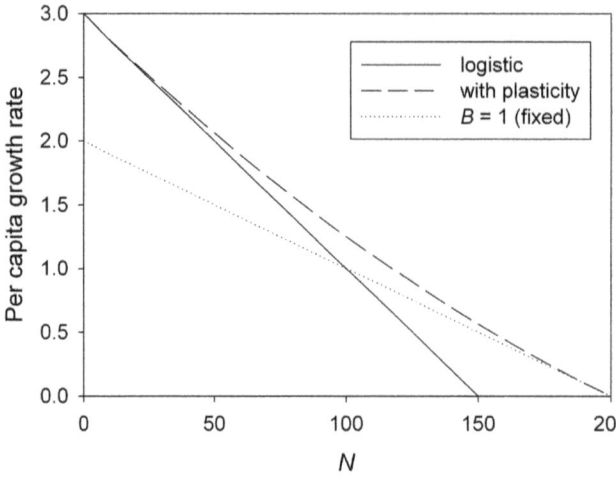

Figure 6.1 Adaptive trait plasticity creates nonlinear density dependence. Per capita growth rates with adaptive plasticity (dashed line) and without (solid line, $B = 0$; dotted line, $B = 1$), for the model in Equations (6.3), (6.5) and (6.9). The dotted curve has B fixed at the optimum for the equilibrium, with plasticity; this leads to lower growth rates at lower N. Parameters in Figures 6.1–6.3 are $r_0 = 3$, $c_0 = 0.02$, $q = 1$, $q' = 0.01$.

dependence in N, and thus its abundance. This changed density then modifies the total interaction strength the focal species exerts on species C.

Equation (6.11) is not logistic growth. Thus, allowing an adaptive, plastic response of the trait to changes in abundance can lead to nonlinear density dependence. Nonlinear density dependence is quite likely for a wide range of reasons (Abrams 2009), and we suggest that shifts in species traits are one such class of reasons. For this example, there is an emergent nonlinearity implying weaker density dependence at higher abundance. The per capita growth rate is shown in Fig. 6.1 with plasticity compared to that with the trait fixed at $B = 0$ and 1 (the equilibrium optimum). Examples of population size trajectories are depicted in Fig. 6.2, comparing this labile population to otherwise similar populations with fixed traits. Figure 6.2 also shows the solution of Equations (6.3–6.4), which is close to the solution of Equation (6.11) for low densities, but approaches the equilibrium more slowly, if $G = 0.2$ (shown); the difference is greater for smaller G and lower for larger G. Larger G implies faster equilibration of B, which therefore should make Equation (6.11) more accurate, because Equation (6.11) assumes instantaneous equilibration of B.

Note that populations with a plastic response have higher equilibrial numbers than do populations with fixed traits (this can be shown to be true in general, and is shown graphically in Fig. 6.3, where the solid line is the isocline for Equation (6.3) and also the equilibrium N as a function of fixed B, while the dashed line is the isocline of Equation (6.4)). In other words, plastic responses that trade off growth rate against direct density dependence tend to increase equilibrial population size. This is a general expectation in models in which fitness depends only upon total population size (Roughgarden 1971).

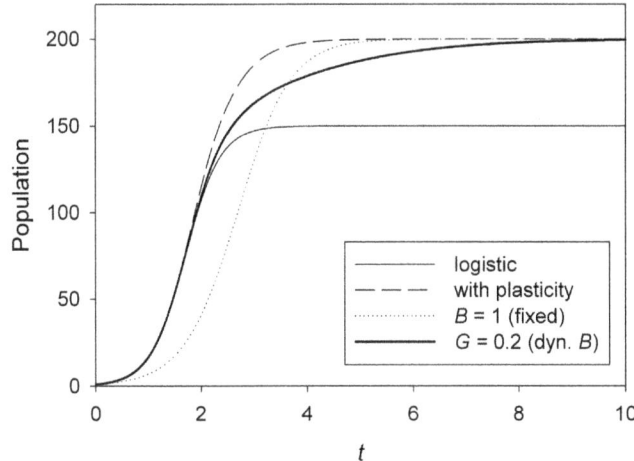

Figure 6.2 Adaptive plasticity can enhance population growth. The figure shows population size as a function of time, with plasticity, and without plasticity with B fixed at 0 (logistic) and 1 (the optimum B with plasticity at equilibrium). The thick solid line is the population size with B given by Equation (6.4) with $G = 0.2$ (and initial $B = 0$). Initial conditions $N = 1$ at $t = 0$.

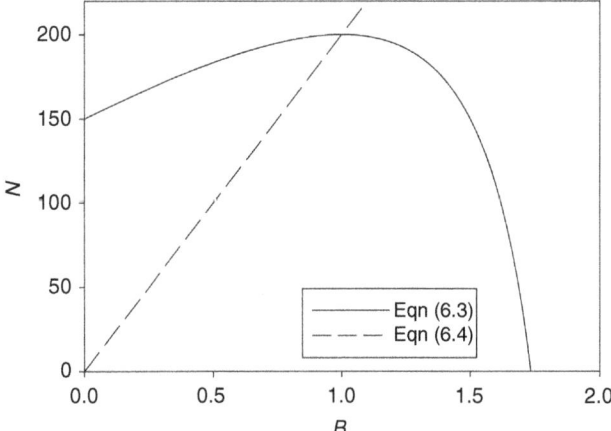

Figure 6.3 Isoclines of Equations (6.3) and (6.4). The isocline of Equation (6.3) is the equilibrium population size as a function of B assuming B is fixed. The Equation (6.4) isocline is given by Equation (6.5), and the equilibrium is the intersection of the isoclines. It can be shown that the equilibrium population size with plasticity is equal to the peak N on the Equation (6.3) isocline (200, at $B = 1$), which is where the slope of this isocline is 0. Therefore, this system does not oscillate as it approaches equilibrium, and if the plastic response lags behind changes in numbers, this does not alter the ultimate equilibrium of the system, or introduce unstable dynamics.

The strength of density dependence s (i.e., the marginal effect on per capita growth of an increase in density) at equilibrium is

$$s = -c_0 + 2\gamma N^* = -\sqrt{c_0^2 - 4\gamma r_0}. \tag{6.13}$$

The trait-mediated response of the species to itself weakens the strength of density dependence. But any factor lowering the maximal growth rate of the species (e.g., a generalist predator acting as a density-independent mortality source; increasing predator density M lowers r_0) reduces density, makes s more negative, and tends to strengthen intraspecific density dependence because of the shift in the trait. If the per capita strength of interspecific competition is constant, and a similar trait-mediated process operates in each of two competing species, this effect could facilitate competitive coexistence (see Holt 1985), mediated through a shift in traits of the competing species. Conversely, mutualists or resources which boost the intrinsic growth rate of the focal species could weaken the strength of density dependence and thereby indirectly hamper competitive coexistence. Similar effects can arise via q and q'.

Discrete-time model of trait-mediated density dependence

Introducing a trait-mediated shift in density dependence did not alter stability in the above model, even if trait equilibration is slow relative to changes in population size. This conclusion is not likely generally true. Consider a species with discrete generations, described by a Ricker model. The model is now

$$N_{t+1} = N_t \omega(N_t, B_t) = N_t \exp\{r(B_t) - c(B_t)N_t\} \tag{6.14}$$

$$\Delta B_t = G\frac{d\ln\omega}{dB_t} = G\left(\frac{dr}{dB_t} - N_t\frac{dc}{dB_t}\right) \tag{6.15}$$

or (equivalently)

$$B_{t+1} = B_t + G\left(\frac{dr}{dB_t} - N_t\frac{dc}{dB_t}\right). \tag{6.16}$$

A local stability analysis (details not shown) assuming Inequality (6.8) holds leads to the stability conditions of $c(B^*)N^* < 2$ and $G(N^*d^2c/dB^2 - d^2r/dB^2) < 2$. For the functional forms in Equation (6.9), these imply $[c_0 - q'^2N^*/(2q)]N^* < 2$ and $Gq < 1$, where N^* is given in Equation (6.12). The first condition is the stability condition for the system, were B_t to track the optimal value for the current N_t. Increasing G makes it harder to meet the second stability condition, so rapid responses can be de-stabilizing.

However, though the stability condition is the same for the two-variable system and the approximation in which B_t tracks its optimum (assuming the second condition is not violated for the two-variable system), the results of instability can markedly differ. Figure 6.4 gives an example. The two-variable system has much bigger excursions in population size than does the one-variable system with optimum B_t (excursions with B_t fixed are in between these two cases). This suggests that plastic responses that allow the

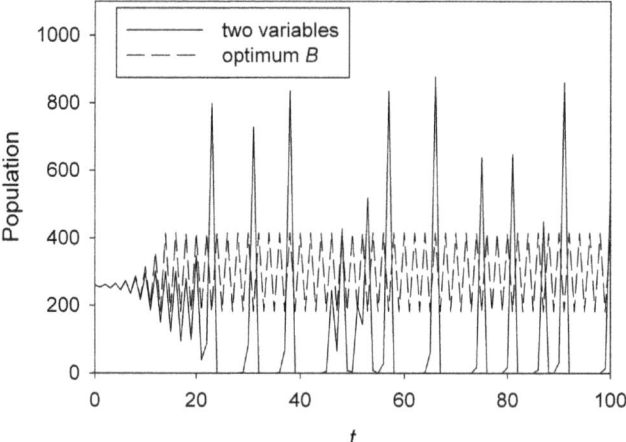

Figure 6.4 Population dynamics for the discrete system in Equations (14–16) with two dynamic variables (N_t and B_t; solid line), and with one variable (N_t) with the trait B_t perfectly tracking the optimum (dashed line). Population size is started 1% above the equilibrium and B is started at the optimum for the equilibrium. Parameters are $r_0 = 4$, $d_0 = 0.022$, $q = 1$, $q' = 0.01$ and $G = 0.5$. The system is unstable in both cases, but lags in the adaptive response in the two-variable case lead to much greater variability in abundance.

population to track quickly its phenotypic optimum moderate the instability inherent in time-lagged, density-dependent systems. However, if adaptation is lagged, instability becomes even more exaggerated. Thus, the existence of plastic or evolutionary trait responses that influence density dependence can be stabilizing, or de-stabilizing, depending upon the rate of response of the trait to changes in density.

Discussion and conclusions

Our overall message is that how trait plasticity influences density dependence is not a sideshow in the story of TMIIs, but essential for a full understanding of this fascinating phenomenon. Our model shows that trait plasticity can generate nonlinear density dependence, so the strength of density dependence will change as interspecific interactions alter abundance. We could have used other systems to illustrate this basic point. For example, in the interaction of predator, prey and parasites, the impact of parasitism upon the vulnerability of the prey (a trait-mediated effect) can strongly influence stability and population dynamics. Fenton and Rands (2006) demonstrate that for trophically transmitted parasites, host manipulation by the parasite so as to facilitate capture can lead to unstable dynamics. We suspect a consideration of trait-mediation will play an increasingly important role in studies of host–parasite interactions.

There may be a methodological implication of these theoretical musings for the empirical question of weighing the relative magnitude of trait-mediated versus density-mediated effects. Consider a system in which a predator imposes mortality on its prey (a density effect), and also induces a change in a prey trait. Both of these then influence a third species, leading to both density-mediated indirect effects (DMIEs) and TMIEs. But now imagine the change in the prey trait also influences how it experiences direct density dependence. This implies an additional shift in prey abundance, which can further impact the third species. So to some extent, there could be trait effects buried within observed indirect density-mediated effects of the predator via this prey species. Teasing these effects apart in experiments might be difficult, in the absence of quantitative models that carefully track all the density-dependent feedbacks at play.

Keeping track of all these feedbacks can be quite a headache, and is often impossible without a formal model in hand. Richard Levins (1975) developed a protocol called 'loop analysis' for analysing many aspects of dynamical systems, which takes account of feedbacks of each species with itself, via pairwise interactions with another species, via three-way interactions, and so forth (hence, the name 'loops'). One can sometimes make inferences about the responses of a system to perturbations based on the qualitative (signed) pattern of interactions. One first describes how a system changes with a system of differential or difference equations. Graphically, the causal flows of the model are represented with a signed digraph, where each node is a variable in the model, and each link represents an interaction. In classical community ecology, nodes are simply species' abundances, but if traits are plastic, traits themselves become nodes. If we for a moment assume that a species' trait influences just a single parameter, and that the system is at equilibrium, the procedures developed by Levins (e.g., see Puccia and Levins 1985, Appendix, pp. 248–250) describe how one can analyse quite generally the impact of a small change in the value of this parameter on the equilibrial levels of each dynamically varying component in the system, and on local stability.

Analytically, even systems with a handful of species can lead to cumbersome expressions using loop analysis, and this may get even worse if traits enter into multiple parameters of the dynamical model: trait-mediated effects in multispecies community models can imply that almost every species interacts with every other species (Abrams 1993). Possibly for reasons such as these, the formalism of loop analysis has not really caught on among ecologists. (Schoener and Spiller, Chapter 2, this volume, still express scepticism.) Nonetheless, applying the formalism of loop analysis, broadened to include traits as variables, can potentially crystallize understanding of the myriad pathways through which trait-mediated effects occur. There has lately been a modest renewal of interest in qualitative analyses (e.g., Montaño-Moctezuma *et al.* 2007; Hosack *et al.* 2009; see Dambacher and Ramos-Jiliberto 2007 for a nice review). We are aware of two papers that explicitly apply this approach to trait-mediated effects. Ramos-Jiliberto and Garay-Narváez (2007) show that induced defences can de-stabilize a one-predator–two-prey module, even though induced defences tend to stabilize each constituent one-predator, one-prey pair, when viewed separately. Garay-Narvaez and Ramos-Jiliberto (2009) extend this approach to argue that in multispecies communities, time delays in induced defences may be de-stabilizing (Fryxell and Lundberg 1998 make a similar point), and that defensive responses aimed at particular predators (specific defences) are more likely stabilizing than are generalized defences. These contributions suggest that qualitative approaches lead to insights even in rather complex systems.

One general message provided by qualitative analyses such as these is that the precise placement and strength of 'self-effects' – density dependence – in webs of interacting species have large and at times surprising consequences for how systems respond to perturbations, and for stability. Understanding the interplay of intraspecific density dependence and trait-mediated interactions is, we suggest, a theme which deserves more concerted attention by both theoreticians and empiricists as this area of research continues to mature.

Acknowledgements

We thank Vlastimil Krivan, an anonymous reviewer, and the editors for their thoughtful comments. We thank the University of Florida Foundation, NSF (grants DEB-0515598 and DEB-0525751) and NIH (grant GM-083192) for support.

References

Abrams, P. A. (1984) Functional responses of optimal foragers. *American Naturalist*, **120**, 382–390.

Abrams, P. A. (1993) Indirect effects arising from optimal foraging. In H. Kawanabe, J. E. Cohen and K. Iwasaki, eds., *Mutualism and Community Organization*. Oxford: Oxford University Press, pp. 255–279.

Abrams, P. A. (1995) Implications of dynamically variable traits for identifying, classifying, and measuring direct and indirect effects in ecological communities. *American Naturalist*, **146**, 112–134.

Abrams, P. A. (2007) Defining and measuring the impact of dynamic traits on interspecific interactions. *Ecology*, **88**, 2555–2562.

Abrams, P. A. (2009) Determining the functional form of density dependence: deductive approaches for consumer–resource systems having a single resource. *American Naturalist*, **174**, 321–330.

Abrams, P. A. (2010) Implications of flexible foraging for interspecific interactions: lessons from simple models. *Functional Ecology*, **24**, 7–17.

Abrams, P. A. and Holt, R. D. (2002) The impact of consumer–resource cycles on the coexistence of competing consumers. *Theoretical Population Biology*, **62**, 281–296.

Abrams, P. A. and Kawecki, T. J. (1999) Adaptive host preference and the dynamics of host–parasitoid interactions. *Theoretical Population Biology*, **56**, 307–324.

Abrams, P. A. and Vos, M. (2003) Adaptation, density dependence, and the abundances of trophic levels. *Evolutionary Ecology Research*, **5**, 1113–1132.

Armstrong, R. A. and McGehee, R. (1976) Coexistence of two competitors on one resource. *Journal of Theoretical Biology*, **56**, 499–502.

Bagchi, R., Swinfield, T., Gallery, R. E. *et al.* (2010) Testing the Janzen-Connell mechanism: pathogens cause overcompensating density dependence in a tropical tree. *Ecology Letters*, **13**, 1262–1269.

Beckerman, A. P. (2005) The shape of things eaten: the functional response of herbivores foraging adaptively. *Oikos*, **110**, 591–601.

Beddington, J. R. (1975) Mutual interference between parasites or predators and its effects on searching efficiency. *Journal of Animal Ecology*, **44**, 331–340.

Berec, L., Eisner, J. and Krivan, V. (2010) Adaptive foraging does not always lead to more complex food webs. *Journal of Theoretical Biology*, **266**, 211–218.

Brashares, J. S., Werner, J. R. and Sinclair, A. R. E. (2010) Social 'meltdown' in the demise of an island endemic: Allee effects and the Vancouver Island marmot. *Journal of Animal Ecology*, **79**, 965–973.

Briggs, C. J. and Hoopes, M. F. (2004) Stabilizing effects in spatial parasitoid-host and predator–prey models: a review. *Theoretical Population Biology*, **65**, 299–315.

Brown, J. S. and Kotler, B. P. (2007) Foraging and the ecology of fear. In D. W. Stephens, J. S. Brown and R. C. Ydenberg, eds., *Foraging: Behavior and Ecology*. Chicago, IL: University of Chicago Press, pp. 437–482.

Cortez, M. H. and Ellner, S. P. (2010) Under-standing rapid evolution in predator–prey interactions using the theory of fast-slow dynamical systems. *American Naturalist*, **176**, E109–E127.

Courchamp, F., Berec, L. and Gascoigne, J. (2009) *Allee Effects in Ecology and Conservation*. Oxford: Oxford University Press.

Creel, S., Christianson, D., Liley, S. and Winnie, Jr, J. A. (2007) Predation risk affects reproductive physiology and demography of elk. *Science*, **315**, 960.

Dambacher, J. M. and Ramos-Jiliberto, R. (2007) Understanding and predicting effects of modified interactions through a qualitative analysis of community structure. *Quarterly Review of Biology*, **82**, 227–250.

DeAngelis, D. L., Goldstein, R. A. and O'Neill, R. V. (1975) A model for trophic interaction. *Ecology*, **56**, 881–892.

Denno, R. F. and Kaplan, I. (2007) Plant-mediated interactions in herbivorous insects: mechanisms, symmetry, and challenging the paradigms of competition past. In T. Ohgushi, T. P. Craig and P. W. Price, eds., *Ecological Communities: Plant Mediation in Indirect Interaction Webs*. Cambridge: Cambridge University Press, pp. 19–50.

Fenton, A. and Rands, S. A. (2006) The impact of parasite manipulation and predator foraging behavior on predator–prey communities. *Ecology*, **87**, 2832–2841.

Fryxell, J. M. and Lundberg, P. (1998) *Individual Behavior and Community Dynamics*. London: Chapman and Hall.

Gaillard, J. M., Festa-Bianchet, M., Yoccoz, N. G., Loison, A. and Toigo, C. (2000) Temporal variation in fitness components and population dynamics of large herbivores. *Annual Review of Ecology and Systematics*, **31**, 367–393.

Garay-Narváez, L. and Ramos-Jiliberto, R. (2009) Induced defenses within food webs: the role of community trade-offs, delayed responses, and defense specificity. *Ecological Complexity*, **6**, 383–391.

Gause, G. F., Smaragdova, N. P. and Witt, A. A. (1936) Further studies of interaction between predators and prey. *Journal of Animal Ecology*, **5**, 1–18

Grimm, V. and Wissel, C. (1997) Babel, or the ecological stability discussions: an inventory and analysis of terminology and a guide for avoiding confusion. *Oecologia*, **109**, 323–334.

Hawlena, D. and Schmitz, O. J. (2010) Physiological stress as a fundamental mechanism linking predation to ecosystem functioning. *American Naturalist*, **176**, 537–556.

Hosack, G. R., Li, H. W. and Rossignol, P. A. (2009) Sensitivity of system stability to model structure. *Ecological Modelling*, **220**, 1054–1062.

Holt, R. D. (1977) Predation, apparent competition, and the structure of prey communities. *Theoretical Population Biology*, **12**, 197–229.

Holt, R. D. (1985) Density-independent mortality, non-linear competitive interactions, and species coexistence. *Journal of Theoretical Biology*, **116**, 479–493.

Horak, P., Tummeleht, L. and Talvik, H. (2006) Predator threat, copulation effort and immunity in male rats (*Rattus norvegicus*). *Journal of Zoology*, **268**, 9–16.

Huisman, J. and Weissing, F. J. (1999) Biodiversity of plankton by species oscillations and chaos. *Nature*, **402**, 407–410.

Ives, A. R. and Dobson, A. P. (1987) Antipredator behavior and the population dynamics of simple predator–prey systems. *American Naturalist*, **130**, 431–447.

Kimbrell, T. and Holt, R. D. (2004) On the interplay of predator switching and prey evasion in determining the stability of predator–prey dynamics. *Israel Journal of Zoology*, **50**, 187–205.

Kondoh, M. (2003) Foraging adaptation and the relationship between food-web complexity and stability. *Science*, **299**, 1388–1391.

Kopp, M. and Gabriel, W. (2006) The dynamic effects of an inducible defense in the Nicholson–Bailey model. *Theoretical Population Biology*, **70**, 43–55.

Kratina, P., Vos, M., Bateman, A. and Anholt, B. R. (2009) Functional responses modified by predator density. *Oecologia*, **159**, 425–433.

Krivan, V. (1997) Dynamical consequences of host-feeding on parasitoid-host dynamics. *Bulletin of Mathematical Biology*, **59**, 809–831.

Krivan, V. (1998) Effects of optimal antipredator behavior of prey on predator–prey dynamics: the role of refuges. *Theoretical Population Biology*, **53**, 131–142.

Krivan, V. (2007) The Lotka–Volterra predator–prey model with foraging-predation risk trade-offs. *American Naturalist*, **170**, 771–782.

Krivan, V. and Schmitz, O. J. (2004) Trait and density mediated indirect interactions in simple food webs. *Oikos*, **107**, 239–250.

Levins, R. (1975) Evolution in communities near equilibrium. In M. Cody and J. M. Diamond, eds., *Ecology and Evolution of Communities*. Cambridge, MA: Harvard University Press, pp. 16–50.

Lochmiller, R. L. and Deerenberg, C. (2000) Trade-offs in evolutionary immunology: just what is the cost of immunity? *Oikos*, **88**, 87–98.

Logofet, D. O. (1993) *Matrices and Graphs: Stability Problems in Mathematical Ecology*. Boca Raton, FL: CRC Press.

Loeuille, N. (2010) Consequences of adaptive foraging in diverse communities. *Functional Ecology*, **24**, 18–27.

May, R. M. (1974) *Stability and Complexity in Model Ecosystems*. Princeton, NJ: Princeton University Press.

Montaño-Moctezuma, G., Li, H. W. and Rossignol, P. A. (2007) Alternative community structures in a kelp-urchin community: a qualitative modeling approach. *Ecological Modelling*, **205**, 343–354.

Mougi, A. and Kishida, O. (2009) Reciprocal phenotypic plasticity can lead to stable predator–prey interaction. *Journal of Animal Ecology*, **78**, 1172–1181.

Murdoch, W. W. and Oaten, A. (1975) Predation and population stability. *Advances in Ecological Research*, **9**, 1–131.

Nowicki, P., Bonelli, S., Barbero, F. and Balletto, E. (2009) Relative importance of density-dependent regulation and environmental stochasticity for butterfly population dynamics. *Oecologia*, **161**, 227–239.

Ohgushi, T. (2005) Indirect interaction webs: herbivore-induced effects through trait change in plants. *Annual Review of Ecology, Evolution, and Systematics*, **36**, 81–105.

Puccia, C. J. and Levins, R. (1985) *Qualitative Modeling of Complex Systems: An Introduction to Loop Analysis and Time Averaging*. Cambridge, MA: Harvard University Press.

Ramos-Jiliberto and Garay-Narváez, L. (2007) Qualitative effects of inducible defenses in trophic chains. *Ecological Complexity*, **4**, 58–70.

Rinaldi, S., Gragnani, A. and De Monte, S. (2004) Remarks on antipredator behavior and food chain dynamics. *Theoretical Population Biology*, **66**, 277–286.

Rosenzweig, M. L. (1974). Aspects of biological exploitation. *Quarterly Review of Biology*, **52**, 371–380.

Roughgarden, J. (1971) Density-dependent natural selection. *Ecology*, **52**, 453–468.

Roughgarden, J. and Feldman, M. W. (1975) Species packing and predation pressure. *Ecology*, **56**, 489–492.

Scheffer, M. (2009) *Critical Transitions in Nature and Society*. Princeton, NJ: Princeton University Press.

Schmitz, O. J. (1992) Exploitation in model food chains with mechanistic consumer-resource dynamics. *Theoretical Population Biology*, **41**, 161–183.

Schmitz, O. J. (2010) *Resolving Ecological Complexity*. Princeton, NJ: Princeton University Press.

Schoener, T. W. (1973) Population growth regulated by intraspecific competition for energy or time: some simple representations. *Theoretical Population Biology*, **4**, 56–84.

Sheriff, M. J., Krebs, C. J. and Boonstra, R. (2009) The sensitive hare: sublethal effects of predator stress on reproduction in snowshoe hares. *Journal of Animal Ecology*, **78**, 1249–1258.

Sinclair, A. R. E. (2003) Mammal population regulation, keystone processes, and ecosystem dynamics. *Philosophical Transactions of the Royal Society of London, Series B*, **348**, 1729–1740.

Skalski, G. T. and Gilliam, J. F. (2001) Functional responses with predator interference: viable alternatives to the Holling type II model. *Ecology*, **82**, 3083–3092.

Urbani, P. and Ramos-Jiliberto, R. (2010) Adaptive prey behavior and the dynamics of intraguild predation. *Ecological Modelling*, **221**, 2628–2633.

Utsumi, S., Ando, Y. and Miki, T. (2010) Linkages among trait-mediated indirect effects: a new framework for the indirect interaction web. *Population Ecology*, **52**, 485–497.

van Baalen, M., Krivan, V., van Rijn, P. C. J. and Sabelis, M. W. (2001) Alternative food, switching predators, and the persistence of predator–prey systems. *American Naturalist*, **157**, 512–524.

Wissinger, S. A., Whiteman, H. H., Denoe, M., Mumford, M. L. and Aubee, C. B. (2010) Consumptive and nonconsumptive effects of cannibalism in fluctuating age-structured populations. *Ecology*, **91**, 549–559.

Ziebarth, N. L., Abbott, K. C. and Ives, A. R. (2010) Weak population regulation in ecological time series. *Ecology Letters*, **13**, 21–31.

CHAPTER SEVEN

Plant effects on herbivore–enemy interactions in natural systems

KAILEN A. MOONEY

Department of Ecology and Evolutionary Biology, University of California – Irvine

and

MICHAEL S. SINGER

Department of Biology, Wesleyan University

Introduction

While it had long been recognized that herbivores are simultaneously influenced by natural enemies (Hairston *et al.* 1960) and plant defences (Fraenkel 1959), Price *et al.* (1980) were among the first to argue forcefully that these dual factors must be considered together. They argued that '[w]e cannot understand the plant–herbivore interaction without understanding the role of enemies. We cannot understand predator–prey interactions without understanding the role of plants' (Price *et al.* 1980, p. 59). This holistic, tritrophic perspective conceptually unites theory from at least three areas of ecological and evolutionary research. First, this tritrophic perspective expands our view on plant defence from one based strictly on the direct defence, to one that also considers the indirect defence of plants by natural enemies (Janzen 1966; Turlings *et al.* 1990), as well as how natural enemies mediate the efficacy of direct defences (Moran and Hamilton 1980; Clancy and Price 1987; Williams 1999; Gassmann and Hare 2005). Second, this tritrophic perspective advances our understanding of the forces shaping the evolution of herbivore host plant choice and diet breadth by incorporating the interactive effects of host plant quality and risk of attack by natural enemies (Bernays 1998; Singer *et al.* 2004a, b). And third, this tritrophic perspective provides a mechanistic framework for understanding the ecological and evolutionary factors that determine the strength of the indirect effects natural enemies have on plant growth, i.e., trophic cascades (Mooney *et al.* 2010).

Tritrophic interactions have received considerable attention in agricultural systems, with numerous studies documenting the effects of crop traits on herbivores and their natural enemies (Hare 1992; Tumlinson *et al.* 1992; Vet and Dicke 1992; Bottrell *et al.* 1998; Turlings and Benrey 1998; Cortesaro *et al.* 2000; Hare 2002; Ode 2006). In contrast, comparatively little is known of the

Trait-Mediated Indirect Interactions: Ecological and Evolutionary Perspectives, eds. Takayuki Ohgushi, Oswald J. Schmitz and Robert D. Holt. Published by Cambridge University Press. © Cambridge University Press 2012.

influences of plant traits on herbivore–enemy interactions from natural systems (e.g., Hare 1992, 2002). Our goals in this review are three-fold. First, where past reviews on this topic have focused on agricultural systems (e.g., Hare 1992, 2002), we give special attention to the evidence for plant variation in herbivore–enemy interactions from natural communities. Second, we position this topic within the framework of trait- and density-mediated indirect interactions. Finally, we consider the evolutionary and ecological implications of plant variation in herbivore–enemy interactions, and we do so with specific reference to the different mechanistic pathways by which such plant effects can occur.

Background
Definitions and terminology

This chapter addresses how plants can vary in the interactions between herbivores and enemies through both trait-mediated indirect interactions (TMIIs) and density-mediated indirect interactions (DMIIs). Definitions of these terms are provided in Chapter 1. The direct and indirect effects we consider thus start with the plant (i.e., the plant is the instigator of the interaction). In this context, the 'effect' of any plant trait on herbivores and enemies can only be considered in reference to some other level of that plant trait. For instance, comparing plants that differ in trichome density would provide an assessment for the effects of leaf trichomes.

We consider all levels of genetic variation, ranging from intraspecific variation (i.e., clones, full- or half-sib families, etc.) within populations to intraspecific genetic variation among populations to variation at higher taxonomic levels (hybrids, species, families, etc.). For simplicity's sake, we use the term plant 'genetic type' to encompass this entire scale of genetic variation, adhering to the common usage of the term 'genotype' in the context of intraspecific variation.

Although we discuss the effects of variable plant traits on 'herbivores' and 'enemies', in all cases we are referring to plant effects on individual herbivores and enemy species. There are certainly instances where variation in plant traits may change herbivore or enemy species composition and, in so doing, influence herbivore–enemy interactions. Yet such dynamics should not be considered a DMII or TMII and are not discussed here. Instead, we limit our discussion to the scenario where the pairwise interaction between a herbivorous species and one enemy species varies among plant genetic types.

Several terms describing the effects of plants on herbivores, enemies and herbivore–enemy interactions must be distinguished. In this chapter, we reserve reference to plants varying in the interactions between herbivores and enemies to those situations where the per capita effects of enemies and herbivores on each other are significantly different among plant genetic types. In other words, a plant effect on an interaction implies a TMII, where plants influence enemy or herbivore traits and thus the interaction between the two. While it can be argued that plant effects on enemy density can involve changes in traits – i.e., foraging behaviours – for the purposes of clarity we classify all plant effects transmitted via enemy density as DMIIs.

A plant's overall influence on enemy suppression of herbivores is in turn inclusive of both TMIIs and DMIIs, referring either to an influence on the herbivore–enemy interaction (a TMII), or to situations where plants vary in enemy density such that the enemy community as a whole has a stronger effect on some plant genetic types than others (a DMII). According to this terminology, two plant genetic types that do not vary with respect to herbi-vore–enemy interactions might still vary with respect to enemy suppression of herbivores if one genetic type supported a higher density of enemies in relation to herbivores (i.e., a higher enemy:herbivore ratio).

Experimental approaches

Documenting plant genetic effects on enemy abundance and herbivore-enemy interactions can be done according to several methodologies, all of which require studying replicate plants of two or more genetic types while controlling for environmental variation. First, the abundance of herbivores and enemies and enemy foraging behaviours can be monitored among plant genetic types (e.g., Johnson and Agrawal 2005; Crutsinger et al. 2006). While this evidence is somewhat circumstantial, variation in enemy density among plant genetic types is at least suggestive of parallel variation in enemy sup-pression of herbivores. Second, more direct evidence comes from cases where plant genetic types can be compared with respect to the rate or frequency of enemy attack, such as rate or frequency of parasitism (e.g., Bailey et al. 2006; Bukovinszky et al. 2008). Third, the strongest evidence comes from experi-mentally manipulating enemy density on multiple plant genetic types and testing for a plant genetic type-by-enemy interaction (Mooney et al. 2010). These differing approaches can also be combined within a single study as, for example, where enemy access to herbivores is manipulated but rates of enemy attack are monitored in open access treatments.

The interpretation of results from the above methodologies is straightforward if herbivore density is similar among plant genetic types, but otherwise requires a more nuanced analytical approach. If plant genetic types vary in their direct resistance to herbivores (i.e., they vary in herbivore density in the absence of enemies) and enemy effects are density dependent (i.e. nonlinear), then tests for a linear association between enemy density (or effect) and herbivore density will lead to the erroneous conclusion that plant genetic type influences predator–herbivore interactions. By way of example, we consider the likely scenario of positive density dependence in enemy recruitment and attack to herbivores; even if enemies acted according to a single density-dependent function, plant genetic types with low herbivore density would have lower rates of enemy recruitment (enemies per herbivore), lower herbivore attack rates and weaker effects of enemy exclusion than plant genetic types with high herbivore densities. Consequently, nonlinearities must be accounted for when plant genetic types vary in direct resistance to herbivores.

Another unique concern arises in considering the results of experiments conducted under largely uncontrolled field conditions. Implicit in the interpretations outlined above is that the same species of herbivores and enemies are interacting in all cases. For instance, enemies might have differing effects on herbivores among plant genetic types due to variation in herbivore species composition. Similarly, the same herbivore species might be suppressed to differing degrees among plant genetic types due to variation in the species composition of the associated enemies. While plant genetic effects on species composition represent an important ecological dynamic, such effects constitute fundamentally different dynamics that should be considered separately from plant mediation of the pairwise interaction between a single herbivore and a single enemy species.

Mechanisms
There are several complementary (not mutually exclusive) means by which plants may influence herbivores, enemies and their interactions with each other (Fig. 7.1). This constellation of pathways can be organized according to whether plants directly influence enemies or herbivores, and whether these

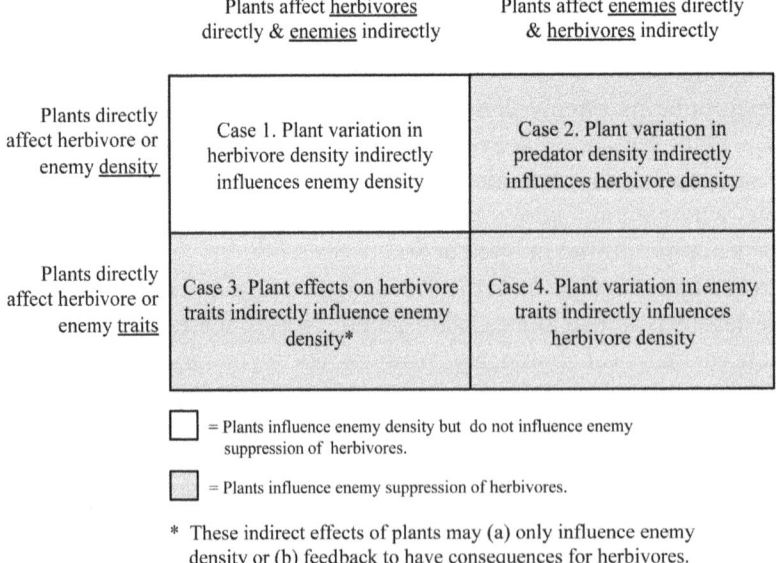

	Plants affect <u>herbivores</u> directly & <u>enemies</u> indirectly	Plants affect <u>enemies</u> directly & <u>herbivores</u> indirectly
Plants directly affect herbivore or enemy <u>density</u>	Case 1. Plant variation in herbivore density indirectly influences enemy density	Case 2. Plant variation in predator density indirectly influences herbivore density
Plants directly affect herbivore or enemy <u>traits</u>	Case 3. Plant effects on herbivore traits indirectly influence enemy density*	Case 4. Plant variation in enemy traits indirectly influences herbivore density

☐ = Plants influence enemy density but do not influence enemy suppression of herbivores.

▨ = Plants influence enemy suppression of herbivores.

* These indirect effects of plants may (a) only influence enemy density or (b) feedback to have consequences for herbivores.

Figure 7.1 Mechanistic pathways by which plant genetic types may vary in their influence on enemy density and enemy suppression of herbivores.

direct effects are on the traits or densities of enemies and herbivores. These direct effects of plants on the density or traits of herbivores may then propagate to enemies indirectly influence. Similarly, the direct effects of plants on the density or traits of enemies may then propagate to influence herbivores indirectly. This classification scheme thus results in four pathways or 'Cases' for how plant genetic types can influence herbivores, enemies and their interactions with each other.

In Case 1, plants directly influence the density of herbivores and this indirectly influences enemy abundance in a DMII. Under this scenario, there is variation among plant genetic types in the densities of both herbivores and enemies, but the interaction between herbivores and enemies remains unchanged. In other words, plant genetic types do not alter the per capita effects of enemies on herbivores. Case 1 thus represents a null model with respect to plant mediation of herbivore suppression by enemies; resources propagate upwards through linear food chains according to a classical bottom-up perspective, but predator suppression of herbivores is not influenced by plant traits. Given the broad evidence for genetic variation in resistance to herbivores (Fritz and Simms 1992), it is likely to be common that enemies vary among plant genetic types in accordance with Case 1.

As an example of Case 1, Bailey et al. (2006) provided evidence for variation in the number of willow galls attacked by birds among genotypes of *Populus angustifolia* and backcross hybrids (from *P.a.* × *P. fremontii*). Here it was determined that plant genetic variation explained most of the variation in gall density, with variation in condensed tannins likely being an important factor. The number of galls attacked by birds was in turn determined almost exclusively by gall abundance. So while most of the variation in the number of galls attacked by birds was determined by plant genetic type, there was no evidence for variation among plant genetic types in the per capita effects of birds on galls. While avian abundance itself was not quantified, this result is still illustrative of density-mediated indirect effects of plants on enemies to the extent that attack rate is taken as a proxy for such counts.

Plants can also have direct influences upon enemy abundance and, in so doing, indirectly influence herbivores via a DMII (Case 2). Under this scenario, the per capita effect of enemies on herbivores is constant, but plant genetic types vary in enemy abundance and thus herbivore suppression. Gassmann and Hare (2005) showed that genetic lines of jimsonweed with glandular trichomes had lower rates of herbivore attack by three generalist predators (lygaeid bugs, lacewing larvae and salticid spiders) as compared to genetic lines with non-glandular trichomes. The reduced predation was due (in part) to glandular trichomes reducing predator residence time. In this way, plant genetic types varied in generalist predator density, and this in turn resulted in an indirect effect on herbivore density.

Plants may also influence enemies through changes in herbivore traits via a TMII (Case 3). Under this scenario, plants indirectly influence enemies by directly affecting either herbivore quality or herbivore susceptibility to attack. Potential scenarios producing this dynamic include the protection herbivores receive from enemies by sequestering plant secondary compounds (Brower *et al.* 1967). In addition, plant resistance can increase herbivore susceptibility to enemies by slowing herbivore development time as outlined in the slow-growth/high-mortality hypothesis (Moran and Hamilton 1980; Price *et al.* 1980; Clancy and Price 1987).

Bukovinszky *et al.* (2008) showed that attack of the aphids *Brevicoryne brassicae* and *Myzus persicae* by parasitoids varied between cultivated and feral strains of Brussels sprouts (*Brassica oleracea* var. *gemmifera*). Structural equation modelling suggested that variation in attack rate was determined in part by plant genetic effects on aphid size. Although this example comes from a cultivated plant species, it nonetheless demonstrates how natural evolutionary processes can generate intraspecific genetic variation that in turn influences herbivore–enemy interactions. Mooney and Agrawal (2008) found genetic variation among milkweed genotypes (full-sib *Asclepias syriaca* families) for ant recruitment to the herbivore *Myzocallis asclepiadis*, an aphid that (unlike other milkweed aphids) is negatively affected and not tended by ants. There was no residual variation among milkweed genotypes for ant abundance when aphid abundance was accounted for, and it thus appears that milkweed did not directly affect ants. These examples thus demonstrate how plant genotypes can vary with enemy density independent of direct effects on the enemy or effects due to variable herbivore density.

Although the indirect effects of plants on enemies outlined in Case 3 may feed back to have consequences for herbivores, this need not be the case. For instance, although Mooney and Agrawal (2008) found genetic variation in milkweed for the number of predatory ants recruited per aphid, this was uncorrelated with the strength of the suppressive effects of those ants on aphids. So here, plant genetic types indirectly influenced enemy abundance without altering the suppressive effects of those enemies on herbivores. In other instances, the indirect effects of plants on predators may in turn result in concomitant variation in the effects of those enemies on herbivores (see the second case study below). For instance, the slow-growth/high-mortality hypothesis is based upon the assumption that plant effects on herbivore growth rate not only influence enemies indirectly, but that these effects on enemies in turn result in higher rates of herbivore suppression (Clancy and Price 1987). So while it is easy to presume that the indirect effects of plants on enemy density should correspond with herbivore suppression, such reciprocal dynamics must be explicitly tested.

Finally, plants may influence the per capita effects of enemies on herbivores in the absence of influencing enemy density via a TMII (Case 4). For example, Ness *et al.* (2009) showed that the carbohydrates provided to ants by extrafloral nectaries can result in more aggressive behaviour of ants towards herbivores, presumably by creating a carbohydrate:protein imbalance. In this way plants may be able to alter enemy foraging behaviour and, in so doing, indirectly influence herbivores by manipulating the per capita effects of enemies on herbivores.

Past works

Since Price *et al.* (1980), the study of tritrophic interactions has progressed rapidly in research on agricultural systems. Studies documenting the effects of the traits of crop species on herbivores and their natural enemies are too numerous to review here, but several such syntheses appear elsewhere (Hare 1992; Tumlinson *et al.* 1992; Vet and Dicke 1992; Bottrell *et al.* 1998; Turlings and Benrey 1998; Cortesaro *et al.* 2000; Hare 2002; Ode 2006). While this large body of work suggests that parallel dynamics might occur in natural systems, the reliability of such inferences is undermined by the fundamental dissimilarities between domesticated and wild plant species, and among the ecological contexts in which those species interact with herbivores and enemies (Hare 1992, 2002).

In contrast to agricultural systems, comparatively little is known of the influences of plant traits on herbivore–enemy interactions from natural systems (e.g., Hare 1992, 2002). In Table 7.1 we list all studies of which we are aware where plants of differing genetic types have been compared with respect to enemy density and herbivore–enemy interactions. Many of the studies included in Table 7.1 were designed specifically to compare the effects of enemies across multiple plant genetic types through predator exclusion (e.g., Forkner and Hunter 2000; Linhart *et al.* 2005; Mooney *et al.* 2010) or by documenting rates of enemy attack (e.g., Gross and Price 1988; Barbosa *et al.* 2001; Bailey *et al.* 2006). But this table also includes studies conducted under the banner of 'community genetics' (Agrawal 2003; Whitham *et al.* 2006; Johnson and Stinchcombe 2007). These studies have tested for the influence of species genotypes (Johnson and Agrawal 2005; Crutsinger *et al.* 2009) or genotypes of interspecific hybrids (Wimp and Whitham 2001) for the structure of the community associated with those genotypes, including enemy abundance and species composition. Heritabilities can then be calculated for the associated community of herbivores and enemies as an extended phenotype of the plant (Shuster *et al.* 2006). Although relatively few studies have been conducted to date, there is an emerging consensus from this work that plant genotype identity can explain much of the variation in arthropod species composition (Wimp and Whitham 2001; Johnson and Agrawal 2005; Crutsinger *et al.* 2009), including the density of enemies.

Several conclusions can be drawn from reviewing these studies. First, plants in natural systems vary both within and among species in ways that influence

Table 7.1 *Studies testing for plant variation in herbivore–enemy interactions in natural systems*

Plant(s)	Herbivore(s)	Enemy(s)	Outcome[1]	Comments	Reference
Interspecific genetic variation					
Box elder (*Acer negundo*) and black willow (*Salix nigra*)	Multiple lepidopteran larvae	Multiple parasitoids	D	Parasitism rates higher on *Acer* than *Salix*.	Barbosa *et al.* 2001
16 milkweed (*Asclepias*) species	An aphid (*Aphis nerii*)	Multiple predatory arthropods	D	Exclusion of predatory arthropods has differing effects on aphids among milkweed species.	Mooney *et al.* 2010
6 milkweed (*Asclepias*) species	An aphid (*Aphis nerii*)	A parasitoid (*Lysiphlebus testaceipes*)	D	Rates of parasitism vary among host plant species.	Helms *et al.* 2004
17 genera/10 families of gymnosperm and angiosperm trees	15 species of lepidopteran larvae from 6 families	Multiple parasitoids	D	Rates of parasitism vary among host plant species.	Lill *et al.* 2002
Fremont cottonwood (*Populus fremontii*), narrowleaf cottonwood (*P. angustifolia*) and hybrids	An aphid (*Chaitophorus populicola*) and other taxa	Multiple predatory and parasitic arthropods	B	Variation among species and hybrids for herbivore and enemy densities, but no statistical tests for variation in enemy recruitment independent of effects due to variable herbivore density.	Wimp and Whitham 2001
2 oak species (*Quercus prinus* and *Q. rubra*)	Different herbivore guilds, some herbivore taxa	Multiple birds	D	Effects of birds differ between host plant species for some herbivore groups in some years.	Forkner and Hunter 2000

2 willows (*Salix phylicifolia* and *S. myrsinifolia*)	Different herbivore guilds, some herbivore taxa	Multiple birds	D	Bird effects on leaf-chewing but not concealed or sap-feeding herbivores vary between host plant species.	Sipura 1999
Horsenettle (*Solanum carolinense*) and groundcherry (*Physalis heterophylla*)	*Tildenia inconspicuella* and *T. georgei* (Lepidoptera: Gelechiidae)	Multiple parasitoids	D	*Solanum* leaf traits result in obligate leaf mining (as compared to facultative on *Physalis*) that in turn results in higher rates of parasitism.	Gross and Price 1988

Interspecific genetic variation

Common milkweed (*Asclepias syriaca*)	An aphid (*Myzocallis asclepiadis*)	An ant (*Formica podzolica*)	D	Negative effects of ants on aphids vary among milkweed full-sib families.	Mooney and Agrawal 2008
Brussels sprouts (*Brassica oleracea* var. *gemmifera*)	Aphids (*Brevicoryne brassicae* and *Myzus persicae*)	Multiple parasitoids	D	Parasitoid and hyper-parasitoid attack varied between cultivated and feral plant populations due to variation in both aphid density and body size.	Bukovinszky et al. 2008
Jimsonweed (*Datura wrightii*)	A beetle (*Lema daturaphila*) and a bug (*Tupiocoris notatus*)	Multiple predatory arthropods	D	Generalist predators spent less time foraging and had weaker effects on herbivores on plant genotypes with glandular trichomes than those without.	Gassmann and Hare 2005

Table 7.1 (cont.)

Plant(s)	Herbivore(s)	Enemy(s)	Outcome[1]	Comments	Reference
Wild cotton (*Gossypium thurberi*)	Multiple species, including a leaf-feeding caterpillar (*Bucculatrix thurberiella*)	An ant (*Forelius pruinosus*)	D	Separate experiments demonstrated genetic variation in extra-floral nectary (EFN) abundance and size and phenotypic effects of EFNs on ants, herbivory and fitness.	Rudgers 2004
Evening primrose (*Oenothera bienis*)	Multiple species	Multiple predatory arthropods	A	No detectable variation among plant genotypes for predator abundance.	Johnson and Agrawal 2005
Cottonwood (*Populus angustifolia*)	Gall-forming aphid (*Pemphigus betae*)	Multiple birds	C	Bird attack of galls followed plant genetic variation in gall density but the rate of bird attack did not vary among genotypes.	Bailey *et al.* 2006
Two willows (*Salix sericea* and *S. eriocephala*)	Leaf-ming moth (*Phyllonorycter salicifoliella*)	Parasitic wasps (Eulophidae)	D	Intraspecific genetic variation for rate of enemy attack on both willows.	Fritz *et al.* 1997
Arroyo willow (*Salix lasiolepis*)	Galling sawfly (*Euura lasiolepis*)	A parasitoid (*Pteromalus* sp.)	D	Host plant clones differed in gall size and smaller galls were parasitized more.	Price and Clancy 1986
Arroyo willow (*Salix lasiolepis*)	Galling sawfly (*Pontania* sp.)	2 parasitoids (*Bracon angelesius* and *Pteromalus* sp.)	D	Galls on host plant clones with faster gall growth and larger galls were more heavily parasitized.	Clancy and Price 1987

Plant	Herbivore	Enemies	Case	Findings	Reference
Arroyo willow (*Salix lasiolepis*)	Leaf-folding sawfly (*Phyllocolpa* sp.)	Multiple predatory and parasitic arthropods	D	Genetic variation in host plant clones for herbivore mortality and parasitism frequency.	Fritz and Nobel 1990
Goldenrod (*Solidago altissima*)	Gall-making flies (*Eurosta solidaginsis*)	Multiple parasitoics and birds	D	Gall size influences rate of parasitism and attack by birds, and there is heritable variation in the plant for gall size.	Weis and Abrahamson 1986
Goldenrod (*Solidago altissima*)	Multiple inquiline herbivores occupying rosette leaf galls	Multiple inquiline predators	B	Goldenrod genotypes varied independently in herbivore and enemy diversity. Tests were not performed on herbivore and enemy density.	Crutsinger et al. 2009
Thyme (*Thymus vulgaris*)	An aphid (*Aphis serpylli*)	Multiple predatory and parasitic arthropods	D	Predator exclusion had different effects on aphids among four genetically controlled monoterpene chemotypes in thyme.	Linhart et al. 2005

[1] Key to study outcomes relative to Cases outlined in Fig. 7.1: A, Did not find plant variation in enemy density or effects of enemies on herbivores (none of Cases 1–4 supported); B, Found plant variation in enemy density (Case 1) but did not seek evidence for plant variation in enemy density independent of herbivore density or for plant variation in enemy suppression of herbivores (Cases 2, 3 and 4 not studied); C, Found plant variation in enemy density due to variation in herbivore density (Case 1) but did not find evidence for plant variation in enemy density independent of herbivore density or for plant variation in enemy suppression of herbivores (Cases 2, 3 and 4 not supported); D, Found plant variation in enemy density independent of herbivore density and/or plant variation in enemy suppression of herbivores (Case 2, 3 or 4 supported).

enemy density and herbivore–enemy interactions; nearly all studies conducted to date document effects of plant genetic type on either enemy abundance or the strength of enemy effects or enemy attack rates.

Second, for most plant taxa, the traits varying among genetic types that are responsible for such effects are unclear. While there are a few instances where traits are implicated (e.g., Gross and Price 1988; Linhart et al. 2005; Bailey et al. 2006), in most cases genetic types are compared but the traits responsible for variation in enemy abundance or enemy suppression of herbivores remain unknown. This is a stark contrast to the agricultural literature that has taken a more trait-based, mechanistic approach (Hare 1992, 2002).

Finally, third, in no instance has a system been exhaustively studied to test for the occurrence or relative importance of the different interaction pathways outlined in Fig. 7.1. Those studies excluding enemies or measuring rates of enemy attack across plant genetic type are able to distinguish between Case 1 and Cases 2, 3 and 4, but in no instance is it clear whether plant mediation of herbivore–enemy interactions is due to effects on enemy density (Case 2), herbivore traits (Case 3) or enemy traits (Case 4). Furthermore, the community genetics studies testing for variation among plant genetic types for enemy abundance (Wimp and Whitham 2001; Johnson and Agrawal 2005; Crutsinger et al. 2009) have failed to test whether enemies simply track plant variation in herbivore resistance (Case 1) or whether plant genetic types vary in enemy abundance after controlling for herbivore density (Cases 2, 3 and 4).

Case studies

We now present two case studies from our own research to provide more detailed examples of the types of experimental and analytical approaches that can be used to study mechanisms of plant mediation of herbivore–enemy interactions in natural systems. We present our interpretation of results in the context of the four cases outlined in Fig. 7.1. Even with this approach, our understanding of these systems is as yet incomplete.

Interspecific variation in predator–herbivore interactions

We illustrate how variation among host-plant species can dictate avian predator–caterpillar interactions. Our collaborative study of a tritrophic forest food web in Connecticut encompasses eight common broadleaf tree taxa in upland forest in Middlesex Co., Connecticut, USA, an assemblage of polyphagous macrolepidopteran larvae, and several taxa of caterpillar-feeding birds (Singer et al. 2012). The seasonality of this study is late spring and early summer (late May–June), when temperate tree foliage is particularly rich in water and utilizable nitrogen (Slanksy and Scriber 1985; Schoonhoven et al. 2005), low in chemical and structural defences (Feeny 1976; Schoonhoven et al. 2005), polyphagous caterpillars are especially abundant (Futuyma and

Gould 1979) and depression of caterpillar densities by bird predation has been detected previously (Holmes *et al.* 1979).

The data presented here are based on a 6-week long field experiment conducted in 2008 in which caterpillar densities on tree branches with bird exclusion bags were compared to those without exclosures (e.g., Holmes *et al.* 1979). The bird exclosures were set up 3 weeks prior to the first sampling of caterpillars, thus capturing effects over the majority of the spring nesting season for many insectivorous bird species in this forest.

Because this study was conducted at the community level, we took great care not to confound host plant associated variation in herbivore–enemy interactions with host-plant variation in caterpillar or bird species composition. Accordingly, we focus here on the dietary generalist caterpillar assemblage, the most abundant of which use all eight experimental tree species. The experimental design includes representatives of each of the tree species in small complete blocks (*c.* 0.25 ha) so that within a block all trees are almost certainly within the foraging range of the same individual birds. Consequently, any host-plant variation in the effects of birds on caterpillars is not confounded with the bird species available to forage on those trees, or with the species composition of caterpillars upon which those birds are foraging.

Our analysis shows a significant influence of host-plant species on the effect size of bird predation of generalist caterpillars. Variation among tree species in generalist caterpillar density was a significant predictor of the strength of bird effects, indicating that variable resistance among trees in turn led to variable rates of bird predation in accordance with Case 1. However, variation in bird predation among tree species persisted after accounting for generalist caterpillar density (estimated from bagged trees) in an analysis of covariance, indicating that density-dependent bird effects were not the sole explanation. The effects of birds after accounting for generalist caterpillar density are depicted in Fig. 7.2a, where the effect sizes of bird predation on generalist caterpillars across tree species are presented as log response ratios (log-transformed proportion of caterpillar density on bird exclusion relative to control trees; Hedges *et al.* 1999).

Having established tree-species variation in bird–caterpillar interactions, we now consider which of Cases 2, 3 and 4 are likely involved. Case 2 can likely be ruled out because the experimental design guarantees that the same individual birds are available to forage on trees of every species. Hence, it is improbable that bird abundance truly varied among tree species in close proximity (Case 2), but rather that the intensity or quality of foraging by birds varied among tree species (Holmes and Robinson 1981).

Distinguishing between Cases 3 and 4 requires information about variation in herbivore and predator phenotypes or behaviours among host-plant

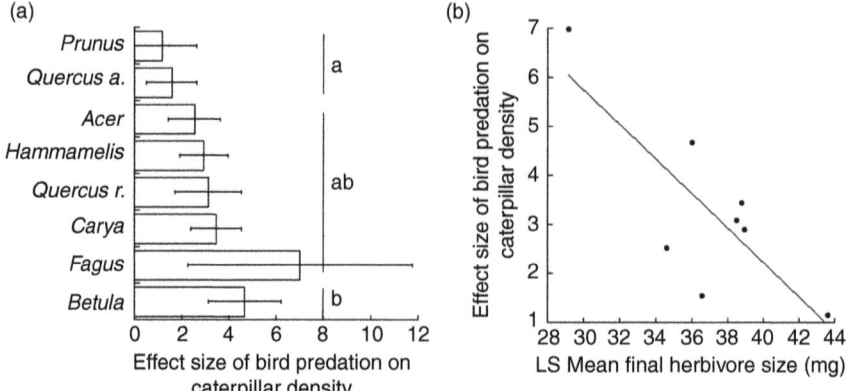

Figure 7.2 Analysis of tree-species variation in the interactions between insectivorous birds and polyphagous (generalist) caterpillars. (a) Variation in least square means (± 95% CI) of the effect size of bird predation on polyphagous caterpillar density among eight tree taxa controlling for tree taxa variation in caterpillar density. The effect sizes are log response ratios, calculated as ln(exclosures/control). Different letters denote different least square means based on a Tukey test. (b) Linear relationship between bird effects on generalist caterpillars (see above) regressed on host plant quality as indicated by caterpillar final size.

species. We have yet to collect data on bird foraging behaviours, precluding a direct test of Case 4. However, we have monitored generalist caterpillar performance on each tree species, estimated as caterpillar final size (pupal or adult mass). Average caterpillar final size is negatively associated with the effect size of bird predation on generalist caterpillars among tree species (Fig. 7.2b). To the extent that larger herbivores also develop more rapidly, which is commonly observed in lepidopteran larvae (e.g., Gripenberg et al. 2010), this pattern is consistent with the slow-growth/high-mortality hypothesis. This hypothesis, however, remains questionable as a general phenomenon due to a mixed empirical record of support (e.g., Moran and Hamilton 1980; Price et al. 1980; Clancy and Price 1987; Haggstrom and Larsson 1995; Williams 1999; Lill and Marquis 2001; Coley et al. 2006; Kaplan et al. 2007). Nevertheless, the slow-growth/high-mortality effect might be particularly expected to occur in our system because a previous analysis of published studies (Williams 1999) showed relatively strong evidence for this hypothesis in studies of predation on externally feeding herbivores.

However, we also acknowledge the possibility that subtle differences among tree species in the relative abundance of various caterpillar species could still determine variation in caterpillar assemblage traits that contribute to the observed differences in the effects of birds. Similarly, although the same individual birds are available to forage on all tree species, it is possible that bird foraging varies in terms of the relative foraging effort each bird species devotes to each tree species (Holmes and Robinson 1981). Clearly, detailed knowledge of the caterpillar and bird species associated with each host-plant species, as well as possible host-plant-related variation in caterpillar and bird traits, will be important to discern mechanisms of indirect effects in this tritrophic community. For example, plant morphology and architecture can influence the likelihood that birds encounter and capture prey (Robinson and Holmes 1982; Marquis and Whelan 1996; Whelan 2001). Similarly, host-plant chemistry can affect a caterpillar's likelihood of being selected as prey by insectivorous birds (Muller *et al.* 2006). Identifying mechanisms of direct and indirect species interactions operating at the scale of communities is an important goal of our research on this forest food web, and of ecology more generally.

Our study of this tritrophic forest food web highlights the importance of phenotypic and genotypic variation among plant species in determining community dynamics and structure. In contrast with many tritrophic studies of species with relatively taxonomically specific interactions, our system includes many taxonomically generalized interactions between plants, herbivores and predators. At face value, it would be tempting to conclude that the tree species, caterpillar species and bird species involved in these generalized interactions are functionally redundant and relatively interchangeable in the community. However, our data support the contrary notion that phenotypic differences among species have significant indirect effects on the magnitude of species interactions, and thus community dynamics and structure. Only by further study of the mechanisms of these direct and indirect interactions will we move toward an understanding of their ecological and evolutionary consequences. This level of understanding is critical for making confident predictions at the community level as natural and anthropogenic perturbations affect the ecosystem.

Case study: intraspecific variation in predator–herbivore interactions

One example of a potential source of intraspecific variation in herbivore–enemy interactions is that of plant sex within dioecious species. It has been proposed that differential investment in reproduction between males and females leads to predictable differences along growth–defence trade-offs (Cornelissen and Stiling 2005). Male plants, with a greater surplus of resources, are expected to be relatively fast growing and poorly defended against herbivores. In contrast, relatively more resource-limited females are expected to be slower growing and invest more in herbivore defences. While plant sex is expected to influence herbivores directly, it is less clear whether this form of intraspecific variation might also have consequences for enemy abundance or the interactions between herbivores and enemies. In other words, how do herbivore–enemy interactions differ based upon the sex of the plant on which they occur? In addition, how do the genetically based influences of plant sex compare to genetic variation within such species more generally?

To address these questions, we have studied the long-lived woody shrub *Baccharis salicifolia*, a dioecious composite (Asteraceae) locally abundant throughout the western United States in riparian areas and wetter upland habitats that can grow to be 3 m in height at maturity. We collected cuttings from 39 individuals in a population occupying the University of California San Joaquin Fresh Water Marsh Reserve in Irvine, CA. Nineteen of these genotypes were male and 20 were female. After rooting 10 to 12 copies of each genotype in potting soil, these plants were transplanted into random locations within a common garden adjacent to the Marsh Reserve in May 2008 (see Mooney *et al.* 2012 for more detail).

In April 2009 we conducted counts on all plants for the most abundant herbivore, the aphid *Uroleucon rudbeckiae*. At this time, we also counted the number of predatory arthropods on each plant, including larval and adult ladybird beetles (Coccinellidae, Coleoptera), syrphid fly larvae (Syrphidae, Diptera), predatory genera of Miridae (Miridae, Hemiptera) and lacewing larvae (Chrysopidae, Neuroptera).

We found variation in aphid and enemy abundance between male and female plants and among plant genotypes (Fig. 7.3a). Genotype means for aphid and enemy abundance spanned 130 fold and 6.4 fold, respectively. Male and female plants also varied significantly, with females having 2.2 fold more aphids and one-third more predators than males. Consequently, variation in herbivore and enemy abundance due to plant sex was a fraction of that seen in the variation among genotypes.

Much of this genetic- and sex-based variation among *Baccharis* plants in enemy abundance was due to enemies responding positively to aphid abundance. In a quadratic regression of genotype means for enemy abundance regressed on aphid abundance, 65% of the variation in enemy abundance among genotypes was explained by genetic variation in aphid abundance. Consequently, the density-dependent response of enemies to variation in herbivore density was responsible for much of the sex- and genotype-based variation in enemy density in accordance with Case 1.

However, the residuals of this quadratic relationship revealed additional statistically significant variation in enemy abundance among *Baccharis* sexes and genotypes (Fig. 7.3b). The differences between males and females were comparatively small and unlikely to be of biological significance. In contrast, the genotypic variation was quite large, ranging from 2.7 fewer predators to 2.3 more predators than predicted by aphid density alone against an overall mean of 4.7 predators per plant. Whether this genetic variation in predator

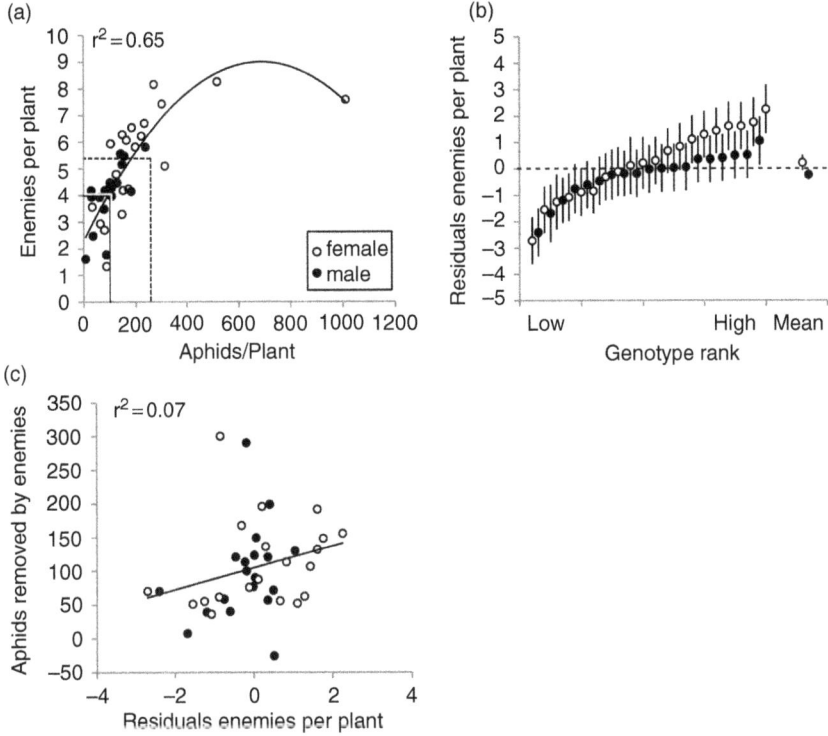

Figure 7.3 Analysis of *Baccharis salicifolia* genetic effects on aphids and enemies.
(a) Means of 39 plant genotypes for enemy abundance related in a quadratic regression
to aphid abundance. Dashed lines indicate mean aphid and enemy abundance for
females, solid lines for males. (b) Plant genotype means ± 1 SE for residual enemy
abundance controlling for aphid abundance with a quadratic regression. Genotypes
are ranked from lowest to highest enemy abundance. Dashed line indicates the density
of predators predicted based upon aphid density. Mean residual abundance is shown
for males and females to the right. (c) The number of aphids removed by enemies (June
manipulative experiment) regressed on residual enemy abundance controlling for
aphid abundance.

density is due to a direct effect of plants on herbivore traits (Case 3) or to a
direct effect of plants on enemy density (Case 2) remains uncertain.

While a plant genetic influence over enemy abundance implies concomitant variation in the suppressive effects of enemies on herbivores, this need
not be the case (see above; Mooney and Agrawal 2008). Consequently, in June
2009 we conducted an enemy exclusion experiment on each plant genotype
to test whether those plants recruiting the most enemies in turn received
stronger herbivore suppression. We randomly selected three plants of each
genotype and added five adult *U. rudbeckiae* to each of two stems on each plant
that were selected to be similar in size. These stems were then enclosed within

mesh bags, with the closed bag serving to exclude enemies while we cut slits in the second bag to allow enemy access. Three weeks later, we counted aphids on each stem. This study is thus somewhat unusual among those listed in Table 7.1 in that it both tested for genetic variation in the density of herbivores and enemies, and then excluded those enemies to test further for genetic variation in their effects on herbivores (but see Mooney and Agrawal 2008).

Aphid abundance differed strongly between *Baccharis* branches open and closed to predators, with mean aphid abundances of 61 ± 5 and 165 ± 12 aphids per branch, respectively. While aphid abundance in this experiment was not significantly influenced by plant genotype, plant sex or their interactions with the effects of enemy exclusion, there was a significant positive genetic correlation between the number of aphids removed by enemies in this June experiment and the residuals of enemy abundance controlling for aphid abundance from the April arthropod counts (one-tailed $P = 0.04$). Simply put, those plant genotypes that recruited more enemies than expected based upon aphid abundance (Fig. 7.3b) in turn received the strongest protection from enemies during the manipulative experiment (Fig. 7.3c). While the strength of this relationship was relatively weak in terms of variance explained ($R^2 = 0.07$), this is due at least in part to the fact that we conducted the manipulative experiment two months after collecting the observational data.

Our study thus shows that *Baccharis* genetic variation in enemy recruitment in turn feeds back to generate a concomitant effect on herbivore suppression. These dynamics suggest that enemies may act as an agent of selection upon plant traits, favouring those genetically controlled and variable plant traits that, while not yet identified, act to increase enemy recruitment and herbivore suppression. While Rudgers (2004) showed that nectar-collecting ants selected for extra-floral nectar production in wild cotton (Table 7.1), these results suggest that similar dynamics may occur amongst plants without obvious structures for enemy recruitment. These findings also demonstrate that a full understanding of plant resistance requires considering not just the pairwise plant–herbivore interaction, but also how plants influence the interactions between herbivores and their enemies (Price *et al.* 1980).

Community and evolutionary consequences
We now turn to considering the consequences of these four classes of effects of plant phenotype on herbivores, enemies and their interactions (Fig. 7.1). These consequences can be both ecological and evolutionary in nature. In addition, these consequences depend on whether the variation in plant phenotypes is genetically based, and when it is genetically based whether this variation is intra- or interspecific in nature.

Intraspecific genetic variation in plant traits

While the genetic basis of anti-herbivore resistance in plants has been long established (Fritz and Simms 1992), there is now increasing evidence that differing genotypes of a single plant species can vary substantially in terms of the species and abundance of all associates, including not only herbivores but also the enemies and parasites of those herbivores (Table 7.1). Yet studies documenting plant genetic effects on community composition have yet to document adequately the mechanisms by which heritable variation in plant traits are working, and it is likely that the net effect of plant genotype acts via some or all of the pathways outlined in Fig. 7.1.

This uncertainty over the mechanism of plant genetic effects on the trophic composition of the associated community in turn renders the evolutionary consequences of such effects for plants ambiguous. While it is of ecological interest to understand how plant genotype identity influences community composition, one implication of such effects is that plants might influence their own fitness by such community-wide effects. But whether this occurs depends on the mechanisms at work. If plant genetic effects operate entirely through effects on herbivore density (Case 1), then the indirect effects of this genetic variation on herbivore enemies would not influence selection on, or the evolution of, plant traits; enemies of herbivores could influence plant fitness under this scenario, but they would not select on plant traits (Inouye and Stinchcombe 2001; Strauss *et al.* 2005). In contrast, if genetically con- trolled plant traits alter enemy suppression of herbivores (Cases 2, 3 or 4) then enemies can act as an agent of selection on plant traits.

Consequently, the mechanisms by which plants influence enemy density determine whether the evolution of herbivore defence in plants is driven entirely by the pairwise plant–herbivore interaction, or whether a consider- ation of genetically based variation in community structure contributes to such evolutionary dynamics. While much work on plant defence has focused strictly on the plant–herbivore interaction, there is an increasing awareness that a more holistic, tritrophic perspective is required (Price *et al.* 1980). If plant resistance mediates herbivore densities but does not influence predator–herbivore interactions, this implies that a reductionist framework can adequately describe the means by which plant defence operates. But to the extent that plant traits also mediate the effects of enemies on herbivores, broadening our perspective to include such dynamics will provide a more accurate and predictive understanding of plant defence (Marquis and Whelan 1996).

Interspecific variation in plant traits

Just as intraspecific genetic variation in plant traits is expected to influence the relative fitness of genotypes, so too will variation among plant species influence the relative performance of those plants. In so doing, these dynamics may feed back to influence plant species composition. The effects of herbivore–enemy interactions for plant species composition have been discussed but little studied. Specifically, it has been proposed that herbivores may influence the competitive balance between fast-growing species, with minimal investment in herbivore defences, and slow-growing but well defended species (Coley et al. 1985; Fine et al. 2004). Enemies may thus indirectly promote the dominance of poorly defended plant species by suppressing herbivores (Polis and Strong 1996). Because plant species vary in strength of benefits received from enemies (Table 7.1), we further propose that enemies should also favour those plant species with traits that maximize enemy suppression of herbivores.

Future directions

Based on the review and discussion just provided, we see four primary areas for future work.

As Table 7.1 makes clear, more studies of more systems are needed. For instance, only 11 species have been investigated for intraspecific variation in enemy density and herbivore–enemy interactions, and four of these are within a single plant family (Salicaceae). In addition, all studies have been conducted in temperate communities and all but one (Johnson and Agrawal 2005) have been conducted with long-lived perennial species.

More work is needed to elucidate the mechanisms by which plant effects occur. Such work should include studies designed to identify the causative plant traits, as well as the interaction pathways by which such traits influence herbivores and enemies (Fig. 7.1). Efforts to identify important plant traits could take a variety of approaches, including plant genetic manipulation (Kessler et al. 2004) and species or genetic correlations between the traits of plant genetic types and plant traits (Mooney et al. 2010). Discriminating among the four mechanistic pathways will require statistical approaches as outlined here (e.g., controlling for herbivore density), as well as data on the traits of herbivores and enemies as they associate with differing plant genetic types. In addition, studies will be required that experimentally control herbivore quality to test directly for plant effects on enemy density (Case 2) and enemy traits (Case 4), and that control for enemy density to test directly for plant effects on herbivore traits (Case 3). Ultimately, a full understanding of these dynamics is likely to require multiple complementary approaches.

A growing body of literature is documenting the community-wide consequences of intraspecific genetic variation (Whitham et al. 2003; Whitham et al. 2006; Johnson and Stinchcombe 2007), including variation in the abundance and species composition of herbivores and enemies associated with differing plant genotypes. One of the proposed implications for this work is that plant influence over community composition may feed back to influence the relative fitness of those genotypes (Whitham *et al.* 2003). Determining the likelihood of this scenario will require establishing whether plant genetic influence over enemies is a case of variable resistance and density-dependent enemy recruitment (Case 1) or plant influence over the strength of herbivore suppression (Cases 2, 3 and 4). In addition, most studies to date have been conducted within relatively homogeneous common gardens, which are likely to overestimate the importance of plant genetic influence in relation to other ecological factors. Consequently, the next step will be to place these interactions in natural ecosystems with realistic levels of habitat and community heterogeneity.

Finally, field experiments are needed to test the importance of herbivores and herbivore–enemy interactions as a determinant of plant-species composition. For instance, while Fine *et al.* (2004) showed that herbivores were likely important for promoting species turnover among tropical soils of differing fertility, the potential roles of herbivore enemies and species-specific herbivore–enemy interactions have yet to be considered.

References

Agrawal, A. A. (2003) Community genetics: new insights into community ecology by integrating population genetics. *Ecology*, **84**, 543–544.

Bailey, J. K., Wooley, S. C., Lindroth, R. L. and Whitham, T. G. (2006) Importance of species interactions to community heritability: a genetic basis to trophic-level interactions. *Ecology Letters*, **9**, 78–85.

Barbosa, P., Segarra, A. E., Gross, P. *et al.* (2001) Differential parasitism of macrolepidopteran herbivores on two deciduous tree species. *Ecology*, **82**, 698–704.

Bernays, E. A. (1998) Evolution of feeding behavior in insect herbivores: success seen as different ways to eat without being eaten. *BioScience*, **48**, 35–44.

Bottrell, D. G., Barbosa, P. and Gould, F. (1998). Manipulating natural enemies by plant variety selection and modification: a realistic strategy? *Annual Review of Entomology*, **43**, 347–367.

Brower, L. P., Van Zandt Brower, J. and Corvino, J. M. (1967) Plant poisons in a terrestrial food chain. *Proceedings of the*

National Academy of Sciences of the United States of America, **57**, 893–898.

Bukovinszky, T., van Veen, F. J. F., Jongema, Y. and Dicke, M. (2008) Direct and indirect effects of resource quality on food web structure. *Science*, **319**, 804–807.

Clancy, K. M. and Price, P. W. (1987) Rapid herbivore growth enhances enemy attack: sublethal plant defenses remain a paradox. *Ecology*, **68**, 733–737.

Coley, P. D., Bateman, M. L. and Kursar, T. A. (2006) The effects of plant quality on caterpillar growth and defense against natural enemies. *Oikos*, **115**, 219–228.

Coley, P. D., Bryant, J. P. and Chapin, F. S. (1985) Resource availability and plant antiherbivore defense. *Science*, **230**, 895–899.

Cornelissen, T. and Stiling, P. (2005) Sex-biased herbivory: a meta-analysis of the effects of gender on plant–herbivore interactions. *Oikos*, **111**, 488–500.

Cortesaro, A., Stapel, J. and Lewis, W. (2000) Understanding and manipulating plant attributes to enhance biological control. *Biological Control*, **17**, 35–49.

Crutsinger, G. M., Cadotte, M. W. and Sanders, N. J. (2009) Plant genetics shapes inquiline community structure across spatial scales. *Ecology Letters*, **12**, 285–292.

Crutsinger, G. M., Collins, M. D., Fordyce, J. A. et al. (2006) Plant genotypic diversity predicts community structure and governs an ecosystem process. *Science*, **313**, 966–968.

Feeny, P. P. (1976) Plant apparency and chemical defense. *Recent Advances in Phytochemistry*, **10**, 1–40.

Fine, P. V. A., Mesones, I. and Coley, P. D. (2004) Herbivores promote habitat specialization by trees in Amazonian forests. *Science*, **305**, 663–665.

Forkner, R. E. and Hunter, M. D. (2000) What goes up must come down? Nutrient addition and predation pressure on oak herbivores. *Ecology*, **81**, 1588–1600.

Fraenkel, G. S. (1959) Raison d'être of secondary plant substances. *Science*, **129**, 1466–1470.

Fritz, R. S. and Nobel, J. (1990) Host plant variation in mortality of the leaf-folding sawfly on the arroyo willow. *Ecological Entomology*, **15**, 25–35.

Fritz, R. S. and Simms, E. L. (1992) *Plant Resistance to Herbivores and Pathogens : Ecology, Evolution, and Genetics*. Chicago, IL: University of Chicago Press.

Fritz, R. S., McDonough, S. E. and Rhoads, A. G. (1997) Effects of plant hybridization on herbivore-parasitoid interactions. *Oecologia*, **110**, 360–367.

Futuyma, D. J. and Gould, F. (1979) Associations of plants and insects in a deciduous forest. *Ecological Monographs*, **49**, 33–50.

Gassmann, A. J. and Hare, J. D. (2005) Indirect cost of a defensive trait: variation in trichome type affects the natural enemies of herbivorous insects on *Datura wrightii*. *Oecologia*, **144**, 62–71.

Gripenberg, S., Mayhew, P. J., Parnell, M. and Roslin, T. (2010) A meta-analysis of preference-performance relationships in phytophagous insects. *Ecology Letters*, **13**, 383–393.

Gross, P. and Price, P. W. (1988) Plant influences on parasitism of two leafminers: a test of enemy-free space. *Ecology*, **69**, 1506–1516.

Haggstrom, H. and Larsson, S. (1995) Slow larval growth on a suboptimal willow results in high predation mortality in the leaf beetle *Galerucella lineola*. *Oecologia*, **104**, 308–315.

Hairston, N. G., Smith, F. E. and Slobodkin, L. G. (1960) Community structure, population control, and competition. *American Naturalist*, **94**, 421–425.

Hare, J. D. (1992) Effects of plant variation on herbivore-natural enemy interactions. In R. S. Fritz and E. L. Simms, eds., *Plant Resistance to Herbivores and Pathogens: Ecology, Evolution, and Genetics*. Chicago, IL: University of Chicago Press, pp. 278–300.

Hare, J. D. (2002) Plant genetic variation in tritrophic interactions. In T. Tscharntke and B. A. Hawkins, eds., *Multitrophic Level Interactions*. Cambridge: Cambridge University Press, pp. 8–43.

Hedges, L. V., Gurevitch, J. and Curtis, P. S. (1999) The meta-analysis of response ratios in experimental ecology. *Ecology*, **80**, 1150–1156.

Helms S. E., Connelly, S. J. and Hunter, M. D. (2004) Effects of variation among plant species on the interaction between a herbivore and its parasitoid. *Ecological Entomology*, **29**, 44–51.

Holmes, R. T. and Robinson, S. K. (1981) Tree species preferences of foraging insectivorous birds in a northern hardwoods forest. *Oecologia*, **48**, 31–35.

Holmes, R. T., Schultz, J. C. and Nothnagle, P. (1979) Bird predation on forest insects: exclosure experiment. *Science*, **206**, 462–463.

Inouye, B. and Stinchcombe, J. R. (2001) Relationships between ecological interaction modifications and diffuse coevolution: similarities, differences, and causal links. *Oikos*, **95**, 353–360.

Janzen, D. H. (1966) Coevolution of mutualism between ants and acacias in Central America. *Evolution*, **20**, 249–275.

Johnson, M. T. J. and Agrawal, A. A. (2005) Plant genotype and environment interact to shape a diverse arthropod community on evening primrose (*Oenothera biennis*). *Ecology*, **86**, 874–885.

Johnson, M. T. J. and Stinchcombe, J. R. (2007). An emerging synthesis between community ecology and evolutionary biology. *Trends in Ecology and Evolution*, **22**, 250–257.

Kaplan, I., Lynch, M. E., Dively, G. P. and Denno, R. F. (2007) Leafhopper-induced plant resistance enhances predation risk in a phytophagous beetle. *Oecologia*, **152**, 665– 675.

Kessler, A., Halitschke, R. and Baldwin, I. T. (2004) Silencing the jasmonate cascade: induced plant defenses and insect populations. *Science*, **305**, 665–668.

Lill, J. T. and Marquis, R. J. (2001) The effects of leaf quality on herbivore performance and attack from natural enemies. *Oecologia*, **126**, 418–428.

Lill, J. T., Marquis, R. J. and Ricklefs, R. E. (2002) Host plants influence parasitism of forest caterpillars. *Nature*, **417**, 170–173.

Linhart, Y. B., Keefover-Ring, K., Mooney, K. A., Breland, B. and Thompson, J. D. (2005) A chemical polymorphism in a multitrophic setting: thyme monoterpene composition and food web structure. *American Naturalist*, **166**, 517–529.

Marquis, R. J. and Whelan, C. (1996) Plant morphology, and recruitment of the third trophic level: subtle and little-recognized defenses? *Oikos*, **75**, 330–334.

Mooney, K. A. and Agrawal, A. A. (2008) Plant genotype shapes ant-aphid interactions: Implications for community structure and indirect plant defense. *American Naturalist*, **168**, E195–E205.

Mooney, K. A., Halitschke, R., Kessler, A. and Agrawal, A. A. (2010) Evolutionary trade-offs in plants mediate the strength of trophic cascades. *Science*, **327**, 1642–1644.

Mooney, K. A., Pratt, R. T. and Singer, M. (2012) The tri-trophic interactions hypothesis: interactive effects on host plant quality, diet breadth and natural enemies on herbivores. *PLoS ONE*, **7**, e34403.

Moran, N. and Hamilton, W. D. (1980) Low nutritive quality as defense against herbivores. *Journal of Theoretical Biology*, **86**, 247–254.

Muller, M. S., McWilliams, S. R., Podlesak, D. *et al.* (2006) Tri-trophic effects of plant defenses: chickadees consume caterpillars based on host leaf chemistry. *Oikos*, **114**, 507–517.

Ness, J. H., Morris, W. F. and Bronstein, J. L. (2009) For ant-protected plants, the best defense is a hungry offense. *Ecology*, **90**, 2823–2831.

Ode, P. J. (2006) Plant chemistry and natural enemy fitness: effects on herbivore and natural enemy interactions. *Annual Review of Entomology*, **51**, 163–185.

Polis, G. A. and Strong, D. R. (1996) Food web complexity and community dynamics. *American Naturalist*, **147**, 813–846.

Price, P. W. and Clancy, K. M. (1986) Interactions among three trophic levels: gall size and parasitoid attack. *Ecology*, **67**, 1593–1600.

Price, P. W., Bouton, C. E., Gross, P. *et al.* (1980) Interactions among three trophic levels: influence of plants on interactions between insect herbivores and natural enemies. *Annual Review of Ecology and Systematics*, **11**, 41–65.

Robinson, S. K. and Holmes, R. T. (1982) Foraging behavior of forest birds: the relationships among search tactics, diet, and habitat structure. *Ecology*, **63**, 1918–1931.

Rudgers, J. A. (2004) Enemies of herbivores can shape plant traits: selection in a facultative ant-plant mutualism. *Ecology*, **85**, 192–205.

Schoonhoven, L. M., Loon, J. J. A. v. and Dicke, M. (2005) *Insect–Plant Biology*, 2nd edition. Oxford: Oxford University Press.

Shuster, S. M., Lonsdorf, E. V., Wimp, G. M., Bailey, J. K. and Whitham, T. G. (2006) Community heritability measures the evolutionary consequences of indirect genetic effects on community structure. *Evolution*, **60**, 991–1003.

Singer, M. S., Carriere, Y., Theuring, C. and Hartmann, T. (2004a) Disentangling food quality from resistance against parasitoids: diet choice by a generalist caterpillar. *American Naturalist*, **164**, 423–429.

Singer, M. C., Farkas, T. E., Skorik, C. M. and Mooney, K. A. (2012) Tri-trophic interactions at a community level: effects of host-plant species quality on bird predation of caterpillars. *American Naturalist*, **179**, 363–374.

Singer, M. S., Rodrigues, D., Stireman, J. O. III and Carriere, Y. (2004b) Roles of food quality and enemy-free space in host use by a generalist insect herbivore. *Ecology*, **85**, 2747–2753.

Sipura, M. (1999) Tritrophic interactions: willows, herbivorous insects and insectivorous birds. *Oecologia*, **121**, 537–545.

Slanksy, F. and Scriber, J. M. (1985) Food consumption and utilization. In G. A. Kerkut and L. I. Gilbert, eds., *Comprehensive Insect Physiology, Biochemistry, and Pharmacology*. Oxford: Pergamon Press, pp. 87–163.

Strauss, S. Y., Sahli, H. and Conner, J. K. (2005) Toward a more trait-centered approach to diffuse (co)evolution. *New Phytologist*, **165**, 81–89.

Tumlinson, J. H., Turlings, T. C. J. and Lewis, W. J. (1992) The semiochemical complexes that mediate insect parasitoid foraging. *Agricultural Zoology Reviews*, **5**, 221–252.

Turlings, T. C. J. and Benrey, B. (1998) Effects of plant metabolites on the behavior and development of parasitic wasps. *Ecoscience*, **5**, 321–333.

Turlings, T. C. J., Tumlinson, J. H. and Lewis, W. J. (1990) Exploitation of herbivore-induced plant odors by host-seeking parasitic wasps. *Science*, **250**, 1251–1253.

Vet, L. E. M. and Dicke, M. (1992) Ecology of infochemical use by natural enemies in a tritrophic context. *Annual Review of Entomology*, **37**, 141–172.

Weis, A. E. and Abrahamson, W. G. (1986) Evolution of host-plant manipulation by gall makers: ecological and genetic factors in the *Solidago–Eurosta* system. *American Naturalist*, **127**, 681–695.

Whelan, C. J. (2001) Foliage structure influences foraging of insectivorous forest birds: an experimental study. *Ecology*, **82**, 219–231.

Whitham, T. G., Bailey, J. K., Schweitzer, J. A. *et al.* (2006) A framework for community and ecosystem genetics: from genes to ecosystems. *Nature Reviews Genetics*, **7**, 510–523.

Whitham, T. G., Young, W. P., Martinsen, G. D. *et al.* (2003) Community and ecosystem genetics: a consequence of the extended phenotype. *Ecology*, **84**, 559–573.

Williams, I. S. (1999) Slow-growth, high-mortality: a general hypothesis, or is it? *Ecological Entomology*, **24**, 490–495.

Wimp, G. M. and Whitham, T. G. (2001) Biodiversity consequences of predation and host plant hybridization on an aphid-ant mutualism. *Ecology*, **82**, 440–452.

CHAPTER EIGHT

The implications of adaptive prey behaviour for ecological communities: a review of current theory

SCOTT D. PEACOR

Department of Fisheries and Wildlife, Michigan State University

and

CLAYTON E. CRESSLER

Department of Ecology and Evolutionary Biology, University of Michigan

Introduction

The overarching goal of community ecology is to address broad problems, from predicting the effects of species' invasion and loss to understanding the processes affecting the diversity, resilience and robustness of ecological systems. Ecological theory addresses these questions through the development of models that examine how species interact within food webs, and how those interactions give rise to the community-level properties we observe in natural systems, such as the relationship between complexity and stability, and the distribution of links across the food web. We argue here that current theory in community ecology is limited in its ability to address these fundamental questions, because it has largely ignored the role of species to modify their traits in response to their environment. This oversight is especially evident from the paucity of theory that considers how the broad problems outlined above are affected by the ubiquitous ability of prey to modify their traits to balance the trade-off between foraging gain and predation risk (Lima and Dill 1990; Lima 1998; Werner and Peacor 2003).

The traditional approach to community ecology is focused largely on linked pairwise interactions between populations, whether through competition, exploitation, or mutualism (May 1973; Pimm 1982; Bender *et al.* 1984; Yodzis 1988; Schoener 1993). These pairwise interactions are treated as building blocks upon which our understanding of larger ecological communities can be built, with the implicit assumption that pairwise interactions are independent of the ecological context in which they are embedded (Wootton 1994; Abrams 1995; Werner and Peacor 2003). A critical evaluation of this assumption began to emerge in the late 1960s. Vandermeer (1969) introduced the notion of 'higher-order interactions' (HOIs) wherein the

Trait-Mediated Indirect Interactions: Ecological and Evolutionary Perspectives, eds. Takayuki Ohgushi, Oswald J. Schmitz and Robert D. Holt. Published by Cambridge University Press. © Cambridge University Press 2012.

interaction between two species is modified by other species in the system. While early attempts to test for HOIs were equivocal (Wilbur 1972; Neill 1974; Werner and Peacor 2003), the idea that species' interactions were not fixed began to gain wider acceptance through the development of foraging theory in the 1970s and 1980s.

Foraging theory was the first formal theory to suggest that the interactions between a consumer and its resources might be altered by ecological context (MacArthur and Pianka 1966; Murdoch 1969; Schoener 1971; Charnov 1976). This theory assumes that consumers optimize their diet depending on the abundance and quality of available resources. Though not reported as such until much later, this was the first ecological theory that included HOIs: a consumer's foraging behaviour, and the strength of the interaction between it and its resources, varied with ecological context (i.e., the abundances of *other* resources in the system). This theory has been widely tested (Stephens and Krebs 1986) and the consequences of flexible foraging by predators for dynamics and structure are now a well-studied problem (McCann and Rooney 2009; Abrams 2010a, b; Beckerman *et al.* 2010).

The insights of foraging theory were extended to consideration of how species might adjust their behaviour to balance the benefits of foraging gain against predation risk through the theoretical work of Peter Abrams (Abrams 1984, 1987, 1992). There is abundant evidence that this expression of phenotypic plasticity is common in many taxa, spanning diverse habitats (Lima 1998; Agrawal 2001). Abrams' work predicted that changes in a species' traits could affect not only the interaction of that species with its own predators and resources, but also impact the interactions among other species in the community. Perhaps the most surprising result of this work is that the indirect effect of one species on another, mediated only through trait change in an intermediate species (so-called trait-mediated indirect effects), can have magnitudes equal to, or even larger than, the magnitude of the direct and indirect interactions arising from consumptive effects (Abrams 1984, 1995). Theory developed since this early work has begun to explore the dynamical and structural consequences of trait modification (reviewed in Bolker *et al.* 2003; Abrams 2010b). Nevertheless, despite the mounting theoretical and empirical evidence suggesting that adaptive prey trait modification (hereafter, APTM) can affect ecological dynamics, APTM is seldom incorporated into general ecological theory (Bolker *et al.* 2003; Abrams 2010b), and we therefore know little about its potential influence on many fundamental ecological problems.

We review the theory addressing the influence of adaptive response of prey to predators on the structure and dynamics of food webs. We provide a broad discussion of the importance of HOIs in ecology in order to provide context for the theory on APTM, and because of the potential congruence between the influences of different HOIs. We then discuss the various approaches and decisions that must be made in order to incorporate APTM into theory. We follow with a review of existing theory on the consequences of APTM for food web structure and dynamics, and highlight how inclusion of APTM can modify canonical predictions regarding apparent and exploitative competition, trophic cascades and the paradox of enrichment, and the relationships among diversity, stability and connectance in food webs. This review of theory provides a basis for an evaluation of the evidence for and against the need to consider APTM to achieve an understanding of ecological communities. Current theory and empirical work strongly suggests that APTM is critical, but current theory and empirical work are not sufficient. We identify key gaps in current theory: the role of environmental variability; the dependence of APTM on system dynamics; and our understanding of the influence of APTM in food webs larger than those typically studied ($N = 3$), and suggest future directions in theory to address these gaps. Finally, we second previous arguments that the most conclusive evidence that APTM is crucial to food web theory will come from studies that integrate theory and experiment, and suggest directions we argue are particularly important in this endeavour.

HOIs in food web theory

An examination of HOIs provides context for the study of APTM, and may also provide clues to the potential influence of APTM. Several disparate mechanisms lead to HOIs (Fig. 8.1). First, one species, denoted an ecological engineer (a conspicuous example being beavers), may modify the environment in a way that affects the interactions between other species (Fig. 8.1a). Second, predator interference may lead to one predator affecting the predation rate of a second predator on a prey (as in the Beddington–DeAngelis functional response, Beddington 1975; DeAngelis *et al.* 1975) (Fig. 8.1b). Third, predator switching occurs when predator preference for a given prey type is affected by the density of a second consumer prey (Fig. 8.1c). Fourth, predators may become satiated by consuming one prey type, affecting consumption rate of a second prey type (Fig. 8.1c). In both these latter cases a HOI results from one prey species affecting the foraging behaviour of a predator and consequently affecting the interaction between the predator and the second prey. However, the different equations used to describe these HOIs have important consequences for model predictions (reviewed below). Finally, modification of prey traits can lead to a variety of higher-order interactions (Fig. 8.1d–g). In a tritrophic food chain, a predator can indirectly affect a resource when the prey modifies traits in response to changes in predation risk (i.e., predator density, Fig. 8.1d). Conversely, a resource can indirectly affect a predator when the consumer modifies traits in response to changes in resource levels (Fig. 8.1e). In two-predator–one-prey systems, a prey response to one predator may affect predator risk to a second predator (Fig. 8.1f). Lastly, there can be

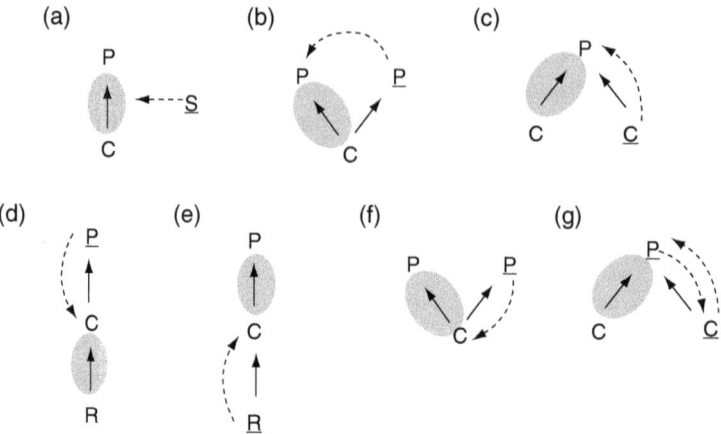

Figure 8.1 Schematic representations of the most commonly studied processes leading to HOIs, in which one species affects the interaction between two others. R, C, P and S represent resources, consumers, predators and species with no trophic significance, respectively. Solid arrows represent trophic consumptive interactions, dotted arrows represent induced trait changes, and the grey ellipses represent the pairwise interaction that is a function of a third species. (a) Predator–consumer interaction is a function of another species (e.g., an ecological engineer) that affects environmental conditions that affect the interaction. (b) Predator–consumer interaction is a function of a second predator in the environment that interferes with the focal predators ability to forage. (c) Predator–consumer interaction is a function of a second consumer's density due to predator satiation or predator switching. In (d–g) HOIs arise from adaptive prey trait modification. (d) A consumer trait-response to reduce predation risk affects the consumer–resource interaction. (e) Consumer trait-response to changes in resource level affects the predator–consumer interaction. (f) A consumer response to a change in one predator's density affects its interaction with a second predator. (g) A consumer modifies its traits to changes in the predator density. In addition to density changes in the consumer as in (c), these trait changes feed back and in turn affect predator traits (foraging preference).

complex interaction between HOIs if both prey and predators can modify their traits in response to changes in density and traits of the other; here prey adjust traits in response to both predator density and foraging behaviour, and the predator's foraging behaviour responds to both the density and traits of prey (Fig. 8.1g).

The different HOIs have received varying amounts of attention in the theoretical literature. Whereas they all have a qualitatively similar representation in ecological theory (i.e., species-pair interactions that are dependent on the density of other species in the food web (Dambacher and Ramos-Jiliberto 2008)), we are unaware of any review that has examined the commonalities and contrasts of their effects. It is clear, however, that the inclusion of some HOIs can strongly influence theoretical predictions. A prominent example is the classic study by McCann *et al.* (1998). In this case, high densities of one resource reduce the consumption rate on a second resource, effectively leading to a sigmoidal (and hence stabilizing) functional response between consumer and resource. While attention to this study has primarily focused on the relative magnitudes of interaction strengths, (i.e., 'weak interactions'), the higher-order nature of the interaction is key to the stabilizing effect found. Similarly, predator switching has also been shown to have large effects on system stability in small food webs (reviewed in Bolker *et al.* 2003; Abrams 2010b) and on basic food web properties including stability and structure in large webs (Kondoh 2003, 2006; Uchida and Drossel 2007).

Terminology

The terms used to describe many HOIs, or subsets thereof, are many and often used differently. 'Trait-mediated indirect effects/interactions' (TMIE, TMII) (Abrams *et al.* 1996; Peacor and Werner 1997, also behaviourally mediated indirect effects, Miller and Kerfoot 1987) occur when one species affects another species by inducing a modification in an intermediate species' phenotype, such as behaviour, morphology and/or life history. This would include all of the HOIs described in Fig. 8.1, except for ecological engineers. This term was created to distinguish them from the indirect interactions typically described in theory, termed 'density-mediated indirect effects' (DMIE), in which one species indirectly affects a second species by changing the density of an intermediate species (Abrams *et al.* 1996). 'Nonconsumptive effects' (NCEs) (Abrams 2008), also termed 'nonlethal effects', occur when the predator causes a TMII through inducing changes in prey traits, as in the interactions depicted in Fig. 8.1d–g. In some of the HOIs, there is adaptive predator behaviour (Fig. 8.1b, c, potentially g), and in some others there is adaptive prey trait modification (Figs. 8.1d–g). The term 'interaction modifications' (Wootton 1994) has been used to encompass the general effect of one species affecting the interaction between two others, and may refer to any of the HOIs in Fig. 8.1. To investigate the literature in the area, all of these terms should be used as keywords.

Approaches to modelling adaptive prey trait modification
Methodology for incorporating traits into models: a specific example

In this section, we review the major approaches that have been taken towards representing APTM in models. As we note above, prey can modify many traits in response to predation risk, including behaviours, such as activity level or habitat choice, morphology, such as the growth of defensive spines, or life history, such as size or reproductive strategy (Tollrian and Harvell

1999). Regardless of the trait under consideration, the fundamental constraint is that modification of the trait involves balancing a fitness trade-off (Bolker et al. 2003). The fitness benefit of the type of trait modification addressed here is increased survivorship due to a reduction in predation risk, but this modification carries a fitness cost such as reduced growth or reproduction (e.g., due to reduced foraging) or increased predation risk from another predator (Fig. 8.1d, f).

The typical approach to incorporating APTM in a model is to define functions relating prey trait expression to both fitness cost and benefit. As we review below, the shapes of these cost and benefit functions are crucial to determining the effects of APTM. To help provide intuition for this consideration, we review a specific model of Abrams (1984). This model examines indirect interactions in a tritrophic chain in which the intermediate consumer (prey) species (C) can adjust its phenotype (T_{op}) to changes in predator (P) and resource (R) density (as in Fig. 8.1d and e).

$$\frac{dR}{dt} = rR\left(1-\frac{R}{K}\right) - a_0 C R T_{op}$$
$$\frac{dC}{dt} = b_0 a_0 C R T_{op} - a_1 P C T_{op}^2 - D_1 C$$
$$\frac{dP}{dt} = b_1 a_1 P C T_{op}^2 - D_2 P. \tag{8.1}$$

Here r and K are the logistic growth parameters, a_i are predation rate parameters, b_i are conversion efficiencies of the resource to prey and prey to predator, and D_i are background mortality rates. T_{op} is a flexible trait, defined as the optimal fraction of time the consumer spends foraging, although this term could be used to describe the expression of other traits that affect consumption rates. In this model, the cost and benefit of modifying T_{op} are specified by the functions describing the dependence of consumer–resource attack rate and predator–consumer attack rate, respectively, on T_{op}. T_{op} is defined as the time spent foraging that maximizes consumer (prey) per capita growth rate, and thus

$$T_{op} = b_0 a_0 R / 2 a_1 P. \tag{8.2}$$

Thus an increase in predation risk will lead to a reduction in the optimal foraging time. Key to the subject of this chapter, using this equation for T_{op} makes the loss term of the per capita resource growth rate equation $a_0 C(b_0 a_0 R / 2 a_1 P)$. Adaptive prey foraging therefore makes the resource growth rate equation a function of the predator density, and therefore an example of a HOI since the resource–consumer interaction is now a function of a third species. Similarly, the predator functional response now depends on resource density, leading to another HOI. The nonlinearities introduced by these HOI underlie the unique and key role that APTM has on the system.

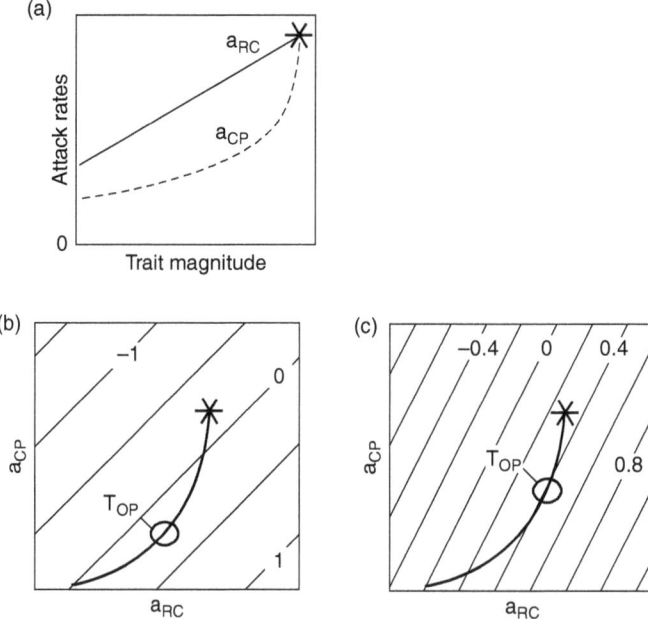

Figure 8.2 Graphical representation of the incorporation of adaptive trait modification in ecological models. These figures correspond to adaptive prey behaviour in a tritrophic chain (Equation (8.1)). (a) A trait change is presented in models as affecting the interaction coefficients between each species pair. The star represents no trait change (e.g., maximum foraging time). A reduction in fraction of time foraging (moving toward origin on x-axis) more strongly reduces the attack rate on the consumer than it reduces the consumer's attack rate on the resource. (b–c) Representation of the consumer growth rate and trait change as a function of the attack rates. The straight lines represent the per capita consumer (prey) growth rate isopleths. Without plasticity, as in traditional theory, the potential attack rates are constrained to a point (corresponding to the star in (a)), whereas with adaptive foraging (trait modification), the potential attack rate combinations are constrained to a line. In (c) the resource density is higher than in (b), making the growth isopleths denser and steeper. T_{op}, the optimal trait modification, occurs at the trait change that maximizes growth rate (indicated by a circle).

To clarify and emphasize the multiple processes and modelling decisions made in this analysis, we use a diagrammatic representation (Fig. 8.2). In conventional theory, the attack rate describing predator–prey interactions is either a constant or a function of one or two of the interacting species. For clarity, we build upon the linear case presented in Equation (8.1). We represent the attack rates (which are proportional to interaction strengths) in a two-dimensional plane, in which the x and y axis represent the consumer–resource and predator–consumer attack rates, respectively. In conventional theory (i.e., no trait modification), the trait magnitude equals a constant and the combination of attack rates would be a single point in this plane, which we indicate by the star. In contrast, with adaptive prey behaviour, the prey can modify the trait magnitude, leading to a range of possible attack rates. Fig. 8.2a illustrates the dependencies shown in Equation (8.1), wherein the consumer–resource and predator–consumer attack rates have a linear and stronger-than-linear dependence on the trait value, respectively. We can translate the two curves in Fig. 8.2a to a single curve illustrating the relationship between the two attack rates, given their functional dependence on the plastic trait (Fig. 8.2b). We can also superimpose the consumer growth rate as a function of the interaction strengths on this figure, as indicated by the parallel isopleths in Fig. 8.2b. T_{op}, as defined by Equation (8.2), occurs at the point on the curve where the growth rate is highest, indicated by the circle (Fig. 8.2b and c). To illustrate the interplay between the prey's trait expression and system dynamics, consider how modifying parameters to lead to a higher resource level will affect T_{op}, which in turn feeds back to affect species abundances. A higher resource level will shift, change the slope, and change the density of the growth rate isopleths, leading to a smaller trait change (i.e., the circle indicating maximum growth rate at the optimal trait change is closer to the star at the trait value in the absence of the trait change). Comparing Fig. 8.2b and c, we see that an increase in resource level increases the per capita growth rate of the prey for both the non-adaptive (compare values at stars) and adaptive (compare values at circles) cases, but that this difference is much smaller in the adaptive case.

Methodology for incorporating trait modification into models: general considerations

As noted above, a key factor in models with APTM is the shapes of the fitness cost and benefit functions, which depend on the assumptions of the modeller. The manner in which cost and benefit functions have been represented in models can be broadly divided into four categories. First, the functional relationship can be explicitly derived from foraging theory (Werner and Anholt 1993; Křivan and Sirot 2004). Second, biologically reasonable relationships that satisfy specific dynamical properties, such as leading to a stable equilibrium, can be chosen (Abrams 1984, 1995, 2003). Third, rather than using explicit equations, only the signs of the derivatives are defined, thereby defining the curvature of the functions (Abrams and Vos 2003). Lastly, equations are used that forgo an explicit representation of traits. In these models, the interaction strength between two species is a direct function of the density of a third species (Arditi et al. 2005; Goudard and Loreau 2008). Whereas the approach of forgoing traits can provide a general framework for studying the effects of HOIs by abstracting out the mechanisms that underlie HOI, it is difficult to relate model results with the natural processes that underlie the HOI.

The shapes of the fitness cost and benefit functions have been shown to be absolutely critical to determining the ecological effects of APTM, and many combinations of linear, nonlinear accelerating, and nonlinear decelerating forms have been explored (Abrams 1984, 1992, 2000, 2004, 2007; Ives and Dobson 1987; Schwinning and Rosenzweig 1990; Abrams and Matsuda 1997; Křivan 1997; Křivan and Sirot 2004). We are unaware of any successful attempts at measuring the shapes of these functions empirically, a situation that has been lamented repeatedly in the literature (Abrams 1995, 2001b, 2008; Bolker *et al.* 2003).We address this gap in the Discussion.

Modelling the dynamics of trait change

Once the trait has been incorporated through the specification of fitness cost and benefit functions, a decision must be made regarding the time-scale of prey response, relative to population dynamics. Many studies have assumed that trait modification is instantaneous, as in the example shown above (Abrams 1984; Křivan 1997, 2007). In such models, the trait is essentially represented as a parameter, as in Equation 1, that instantaneously takes on the value that maximizes fitness or maintains an ideal free distribution (Abrams 1984; Schwinning and Rosenzweig 1990). However, this assumption strikes many as too unrealistic (Bolker *et al.* 2003; Abrams 2010b). The most common alternative formulation borrows an approach from quantitative genetics and assumes that the rate of trait change is determined by the steepness of the fitness gradient with respect to the trait (Abrams *et al.* 1993; Abrams 1999, 2001a). A second alternative formulation specifies the rate of trait change as a function of the environment, e.g., predator density, without any reference to fitness (Vos *et al.* 2004a, b). In the two alternative formulations, the trait becomes a state variable with an explicit dependence upon the environment (e.g., the densities of other species), and trait dynamics and population dynamics occur simultaneously. The difference between assuming instantaneous trait change versus specifying a dynamical equation for the trait is not trivial, as a number of studies have shown that different assumptions can produce qualitatively different effects; for example, instantaneous change is often stabilizing, whereas trait change proportional to the fitness gradient can be de-stabilizing (Abrams 1999, 2001a, 2003; Ma *et al.* 2003).

Theory must also consider prey perception of environmental change. Most studies assume that prey trait modification is 'perfect', that is, APTM always increases fitness. However, studies that have introduced error into trait modification, through e.g., imperfect information or limited perception, have shown that this can have large effects on system dynamics (Luttbeg and Schmitz 2000; Abrams and Matsuda 2003; Kimbrell and Holt 2004; Abrams 2007). For example, both Luttbeg and Schmitz (2000) and Kimbrell and Holt

(2004) show that the degree to which prey are able to perceive their environment has important consequences for system persistence.

Adaptive prey trait modification in food web theory

In this section, we review theory on how APTM affects long-term dynamics in small $(N \leq 3)$ and large food webs, and identify mechanisms that have consistent effects on food web dynamics. First, we briefly review results from two-species predator–prey systems to provide some context for the considerations of larger webs. We then consider the three basic three-species food web topologies, laid out above, that include APTM (Fig. 8.1d–g). These topologies are the focus of most theoretical studies because they are the simplest systems that can exhibit indirect effects: in one-predator–two-prey systems, the focus is on apparent competition between the prey; in two-predator–one-prey systems, the focus is on exploitative competition between the predators; in tritrophic chains, the focus is on trophic cascades and the paradox of enrichment. For each topology, we address the mechanisms that induce stability or instability, with a special focus on the ecologically relevant indirect effects observed in traditional ecological models. Finally, we review the few studies that have incorporated APTM into larger food webs.

APTM in predator-prey models

Ives and Dobson (1987) provided one of the first analyses of long-term dynamics in systems with APTM, and suggested that prey trait modification increased stability. Since this paper, studies have altered assumptions about the shapes of the fitness cost and benefit functions and the timing of trait change. In general, instantaneous responses to predation tend to be stabilizing (Ives and Dobson 1987; Křivan 1997, 2007; Cressman et al. 2004), whereas incorporating lags can be de-stabilizing. When predation risk increases at a decreasing rate with increasing trait value (i.e., predation risk saturates at high levels of the trait), the effect is often de-stabilizing (Abrams et al. 1993; Abrams and Matsuda 1997). This is especially true when the trait is incorporated into the Type II functional response of a predator. De-stabilization results from 'chase' cycles generated by the interaction between prey behaviour and predator satiation: when prey are abundant, predators are satiated; this leads to a reduction in prey defence, which leads to an increase in predation and a declining prey population, which causes prey to increase defence levels. As long as prey and/or predator behavioural responses are not instantaneous, this can lead to unstable population dynamics, though it may not alter the effects on system persistence (Abrams 2000, 2003). However, when risk increases linearly or superlinearly with increasing trait value, this tends to stabilize dynamics (Ives and Dobson 1987; Křivan 1997, 2007; Křivan and Sirot 2004; Cressman et al. 2004). Coevolution of prey and predator foraging traits also tends to increase stability (Křivan 2007; Mougi and Nishimura 2008).

APTM in tritrophic food chains

The tritrophic chain topology (Fig. 8.1d, e) was studied by Abrams (1984) in the first paper introducing the notion of indirect effects arising from prey behaviour (i.e., TMIEs). Abrams documented that interactions arising from trait modification could be of larger magnitude than those arising from consumptive effects (DMIE). Trophic cascades and the paradox of enrichment have been the focus of most of the dynamic studies since that time.

There has been considerable theoretical interest in the effect of APTM on the magnitude of trophic cascades, perhaps motivated by empirical work suggesting that TMII may be the most important cause of such cascades (Schmitz *et al.* 2004). Traditional ecological theory (Oksanen *et al.* 1981) predicts that density-mediated trophic cascades will have consistent effects on species in the chain: in particular, enriching the system will have no effect on the middle consumer, because any increase in prey abundance will be countered by an increase in predator abundance. However, several theoretical studies have shown that APTM in the middle consumer can lead to an increased abundance of the middle species in response to enrichment (Abrams and Vos 2003; Křivan and Schmitz 2004; Křivan and Sirot 2004; Vos *et al.* 2004a, b; Ramos-Jiliberto *et al.* 2008). Abrams and Vos (2003) use a very general framework to show that, depending on the shapes of the fitness cost and benefit functions, each species' equilibrium abundance can change in either a positive or negative direction, emphasizing that including APTM leads to model results that contrast with traditional theory (see also Abrams and Matsuda 2005; Abrams 2009).

The paradox of enrichment (Rosenzweig 1971) describes how increasing the carrying capacity of the basal species de-stabilizes the dynamics of a tritrophic system; increasing basal carrying capacity leads first to cycles, and ultimately to chaotic dynamics (Hastings and Powell 1991). Several studies have shown that APTM tends to be stabilizing (Fryxell and Lundberg 1998; Rinaldi *et al.* 2004; Vos *et al.* 2004a; Ramos-Jiliberto *et al.* 2008); APTM reduces the amplitude of fluctuations, preventing the transition to chaos and stabilizing the persistence of the system.

APTM in two-predator–one-prey webs

A dominant paradigm in ecology is that two predators cannot coexist on a single prey (Armstrong and McGehee 1980). However, Matsuda *et al.* (1993) showed that coexistence was possible if APTM reduced susceptibility to one predator, while enhancing susceptibility to the other, i.e., if trait modification is predator-specific. This specificity generates a mutualistic TMII between the predators that opposes the density-mediated competitive interaction, enhancing coexistence of all three species. Generalized defensive traits that are equally effective against both predators, on the other hand, promote instability and reduce the likelihood of coexistence. Matsuda *et al.* (1994, 1996) and Kondoh (2007) have extended this finding to show that predator-specific trait modification is stabilizing, whereas generalized trait modification is de-stabilizing, even when this topology is embedded in more complicated topologies.

APTM in one-predator–two-prey webs

One-predator–two-prey webs were the focus of much of the early theory on the effects of predator foraging behaviour on ecological dynamics (Murdoch 1969), and recent theory has considered how the inclusion of APTM with predator behaviour can alter standard predictions. In traditional ecological theory, one-predator–two-prey systems are characterized by the negative DMII between the two prey, known as apparent competition, that is expected to lead to competitive exclusion, analogous to exploitative competition (Holt *et al.* 1994). However, if predators alter their foraging behaviour in response to prey density, either through predator switching, diet choice or satiation, the intensity of apparent competition tends to be reduced, permitting coexistence of all three species (see reviews in Fryxell and Lundberg 1998; Bolker *et al.* 2003; Abrams 2010a).

Abrams and Matsuda (1993) were the first to explore the consequences of adaptive foraging in both prey and predators. They showed that this interaction could produce either positive switching (predator preference for most abundant prey, as in predator switching models) or negative switching (predator preference for the less abundant prey), but that the overall result was a mutualistic TMII between the prey. Later studies using similar methodologies (i.e., ordinary differential equation models), but with different assumptions about shape of fitness cost and benefit functions, have confirmed this basic result. In general, APTM tends to reduce the intensity of apparent competition between the prey, making coexistence more likely (Abrams and Matsuda 1996; Abrams *et al.* 1998; Abrams 2000; Yamauchi and Yamamura 2005). Whether APTM is also stabilizing seems to depend upon modelling assumptions related to the shapes of the fitness cost and benefit functions and the relative speed of adaptation (Abrams 2000). Studies of this topology using individual-based modelling further suggest that APTM can lead to stable coexistence. Kimbrell and Holt (2004) allowed prey to evolve modified behaviour to predators (sensitivity) and predators to evolve preference behaviour. They showed the prey sensitivity alone was stabilizing, and that if both prey and predator traits were adaptive, the system was more stable than a system without adaptation.

APTM in larger systems

There have been very few studies of APTM in larger food webs (Matsuda *et al.* 1994, 1996; Dambacher and Ramos-Jiliberto 2007; Kondoh 2007). Kondoh (2007) examined the influence of APTM on community-level stability metrics in food webs of 3–12 species with multiple trophic levels and different levels of connectivity. In this model, prey could modify a trait to reduce predation risk, at the cost of reduced reproduction (for non-basal species) or reduced intrinsic growth rate (for basal species). Trait modification was either general (reduced risk from all predators) or specific (reducing risk from one predator led to increased risk to other predators). When models were constrained to two trophic levels, APTM led to HOIs of the form shown in Fig. 8.1f and g; for more than two trophic levels, all of the HOI caused by APTM shown in Fig. 8.1 are possible. Predator-specific APTM led to increased community persistence as the fraction of adaptive species or the rate of adaptation was increased. Further, a positive relationship between the number of connections and the stability of the food web was observed in simulations with few species, while a unimodal connectance–stability relationship was observed for webs with many species. In contrast, generalized trait modification decreased persistence and led to negative connectance–stability relationships. The mechanism underlying these effects was a trait-mediated 'rescue effect' explored by Matsuda and colleagues (Matsuda et al. 1993, 1994, 1996). In this case, when a prey adapts its trait to a more abundant predator, this increases susceptibility to rarer predators, thereby increasing those predators' growth rates and 'rescuing' them. The decrease in stability as connectance increases in large webs is attributed to increased competition between predators for the same prey. However, if predators were also allowed to forage adaptively, as in Matsuda et al. (1993, 1994, 1996), this pattern might change, as adaptive foraging by predators can generate positive complexity–stability relationships (Kondoh 2003, 2006; Uchida and Drossel 2007). Furthermore, the contribution of the HOI caused by APTM in a tritrophic interaction (Fig. 1d, e) was not discussed, and therefore the contribution was not clear.

Discussion

We have examined adaptive prey traits within the broader context of higher-order interactions and reviewed the most common approaches to incorporating APTM into population models and the theoretical literature concerning the effects of APTM on long-term ecological dynamics. Many studies show that APTM tends to increase system persistence (the coexistence of all species). The qualitative effects of one species on another (for example, the sign of the interaction between them) are often modified by the inclusion of APTM; this is often the mechanism leading to coexistence (Abrams and Vos 2003). APTM is also frequently predicted to have a stabilizing effect on dynamics, preventing large amplitude fluctuations (Abrams 2000; Vos et al. 2004a; Yamauchi and Yamamura 2005; Ramos-Jiliberto et al. 2008). However, across simple predator–prey, one-predator–two-prey and tritrophic systems, it has been predicted that APTM can de-stablize system dynamics if the predator has a Type II functional response (Abrams and Matsuda 1997; Abrams 2000, 2003, 2007).

This is a result of 'chase' cycles generated by the interaction between prey behaviour and predator satiation: when prey are abundant, predators are satiated; this leads to a reduction in prey defence, which leads to an increase in predation and a declining prey population, which causes prey to increase defence levels. As long as prey and/or predator behavioural responses are not instantaneous, this can lead to unstable population dynamics, though it may not alter the effects on system persistence (Abrams 2000, 2003).

The current body of theory therefore elucidates important considerations for the incorporation of APTM in theory, and makes clear that APTM can strongly influence model predictions. However, crucial issues remain. In particular, the overarching question of whether APTM is necessary to understand ecological systems has not been definitively answered. We suggest that the evidence from existing theory and experiment is strongly suggestive, but nevertheless equivocal. The current theory has not included factors that are intrinsic to the evolution and expression of APTM and therefore potentially omits processes crucial to understanding their influence in food webs, including environmental stochasticity and trait expression based on system dynamics. Furthermore, APTM studies have largely been restricted to small webs, and it is unclear how the results of these studies will scale up to larger webs. We next discuss these needs in more detail and suggest future directions for addressing them. Finally, we argue that the strongest case for the importance of APTM will be made by studies that bridge the gap between theory and experiment, and discuss particularly profitable directions in this area.

Evidence, pro and con, for the need to include APTM in ecological theory

The broad goals of ecological theory are to understand dynamical and structural properties of food webs that affect basic properties including stability, ecosystem function, diversity and the cascading effects of species introductions and removals. Empirical work strongly indicates that APTM affects the very nature of the interactions between species that forms the cornerstone of ecological theory. A large body of empirical work shows that APTM contributes substantially to the net effect of the predator (reviewed in Werner and Peacor 2003; Schmitz *et al.* 2004; Miner *et al.* 2005; Ohgushi 2005; Creel and Christianson 2008; Heithaus *et al.* 2008; Peacor and Werner 2008; Peckarsky *et al.* 2008). Further, empirical work has supported some counterintuitive theoretical predictions such as a predator having a positive effect on its prey's growth rate (Abrams 1987; Peacor 2002). However, the great majority of empirical studies have addressed short-term effects on fitness correlates, rather than long-term effects on demographic properties including population dynamics, stability and abundance. We address this disconnect between theory and empirical work below, in the section 'Bridging the gap between theory and experiment.'

We argue that the theoretical evidence for the need to include APTM is equivocal: theory to date suggests that APTM should have a strong influence, but there are also crucial elements missing from current theory.

There are two arguments stemming from theory that suggest APTM is crucial to food web theory. First, given the importance of other HOIs in ecological theory (reviewed in section 'HOIs in food web theory: various mechanisms including APTM'), it is plausible that the HOIs arising from APTM will be similarly influential. Second, the general conclusion of all of the studies summarized here is that APTM appears to have very large effects on system dynamics. In addition to altering the net influence of one species on another, theory predicts that APTM can even change the sign of the inter-action between them, and can affect food web properties, such as stability and connectance (Matsuda *et al.* 1994, 1996; Kondoh 2007).

Despite the suggestion that APTM will be important, there are a number of reasons to temper this suggestion. First, the results summarized above are highly dependent upon the assumptions made about the shapes of the fitness cost and benefit functions and the timing of defence expression, and there is very little empirical work that could guide modellers (Abrams 2010b). Second, current theory has largely failed to consider a number of aspects of APTM that are crucial to its role in natural systems. Finally, ecological theory of APTM has almost been exclusively restricted to the study of simple three-species sys-tems, but the translation of these results to large food webs is tenuous. We discuss these three issues in the next section.

Future needs for theory of APTM
Environmental stochasticity
Despite the recognition that a variable environment is fundamental to the evolution and expression of phenotypic plasticity, including APTM, exist-ing theory examining the influence of APTM has largely ignored environ-mental stochasticity. This stochasticity can be caused by spatial variability faced by individuals from the same population, or temporal variability due, in large part, to abiotic factors. A winter freeze kills fish in some ponds, differentially and dramatically altering predation risk for many species. An invasive species modifies the environment as an ecological engineer, or has strong influence on a predator's dynamics, strongly influencing predator-induced effects on prey. It is this type of environmental variabil-ity that empiricists are familiar with, and that most empirical studies simulate. However, there are almost no studies of APTM that include such exogenously driven variability. Indeed, not only is stochastisity excluded, but most theory on the effects of APTM is conducted at equili-brium (e.g., Abrams 1984, 1992, 1995; Křivan 1997; Abrams and Vos 2003; Křivan and Sirot 2004).

The incongruence of the importance of environmental stochasticity as a driver of APTM and its absence in ecological models raises concern as to whether existing theory represents APTM accurately enough to give meaningful predictions. Take, for example, studies that examine whether the addition of APTM is stabilizing. We know from long-term observational studies that environmental variation can have large effects on species densities (Werner *et al.* 2007a, b), suggesting that systems may rarely be near equilibrium. Moreover, environmental stochasticity can interact with nonlinearities in ecological models (such as those introduced by APTM) to produce dynamics that would not be predicted through a study of the underlying deterministic model (Coulson *et al.* 2001; Rohani *et al.* 2002). Therefore, analysis done near equilibrium, with the associated assumptions required, may not be relevant. Similar issues may exist in studies that compare the relative magnitudes of DMIEs and TMIEs near equilibrium using the approach, outlined above, of perturbing the system away from equilibrium (Abrams 1984). Small perturbations made near equilibrium may not be representative of the time-integrated influence that APTM has under varying conditions far from equilibrium. Furthermore, stochasticity has the ability to modify the strength of ecological interactions. For example, whereas stochasticity could weaken indirect effects, due to stochastic extinction 'short-circuiting' chains of interaction (Schoener 1993), studies of apparent competition have shown that environmental stochasticity can actually amplify the strength of DMIEs (Holt and Barfield 2003; Brassil 2006).

There are at least three distinct approaches to modelling environmental stochasticity that could potentially be used to examine ATBP. The first adds stochasticity to an existing deterministic model. Coulson *et al.* (2004) review a number of different treatments of stochasticity in these cases, such as adding stochasticity as 'noise' around the deterministic dynamics or in such a way that it can interact with nonlinearities in the model. The second approach treats one of the populations in the system as a stochastic variable (Peacor *et al.* 2006). The third approach uses individual-based models, wherein all of the species are dynamically linked, but the encounters between individuals are stochastic (Luttbeg and Schmitz 2000; Ovadia and Schmitz 2004). There are trade-offs between these different approaches. The first approach can uncover important feedbacks between stochasticity and deterministic dynamics (Coulson *et al.* 2001; Rohani *et al.* 2002), but is challenging to formulate and analyse correctly. The second approach is easier to formulate, but does not include potential feedbacks between populations, as one variable is no longer impacted by others in the system. The third approach is computationally intensive and connecting model predictions to model assumptions can be challenging.

We are aware of only a few studies that have considered environmental stochasticity. In one theoretical study, predator dynamics are modelled as a forcing function independent of prey dynamics and predator presence varies stochastically (Peacor et al. 2006, 2007). APTM strongly influenced system stability by a mechanism that required strong fluctuation in predator density that would not be represented in equilibrium studies. While this study is suggestive, we argue that the inclusion of environmental stochasticity is required for the theory on APTM from first principles, as environmental stochasticity is the dynamical process underpinning APTM.

Adaptive prey response to system dynamics

The existing theory of APTM typically assumes that prey alter traits to maximize instantaneous fitness (Abrams 2010b). Implicit in this assumption is that prey response depends only on the current environmental state. This has the strength of allowing trait change to happen on the same timescale as system dynamics. However, it does not account for the extensive evidence from behavioural ecology demonstrating that prey response is shaped by myriad aspects of both individual state and the environment, including aspects that depend on time horizons of predation risk, life history strategy and length of the growing season (Abrams 1991; Rowe and Ludwig 1991; Lima and Bednekoff 1999; Van Buskirk et al. 2002; Pecor and Hazlett 2003). An explicit example of this is the adaptive response of organisms inhabiting ephemeral ponds that must metamorphose before emergence. In this system, the APTM strategy will depend on pond duration (Relyea 2007), so that species or populations in pond types with different durations will respond differently to identical resource and predator densities. From this example, it is clear that assuming APTM depends only on current environmental state can make misleading predictions. Instead, theory must consider how APTM is shaped by natural selection to respond to the dynamics of the system over longer time horizons. There are currently very few studies that optimize APTM over the lifespan while simultaneously allowing APTM to affect system dynamics, likely because the inherent feedback between trait change and the dynamics of the environment makes this a very challenging optimization.

One approach to this problem finds the optimal APTM strategy using iterative techniques, such as state variable dynamic modelling (Mangel and Clark 1989), genetic algorithms (Luttbeg and Sih 2004; Cressler et al. 2010) or neural nets (Strand et al. 2002). All of these techniques determine the set of 'rules' that maximize expected fitness over a specified time horizon (e.g., the lifespan of an individual or the length of a season). The ecological consequences of this strategy are then found by inserting these rules into a dynamic model. The strength of these approaches is that optimal rules can depend upon any number of factors, including time horizons, individual state or future environmental state. However, it can be difficult (but not impossible) to incorporate population- and community-level effects into these models (Mangel and Clark 1989). Luttbeg and Schmitz (2000) use a state variable dynamic model to determine optimal prey foraging behaviour, and the consequences of flexible behaviour and imperfect information on system dynamics. In their model, behaviour determined prey fecundity and predator-dependent mortality, and was influenced by the prey's perception of their environment. Prey behaviour fed back on system dynamics through its effect on predator foraging success and reproduction. Given a constraint set (fixed/flexible behaviour/perception), the behavioural strategy that maximized reproductive output was determined and then integrated into a dynamical model of the abundance of the predator and prey over multiple seasons (for another example using this approach, see Takimoto 2003).

A second approach to incorporate dynamics-based APTM uses computational optimization algorithms to find ('evolve') the optimal foraging strategy of individual foragers (Kimbrell and Holt 2004; Luttbeg and Sih 2004; Peacor et al. 2006, 2007). Each individual in the population initially possesses a unique strategy. These individuals are then placed into a dynamic environment (often using an individual-based model) and allowed to grow, reproduce, and die, each according to its strategy. Reproduction between strategies creates new strategy variants, as in evolution, and over time, the biased reproduction of more fit individuals leads to a strategy with high fitness (Holland 1992). The difference between this approach and the one illustrated in the previous paragraph is that here optimization and dynamics occur simultaneously and population- and community-level dynamics are an emergent property of the foraging strategy that evolves. For example, Peacor et al. (2007) compare the optimal behaviour of a focal prey in a predator–prey system and in a one-predator–two-prey system. In the latter system, the focal prey cannot perceive the presence of the competitor, but base their behaviour only on resource and predator densities. The growth rate and the optimal behavioural response of the focal prey to identical environmental conditions differ between the two systems due to the long-term effects the competitor has on resource levels. Although not presented, the optimal behaviour was also a function of the time period of predator presence and absence (Peacor, unpublished data). Using a similar approach, Kimbrell and Holt (2004) found that the sensitivity of the prey to predation risk strongly affected the stability of a one-predator–two-prey system.

Scaling up insights from small webs to large webs
Ultimately, we need to understand the effect of APTM in large food webs to understand their influence on fundamental ecological questions. The majority of studies have, however, focused on low dimensional systems, which can represent 'community modules' (Holt 1997) of strongly interacting species that may be useful as building blocks in assembling complex communities (Bascompte and Mélian 2005; Stouffer et al. 2007). Thus, we are closer to an

understanding of the full complexity of ecological systems if the effects observed in small webs persist when these webs are embedded in more complex networks. Existing theory gives conflicting answers to this question. For example, small webs exhibiting chaotic dynamics are stabilized by inclusion into larger webs (Fussmann and Heber 2002). Similarly, unstable trophic modules are stabilized when connected to stable trophic modules in a model of a Caribbean food web (Kondoh 2008). On the other hand, keystone species effects are maintained when keystone systems are embedded into more complex webs (Brose *et al.* 2005).

The attempt to scale inferences from small food web studies of APTM to larger food webs is complicated by two related challenges: the superposition of webs and the interaction of APTM with other HOIs. As reviewed herein, current theory has shown that APTM can have a wide variety of effects, depending on the food web structure and the model assumptions. This makes inference difficult when multiple community modules are superimposed on one another. Consider, for example, the food web shown in Fig. 8.3a. Predator P1 induces APTM in consumer C1. Thus, we can consider at least two modules: the

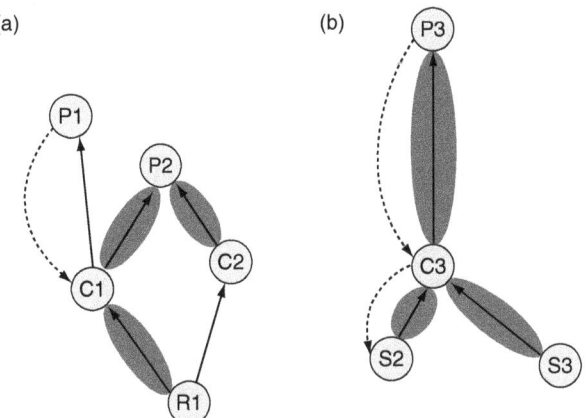

Figure 8.3 An heuristic diagram of a large food web. Panel (a) depicts the trait-mediated indirect effects arising from adaptive change in consumer C1's foraging behaviour. This modified C1's interaction with its resource, R1. It also affects the interaction of C1 with its second predator, P2. Depending on the shapes of the cost and benefit functions, it is possible for either, both or neither of the modules P1–C1–R1 and P1–P2–C1 to be stable, thereby making it difficult to predict the stability of the P1–P2–C1–R1 system. Panel (b) shows the indirect effects arising from the interaction of multiple APTMs: species S2 can adjust its traits in response to C3, and C3 can adjust its traits in response to both P3 and S2. This leads to trait-mediated indirect effects of P3 on S2 and P3 on S3 (Fig. 8.1d), S2 and S3 on P3 (Fig. 8.1e), and S2 on S3 (Fig. 8.1g). As above, it is impossible to predict, a priori, what effect these TMIIs will have on the overall stability of the P3–C3–S2–S3 system. See colour plate section.

tritrophic interaction between P1, C1 and R1, and the two-predator–one-prey interaction between P1, P2 and C1. The APTM of C1 generates a positive TMII between P1 and R1, an effect that has been shown to be stabilizing (Vos *et al.* 2004a; Ramos-Jiliberto 2008). However, if this APTM is effective against predator P2 as well, the P1–P2–C1 system is predicted to be unstable (Matsuda *et al.* 1993). It is unclear whether the P1–P2–C1–R1 system will be stable or unstable. Furthermore, given the ubiquity of APTM, in any given community there will likely be more than one species reacting to the environment through modified traits and therefore multiple interacting HOIs (Fig. 8.3b; Abrams 2010b). This can produce similar problems with inference to those presented by a single instance of APTM leading to multiple TMIIs. Moreover, work by Abrams (1992) on a four-species chain with two species employing APTM has shown the possibility of highly nonintuitive outcomes.

The inclusion of other processes leading to HOIs in large webs is also limited but informative. The emerging consensus is that they are critical to food web structure, as has been found for predator satiation and predator switching (reviewed in 'HOIs in food web theory: various mechanisms including APTM'). We simply need more work incorporating APTM into models of larger food webs to know if the same will be true for prey trait modification.

Existing studies provide contrasting approaches to incorporating APTM into food web models that can be used as starting points for further theory. The study of Kondoh (2007) is a straightforward extension of the most common approaches reviewed above for small webs, wherein traits are explicitly considered and each species' traits change in the direction of increasing fitness. This study fixed the properties of the web, such as the number of species and connectance, and then studied the dynamics. Arditi *et al.* (2005) and Goudard and Loreau (2008), on the other hand, considered only generalized HOIs, where pairwise interaction strengths depend upon the densities of other species in the web, and take a food web assembly approach, so that the properties of the web are not fixed. The general framework presented in these studies could be adapted to investigate specific APTM, for example by constraining the possible signs of the interaction modification and/or including HOI in specific links (see Dambacher and Ramos-Jiliberto 2007).

Bridging the gap between theory and experiment

There is a large gap between theoretical and empirical work in the study of the effects of APTM (Bolker *et al.* 2003; Abrams 2010b). There are presently hundreds of studies in diverse systems that have demonstrated direct effects of APTM on prey demographics and indirect effects (i.e., TMII) on species interacting with the prey (reviewed in Werner and Peacor 2003; Schmitz *et al.* 2004; Miner *et al.* 2005; Ohgushi 2005; Peacor and Werner 2008). Many new avenues are being explored, including how environmental context (such as refuges: Grabowski 2004; Trussell *et al.* 2006a, or resource dynamics and experimental duration: Turner 2004; Werner and Peacor 2006) can affect the relative magnitude of TMIIs, how APTM can affect ecosystem processes (Trussel *et al.* 2006b; Schmitz *et al.* 2008) and the feedback effects of prey and predator responses to each other (Lima 2002; Luttbeg and Sih 2004). However, despite the profusion of activity in this area, theoreticians' calls for empirical measurement of certain relationships, and for more integration between theory and empirical work, have not been addressed. We are aware of several ongoing investigations moving in this direction (Trussell and Luttbeg, pers. comm., Vandermeer and Baskerville, pers. comm.). We repeat the call of earlier reviews for the need for integration of theory and empirical work here (Abrams 1995, 2001b, 2008, 2010b; Bolker *et al.* 2003).

One critical need is empirical measurement of the relationship between trait expression and fitness costs and benefits (Fig. 8.2). The review above makes clear that varying the shapes of the cost and benefit functions strongly influences theoretical predictions of the effects of APTM (Abrams 1984, 1992; Abrams and Vos 2003; Křivan and Sirot 2004), suggesting an accurate empirical description is necessary. Many biologically reasonable shapes have been proposed (Werner and Anholt 1993; Abrams 2003; Křivan and Sirot 2004), and, theoretically, differentiating between these shapes should be possible based on the results of short-term experiments (Bolker *et al.* 2003). However, surprisingly, we are not aware of a single study that has documented these relationships. A number of studies have shown how different levels of predation risk lead to continuous changes in trait expression (reviewed in Peacor and Werner 2008), but this does not yield the information required. A clear impediment is that the fitness cost curve is inherently difficult to measure; predator presence will affect prey traits, but predation rates must be measured as a function of trait values. There are likely creative ways around this problem, at least in some systems. For example, McCoy and Bolker (2008) quantified the effect of prey size and experience on predator functional response. While not explicitly measuring the shape of the fitness cost and benefit functions (because only two levels of trait modification were considered), the authors do examine the interaction of multiple factors on the strength of ecological interactions and their methodology could potentially be extended to consider cost and benefit functions explicitly.

In addition to the need to measure the functional relationship between trait modification and fitness costs and benefits relationship, is the measurement of the magnitude of the predator effect on consumer (prey)–resource interaction. This parameter is critical to any theory on APTM (representing the predator effect on the second and first terms in the resource and consumer growth rates in Equation (8.1), respectively). In an analysis of empirical studies, Peacor and Werner (2004a) found that induced effect of predators on consumer (prey) traits commonly modifies the consumer–resource attack rates, a component of interaction strength, in the range of 20–90%. These magnitudes are clearly of an order to be important to dynamics. It would be helpful if empiricists measured the effect of trait modifications on interaction strengths alongside studies of indirect interactions and other responses; e.g., did the large indirect effects reported in a given empirical study arise from a 10% or 50% change in the interaction strength? We are aware of only a few studies that address this problem (McPeek and Peckarsky 1998; Peacor and Werner 2004a; Pangle *et al.* 2007; Creel and Christianson 2008). Furthermore, theoretical studies also seldom investigate how much the interaction strengths varied to produce the results reported. Did they require a 99% change in interaction strengths, which are likely biologically implausible? Attention to this factor by both empiricists and theorists would help to close the gap between empirical and theoretical studies.

We second the call by Bolker *et al.* (2003) in their review of theory of TMIIs to integrate theory with empirical studies of the influence of APTM in food webs. There are a number of studies that have begun to move in this direction. The model of Vos *et al.* (2004a, b) was explicitly parameterized from laboratory data and was studied in an attempt to explain a pattern observed in experimental data (Verschoor *et al.* 2004). The individual-based model of Schmitz and colleagues (Schmitz 2000; Ovadia and Schmitz 2004) was calibrated from a mechanistic understanding of a particular field system (Beckerman *et al.* 1997). The short-term model output was then compared against experimental data before using the model to make predictions over longer time periods.

What is needed now is a comprehensive programme that combines theory, laboratory studies and field studies, as has been successfully applied to the study of other processes (Werner 2001). Laboratory (or very controlled field) studies will provide an understanding of the adaptive prey responses and enable the development of theory to form predictions for experiments with multispecies assemblages. Field experiments can then be used to test these predictions. An integrative process between laboratory studies, modelling and field work will yield the manner in which APTM affects the system dynamics. Studies of the context dependence of TMIEs of predators on resources (Peacor and Werner 2000, 2004b; Turner 2004) may be considered an example of a first step in such a research programme. For example, Peacor and Werner (2000) examined the effect of predation risk on an assemblage of tadpole species and size classes. Laboratory experiments yielded information on the APTM of the different tadpole species and size classes. They used this information to build a very simple model that accurately predicted the effect of predator-induced modification of prey behaviour on competition, and predicted how environmental factors (such as growing versus nongrowing resources) influenced the magnitude of predator-induced changes in prey growth rates. This validation of short-term predictions is an important step, as the results are complex and require a model to help interpret them (Ovadia and Schmitz 2004). For example, under certain conditions, predator-induced changes in prey behaviour may have no effect on the prey's growth rate, but a strong indirect effect on the growth rate of the prey's competitors (Peacor and Werner 2000). Ultimately, what is required are quantitative predictions of the influence of APTM over longer (multigenerational) timescales in systems with many species. Experiments in controlled venues (e.g., mesocosms) will be helpful to develop this approach and yield insights into the processes, but ultimately, field-based studies using this approach will have the highest impact.

An alternative approach to integrating theory and data is the re-examination of existing time series data. A number of 'classic' ecological patterns, such as the cycles between lynx and hare abundances in the Arctic and cycles in Scandinavian voles, occur in systems in which APTM is hypothesized to be operating (Sundell et al. 2004; Peckarsky et al. 2008). These time series have been the subject of extensive modelling efforts, and so, for many of these systems, parameterized models already exist (Turchin 2000; King and Schaffer 2001). A profitable direction may be to revisit some of these time series and quantitatively test whether the inclusion of APTM into existing models is able to better explain the observed patterns. Such an indirect test was performed in the study of Dambacher and Ramos-Jiliberto (2007), who showed that by extending a previously published model for a lake food web by including APTM, they were better able to predict observed patterns of clear water phases.

Conclusions

A principal goal of our review has been to address whether the adaptive response of a species to predation risk, herein denoted adaptive prey trait modification, is required for ecological theory to address adequately fundamental questions in community ecology, such as the relationship between stability and diversity, community resilience and robustness, and predicting the effects of species' invasion and loss. We argue that the current theoretical evidence is suggestive, but not definitive. This assessment is based on the exclusion of important processes intrinsic to APTM and the limited amount of theory in larger systems. On the other hand, APTM will likely strongly influence theory based on first principles: empirical studies have shown that the basic building blocks of food web theory, pairwise species interactions, are greatly impacted by changes in food web context and the consequent TMIIs observed in many systems are testament to the influence of APTM. The underlying mechanism for such TMIIs, phenotypic plasticity in traits involved in fitness trade-offs, has been documented in hundreds of species. Further, existing theory in simple systems supports the supposition that APTM is critical by demonstrating that its inclusion has dramatic effects on standard outcomes in each of the classic configurations examined, and inclusion of other forms of HOIs has large impacts on theoretical predictions (Kondoh 2003; Uchida *et al.* 2007; McCann and Rooney 2009; Abrams 2010a; Beckerman *et al.* 2010). Yet despite this evidence, even theoretical studies that demonstrate the important of interaction strength in dynamical models (e.g., Neutel *et al.* 2002; Pawar 2009) are not considering the potential implications of the dynamic nature of these interaction strengths introduced by APTM (Kondoh 2007). We therefore argue that theory in large, complex systems that include APTM is needed, and that there is much to be garnered to inform such studies from present studies in small systems. What is most essential is much tighter integration of theory and data than has been achieved heretofore. Many of the above-mentioned gaps in current theory will likely be addressed by coupling system-specific models with empirical studies that examine systems with more (i.e., $N > 3$) species over long timescales and measure relevant parameters and functional shapes needed for close integration with ecological theory. Such a research programme will go a long way towards answering the question of whether APTM is critical to the structure and dynamics of ecological systems, and given the overwhelming evidence of the potential importance of APTM to fundamental ecosystems properties, we expect much empirical and theoretical activity in this area in the coming decades. We second the call of previous reviews (Bolker *et al.* 2003; Abrams 2010b) for these activities to be done in tandem.

Acknowledgements

We thank E. Baskerville, J. Vonesh, B. Luttbeg, the Aaron King laboratory, and the Earl Werner laboratory for helpful comments on this manuscript. Scott Peacor was supported by the National Science Foundation (OCE-0826020) during the preparation of this manuscript. SDP acknowledges support from the Michigan Agricultural Experiment Station.

References

Abrams, P. (1984) Foraging time optimization and interactions in food webs. *American Naturalist*, **124**, 80–96.

Abrams, P. (1987) Indirect interactions between species that share a predator: varieties on a theme. In W. Kerfoot and A. Sih, eds., *Predation: Direct and Indirect Impacts on Aquatic Communities*. Hanover, NH: University Press of New England, pp. 38–54.

Abrams, P. (1991) Life-history and the relationship between food availability and foraging effort. *Ecology*, **72**, 1242–1252.

Abrams, P. (1992) Predators that benefit prey and prey that harm predators: unusual effects of interacting foraging adaptations. *American Naturalist*, **140**, 573–600.

Abrams, P. (1995) Implications of dynamically variable traits for identifying, classifying, and measuring direct and indirect effects in ecological communities. *American Naturalist*, **146**, 112–134.

Abrams, P. (1999) The adaptive dynamics of consumer choice. *American Naturalist*, **153**, 83–97.

Abrams, P. (2000) The impact of habitat selection on the spatial heterogeneity of resources in varying environments. *Ecology*, **81**, 2902–2913.

Abrams, P. (2001a) Modelling the adaptive dynamics of traits involved in inter- and intraspecific interactions: an assessment of three methods. *Ecology Letters*, **4**, 166–175.

Abrams, P. (2001b) Describing and quantifying interspecific interactions: a commentary on recent approaches. *Oikos*, **94**, 209–218.

Abrams, P. (2003) Can adaptive evolution or behaviour lead to diversification of traits determining a trade-off between foraging gain and predation risk? *Evolutionary Ecology Research*, **5**, 653–670.

Abrams, P. (2004) Trait-initiated indirect effects due to changes in consumption rates in simple food webs. *Ecology*, **85**, 1029–1038.

Abrams, P. (2007) Habitat choice in predator–prey systems: spatial instability due to interacting adaptive movements. *American Naturalist*, **169**, 581–594.

Abrams, P. (2008) Measuring the impact of dynamic antipredator traits on predator–prey-resource interactions. *Ecology*, **89**, 1640–1649.

Abrams, P. (2009) Adaptive changes in prey vulnerability shape the response of predator populations to mortality. *Journal of Theoretical Biology*, **261**, 294–304.

Abrams, P. (2010a) Quantitative descriptions of resource choice in ecological models. *Population Ecology*, **52**, 47–58.

Abrams, P. (2010b) Implications of flexible foraging for interspecific interactions: lessons from simple models. *Functional Ecology*, **24**, 7–17.

Abrams, P. and Matsuda, H. (1996) Positive indirect effects between prey species that share predators. *Ecology*, **77**, 610–616.

Abrams, P. and Matsuda, H. (1997) Prey adaptation as a cause of predator–prey cycles. *Evolution*, **51**, 1742–1750.

Abrams, P. and Matsuda, H. (2005) The effect of adaptive change in the prey on the dynamics of an exploited predator population. *Canadian Journal of Fisheries and Aquatic Sciences*, **62**, 758–766.

Abrams, P. and Vos, M. (2003) Adaptation, density dependence and the responses of trophic level abundances to mortality. *Evolutionary Ecology Research*, **5**, 1113–1132.

Abrams, P., Holt, R. and Roth, J. (1998) Apparent competition or apparent mutualism? Shared predation when populations cycle. *Ecology*, **79**, 201–212.

Abrams, P., Matsuda, H. and Harada, Y. (1993) Evolutionarily unstable fitness maxima and stable fitness minima of continuous traits. *Evolutionary Ecology*, **7**, 465–487.

Abrams, P., Menge, B., Mittelbach, G., Spiller, D. and Yodzis, P. (1996) The role of indirect effects in food webs. In G. Polis and K. Winemiller, eds., *Food Webs: Dynamics and Structure*. New York: Chapman and Hall, pp. 371–395.

Agrawal, A. (2001) Phenotypic plasticity in the interactions and evolution of species. *Science*, **294**, 321–326.

Arditi, R., Michalski, J. and Hirzel, A. (2005) Rheagogies: modeling non-trophic effects in food webs. *Ecological Complexity*, **2**, 249–258.

Armstrong, R. and McGehee, R. (1980) Competitive exclusion. *American Naturalist*, **115**, 151–170.

Bascompte, J. and Mélian, C. (2005) Simple trophic modules for complex food webs. *Ecology*, **86**, 2868–2873.

Beckerman, A., Petchey, O. and Morin, P. (2010) Adaptive foragers and community ecology: linking individuals to communities and ecosystems. *Functional Ecology*, **24**, 1–6.

Beckerman, A., Uriarte, M. and Schmitz, O. (1997) Experimental evidence for a behavior-mediated trophic cascade in a terrestrial food chain. *Proceedings of the National Academy of Sciences of the United States of America*, **94**, 10735–10738.

Beddington, J. (1975) Mutual interference between parasites or predators and its effect on searching efficiency. *Journal of Animal Ecology*, **44**, 331–340.

Bender, E., Case, T. and Gilpin, M. (1984) Perturbation experiments in community ecology: theory and practice. *Ecology*, **65**, 1–13.

Bolker, B., Holyoak, M., Křivan, V., Rowe, L. and Schmitz, O. (2003) Connecting theoretical and empirical studies of trait-mediated interactions. *Ecology*, **84**, 1101–1114.

Brassil, C. (2006) Can environmental variation generate positive indirect effects in a model of shared predation? *American Naturalist*, **167**, 43–54.

Brose, U., Berlow, E. and Martinez, N. (2005) Scaling up keystone effects from simple to complex ecological networks. *Ecology Letters*, **8**, 1317–1325.

Charnov, E. (1976) Optimal foraging, marginal value theorem. *Theoretical Population Biology*, **9**, 129–136.

Coulson, T., Catchpole, E., Albon, S. *et al.* (2001) Age, sex, density, winter weather, and population crashes in Soay sheep. *Science*, **292**, 1528–1531.

Coulson, T. Rohani, P. and Pascual, M. (2004) Skeletons, noise, and population growth: the end of an old debate? *Trends in Ecology and Evolution*, **19**, 359–364.

Creel, S. and Christianson, D. (2008) Relationships between direct predation and risk effects. *Trends in Ecology and Evolution*, **23**, 194–201.

Cressler, C., King, A. and Werner, E. (2010) Interactions between behavioral and life-history trade-offs in the evolution of integrated predator-defense plasticity. *American Naturalist*, **176**, 276–288.

Cressman, R., Křivan, V. and Garay, J. (2004) Ideal free distributions, evolutionary games, and population dynamics in multiplespecies environments. *American Naturalist*, **164**, 473–489.

Dambacher, J. and Ramos-Jiliberto, R. (2007) Understanding and predicting effects of modified interactions through a qualitative analysis of community structure. *Quarterly Review of Biology*, **82**, 227–250.

DeAngelis, D., Goldstein, R. and O'Neill, R. (1975) A model for tri-trophic interaction. *Ecology*, **56**, 881–892.

Fryxell, J. and Lundberg, P. (1998) *Individual Behavior and Community Dynamics*. Heidelberg, Germany: Springer.

Fussmann, G. and Heber, G. (2002) Food web complexity and chaotic population dynamics. *Ecology Letters*, **5**, 394–401.

Goudard, A. and Loreau, M. (2008) Nontrophic interactions, biodiversity, and ecosystem functioning: an interaction web model. *American Naturalist*, **171**, 91–106.

Grabowski, J. (2004) Habitat complexity disrupts predator–prey interactions but not the trophic cascade on oyster reefs. *Ecology*, **85**, 995–1004.

Hastings, A. and Powell, T. (1991) Chaos in a three-species food-chain. *Ecology*, **72**, 896–903.

Heithaus, M., Frid, A., Wirsing, A. and Worm, B. (2008) Predicting ecological consequences of marine top predator declines. *Trends in Ecology and Evolution*, **23**, 202–210.

Holland, J. H. (1992) *Adaptation in Natural and Artificial Systems: An Introductory Analysis with Applications to Biology, Control, and Artificial Intelligence*. Boston, MA: The MIT Press.

Holt, R. (1997) Community modules. In A. Gange and V. Brown, eds., *Multitrophic Interactions in Terrestrial Ecosystems*. Cambridge: Cambridge University Press.

Holt, R. and Barfield, M. (2003) Impacts of temporal variation on apparent competition and coexistence in open ecosystems. *Oikos*, **101**, 49–58.

Holt, R., Grover, J. and Tilman, D. (1994) Simple rules for interspecific dominance in systems with exploitative and apparent competition. *American Naturalist*, **144**, 741–771.

Ives, A. and Dobson, A. (1987) Antipredator behavior and the population-dynamics of simple predator–prey systems. *American Naturalist*, **130**, 431–447.

Kimbrell, T. and Holt, R. (2004) On the interplay of predator switching and prey evasion in determining the stability of predator–prey dynamics. *Israel Journal of Zoology*, **50**, 187–205.

King, A. and Schaffer, W. (2001) The geometry of a population cycle: a mechanistic model of snowshoe hare demography. *Ecology*, **82**, 814–830.

Kondoh, M. (2003) Foraging adaptation and the relationship between food web complexity and stability. *Science*, **299**, 1388–1391.

Kondoh, M. (2006) Does foraging adaptation create the positive complexity-stability relationship in realistic food-web structure? *Journal of Theoretical Biology*, **238**, 646–651.

Kondoh, M. (2007) Anti-predator defence and the complexity-stability relationship of food webs. *Proceedings of the Royal Society of London, Series B*, **274**, 1617–1624.

Kondoh, M. (2008) Building trophic modules into a persistent food web. *Proceedings of the National Academy of Sciences of the United States of America*, **105**, 16631–16635.

Křivan, V. (1997) Dynamic ideal free distribution: effects of optimal patch choice on predator–prey dynamics. *American Naturalist*, **149**, 164–178.

Křivan, V. (2007) The Lotka-Volterra predator–prey model with foraging-predation risk trade-offs. *American Naturalist*, **170**, 771–782.

Křivan, V. and Schmitz, O. (2004) Trait and density mediated indirect interactions in simple food webs. *Oikos*, **107**, 239–250.

Křivan, V. and Sirot, E. (2004) Do short-term behavioural responses of consumers in tritrophic food chains persist at the population time-scale? *Evolutionary Ecology Research*, **6**, 1063–1081.

Lima, S. (1998) Stress and decision making under the risk of predation: recent developments from behavioral, reproductive, and ecological perspectives. *Stress and Behavior*, **27**, 215–290.

Lima, S. (2002) Putting predators back into behavioral predator–prey interactions. *Trends in Ecology and Evolution*, **17**, 70–75.

Lima, S. and Bednekoff, P. (1999) Temporal variation in danger drives antipredator behavior: the predation risk allocation hypothesis. *American Naturalist*, **153**, 649–659.

Lima, S. and Dill, L. (1990) Behavioral decisions made under the risk of predation: a review and prospectus. *Canadian Journal of Zoology*, **68**, 619–640.

Luttbeg, B. and Schmitz, O. (2000) Predator and prey models with flexible individual behavior and imperfect information. *American Naturalist*, **155**, 669–683.

Luttbeg, B. and Sih, A. (2004) Predator and prey habitat selection games: the effects of how prey balance foraging and predation risk. *Israel Journal of Zoology*, **50**, 233–254.

Ma, B., Abrams, P. and Brassil, C. (2003) Dynamic versus instantaneous models of diet choice. *American Naturalist*, **162**, 668–684.

MacArthur, R. and Pianka, E. (1966) On optimal use of a patchy environment. *American Naturalist*, **100**, 603–610.

McCann, K. and Rooney, N. (2009) The more food webs change, the more they stay the same. *Philosophical Transactions of the Royal Society of London, Series B*, **364**, 1789–1801.

McCann, K., Hastings, A. and Huxel, G. (1998) Weak trophic interactions and the balance of nature. *Nature*, **395**, 794–798.

McCoy, M. and Bolker, B. (2008) Trait-mediated interactions: influence of prey size, density and experience. *Journal of Animal Ecology*, **77**, 478–486.

Mangel, M. and Clark, C. W. (1989) *Dynamic Modeling in Behavioral Ecology*. Princeton, NJ: Princeton University Press.

Matsuda, H., Abrams, P. and Hori, H. (1993) The effect of adaptive antipredator behavior on exploitative competition and mutualism between predators. *Oikos*, **68**, 549–559.

Matsuda, H., Hori, M. and Abrams, P. (1994) Effects of predator-specific defense on community complexity. *Evolutionary Ecology*, **8**, 628–638.

Matsuda, H., Hori, M. and Abrams, P. (1996) Effects of predator-specific defence on biodiversity and community complexity in two-trophic-level communities. *Evolutionary Ecology*, **10**, 13–28.

May, R. M. (1973) *Stability and Complexity in Model Ecosystems*. Princeton, NJ: Princeton University Press.

McPeek, M. and Peckarsky, B. (1998) Life histories and the strengths of species interactions: combining mortality, growth, and fecundity effects. *Ecology*, **79**, 867–879.

Miller, T. and Kerfoot, W. (1987) Redefining indirect effects. In W. Kerfoot and A. Sih, eds., *Predation: Direct and Indirect Impacts on Aquatic Communities*. Hanover, NH: University Press of New England, pp. 33–37.

Miner, B., Sultan, S., Morgan, S., Padilla, D. and Relyea, R. (2005) Ecological consequences of phenotypic plasticity. *Trends in Ecology and Evolution*, **20**, 685–692.

Mougi, A. and Nishimura, K. (2008) The paradox of enrichment in an adaptive world. *Proceedings of the Royal Society of London, Series B*, **275**, 2563–2568.

Murdoch, W. (1969) Switching in general predators: experiments on predator specificity and stability of prey populations. *Ecological Monographs*, **39**, 335–354.

Neill, W. (1974) Community matrix and interdependence of competition coefficients. *American Naturalist*, **108**, 399–408.

Neutel, A., Heesterbeek, J. and de Ruiter, P. (2002) Stability in real food webs: weak links in long loops. *Science*, **296**, 1120–1123.

Ohgushi, T. (2005) Indirect interaction webs: herbivore-induced effects through trait change in plants. *Annual Review of Ecology, Evolution, and Systematics*, **36**, 81–105.

Oksanen, L., Fretwell, S., Arruda, J. and Niemela, P. (1981) Exploitation ecosystems in gradients of primary productivity. *American Naturalist*, **118**, 240–261.

Ovadia, O. and Schmitz, O. (2004) Scaling from individuals to food webs: the role of size-dependent responses of prey to predation risk. *Israel Journal of Zoology*, **50**, 273–297.

Pangle, K., Peacor, S. and Johannsson, O. (2007) Large nonlethal effects of an invasive invertebrate predator on zooplankton population growth rate. *Ecology*, **88**, 402–412.

Pawar, S. (2009) Community assembly, stability and signatures of dynamical constraints on food web structure. *Journal of Theoretical Biology*, **259**, 601–612.

Peacor, S. (2002) Positive effect of predators on prey growth rate through induced modifications of prey behaviour. *Ecology Letters*, **5**, 77–85.

Peacor, S. and Werner, E. (1997) Trait-mediated indirect interactions in a simple aquatic food web. *Ecology*, **78**, 1146–1156.

Peacor, S. and Werner, E. (2000) Predator effects on an assemblage of consumers through induced changes in consumer foraging behavior. *Ecology*, **81**, 1998–2010.

Peacor, S. and Werner, E. (2004a) How dependent are species-pair interaction strengths on other species in the food web? *Ecology*, **85**, 2754–2763.

Peacor, S. and Werner, E. (2004b) Context dependence of nonlethal effects of a predator on prey growth. *Israel Journal of Zoology*, **50**, 139–167.

Peacor, S. and Werner, E. (2008) Nonconsumptive effects of predators and trait-mediated indirect interactions. *Encyclopedia of Life Sciences*, Wiley Online Library.

Peacor, S., Allesina, S., Riolo, R. and Hunter, T. (2007) A new computational system, DOVE (Digital Organisms in a Virtual Ecosystem), to study phenotypic plasticity and its effects in food webs. *Ecological Modelling*, **205**, 13–28.

Peacor, S., Allesina, S., Riolo, R. and Pascual, M. (2006) Phenotypic plasticity opposes species invasions by altering fitness surface. *PLoS Biology*, **4**, 2112–2120.

Peckarsky, B., Abrams, P., Bolnick, D. *et al.* (2008) Revisiting the classics: considering nonconsumptive effects in textbook examples of predator–prey interactions. *Ecology*, **89**, 2416–2425.

Pecor, K. and Hazlett, B. (2003) Frequency of encounter with risk and the trade-off between pursuit and antipredator behaviors in crayfish: a test of the risk allocation hypothesis. *Ethology*, **109**, 97–106.

Pimm, S. L. (1982) *Food Webs*. Chicago, IL: University of Chicago Press.

Ramos-Jiliberto, R., Mena-Lorca, J., Flores, J. and Morales-Alvarez, W. (2008) Role of inducible defenses in the stability of a tri-trophic system. *Ecological Complexity*, **5**, 183–192.

Relyea, R. (2007) Getting out alive: how predators affect the decision to metamorphose. *Oecologia*, **152**, 389–400.

Rinaldi, S., Gragnani, A. and De Monte, S. (2004) Remarks on antipredator behavior and food chain dynamics. *Theoretical Population Biology*, **66**, 277–286.

Rohani, P., Keeling, M. and Grenfell, B. (2002) The interplay between determinism and stochasticity in childhood diseases. *American Naturalist*, **159**, 469–481.

Rosenzweig, M. L. (1971) Paradox of enrichment: destabilization of exploitation ecosystems in ecological time. *Science*, **171**, 385–387.

Rowe, L. and Ludwig, D. (1991) Size and timing of metamorphosis in complex life-cycles: time constraints and variation. *Ecology*, **72**, 413–427.

Schmitz, O. (2000) Combining field experiments and individual-based modeling to identify the dynamically relevant organizational scale in a field system. *Oikos*, **89**, 471–484.

Schmitz, O., Grabowski, J., Peckarsky, B. *et al.* (2008) From individuals to ecosystem function: toward an integration of evolutionary and ecosystem ecology. *Ecology*, **89**, 2436–2445.

Schmitz, O., Křivan, V. and Ovadia, O. (2004) Trophic cascades: the primacy of trait-mediated indirect interactions. *Ecology Letters*, **7**, 153–163.

Schoener, T. (1971) Theory of feeding strategies. *Annual Review of Ecology and Systematics*, **2**, 369–404.

Schoener, T. (1993) On the relative importance of direct versus indirect effects in ecological communities. In H. Kawanabe, J. Cohen and K. Iwasaki, eds., *Mutualisms and Community*

Organization. Oxford: Oxford University Press, pp. 365–411.

Schwinning, S. and Rosenzweig, M. (1990) Periodic oscillations in an ideal-free predator–prey distribution. *Oikos*, **59**, 85–91.

Stephens, D. and Krebs, J. (1986) *Foraging Theory*. Princeton, NJ: Princeton University Press.

Stouffer, D., Camacho, J., Jiang, W. and Amaral, L. (2007) Evidence for the existence of a robust pattern of prey selection in food webs. *Proceedings of the Royal Society of London, Series B*, **274**, 1931–1940.

Strand, E., Huse, G. and Giske, J. (2002) Artificial evolution of life history and behavior. *American Naturalist*, **159**, 624–644.

Sundell, J., Dudek, D., Klemme, I., *et al.* (2004) Variation in predation risk and vole feeding behaviour: a field test of the risk allocation hypothesis. *Oecologia*, **139**, 157–162.

Takimoto, G. (2003) Adaptive plasticity in ontogenetic niche shifts stabilizes consumer–resource dynamics. *American Naturalist*, **162**, 93–109.

Tollrian, R. and Harvell, C. D. (1998) *The Ecology and Evolution of Inducible Defenses*. Princeton, NJ: Princeton University Press.

Turner, A. (2004) Non-lethal effects of predators on prey growth rates depend on prey density and nutrient additions. *Oikos*, **104**, 561–569.

Trussell, G., Ewanchuk, P. and Matassa, C. (2006a) Habitat effects on the relative importance of trait- and density-mediated indirect interactions. *Ecology Letters*, **9**, 1245–1252.

Trussell, G., Ewanchuk, P. and Matassa, C. (2006b) The fear of being eaten reduces energy transfer in a simple food chain. *Ecology*, **87**, 2979–2984.

Turchin, P. (2000) Living on the edge of chaos: population dynamics of Fennoscandian voles. *Ecology*, **81**, 3099–3116.

Uchida, S. and Drossel, B. (2007) Relation between complexity and stability in food webs with adaptive behavior. *Journal of Theoretical Biology*, **247**, 713–722.

Uchida, S., Drossel, B. and Brose, U. (2007) The structure of food webs with adaptive behaviour. *Ecological Modelling*, **206**, 263–276.

Van Buskirk, J., Müller, C., Portmann, A. and Surbeck, M. (2002) A test of the risk allocation hypothesis: tadpole responses to temporal change in predation risk. *Behavioral Ecology*, **13**, 526–530.

Vandermeer, J. (1969) Competitive structure of communities: an experimental approach with protozoa. *Ecology*, **50**, 362–370.

Verschoor, A., Vos, M. and van der Stap, I. (2004) Inducible defences prevent strong population fluctuations in bi- and tri-trophic food chains. *Ecology Letters*, **7**, 1143–1148.

Vos, M., Kooi, B., DeAngelis, D. and Mooij, W. (2004a) Inducible defences and the paradox of enrichment. *Oikos*, **105**, 471–480.

Vos, M., Verschoor, A., Kooi, B., Wackers, F., DeAngelis, D. and Mooij, W. (2004b) Inducible defenses and trophic structure. *Ecology*, **85**, 2783–2794.

Werner, E. (2001) Ecological experiment and a research program in community ecology. In W. Resetarits and J. Bernardo, eds., *Experimental Ecology: Issues and Perspectives*, 1st edn. Oxford: Oxford University Press, pp. 3–26.

Werner, E. and Anholt, B. (1993) Ecological consequences of the trade-off between growth and mortality rates mediated by foraging activity. *American Naturalist*, **142**, 242–272.

Werner, E. and Peacor, S. (2003) A review of trait-mediated indirect interactions in ecological communities. *Ecology*, **84**, 1083–1100.

Werner, E. and Peacor, S. (2006) Lethal and nonlethal predator effects on an herbivore guild mediated by system productivity. *Ecology*, **87**, 347–361.

Werner, E., Skelly, D., Relyea, R. and Yurewicz, K. (2007a) Amphibian species richness across environmental gradients. *Oikos*, **116**, 1697–1712.

Werner, E., Yurewicz, K., Skelly, D. and Relyea, R. (2007b) Turnover in an amphibian metacommunity: the role of local and regional factors. *Oikos*, **116**, 1713–1725.

Wilbur, H. (1972) Competition, predation, and structure of *Ambystoma–Rana sylvatica* community. *Ecology*, **53**, 3–21.

Wootton, J. (1994) The nature and consequences of indirect effects in ecological communities. *Annual Review of Ecology and Systematics*, **25**, 443–466.

Yamauchi, A. and Yamamura, N. (2005) Effects of defense evolution and diet choice on population dynamics in a one-predator-two-prey system. *Ecology*, **86**, 2513–2524.

Yodzis, P. (1988) The indeterminacy of ecological interactions as perceived through perturbation experiments. *Ecology*, **69**, 508–515.

CHAPTER NINE

Community consequences of phenotypic plasticity of terrestrial plants: herbivore-initiated bottom-up trophic cascades

TAKAYUKI OHGUSHI

Center for Ecological Research, Kyoto University

Introduction

Much of earth's biodiversity is composed of species that feed on plants, and in turn these herbivores are the prey base for predatory species. A high diversity of herbivores may support a high diversity of parasites and predators, thus potentially allowing the diversity of plants to cascade upwards to higher trophic levels (Hunter and Price 1992). It is well accepted that increasing the species diversity and/or functional diversity of terrestrial plants leads to greater species diversity of herbivorous and predacious arthropods (Haddad *et al.* 2009; Scherber *et al.* 2010). Moreover, studies of community genetics have shown that genetic variation within a plant species or hybrid zone can greatly influence the species richness and abundance of arthropods associated with the plant (Whitham *et al.* 2006).

It is important to recognize that the herbivore-induced phenotypic plasticity of plants can generate plant-based resource variation, and that these phenotypic variations within a plant species can potentially have a strong bottom-up effect on the community structure and biodiversity of arthropods (Ohgushi 2005). Different plant phenotypes offer distinct niches enabling arthropods to coexist, which leads to the hypothesis that a plant population with high phenotypic diversity would offer a greater variety of niches for arthropods than a population with low phenotypic diversity, and this would result in a greater diversity and an altered community structure of arthropods.

While plant phenotypic plasticity in response to herbivory is a ubiquitous phenomenon in nature (Karban and Baldwin 1997) and there is a growing body of evidence demonstrating plant-mediated indirect effects through herbivore-induced trait changes in plants (Ohgushi 2005, 2007; Ohgushi *et al.* 2007; Kaplan and Denno 2007), the search for community-wide consequences of herbivore-induced changes in plant traits has only recently begun (Ohgushi

Trait-Mediated Indirect Interactions: Ecological and Evolutionary Perspectives, eds. Takayuki Ohgushi, Oswald J. Schmitz and Robert D. Holt. Published by Cambridge University Press. © Cambridge University Press 2012.

2005, 2008; Poelman *et al.* 2008). Thus, we know remarkably little about the consequences of herbivore-induced plant phenotypic plasticity for community properties and biodiversity of higher trophic levels.

However, the ubiquity of phenotypic plasticity of plants caused by associated herbivores strongly suggests that community-wide bottom-up cascading effects should be prevalent and important in structuring plant-based communities in a wide range of ecosystems, in addition to top-down trophic cascades. The paucity of evidence for such effects is simply due to a lack of study.

In this chapter, I will explore how herbivore-generated phenotypic changes in terrestrial plants initiate bottom-up cascading effects, leading to the alteration of community structure and biodiversity of not only herbivores but also their natural enemies. My goal is to encourage ecologists strongly to meet the challenge of this novel topic by addressing the importance and ubiquitous nature of bottom-up cascading effects mediated via plant traits in multitrophic systems. This chapter is not intended as a comprehensive review of the topic. In an effort to inspire future research, I intend to highlight the area that represents the most exciting recent ecological advances, and to draw attention to questions that are still largely unexplored.

Top-down effects of herbivores initiate bottom-up cascades through phenotypic plasticity of plants

In the past two decades, we have learned much about top-down and bottom-up effects in various multitrophic systems. Among such effects, trophic cascades are well-known indirect effects that significantly determine abundance and/or productivity in terrestrial and aquatic ecosystems, which can govern community structure and ecosystem functioning (Terborgh and Estes 2010). Early research in terrestrial systems focused mainly on top-down trophic cascades, which are the indirect, positive effects of predators on plant biomass via herbivores (Pace *et al.* 1999; Shurin *et al.* 2002). More recently, bottom-up trophic cascading effects that are the indirect effects of plants on predators mediated through herbivores have received increasing attention (Bukovinszky *et al.* 2008; Haddad *et al.* 2009; Scherber *et al.* 2010).

It should be noted that the top-down and bottom-up effects interact with each other via changes in the traits or abundance of plants and top predators. Bottom-up effects may alter the sign and strength of top-down cascading effects, and vice versa. For instance, bottom-up effects that vary with changes in growth, quality and structure of plants alter the strength and signs of top-down effects or prey–predator interactions (Forkner and Hunter 2000; Denno *et al.* 2002; Pearson 2010). However, previous studies have concentrated chiefly on how either top-down or bottom-up forces regulate the abundance or productivity of each trophic level independently, and biodiversity research on bottom-up effects of diverse plant species has focused on plant-initiated bottom-up effects but has not included the top-down force as an initiator of bottom-up effects (i.e., herbivore-initiated bottom-up cascades). Since there is increasing evidence suggesting that herbivore-induced changes in plant traits as a result of top-down herbivory have the potential to link the top-down effects of herbivores and the plant-mediated bottom-up effects (Ohgushi *et al.* 2007), we should consider the link between bottom-up and top-down forces to understand how plant–herbivore interactions function in multitrophic systems.

Phenotypic plasticity alters direct and indirect interactions in a community, and thus provides a basis for trait-mediated indirect interactions (TMIIs) among members in an ecological community, generating variation in the strength and direction of interactions (Miner *et al.* 2005). Previous studies on phenotypically plastic responses mainly examined patterns of expression and their adaptive nature, and therefore the community and biodiversity consequences of phenotypic plasticity remain largely unknown. More recently, ecologists have begun to address how phenotypic plasticity governs patterns and processes of trophic interactions, community organization, ecosystem functioning and biodiversity (Werner and Peacor 2003; Ohgushi 2005; Schmitz 2008).

Exploring multiple interaction linkages that are generated by herbivore-induced plant phenotypes has the potential to link phenotypic expression to ecological community. To understand effects of bottom-up cascades on higher trophic levels, we should focus on community-level impacts of phenotypically plastic responses of plants. Note that herbivore-initiated bottom-up effects can be divided into direct effects of a plant on a predator without herbivore mediation (i.e., herbivore–plant–predator effect) and indirect effects of a plant on a predator with herbivore mediation (i.e., herbivore–plant–herbivore–predator effect) (Fig. 9.1). In the former (herbivore–plant–predator effect), a herbivore impacts a plant trait, which directly affects the response of predators without herbivore mediation, such as when an induced plant volatile acts as an attractant to predators or parasitoids (a synomone; Dicke and Baldwin 2010) or when modifications of the host plant are made by galling arthropods or sap exudation is induced by boring insects, both of which attract natural enemies (Marquis and Lill 2007; Yoshimoto and Nishida 2008). The latter (herbivore–plant–herbivore–predator effect) is a herbivore-initiated bottom-up trophic cascade, and it involves a herbivore to plant to herbivore to predator interaction. A herbivore alters a plant phenotype, and this in turn modifies the resource condition of one or more herbivores, which then influences the response of predators; the impact of the initial herbivore is transmitted from one trophic level to another step by step. Such multi-level pathways: herbivore-initiated bottom-up trophic cascades, will be chiefly discussed in this review. Determining the patterns and the mechanisms producing herbivore-initiated bottom-up trophic cascades will answer key ecological questions about multitrophic level interactions:

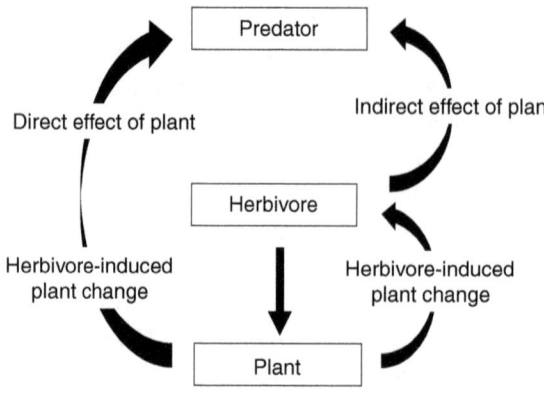

Figure 9.1 Two types of herbivore-induced bottom-up effects: a direct effect of trait changes in plant on predator, and an indirect effect of trait changes in plant on predator via changes in herbivore, i.e., bottom-up trophic cascade. Note that herbivore-induced bottom-up effects are divided into a direct effect of plant on predator without herbivore mediation (i.e., herbivore–plant–predator effect) and an indirect effect of plant on predator with herbivore mediation (i.e., herbivore–plant–herbivore–predator effect).

1. How do top-down effects of herbivores initiate bottom-up effects?
2. To what extent are plastic responses of plants following herbivore attack transmitted throughout a whole community of higher trophic levels?
3. What are the biodiversity and community consequences of the phenotypic plasticity of terrestrial plants?

Herbivore-initiated bottom-up cascading effects on arthropod communities

Since most studies on herbivore-initiated bottom-up effects have focused on a single species at each trophic level, there are very limited data on the community-level impacts on herbivores and predators (Table 9.1). I divided herbivore-generated plant phenotypes into six categories: induction of resistance or susceptibility by secondary metabolites, decrease or increase in nutritional quality, induced regrowth and ecosystem engineering. Although studies of these effects are still rare and only enable one to draw very limited general conclusions, several trends have emerged from the existing data: induced resistance or a decrease in nutritional quality result in a decrease in abundance and species richness of herbivores. On the other hand, the induction of susceptibility by secondary metabolites, an increase in nutritional quality, induced regrowth and ecosystem engineering mostly lead to an increase in abundance and species richness of herbivores. Note that herbivore-initiated effects on the abundance and species richness of predators have been shown to be strongly correlated with those of herbivores. Twelve out of 14 case studies that examined both herbivores and predators showed that herbivore-generated plant phenotypes had the same direction (positive or negative) of effects on the community properties of

Table 9.1 *Studies on community-level effects of herbivore-induced plant responses on herbivore and/or predator communities*

Category	Plant response to herbivory	Herbivore-induced plant phenotype	Effect on herbivore communities			Effect on predator/parasitoid communities			References
			Abundance	Species richness	Community composition	Abundance	Species richness	Community composition	
1	Induction of resistance by secondary metabolites	Increase in secondary substances (protease inhibitor, condensed tannin, glucosinolates)	Negative	n/a	n/a	n/a	n/a	n/a	Wold and Marquis (1997), Thaler et al. (2001), Faeth (1986), Poelman et al. (2010)
			Negative	Negative	n/a	n/a	n/a	n/a	Kessler et al. (2004)
		Increase in protease inhibitor	Negative for flea beetle and pirate bug, no effect on aphids	n/a	n/a	Positive for ladybirds, no effect on spiders	n/a	n/a	Rodriguez-Saona et al. (2005)
2	Induction of susceptibility by secondary metabolites	Increase in glucosinolates	Positive	n/a	n/a	n/a	n/a	n/a	Poelman et al. (2010)
3	Decrease in nutritional quality	Decrease in nitrogen	Negative	n/a	n/a	n/a	n/a	n/a	West (1985)
		Decrease in shoot growth	Negative	Negative	n/a	n/a	n/a	n/a	Waltz and Whitham (1997)
		Decrease in quality (?)	Negative	Negative	n/a	Negative	Negative	Change	González-Megías and Gómez (2003)
4	Increase in nutritional quality	Increase in leaf quality (?)	Positive	Positive	n/a	n/a	n/a	n/a	Bailey and Whitham (2002)

Table 9.1 (*cont.*)

Category to herbivory	Plant response Herbivore-induced plant phenotype	Effect on herbivore communities			Effect on predator/parasitoid communities			References
		Abundance	Species richness	Community composition	Abundance	Species richness	Community composition	
	Increase in plant quality (?)	Positive	Positive	n/a	Positive	n/a	n/a	Dickson and Whitham (1996), Waltz and Whitham (1997)
5 Induced regrowth	Increase in leaf nitrogen	Positive	Positive	Change	Positive	Positive	No change	Katayama et al. (2011)
	Shoot regrowth	Positive	n/a	n/a	n/a	n/a	n/a	Pilson (1992), Hunter (1987)
	Shoot regrowth	Positive	Positive	n/a	n/a	n/a	n/a	Tscharntke (1999)
	Shoot regrowth, increased quality (nitrogen, water content, leaf size, chlorophyll)	Positive	n/a	n/a	Positive	n/a	n/a	Danell and Huss-Danell (1985)
	Shoot regrowth, increased quality (nitrogen)	Positive	Positive	n/a	Positive	Positive	n/a	Nakamura et al. (2006)
	Shoot regrowth, increased quality (nitrogen)	Positive	Positive	Change	Positive	Positive	Change	Utsumi and Ohgushi (2009)
	Shoot regrowth, increased quality (nitrogen)	Positive	Positive	Change	No effect	No effect	Change	Utsumi et al. (2009)
6 Ecosystem engineering	Leaf shelter	n/a	Positive	n/a	n/a	n/a	n/a	Lill and Marquis (2003)
	Leaf shelter	Positive	n/a	n/a	Positive	n/a	n/a	Fournier et al. (2003)
	Leaf shelter	Positive	n/a	Change	Positive	n/a	Change	Lill and Marquis (2004)
	Leaf shelter	Positive	Positive	n/a	Positive	Positive	n/a	Martinsen et al. (2000), Kagata and Ohgushi (2004)
	Leaf gall	Positive	Positive	Change	Positive	Positive	Change	Bailey and Whitham (2003)

both herbivores and predators. Next, I will discuss data collected by my research group to illustrate how changes induced in plant phenotypes determine arthropod community structure and biodiversity.

Bottom-up trophic cascade initiated by stem-borer in a willow system

Compensatory growth is a common response of plants to herbivory (Table 9.1, Category 5) and indirectly influences preference and performance, and thus the abundance of late emerging insects.

We conducted field experiments to examine the community-level impacts of willow regrowth in response to damage by a stem-boring larval swift moth, *Endoclita excrescence*, on herbivorous and predaceous arthropods on three willow species, *Salix gilgiana*, *S. eriocarpa* and *S. serissaefolia* (Utsumi and Ohgushi 2009). Moth boring induced the growth of lateral shoots, which grew more rapidly and had leaves with significantly higher nitrogen and water contents than leaves on uninduced shoots. As a result, the moth-boring indirectly altered the herbivore and predator arthropod community structure at the shoot level within the tree (Fig. 9.2). The relative abundance of herbivorous insects was 3.4-, 2.0- and 1.6-fold greater on the newly emerged lateral shoots than on control shoots in *S. gilgiana*, *S. eriocarpa* and *S. serissaefolia*, respectively. Similarly, the species richness was 1.4–1.9 times greater on the lateral shoots than on control shoots. Shoot regrowth had different effects on the densities of different feeding groups. Specialist chewers and sap-feeders had higher densities on the new lateral shoots, but generalist chewers and sap-feeders did not increase. The overall abundance and species richness of predators also increased on the lateral shoots. The relative abundance of predators was 2.9- and 1.9-fold greater on the newly emerged lateral shoots than on controls in *S. gilgiana* and *S. eriocarpa*, respectively. In each willow species, predator species richness was 2.0–6.5 times greater on the lateral shoots than on controls. The responses of predators differed among willow species. On *S. gilgiana*, ant density was significantly increased on the lateral shoots, but not spider density. On *S. eriocarpa*, spider density was significantly increased on the lateral shoots, but not ant density. Furthermore, PCA ordination revealed that the moth-boring and willow species differentially affected the community composition of arthropods.

In addition to the shoot-level response, we examined the response of the entire arthropod community structure at the tree level (Utsumi *et al.* 2009). Three herbivory treatments and a control were initiated in a common garden: trees damaged by a swift moth caterpillar, artificial cutting of 25% and 100% of stems and control trees without herbivory. The natural and artificial herbivory enhanced shoot production, and resulted in full compensation for

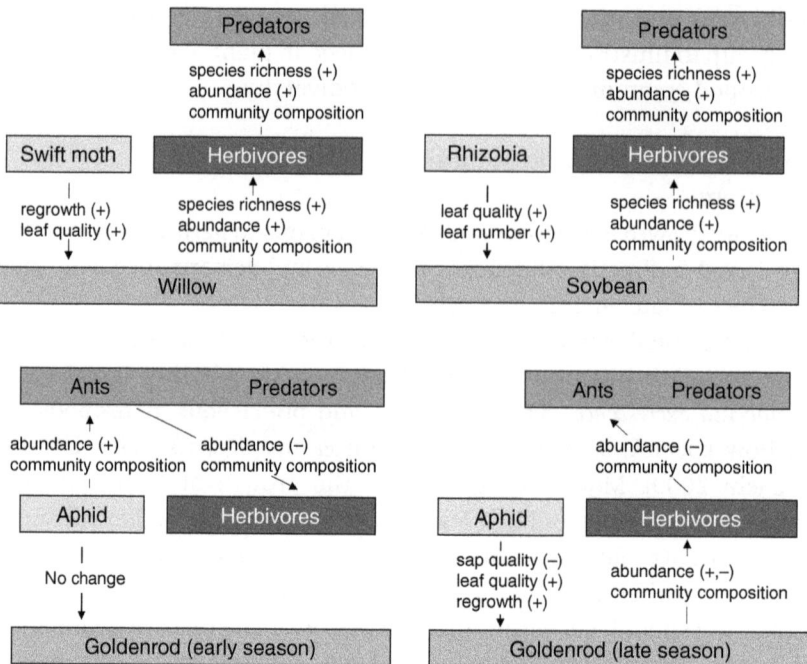

Figure 9.2 Diagrams of bottom-up trophic cascades initiated by herbivores and microbial symbionts. Bottom-up trophic cascading effects initiated by swift moth caterpillars on willow, underground rhizobia on soybean, and aphids on goldenrod in early and late seasons. See colour plate section.

biomass loss by herbivory. The relative abundance of herbivorous insects in the bored and 25%-cut treatments was 1.4–1.7 times greater than that in controls in all the willow species. The moth-boring and 25%-cut treatments significantly increased herbivore species richness by 1.3–1.6 times compared to the controls. Predator species richness and relative abundance were 1.5–1.4 times greater in bored treatments than in controls in *S. gilgiana*. Ordination analysis indicated that the moth-boring and artificial cutting treatments significantly altered the community composition of herbivorous species. Changes in the community structure of herbivores were likely due to changes in plant quality that were determined by the intensity of herbivore damage. In fact, the moth-boring and artificial cutting treatments yielded significantly greater variance and mean of foliar nitrogen than the control, but there were no differences in aboveground biomass among treatments. The community composition of predator species was also significantly different among treatments. Thus, moth-boring has community-wide impacts on arthropod assemblages at both the shoot and tree levels via induced shoot growth in these three willow species.

Bottom-up trophic cascade initiated by aphid in a goldenrod system

We found herbivore-initiated bottom-up cascading effects not only in woody plants but also in herbaceous plants, such as the tall goldenrod, *Solidago altissima* (Ando and Ohgushi 2008; Ando *et al.* 2011). Tall goldenrod was introduced from the United States 100 years ago, and has become widely distributed in Japan. Its common insect herbivores are aphids, leafhoppers and geometrid moth caterpillars in spring and early summer, and grasshoppers and soft scales in autumn. The most abundant aphid, *Uroleucon nigrotuberculatum*, occurs from late June to late July. The native ant *Formica japonica* is frequently observed feeding on scattered aphid honeydew on leaves. Aphid infestation did not change shoot growth, foliar quality of nitrogen or water content in the early part of the growing season, but it increased foliar nitrogen and stimulated lateral shoot growth in autumn, when aphids were no longer present (Table 9.1, Categories 4 and 5). Ants attracted to honeydew of the aphids and scale insects affected other herbivores in the early and late part of the growing season, respectively.

An aphid removal experiment revealed indirect effects of the aphid on abundance, species richness and community composition of herbivores and predators through changes in ant abundance in the early (aphid present) and late (aphid absent) season in a different way (Fig. 9.2). Although aphids did not affect the species richness of either herbivores or predators, aphid removal resulted in a 174% increase in the abundance of herbivores, and a 53% decrease in predators in the early season. Leafhopper and the geometrid moth caterpillars were eight times more abundant on the aphid-free plants, due to the decreased number of ants. The increased density of predators on aphid-infested plants was due to increased numbers of ants. Ordination analysis showed that the community composition of both herbivores and predators differed significantly between aphid-infested and aphid-free plants. In late September, when the aphids were absent, the species richness of herbivores and predators was not affected by aphid colonization early in the season. In contrast, the abundance of herbivores and predators was decreased by 22.9% and 46.6% on aphid-exposed plants, respectively. The density response differed between the soft scale and the grasshopper. Aphid colonization in the early season resulted in 4.0-fold decrease of soft scales, probably because of a decline of phloem quality in aphid-exposed plants. As a result, the number of ants that tended soft scales decreased significantly. Conversely, the grasshopper density on aphid-exposed plants was 3.3-fold greater than that on aphid-free plants. The increase in grasshopper density on aphid-exposed plants was due to increased leaf nitrogen and the decreased negative impact of ants associated with scale insects. Thus, aphids negatively affected the soft scales but positively affected the grasshoppers in the late part of the growing season. As a result, the community composition of both herbivores and predators differed significantly between aphid-infested and aphid-free plants.

Thus, the bottom-up cascade effects initiated by aphids appeared to have occurred in the late season when aphids were absent because induced changes in plant traits occurred with a considerable delay (Fig. 9.2). This indicates that herbivore-induced plant changes can initiate trophic cascades in a temporally separated community, which is likely to be much more common than is currently appreciated. On the other hand, we did not detect such bottom-up cascades in the early season when aphids were present, because aphids did not change plant traits. Rather, the increased number of ants attracted to aphid honeydew had a significantly negative impact on the abundances of co-occurring herbivores. Note that the intensity of the ant-mediated indirect effects on other insects is largely dependent on the herbivore-induced changes in plant quality and availability of leaf shelters constructed by ecosystem engineers, both of which strongly influence the abundance of honeydew-producing insects (Wimp and Whitham 2007).

Bottom-up trophic cascade initiated by belowground microbe in a soybean system

Most terrestrial plants harbour microbial symbionts in some form, and symbiotic microbes such as mycorrhiza and endophytes have strong impacts on arthropod communities on plants associated with these microbial symbionts (Chaneton and Omacini 2007; Rudgers and Clay 2008). These symbiotic microbes may exert bottom-up indirect effects transmitted to higher trophic levels through a modification of host-plant quality and quantity. Although symbiotic microbes are not herbivores and thus this system is not an example of herbivore-initiated bottom-up cascading effects, these microbes can affect the quantity and quality of plant traits, through responses of plant traits similar to those following herbivore attack. Thus, consideration of microbial symbiont-initiated bottom-up effects provides insights useful for understanding herbivore-initiated bottom-up trophic cascades.

We conducted a common garden experiment to explore the effects of rhizobia, nitrogen-fixing soil bacteria on aboveground arthropod communities on soybean, using a root-nodulating strain and its non-nodulating mutant (Katayama et al. 2011). The root-nodulating soybean strain (R+) grew larger, and it had a greater number of leaves and increased foliar nitrogen than the non-nodulating mutant (R–) (Table 9.1, Category 4). Both species richness and abundance of herbivores on R+ plants were significantly greater than those on R– plants (Fig. 9.2). The species richness of sap feeders and leaf chewers on R+ plants was increased by 1.7-fold and 2.9-fold, and their abundance was increased by 1.3-fold and 2.4-fold, respectively. Likewise, R+ plants had greater species richness and relative abundance of predators than R– plants. Predators on R+ plants had a 1.4-fold and 1.8-fold increase in species richness and abundance, respectively. In addition, the herbivore community composition on R+ plants significantly differed from that on R− plants. Predator species richness increased with increasing species richness and abundance of herbivores, and the relative abundance of predators increased with herbivore species richness, indicating that the species richness and relative abundance of predators are largely dependent on those of herbivores. Thus, belowground rhizobia have a significant community-wide impact on aboveground arthropods on plants.

Herbivore-initiated bottom-up cascading effects in other systems

I have summarized our data demonstrating how phenotypic changes in plants by herbivores and underground symbiotic microbes initiate community-level bottom-up trophic cascades. Biodiversity research has chiefly focused on species diversity and/or functional group diversity (Haddard *et al.* 2009) or genetic diversity (Hughes *et al.* 2008) of plants as a primary cause of the diversity and community structure of higher trophic levels. In contrast, community-wide impacts of herbivore-induced plant phenotypes on predators and/or parasitoids mediated by herbivores have received little attention. Therefore, I have included studies on the effects of herbivore-induced plant traits on herbivore communities alone (Table 9.1).

Induced plant resistance or susceptibility mediated by secondary metabolites

One very common and well-known plant response against herbivory is induced resistance (Table 9.1, Category 1). Although there is an increasing body of evidence on indirect interactions between herbivores mediated by induced plant resistance, surprisingly few studies have attempted to examine their community consequences, such as effects on species richness and relative abundance. Kessler *et al.* (2004) found that genetically modified (lipoxygenase-deficient) tobacco that lacked induced resistance received greater herbivory than wild tobacco with induced resistance. In addition, two new species (a leafhopper and a cucumber beetle) attacked non-defended plants, suggesting that induced defences may shift the species richness and composition of the herbivore community. In their herbivore-removal experiment on white oak, Wold and Marquis (1997) compared damaged and intact leaves within a tree and showed that early-season insect herbivory increased foliar tannin and decreased nitrogen late in the growing season, and that the lower leaf quality decreased herbivore density in the late season. Likewise, Thaler *et al.* (2001) reported the effects of using elicitors to induce resistance in tomato on the performance of armyworms, flower thrips, flea beetles and aphids. Application of jasmonic acid induced proteinase inhibitors and polyphenol oxidase, which enhance plant resistance. The numbers of these herbivores from a range of different feeding guilds were reduced by 50–75% on the jasmonic-acid-treated plants.

In contrast, Poelman *et al.* (2010) reported the positive effects of secondary substances of white cabbage induced by feeding of white butterfly caterpillars on the overall abundance of subsequent herbivores (Table 9.1, Category 2). This occurred because specialist herbivores significantly increased on the induced plants. Induced glucosinolates have contrasting effects on generalist and specialist herbivores, and they strongly attract specialist herbivores by providing oviposition stimuli. Also, other secondary metabolites may positively affect specialist herbivores. Thus, induced secondary substances are more likely to result in an increase in susceptibility to specialist herbivores, thereby enhancing the overall abundance of the specialist-dominated community.

Induced changes in plant nutritional quality

As can be seen in the soybean and goldenrod systems, increased nutritional quality can enhance biodiversity components of higher trophic levels (Table 9.1, Category 4). By removing a gall-forming aphid and a leaf-feeding beetle from cottonwoods, Waltz and Whitham (1997) observed the effects of both the aphid and the leaf beetle on communities of herbivores and predators including ants and spiders. They found opposite impacts of the removal. Aphid removal decreased species richness and abundance by 32% and 55%, while beetle removal increased species richness and abundance by 120% and 75%, respectively. Aphids caused less immediate damage to the plant because they feed from phloem; indeed, they may have improved leaf quality, which makes galled leaves more attractive to other herbivores. While aphids consumed large volumes of photosynthate, 90% was excreted in the form of honeydew, which subsequently supported a diverse fauna. In contrast, beetles consumed 33% of the foliage of juvenile ramets and negatively affected subsequent plant growth (such as shoot length), which significantly reduced the plant suitability for both beetles and other community members. Thus, reduced nutritional quality of plants can have a negative impact on the arthropod community (Table 9.1, Category 3). Similarly, González-Megías and Gómez (2003) demonstrated that a leaf beetle feeding on flowers, fruits and vegetative tissues had a great, negative impact on the structure of the arthropod community on a cruciferous shrub. The leaf beetle removal increased the abundances of sap-suckers, flower-feeders, folivores and predators, resulting in an altered community composition and an increase in species richness. They suggested that this large, negative impact of the leaf beetle on community structure was caused not only by direct consumption but also by reduced nutritional quality due to the beetle feeding.

Damage-induced regrowth

Herbivores alter plant architecture in diverse ways that range from superficial modifications to transformations of the entire plant form. Not only woody but also herbaceous plants have the ability to regrow in response to the mechanical damage caused by most feeding modes of insect herbivores. Meristem destruction due to herbivory usually results in the development of dormant buds adjacent or basal to the site of mechanical damage. In spite of its ubiquity, little attention has been paid to the community-level consequences of compensatory responses (Table 9.1, Category 5). However, as our studies on the willow system illustrated, the damage-induced regrowth in response to herbivory can result in a large community-wide impact through increasing plant susceptibility to later-attacking insect herbivores. This is because the damage-induced regrowth produces rapidly growing tissues that provide a highly nutritional resource to herbivores (Price 1991), in effect juvenilizing plant parts. Likewise, enhanced lateral shoot production of a common reed induced by stem-boring moth larvae had community-level effects on subsequent insect herbivores (Tscharntke 1999). Production of new lateral shoots with nutrient-rich tissue significantly affected the community structure by increasing the abundance of six species of gall-making insects. In addition to the regrowth induced by insect herbivory, plant regrowth after mammalian browsing is also very common (Gómez and González-Megías 2007). Mammalian browsing often indirectly affects herbivorous insects. For example, Danell and Huss-Danell (1985) found that herbivorous insects, including aphids, psyllids, leaf miners and leaf gallers, were more abundant on birch trees previously browsed upon by moose than on unbrowsed trees. Browsed trees produced larger leaves with more nitrogen and chlorophyll, and this improved leaf quality, resulting in higher densities of herbivorous insects. Likewise, Bailey and Whitham (2002) documented the effects of elk browsing on the regeneration of aspen, and arthropod species richness and abundance.

Provision of habitats by insect ecosystem engineers

The structures produced by common insect herbivore guilds, including gall makers, leaf rollers, case bearers, stem borers and leaf miners, provide new habitats for other herbivores and/or natural enemies by adding a wide range of shelters (Table 9.1, Category 6). Ecosystem engineering occurs not because of direct responses of plants to herbivory. Instead, it results from structural alterations of plants manipulated by engineers. This contrasts with the trait mediation by insects discussed above, where herbivory induces responses of the plant itself. Such engineers have the potential to influence greatly the species richness, abundance and community structure by providing new habitats for secondary users of not only herbivores but also predators (Marquis and Lill 2007). Thus, insect ecosystem engineers belong to the category of herbivores that initiate bottom-up trophic cascades by adding a wide variety of constructs to plants, which is a primary focus of this review. For instance, Martinsen *et al.* (2000) demonstrated a community-level impact of a leaf-rolling moth on arthropod herbivores and predators on cottonwoods. Leaf roll removal caused a 3-fold decline in herbivore species richness and an 8-fold decline in predator species richness, a 2.5-fold decline in herbivore abundance and a 25-fold decline in predator abundance. Conversely, leaf-roll addition led to a 3.5-fold increase in both herbivore and predator species richness, a 17-fold increase in herbivore abundance and a 3-fold increase in predator abundance. Likewise, Bailey and Whitham (2003) found that a leaf-galling sawfly providing leaf shelters increased the species richness and relative abundance of the arthropod community on aspen. Arthropod species richness and relative abundance increased with increasing gall density. More than 90% of all individuals of five herbivore species and six predator species were found within galls. As a result, arthropod community composition differed significantly when galls were present compared to when they were absent.

Herbivore responses to herbivore-induced plant phenotypes

How do herbivore-induced plant phenotypes alter the biodiversity and community structure of higher trophic levels? Previous studies on the effects of species diversity or functional diversity of plants on the diversity of arthropod consumers have suggested several mechanisms responsible for the increases in species richness and abundance of herbivorous arthropods (Siemann 1998; Haddad *et al.* 2009). First, many herbivores exhibit some degree of feeding specialization, and thus a diverse plant community should provide a greater diversity of resources for a greater number of herbivore species. Second, diverse plant communities are more productive than simple plant communities. Higher productivity provides a greater quantity of resources for consumers, thereby increasing consumer abundance and species richness. Third, higher productivity increases the abundance of rare resources or combinations of resources and conditions that are required by specialist species, allowing scarce resources to become abundant enough to support additional species. Although these scenarios are considered to provide explanations of underlying mechanisms relevant to why plant species diversity or functional diversity has a strong impact on the diversity and abundance of herbivores, they would be applicable in the case of intraspecific phenotypic variations in plants caused by a wide variety of herbivores. Similarly, investigators of community genetics have emphasized that intraspecific genetic variations in a plant species or in hybrid zones can have important consequences for the community organization of higher trophic levels (Whitham *et al.* 2006). Intraspecific genetic variation and hybridization may be important factors shaping the diversity and structure of communities, because plant genotypes vary in their resistance and susceptibility to multiple species of herbivores based on resource quality (Johnson et al. 2006).

An understanding of the mechanisms responsible for bottom-up effects initiated by inter- or intraspecific variations in the quality of terrestrial plants may offer the potential to draw strong inferences about how herbivore-induced plant changes generate bottom-up cascading effects. Herbivore-generated plant phenotypes produce a great deal of quantitative and qualitative resource variations for herbivores both within and among plants. Ohgushi (2005) argued that plant responses to herbivores play an important role in increasing resource heterogeneity, thereby creating niches for insect herbivores, allowing more species to coexist. In the willow system, the compensatory regrowth induced by the stem borer increased the level of nutritional quality and its variation among shoots or trees. The result was that species richness and relative abundance of plant-associated arthropods increased, which in turn altered community composition at both the shoot and tree levels. Several recent reviews have demonstrated how herbivore-induced phenotypic changes in plants affect preference and/or performance of temporally or spatially separated herbivore species (Ohgushi 2005; Kaplan and Denno 2007), and this provides a mechanistic basis for how herbivory alters the biodiversity and community structure of arthropod herbivores. This review highlights the fact that different plant responses to herbivores will produce variation in divergent responses regarding herbivore species richness and abundance. Induced resistance tends to have negative effects on both species richness and abundance, while damage-induced regrowth and ecosystem engineering have positive effects. Induced changes in nutritional quality will affect the components of biodiversity.

The specificity of herbivore inducers that alter plant phenotypes and herbivore receivers that are impacted by the induced plant can determine the strength and sign of cascading effects of herbivore-induced plant traits. Some herbivores induce changes in plant phenotypes that affect only a small number of other species, while others induce profound changes in phenotypes that affect many other species (Poelman et al. 2008). Because of their strong effects on plant phenotypes, specialist herbivores are often strong drivers of arthropod community structure through induced plant responses, while generalists generally do not induce such strong effects. The specificity of inducers is also influenced by feeding guild (Waltz and Whitham 1997; Rodriguez-Saona et al. 2005) and species identity (Agrawal 2000; Van Zandt and Agrawal 2004). The feeding mode may affect the strength of induced responses in a different way. As already mentioned, aphids and leaf beetles on cottonwoods had opposite impacts on the species richness and abundance of arthropod herbivores and predators (Waltz and Whitham 1997).

The specificity of receivers is dependent on the differential effects of a plant response to different inducers. This occurs when two or more species exhibit different preference or performance on plants expressing a given induced phenotype. The specificity of receivers is largely determined by a herbivore's feeding guild (Bailey and Whitham 2002; Koricheva *et al.* 2009), diet breadth (Martinsen *et al.* 1998; Poelman *et al.* 2010) and species identity (Van Zandt and Agrawal 2004). Generalist herbivores tend to avoid induced plants, but specialists across a range of feeding guilds often both prefer and perform better on induced plants (Poelman *et al.* 2008). This pattern of specificity of receivers was also supported by our manipulative experiment in the willow system. Specialist chewers and sap-feeders increased in response to the induced lateral shoots, but generalist chewers and sap-feeders did not. Thus, the combination of differential plant responses to different herbivores and the differential responses of arthropods to the changes in plant phenotypes induced by herbivores would shape the community structure of arthropod herbivores.

Predator responses to changes in properties of herbivores

How are the changes in herbivores caused by plant responses to herbivore initiators transmitted onwards to predators to influence their biodiversity and community structure? Hunter and Price (1992) suggested that a high diversity of herbivores supports a high diversity of predators and parasites, thus potentially allowing the diversity of plants to cascade upwards to higher trophic levels. Also, heterogeneity and productivity of herbivorous prey can determine the abundance, species richness and distribution of predators. In a long-term biodiversity experiment in 82 sown grassland plots, Scherber *et al.* (2010) found that plant species richness had highly significant overall effects on the abundance and species richness of insect herbivores and their natural enemies. The species richness and abundance of natural enemies increased with increasing species richness and abundance of herbivores that resulted from an increase in plant species richness. Also, the densities of arthropod predators are significantly correlated with the densities of leaf chewers, phloem feeders and leaf miners among oak saplings (Forkner and Hunter 2000). Likewise, Cardinale *et al.* (2006) found an interdependence of predator species richness and prey abundance in a system of predatory ladybird beetles and pea aphids on alfalfa. Predator species richness had a positive relationship with aphid density, because predators aggregate and remain longer in higher aphid density patches. In our study of the bottom-up trophic cascades initiated by rhizobia, predator richness increased with increasing herbivore species richness and abundance. Thus, these examples suggest that the overall density and/or diversity of predators would respond to those of herbivores. Twelve out of 14 case studies that examined both herbivores and predators suggested that species richness and/or relative abundance of predators were positively associated with those of herbivores (Table 9.1). Increases in diversity of plant species or genotypes, typically allowing herbivore diversity to increase, also support predator biodiversity (Crutsinger *et al.* 2006; Haddad *et al.* 2009). The increase in abundance and heterogeneity of prey herbivores increases the number of niches allowing predators to coexist. This indicates that plant–herbivore interactions can indirectly control predator biodiversity.

Herbivore-induced resistance in plants often decreases performance of not only insect herbivores but also their natural enemies through reduced quality of prey herbivores. For example, Rodriguez-Saona *et al.* (2005) found that damage to tomato plants by beet armyworm induced a three-fold higher activity of proteinase inhibitors that decrease caterpillar quality. As a result, the caterpillar-induced responses of tomatoes indirectly prolonged the developmental time, increased the mortality, and reduced the pupal mass of a parasitoid wasp. Similarly, Havill and Raffa (2000) found that induced defences in poplar trees decreased the quality of gypsy moth caterpillars, resulting in reduced reproductive output of a parasitoid wasp. Generalist parasitoids may be more negatively affected than specialists by plant defence chemistry, because specialist parasitoids have evolved better capacities for metabolizing or tolerating high levels of plant toxins (Barbosa *et al.* 1991; Harvey *et al.* 2005). On the other hand, some specialist herbivores can use plant secondary compounds in defence against their natural enemies (Pasteels *et al.* 1988; Denno *et al.* 1990). This suggests that the induced defence chemicals of plants increase the defences of prey species, thereby indirectly reducing the abundance of natural enemies. In contrast, previous herbivory lowers an encapsulation response of the cabbage white butterfly larvae to its parasitoid wasp, because it reduces larval weight (Bukovinszky *et al.* 2009), indicating that induced secondary compounds may reduce the immune defence of larvae against parasitoids.

Changes in nutritional quality of plants can also indirectly affect the performance of generalist predators by changing the nutritional quality of prey herbivores. For instance, Kagata *et al.* (2005) examined how improved leaf quality due to shoot regrowth produced by artificial cutting of willows influenced the performance of a leaf beetle and its predatory ladybird beetle. Performance in terms of survival rate, developmental time and adult mass of the leaf beetle was higher when the beetle fed on leaves of cut trees compared to uncut trees. Predator performance measured by development time and adult mass was also significantly improved when the predators fed on beetle larvae that had fed on leaves of cut trees. Furthermore, host-plant quality can indirectly shift the abundance and community structure of parasitoids through their insect hosts. Bukovinszky *et al.* (2008) compared the community structure of primary and secondary parasitoids of aphids that feed on feral and domesticated Brussels sprouts. Feral plants had higher nutritional quality and a higher density of aphids with larger body size, which in turn resulted in a higher species richness of parasitoids, a greater abundance of parasitized aphids and larger parasitoid offspring.

Biodiversity consequences of herbivore-initiated bottom-up cascading effects

Here, I will focus on two important biodiversity properties as an outcome of herbivore-initiated bottom-up trophic cascades: community composition and interaction diversity, both of which have received less attention than species richness and relative abundance in biodiversity research. Changes in community composition may result in changes in the relative abundance of dominant species or keystone species, and their interactions with other species, which may dramatically alter community dynamics. As we saw in the willow and goldenrod systems, the arthropod community compositions on damaged plants significantly changed compared to that on undamaged plants even when species richness or relative abundance remained unaltered. Community composition consists of not only species composition but also network topology. For example, Brussels sprouts with higher nutritional quality supported the increased link density and connectance of aphid–primary parasitoid and primary–secondary parasitoid links (Bukovinszky et al. 2008).

Interaction diversity, determined by the number and type of interactions, has a great impact on the species richness, abundance and stability of communities (Melián et al. 2009; Petchey et al. 2010). It is critical to include different types of interactions such as non-trophic and facilitative interactions in order to gain better understanding of how ecological communities are structurally organized (Borer et al. 2002; Callaway 2007; Goudard and Loreau 2008). For example, Bailey and Whitham (2007) examined the effects of beaver felling of trees and gall construction by leaf-galling sawfly on the arthropod community on cottonwoods, and found that arthropod diversity increased as the interaction diversity increased via indirect interactions of these two dominant keystone herbivores. Herbivores can generate various responses in plants that increase the diversity of plant phenotypes, thereby indirectly altering the preference and performance of other community members. In this context, herbivore-generated phenotypes in plants offer an important mechanistic basis for an increase in the diversity of interactions by providing new niches to allow more species to coexist, particularly non-trophic, facilitative and TMIIs, all of which have rarely been considered in traditional food webs. For example, there is four times more interaction diversity in indirect interaction webs, which include herbivore-induced phenotypic changes in plants, on willows and goldenrods, compared to the number of interactions embedded in food webs. This leads to a 3.8–4.8 times greater link density of species interactions on willows and goldenrods, and a 4.3-fold increase in herbivore diversity on willows (Ohgushi 2008).

Spatial and temporal resource mosaics created by herbivore-generated plant phenotypes

Herbivore-generated plant phenotypes can increase the plant-based resource variation, and as different plant phenotypes offer distinct niches for arthropods, the phenotypic variation within a plant potentially has a strong bottom-up effect on the biodiversity of associated arthropods. It should be noted that the effects of herbivore-induced plant phenotypes on the species diversity of higher trophic levels largely depend on spatial and temporal variations in the altered phenotypes at the individual plant level. Recent studies have highlighted the substantial within-plant heterogeneity in quality (Orians and Jones 2001; Roslin *et al.* 2006). In this context, the plant phenotypes induced by herbivore attack create a resource mosaic consisting of two types of resources, such as resistant versus non-resistant tissue or regrowth versus non-regrowth foliage, which is available to herbivores at the whole-plant level. Thus, a herbivore-generated resource mosaic that increases the phenotypic variance in available plant traits would foster a greater diversity of herbivores and/or their natural enemies. When all of the plants are of one type, we can predict a lower total diversity than when plants contain both types. This suggests that the herbivore diversity would reach a peak at intermediate levels of herbivore-generated plant phenotypes. In the case of the herbivore-induced regrowth in the willow system, the stem-borer produced a mosaic of regrowth (new) and non-regrowth (old) foliage within a tree. The species richness and relative abundance of herbivorous arthropods on bored and 25% cut trees, both of which had a mixture of both foliage types, were significantly greater than those on control trees (with only non-regrowth foliage) and 100% cut trees (with only regrowth foliage). Also, predators increased in species richness and relative abundance on bored and 25% cut trees. The within-tree variation in foliar nitrogen was highest in bored and 25% cut trees, and lowest in control and 100% cut trees. These results strongly support the above hypothesis that the herbivore diversity would reach a peak at intermediate levels of herbivore-generated plant phenotypes. This scenario would be applicable to the species diversity of late-season herbivores that potentially feed on either or both herbivore-induced and non-induced plant tissues generated as a result of early-season herbivory at a time of year when new tissue typically is not produced, at least in seasonal environments. For plants that have indeterminate growth, producing new foliage throughout the season in tropical plants and many herbs, one would predict less of a response from the herbivore community to new tissues than for temperate woody plants that normally have just a single flush.

Conclusions and future directions

This review has shown various patterns and underlying mechanisms of bottom-up trophic cascading effects generated by plant phenotypic responses to herbivores. Although there are still few studies from which to draw general conclusions, several trends emerge from the available data I have reviewed here.

1. Top-down effects of herbivores on plants generating bottom-up cascading effects on the community properties and biodiversity of higher trophic levels should be common and ubiquitous in a wide range of terrestrial systems, in contrast to the prevailing view that community-level trophic cascades are rare in terrestrial systems.
2. The variation in responses of species richness and relative abundance in herbivore communities is largely dependent on the different ways in which plants respond to herbivory. Damage-induced regrowth and ecosystem engineering tend to enhance herbivore species richness and abundance, whereas increased plant resistance tends to decrease these components of biodiversity in herbivores. Nutritional change is variable in its effects on biodiversity, depending on whether the trait changes induced have a positive or negative influence on resource quality for herbivores.
3. The species richness and/or abundance of predators tend to be positively associated with those of herbivores, strongly suggesting that plants have indirect effects on predator communities via changes in herbivores.
4. Herbivore-generated plant phenotypes can initiate bottom-up cascading effects in temporally or spatially separated communities, as illustrated by the goldenrod and soybean systems.
5. The specificity of herbivore initiators and receivers regarding their feeding breadth or guild, combined with differences in trait changes of plants following herbivory, could alter the strength and consequences of the herbivore-initiated bottom-up trophic cascades.

The results of the studies reviewed here can be used to generate new hypotheses to explain how the alteration of plant phenotypes generated by a wide range of herbivores can initiate important bottom-up cascading effects in multitrophic systems. This review also illustrates the connection between the top-down effects of herbivores and the bottom-up effects of induced plant phenotypes, which allows certain keystone herbivores to change dramatically plant traits producing direct and indirect effects that cascade upwards through their own trophic level and beyond. More research on the mechanisms by which herbivores induce TMIEs in terrestrial plants will be essential in order to integrate the top-down and bottom-up forces of plants to gain a more exact understanding of how multitrophic-level interactions shape ecological communities and determine biodiversity. More attention should be paid to the community composition and interaction diversity, in particular to non-trophic and facilitative interactions, caused by TMIEs of plants following herbivore attack. Explorations of the spatial and temporal resource mosaics within a plant will be needed to understand how plant phenotypes, induced by an intermediate level of herbivory, promote the biodiversity of herbivores and their natural enemies. Addressing these key issues will greatly benefit from multidisciplinary approaches that include studies at different levels of biological organization, from phenotypic expression to community organization and biodiversity maintenance in a wide range of ecosystems.

References

Agrawal, A. A. (2000) Specificity of induced resistance in wild radish: causes and consequences for two specialist and two generalist caterpillars. *Oikos*, **89**, 493–500.

Ando, Y. and Ohgushi, T. (2008) Ant- and plant-mediated indirect effects induced by aphid colonization on herbivorous insects on tall goldenrod. *Population Ecology*, **50**, 181–189.

Ando, Y., Utsumi, S. and Ohgushi, T. (2011) Community-wide impact of an exotic aphid on an introduced tall goldenrod. *Ecological Entomology*, **36**, 643–653.

Bailey, J. K. and Whitham, T. G. (2002) Interactions among fire, aspen, and elk affect insect diversity: reversal of a community response. *Ecology*, **83**, 1701–1712.

Bailey, J. K. and Whitham, T. G. (2003) Interactions among elk, aspen, galling sawflies and insectivorous birds. *Oikos*, **101**, 127–134.

Bailey, J. K. and Whitham, T. G. (2007) Biodiversity is related to indirect interactions among species of large effect. In T. Ohgushi, T. P. Craig and P. W. Price, eds., *Ecological Communities: Plant Mediation in Indirect Interaction Webs*. Cambridge: Cambridge University Press, pp. 306–328.

Barbosa, P., Gross, P. and Kemper, J. (1991) Influence of plant allelochemicals on the tobacco hornworm and its parasitoid, *Cotesia congregate. Ecology*, **72**, 1567–1575.

Borer, E. T., Anderson, K., Blanchette, C. A. *et al.* (2002) Topological approaches to food web analyses: a few modifications may improve our insights. *Oikos*, **99**, 397–401.

Bukovinszky, T., Poelman, E. H., Gols, R. *et al.* (2009) Consequences of constitutive and induced variation in plant nutritional quality for immune defence of a herbivore against parasitism. *Oecologia*, **160**, 299–308.

Bukovinszky, T., van Veen, F. J. F., Jongema, Y. and Dicke, M. (2008) Direct and indirect effects of resource quality on food web structure. *Science*, **319**, 804–807.

Callaway, R. M. (2007) *Positive Interactions and Interdependence in Plant Communities*. Dordrecht, The Netherlands: Springer.

Cardinale, B. J., Weis, J. J., Forbes, A. E., Tilmon, K. J. and Ives, A. R. (2006) Biodiversity as both a cause and consequence of resource availability: a study of reciprocal causality in a predator–prey system. *Journal of Animal Ecology*, **75**, 497–505.

Chaneton, E. J. and Omacini, M. (2007) Bottom-up cascades induced by fungal endophytes in multitrophic systems. In T. Ohgushi, T. P. Craig and P. W. Price, eds., *Ecological Communities: Plant Mediation in Indirect Interaction Webs*. Cambridge: Cambridge University Press, pp. 164–187.

Crutsinger, G. M., Collins, M. D., Fordyce, J. A.
et al. (2006) Plant genotypic diversity
predicts community structure and governs
an ecosystem process. Science, **313**, 966–968.

Danell, K. and Huss-Danell, K. (1985) Feeding by
insects and hares on birches earlier affected
by moose browsing. Oikos, **44**, 75–81.

Denno, R. F., Gratton, C., Peterson, M. A. et al.
(2002) Bottom-up forces mediate natural-
enemy impact in a phytophagous insect
community. Ecology, **83**, 1443–1458.

Denno, R. F., Larsson, S. and Olmstead, K. L. (1990)
Role of enemy-free space and plant quality
in host-plant selection by willow beetles.
Ecology, **71**, 124–137.

Dicke, M. and Baldwin, I. T. (2010) The
evolutionary context for herbivore-induced
plant volatiles: beyond the 'cry for help'.
Trends in Plant Science, **15**, 167–175.

Dickson, L. L. and Whitham, T. G. (1996)
Genetically-based plant resistance traits
affect arthropods, fungi, and birds. Oecologia,
106, 400–406.

Faeth, S. H. (1986) Indirect interactions between
temporally separated herbivores mediated
by the host plant. Ecology, **67**, 479–494.

Forkner, R. E. and Hunter, M. D. (2000) What goes
up must come down? Nutrient addition and
predation pressure on oak herbivores.
Ecology, **81**, 1588–1600.

Fournier, V., Rosenheim, J. A., Brodeur, J.,
Laney, L. O. and Johnson, M. W. (2003)
Herbivorous mites as ecological engineers:
indirect effects on arthropods inhabiting
papaya foliage. Oecologia, **135**, 442–450.

Gómez, J. M. and González-Megías, A. (2007)
Trait-mediated indirect interactions,
density-mediated indirect interactions, and
direct interactions between mammalian
and insect herbivores. In T. Ohgushi,
T. P. Craig and P. W. Price, eds., Ecological
Communities: Plant Mediation in Indirect
Interaction Webs. Cambridge: Cambridge
University Press, pp. 104–123.

González-Megías, A. and Gómez, J. M. (2003)
Consequences of removing a keystone
herbivore for the abundance and diversity of

arthropods associated with a cruciferous
shrub. Ecological Entomology, **28**, 299–308.

Goudard, A. and Loreau, M. (2008) Nontrophic
interactions, biodiversity, and ecosystem
functioning: an interaction web model.
American Naturalist, **171**, 91–106.

Haddad, N. M., Crutsinger, G. M., Gross, K. et al.
(2009) Plant species loss decreases arthropod
diversity and shifts trophic structure. Ecology
Letters, **12**, 1029–1039.

Harvey, J. A., van Nouhuys, S. and Biere, A. (2005)
Effects of quantitative variation in
allelochemicals in Plantago lanceolata on
development of a generalist and a specialist
herbivore and their endoparasitoids. Journal
of Chemical Ecology, **31**, 287–302.

Havill, N. P. and Raffa, K. F. (2000) Compound
effects of induced plant responses on insect
herbivores and parasitoids: implications for
tritrophic interactions. Ecological Entomology,
25, 171–179.

Hughes, A. R., Inouye, B. D., Johnson, M. T. J.,
Underwood, N. and Vellend, M. (2008)
Ecological consequences of genetic
diversity. Ecology Letters, **11**, 609–623.

Hunter, M. D. (1987) Opposing effects of spring
defoliation on late season oak caterpillars.
Ecological Entomology, **12**, 373–382.

Hunter, M. D. and Price, P. W. (1992) Playing
chutes and ladders: heterogeneity and the
relative roles of bottom-up and top-down
forces in natural communities. Ecology, **73**,
724–732.

Johnson, M. T. J., Lajeunesse, M. J. and
Agrawal, A. A. (2006) Additive and
interactive effects of plant genotypic
diversity on arthropod communities and
plant fitness. Ecology Letters, **9**, 24–34.

Kagata, H. and Ohgushi, T. (2004) Leaf miner as a
physical ecosystem engineer: secondary use
of vacant leaf mines by other arthropods.
Annals of Entomological Society of America, **97**,
923–927.

Kagata, H., Nakamua, M. and Ohgushi, T. (2005)
Bottom-up cascade in a tri-trophic system
different impacts of host-plant regeneration
on performance of a willow leaf beetle and

its natural enemy. *Ecological Entomology*, **30**, 58–62.

Kaplan, I. and Denno, R. F. (2007) Interspecific interactions in phytophagous insects revisited: a quantitative assessment of competition theory. *Ecology Letters*, **10**, 977–994.

Karban, R. and Baldwin, I. T. (1997) *Induced Responses to Herbivory*. Chicago, IL: The University of Chicago Press.

Katayama, N., Zhi Qi Zhang, Z. Q. and Ohgushi, T. (2011) Community-wide effects of belowground rhizobia on aboveground arthropods. *Ecological Entomology*, **36**, 43–51.

Kessler, A., Halitschke, R. and Baldwin, I. T. (2004) Silencing the jasmonate cascade: induced plant defenses and insect populations. *Science*, **305**, 665–668.

Koricheva, J., Gange, A. C. and Jones, T. (2009) Effects of mycorrhizal fungi on insect herbivores: a meta-analysis. *Ecology*, **90**, 2088–2097.

Lill, J. T. and Marquis, R. J. (2003) Ecosystem engineering by caterpillars increases insect herbivore diversity on white oak. *Ecology*, **84**, 682–690.

Lill, J. T. and Marquis, R. J. (2004) Leaf ties as colonization sites for forest arthropods: an experimental study. *Ecological Entomology*, **29**, 300–308.

Marquis, R. J. and Lill, J. T. (2007) Effects of arthropods as physical ecosystem engineers on plant-based trophic interaction webs. In T. Ohgushi, T. P. Craig and P. W. Price, eds., *Ecological Communities: Plant Mediation in Indirect Interaction Webs*. Cambridge: Cambridge University Press, pp. 246–274.

Martinsen, G. D., Driebe, E. M. and Whitham, T. G. (1998) Indirect interactions mediated by changing plant chemistry: beaver browsing benefits beetles. *Ecology*, **79**, 192–200.

Martinsen, G. D., Floate, K. D., Waltz, A. M., Wimp, G. M. and Whitham, T. G. (2000) Positive interactions between leafrollers and other arthropods enhance biodiversity on hybrid cottonwoods. *Oecologia*, **123**, 82–89.

Melián, C. J., Bascompte, J., Jordano, P. and Křivan, V. (2009) Diversity in a complex ecological network with two interaction types. *Oikos*, **118**, 122–130.

Miner, B. G., Sultan, S. E., Morgan, S. G., Padilla, D. K. and Relyea, R. A. (2005) Ecological consequences of phenotypic plasticity. *Trends in Ecology and Evolution*, **20**, 685–692.

Nakamura, M., Kagata, H. and Ohgushi, T. (2006) Trunk cutting initiates bottom-up cascades in a tri-trophic system: sprouting increases biodiversity of herbivorous and predaceous arthropods on willows. *Oikos*, **113**, 259–268.

Ohgushi, T. (2005) Indirect interaction webs: herbivore-induced effects through trait change in plants. *Annual Review of Ecology, Evolution, and Systematics*, **36**, 81–105.

Ohgushi, T. (2007) Nontrophic, indirect interaction webs of herbivorous insects. In T. Ohgushi, T. P. Craig and P. W. Price, eds., *Ecological Communities: Plant Mediation in Indirect Interaction Webs*. Cambridge: Cambridge University Press, pp. 221–245.

Ohgushi, T. (2008) Herbivore-induced indirect interaction webs on terrestrial plants: the importance of non-trophic, indirect, and facilitative interactions. *Entomologia Experimentalis et Applicata*, **128**, 217–229.

Ohgushi, T., Craig, T. P. and Price, P. W. (2007) *Ecological Communities: Plant Mediation in Indirect Interaction Webs*. Cambridge: Cambridge University Press.

Orians, C. M. and Jones, C. G. (2001) Plants as resource mosaics: a functional model for predicting patterns of within-plant resource heterogeneity to consumers based on vascular architecture and local environmental variability. *Oikos*, **94**, 493–504.

Pace, M. L., Cole, J. J., Carpenter, S. R. and Kitchell, J. F. (1999) Trophic cascades revealed in diverse ecosystems. *Trends in Ecology and Evolution*, **14**, 483–488.

Pasteels, J. M., Rowell-Rahier, M. and Raupp, M. J. (1988) Plant-derived defense in chrysomelid beetles. In P. Barbosa and D. K. Letourneau, eds., *Novel Aspects of Insect-Plant Interactions*. New York: John Wiley and Sons, pp. 235–272.

Pearson, D. E. (2010) Trait- and density-mediated indirect interactions initiated by an exotic invasive plant autogenic ecosystem engineer. *American Naturalist*, **176**, 394–403.

Petchey, O. L., Morin, P. J. and Olff, H. (2010) The topology of ecological interaction networks: the state of the art. In H. A. Verhoef and P. J. Morin, eds., *Community Ecology: Processes, Models, and Applications*. New York: Oxford University Press, pp. 7–22.

Pilson, D. (1992) Aphid distribution and the evolution of goldenrod resistance. *Evolution*, **46**, 1358–1372.

Poelman, E. H., van Loon, J. J. A. and Dicke, M. (2008) Consequences of variation in plant defense for biodiversity at higher trophic levels. *Trends in Plant Science*, **13**, 534–541.

Poelman, E. H., van Loon, J. J. A., van Dam, N. M., Vet, L. E. M. and Dicke, M. (2010) Herbivore-induced plant responses in *Brassica oleracea* prevail over effects of constitutive resistance and result in enhanced herbivore attack. *Ecological Entomology*, **35**, 240–247.

Price, P. W. (1991) The plant vigor hypothesis and herbivore attack. *Oikos*, **62**, 244–251.

Rodriguez-Saona, C., Chalmers, J. A., Raj, S. and Thaler, J. S. (2005) Induced plant responses to multiple damagers: differential effects on an herbivore and its parasitoid. *Oecologia*, **143**, 566–577.

Roslin, T., Gripenberg, S., Salminen, J.-P. et al. (2006) Seeing the trees for the leaves – oaks as mosaics for a host-specific moth. *Oikos*, **113**, 106–120.

Rudgers, J. A. and Clay, K. (2008) An invasive plant-fungal mutualism reduces arthropod diversity. *Ecology Letters*, **11**, 831–840.

Scherber, C., Eisenhauer, N., Weisser, W. W. et al. (2010) Bottom-up effects of plant diversity on multitrophic interactions in a biodiversity experiment. *Nature*, **468**, 553–556.

Schmitz, O. J. (2008) Effects of predator hunting mode on grassland ecosystem function. *Science*, **319**, 952–954.

Shurin, J. B., Borer, E. T., Seabloom, E. W. et al. (2002) A cross-ecosystem comparison of the strength of trophic cascades. *Ecology Letters*, **5**, 785–791.

Siemann, E. (1998) Experimental tests of effects of plant productivity and diversity on grassland arthropod diversity. *Ecology*, **79**, 2057–2070.

Terborgh, J. and Estes, J. A. (2010) *Trophic Cascades: Predators, Prey, and the Changing Dynamics of Nature*. Washington DC: Island Press.

Thaler, J. S., Stout, M. J., Karban, R. and Duffey, S. S. (2001) Jasmonate-mediated induced plant resistance affects a community of herbivores. *Ecological Entomology*, **26**, 312–324.

Tscharntke, T. (1999) Insects on common reed (*Phragmites australis*): community structure and the impact of herbivory on shoot growth. *Aquatic Botany*, **64**, 399–410.

Utsumi, S. and Ohgushi, T. (2009) Community-wide impacts of herbivore-induced plant regrowth on arthropods in a multi-willow species system. *Oikos*, **118**, 1805–1815.

Utsumi, S., Nakamura, M. and Ohgushi, T. (2009) Community consequences of herbivore-induced bottom-up trophic cascades: the importance of resource heterogeneity. *Journal of Animal Ecology*, **78**, 953–963.

Van Zandt, P. A. and Agrawal, A. A. (2004) Specificity of induced plant responses to specialist herbivores of the common milkweed *Asclepias syriaca*. *Oikos*, **104**, 401–409.

Waltz, A. M. and Whitham, T. G. (1997) Plant development affects arthropod communities: opposing impacts of species removal. *Ecology*, **78**, 2133–2144.

Werner, E. E. and Peacor, S. D. (2003) A review of trait-mediated indirect interactions in ecological communities. *Ecology*, **84**, 1083–1100.

West, C. (1985) Factors underlying the late seasonal appearance of the lepidopterous leaf-mining guild on oak. *Ecological Entomology*, **10**, 111–120.

Whitham, T. G., Bailey, J. K., Schweitzer, J. A. *et al.* (2006) A framework for community and ecosystem genetics: from genes to ecosystems. *Nature Reviews Genetics*, **7**, 510–523.

Wimp, G. M. and Whitham, T. G. (2007) Host plants mediate aphid-ant mutualisms and their effects on community structure and diversity. In T. Ohgushi, T. P. Craig and P. W. Price, eds., *Ecological Communities: Plant Mediation in Indirect Interaction Webs.* Cambridge: Cambridge University Press, pp. 275–305.

Wold, E. N. and Marquis, R. J. (1997) Induced defense in white oak: effects on herbivores and consequences for the plant. *Ecology*, **78**, 1356–1369.

Yoshimoto, J. and Nishida, T. (2008) Plant-mediated indirect effects of carpenterworms on the insect communities attracted to fermented tree sap. *Population Ecology*, **50**, 25–34.

CHAPTER TEN

Model-based, response-surface approaches to quantifying indirect interactions

TOSHINORI OKUYAMA

Department of Entomology, National Taiwan University

and

BENJAMIN M. BOLKER

Department of Mathematics and Statistics and Department of Biology,
McMaster University

Introduction

Community dynamics are determined by demographic processes such as reproduction and predation. These demographic processes are influenced in turn by individual traits. While earlier community models assumed that individual traits were static (e.g., Rosenzweig and MacArthur 1963), in the past decade community studies have begun to focus on flexible traits (see Abrams 2010, for review). For example, the Holling type II functional response model describes the predation rate as $aR/(1 + ahR)$ where a is the attack rate, h is the handling time, and R is the prey (resource) density. The original derivation of the model assumed that predators' search behaviour and handling behaviour were static (Holling 1959), and thus could be represented by the static parameters. However, because real individuals' behaviours are flexible, their traits may vary with changing ecological conditions. For example, the antipredator behaviour of a consumer may interfere with its searching behaviour, thus reducing its foraging (attack) rate a as it responds to increasing predator density (Lima and Dill 1990; Persons and Rypstra 2001; Werner and Peacor 2003; Caro 2005). Similarly, handling time may also change with resource density (Cook and Cockrell 1978; Giller 1980; Samu 1993). It is now well accepted that considering flexible traits in ecological studies is both justified by the results of many experimental studies (Werner and Peacor 2003) and important for understanding community dynamics (Fryxell and Lundberg 1998; Bolker *et al.* 2003).

Flexible traits generate a variety of indirect interactions. (This is not their only important effect: they can also modify the dynamics of simple communities consisting of only two species without indirect interactions, e.g., Křivan (2007).) Indirect interactions that are driven by changes in the traits of intermediate species are called *trait-mediated indirect interactions* (TMIIs),

Trait-Mediated Indirect Interactions: Ecological and Evolutionary Perspectives, eds. Takayuki Ohgushi, Oswald J. Schmitz and Robert D. Holt. Published by Cambridge University Press. © Cambridge University Press 2012.

while those that are instead driven by changes in the density of the intermediate species are called *density-mediated indirect interactions* (DMIIs) (Abrams 1995; Abrams *et al.* 1996). DMIIs were traditionally considered to drive such classic examples of indirect interactions as trophic cascades and keystone predation, but the importance of TMIIs in these and other large-scale phenomena is increasingly acknowledged (Schmitz *et al.* 2004). Despite their importance, however, the specific roles of each indirect interaction are relatively poorly understood.

While ecologists are primarily interested in TMIIs because of their potential large-scale effects (e.g., the persistence or coexistence of species in natural communities over many generations), the majority of empirical studies of indirect interactions to date have aimed at quantifying the existence or *relative* importance of indirect interactions (Okuyama and Bolker 2007). As is appropriate for this stage of research (i.e., confirming the existence of an ecological factor, rather than trying to quantify its effects at the community scale), the typical experimental design (see below) has been optimized for detecting significant effects of TMIIs and comparing their strength with that of DMIIs under a particular set of conditions, rather than for quantifying the interactions in a way that could be generalized outside the experimental setting. (Although some studies have extended the search for TMIIs to a larger scale by directly exploring effects at the ecosystem level (Thomas *et al.* 1998; Schmitz 2010), they have still failed to provide a general quantitative framework for estimating parameters that would allow extrapolation.) In previous work (Okuyama and Bolker 2007), we pointed out that the standard metrics for the strengths of density- and trait-mediated interactions cannot in fact be integrated in any straightforward way into standard quantitative models for the dynamics of ecological communities.

Here, we provide further demonstrations of the context dependence of these metrics and, more importantly, follow up on a suggestion made there to 'build a simple model of resource dynamics, predation rate, and forager responses and parameterize it from the system'. We first review the standard experimental design that has been used in most studies that have attempted to quantify the strength of (as opposed to simply detecting) TMII. We then discuss problems that emerge from the standard approach in the case of a three-species food chain, using a simulation model for illustration. Finally, we use an explicitly model-based approach to demonstrate that one can indeed parameterize the functions that describe trait-mediated interactions, in a way suitable for extrapolating to longer-term dynamics. While the approach we suggest is a straightforward adaptation of well-known experimental designs and analytical methods (Inouye 2001; Hall *et al.* 2007), it is qualitatively different from current approaches in the TMII literature, and could open new avenues in TMII research.

Common experimental design

The most common community module (Holt 1997) considered in experiments aimed at quantifying TMII is the three species linear food chain, predator–consumer–resource. This system typically displays indirect mutualisms between the predator and the resource: the removal of consumers by predators increases the growth rate or density of the resource (i.e., inducing a classic (DMII) trophic cascade), while the reduction of consumer foraging rates in order to balance their risks of predation and starvation induces a trait-mediated trophic cascade (Schmitz et al. 2004).

Most laboratory studies of indirect interactions use a standard experimental design with three treatments: no-predator, true-predator and threat-predator treatment (see Okuyama and Bolker 2007, for review). We denote R_t, N_t and P_t to describe the density of the resource, consumer and predator at time t, respectively. In each treatment, a known initial density of resource (R_0), consumers (N_0) and predators (P_0) are added to an experimental micro- or mesocosm. In the no-predator treatment, predators are not introduced ($P_0 = 0$), and thus there are no indirect interactions between the predator and the resource. In the threat predator treatment, consumers can sense the presence of predators but not be attacked by them (e.g., caged predators are present or predator cues are added to the environment) (Peacor and Werner 2001). Thus, the consumers behave as though predators are present, inducing a TMII. In the true predator treatment, predators that can kill the consumers are present, $P_0 > 0$, so that both DMIIs and TMIIs occur. These experiments are normally run for a fixed experimental duration T, and the densities at the end of the experiment (R_T, N_T and P_T) are recorded. In the typical case, predator densities are constant (e.g., the experiment is too short to include predator reproductions or deaths), and the primary response of interest is the density of surviving resources, R_T. If we denote $[P_t]$ as the number of threat predators (e.g., number of caged predators), we can summarize the common experimental design as follows:

$$(R_0, N_0, P_0 = 0) \rightarrow (R_T^{no}, N_T^{no}, P_T^{no}) \qquad \text{No predator}$$
$$(R_0, N_0, P_0 = p) \rightarrow (R_T^{true}, N_T^{true}, P_T^{true}) \qquad \text{True predator}$$
$$(R_0, N_0, [P_0] = p) \rightarrow (R_T^{threat}, N_T^{threat}, P_T^{threat}) \quad \text{Threat predator}$$

where $p > 0$. We compare $(R_T^{no}, R_T^{true}, R_T^{threat})$ to quantify TMII and DMII (superscripts denote the treatment). For example, both R_T^{no} and R_T^{threat} exclude the density effect of the predator on the consumer, and thus comparing them gives the strength of an isolated TMII (e.g., Luttbeg et al. 2003). We have previously discussed some of the challenges associated with this method of comparing the strengths of TMII and DMII, and particularly of connecting these results to reasonable mechanistic models of demography and behaviour (Okuyama and Bolker 2007).

In Okuyama and Bolker (2007), we were unable to come up with any consistent match between these metrics and the parameters of simple mechanistic models describing the community dynamics over a fixed experimental period (and we have seen no evidence that other researchers have succeeded since). One indication that these problems are actually unsolvable is that, as we illustrate below, the results from the standard experimental design are specific to initial densities (R_0, N_0, P_0) and the experimental duration T (Okuyama and Bolker 2007; Abrams 2008). While we and Abrams have noted these problems, strategies for dealing with them have not been discussed explicitly.

Problems

We will use the simple three-species linear food chain discussed above to illustrate the problems. We assume that the predator and consumer exhibit Type II functional responses including flexible traits. For simplicity, we also assume only the consumer is flexible. The consumer's foraging rate on the resource is

$$\frac{a_N C_\alpha R}{1 + a_N C_\alpha h_N R} \tag{10.1}$$

where a_N, h_N and C_α are the attack rate, handling time and flexible trait (e.g., foraging effort) of the consumer, respectively. Similarly, the predator's predation rate on the consumer is

$$\frac{a_P C_\beta N}{1 + a_P C_\beta h_P N} \tag{10.2}$$

where a_P and h_P are the attack rate and handling time of the predator on the consumer, and C_β describes the flexible trait. C_α and C_β may be different because the trait may differentially affect the cost (predation risk) and the benefit (resource intake). For concreteness, we assume (as is often observed) a trade-off between foraging rate and predation risk, i.e. that C_α and C_β are both increasing functions of activity rate or some similar trait. As the consumer increases its trait expression, its resource intake rate increases through increasing C_α (Equation (10.1)), but its predation risk simultaneously increases through increasing C_β (Equation (10.2)). While static trait models assume that C_α and C_β are constant, flexible trait models often consider adaptive traits that maximize the expected fitness of the forager either instantaneously (e.g., Stephens and Krebs 1986; Abrams 1992; Křivan and Sirot 2004) or with a time delay (Abrams 2001; Kondoh 2003).

We use a more phenomenological formulation that simply makes C_α and C_β functions of predator density:

$$C_\alpha = \exp(\alpha P) \tag{10.3}$$

$$C_\beta = \exp(\beta P). \tag{10.4}$$

In our example, consumers respond to predator presence by reducing their foraging rate, reducing their predation risk ($\beta < 0$) but also interfering with foraging success ($\alpha < 0$: Křivan and Schmitz 2004). Many other behavioural adjustments, particularly responses of consumers to changes in resource or conspecific (consumer) density, or interactions between the effects of these densities, can easily be included in this framework. For example, we could represent the reduction of consumer antipredator behaviour in the presence of high densities of (risk-diluting) conspecifics as $C_\alpha = \exp\{(\alpha_P - \gamma_{NP} N)P\}$, where α_P is the level of antipredator response at low conspecific densities, decreasing at a rate γ_{NP} per unit conspecific density. The studies that have explored this behavioural response (e.g., Peacor 2003; McCoy 2007) have focused on measuring the trait response. In contrast, the approach we suggest below characterizes the actual effect of the trait response on consumer foraging and predation risk. This approach is more difficult than measuring traits directly, but provides a more valuable endpoint – the actual effects of the trait modification on the community dynamics. As another extension, cooperative antipredator behaviours could be characterized by $C_\beta = \exp(\beta_P P - \beta_N N)$, so that increasing conspecific density lowered predator attack rate. Here, however, we will discuss only the general approach to the system described by Equations (10.1) and (10.4).

As discussed above, the common experimental design records the change in the number of species in a given time from predetermined initial densities of each species (R_0, N_0 and P_0). The expectation is that comparing total resource consumption from the different treatments will quantify the relative strength of indirect interactions in a way that is generally applicable at least to the particular organisms in question. Figure 10.1 shows the dependence of the usual TMII index ($\gamma_T = R_T^{\text{threat}} - R_T^{\text{no}}$), and the TMII/DMII ratio (γ_T/γ_D, where $\gamma_D = R_T^{\text{true}} - R_T^{\text{no}} - \gamma_T$), on the experimental duration and starting consumer densities. For $N_0 = 10$, the TMII index increases from ≈ 5 to 20 as experimental duration increases, while the TMII/DMII ratio drops from ≈ 1 to 0.3. For $N_0 = 20$ the TMII index peaks at intermediate experimental durations, while the TMII/DMII ratio again decreases monotonically.

Some of the effects shown in Fig. 10.1 are driven by depletion. If we run an experiment long enough, the consumers will have time to deplete all of the resources even in the threat-predator treatment, making the no- and threat-predator treatments equivalent – thus the estimate of TMII will decrease for very long durations. One could say that a sensible experimenter would never run an experiment long enough for most of the resources to be consumed. However, looking at Fig. 10.1, it is far from clear what the 'correct' experimental design would be (and theory offers no guidance): if one runs the experiment for too short a time with low consumer densities, one also sees low estimates of TMII, and the TMII/DMII ratio drops with increasing experimental duration regardless of the consumer density. Furthermore, for short

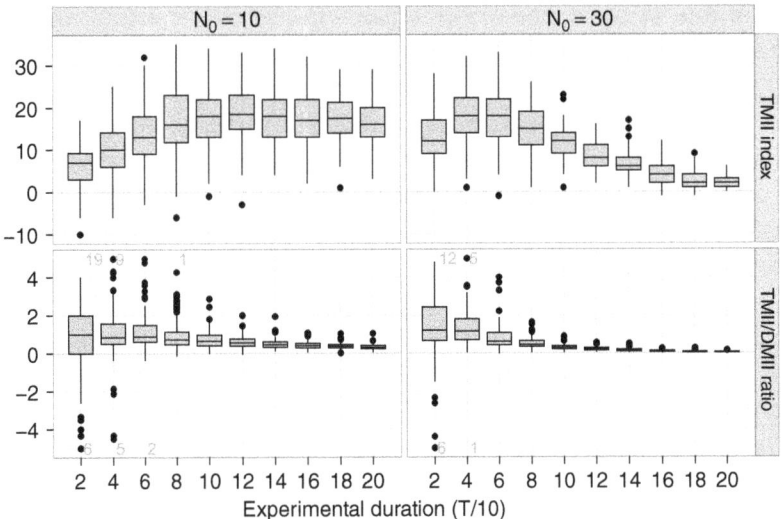

Figure 10.1 Relationship among the initial consumer densities (N_0), experimental duration (T), and TMII index and TMII/DMII ratio (see text for index and ratio definitions). Boxplots and points represent 100 simulations for each combination of input parameters. Estimation of the TMII/DMII ratio is highly unstable for short experimental durations. Values with $|\gamma_T/\gamma_D| > 5$ were excluded from the plot; the numbers of points above and below the lower limits for each experimental duration are noted in grey. Other parameter values: {$R_0 = 100$, $P_0 = 5$, $a_N = 0.001$, $a_P = 0.01$, $h_N = 0.5$, $h_P = 0.5$, $\alpha = -0.1$, $\beta = -0.1$}.

experimental durations the estimates of the TMII/DMII ratio are extremely unstable, suggesting the same general trade-off that occurs in functional response experiments between short experiments that do not provide enough information on attack rates for reliable estimation and long experiments that are subject to depletion effects (Juliano and Williams 1987).

Thus, the quantification of indirect effects is highly specific to the details of a particular experiment – not just the organisms used and the specific manipulations used to provide predator cues, but also the initial densities and experimental duration. In other words, R_T, N_T and P_T depend on R_0, N_0, P_0 and T. It is difficult to make any generalizations (other than rejecting a null hypothesis that $C_\alpha = C_\beta = 0$) from experimental designs that look only at the terminal densities from two predator-density treatments.

Parallels with static-trait communities

The problems we discussed above are not specific to studies quantifying TMII. Experimental results in any subfield of community ecology, particularly those that attempt to quantify ecological processes rather than reject null hypotheses, are likely to be specific to the details of experiments such as the treatment levels and experimental duration. For example, many studies have examined the predation rate of consumer–resource pairs (in the classic setting where species' traits are assumed to be static) (Hassell 1978; Jeschke *et al.* 2004). These studies use experimental designs similar to those discussed above, except that there are only two species under consideration (in our terminology, $P_0 = 0$; because traits are assumed to be static, $\alpha = \beta = 0$). Therefore, the experimental data from those studies (i.e., number of resource consumed) are specific to the initial densities as well as the experimental duration. Nonetheless, as suggested above, these data can be still useful if we are interested in testing for the *existence* of certain effects, i.e., testing null hypotheses such as the occurrence of consumption (e.g., does cannibalism occur?) (Rosenthal and Platts 1990; Caldwell and Araújo 1998), or differences in predation rates between different conditions or levels of a factor such as temperature (Sanchez-Salazar *et al.* 1987; Perdikis *et al.* 1990). As discussed above, previous studies of TMII have had similar objectives of showing the existence of TMII, or that the strength of TMII varies under different conditions (e.g., Trussell *et al.* 2006).

For communities without flexible traits, theoreticians have identified functional response as a central phenomenon connecting predation and community dynamics. If we want to understand the dynamics of communities, we need to be able to predict predation rate at all possible population densities – the functional response. Conducting experiments at many different levels of density is necessary in order to characterize the pattern (Inouye 2001), but ultimately we also need a model (at the very least, some assumption about the form of the functional response curve) in order to interpolate between experimental levels. Using one of the many simple, mechanistically derived classes of functional response models (Turchin 2003) allows us to incorporate the experimental duration T into the estimation procedure so that the results are independent of the experimental duration and can be extrapolated to other experimental durations (given that the assumptions of the model are sufficiently realistic) (Juliano and Williams 1987; Bolker 2008).

The same reasoning applies to TMII studies. In order to connect TMII and community dynamics, we need to know how TMII changes with varying environmental conditions (e.g., population densities). As we show below, assuming flexible trait models (e.g., Equations (10.3) and (10.4)) will allows us to interpolate between observed responses and characterize a response function or surface for the strength of TMII as a function of population densities. However, the classes of models for flexible traits are less well developed theoretically. Here we have assumed simple exponential forms for the effects of foraging effort on consumption and predation risk (Equations (10.3) and (10.4)), but there are many different ways to model flexible traits, the details of which may affect inferences about community level dynamic.

This issue, too, has a parallel in studies of communities with static traits. Even for traditional static-trait models, there are many ways to model ecological processes (specifically, the functional response), and some of the detailed differences can be important for community dynamics (Fussmann and Blasius, 2005). However, because theoreticians have provided candidate models to be tested, empirical studies have attempted to tease apart some of the models (Jost and Ellner 2000; Fussman et al. 2005; Schenk et al. 2005; Saha et al. 2007; Kratina et al. 2009; Hauzy et al. 2010), and useful debates continue (Abrams and Ginzburg 2000; Fussman et al. 2007; Jensen et al. 2007). Even in the absence of a general consensus about the appropriate form of functional response, clear initial guidance (i.e. candidate models) has allowed ecologists to study many systems and refine the models. No such guidance is currently available for flexible trait models.

Part of the problem may lie in the broad diversity of theoretical approaches to communities with flexible traits. For example, some studies (cited above) have used optimal foraging approaches (with or without time delays in trait responses), while others have assumed a specific functional form (e.g., Křivan and Schmitz 2004; Rudolf 2007). In optimal foraging models, the functional form of the flexible trait is determined by the fitness costs, which depend on the fitness currency (usually an assumed trade-off between growth and future fecundity) and other aspects of the model such as the functional response. In theoretical studies, simple linear functional responses are often assumed for simplicity (for example, when the predator functional response is Holling Type II, the dilution effect induces games among the consumers – the fitness of a consumer depends on the traits of its conspecifics – and a unique optimal solution may not exist (Okuyama 2006)). To some degree, theoretical studies have been driven by mathematical tractability and elegance rather than by the ease of empirical evaluation; the great variety of theoretical models can also make it difficult to identify candidate models.

Solutions: model-based analysis of response-surface designs

We have discussed the problems and the importance of the model-based approach. Here, we will describe some examples to illustrate the methodology and the utility of the approach. First, we will discuss a method to characterize flexible trait models, and then discuss how the models can be used to estimate TMIIs.

Characterizing flexible trait models

One straightforward solution that will allow experimentalists and theoreticians to move beyond the current, problematic approaches is to fit parametric, dynamical models. Historically, it has been difficult for researchers to use general, flexible models to quantify the results of experiments because only a few simple, special cases have analytical solutions that can be used in standard nonlinear regression programs (e.g., the Holling family of functional responses, or the Rogers random predator equation (Rogers 1972), in the static-trait example). We know of no such examples where one can derive a simple equation for the expected density of resource remaining (or total resource consumed) after time T as a function of the functional response parameters (a_N, a_P, h_N, h_P), trait responses (α, β) and starting conditions. However, with modern computational power it is reasonably straightforward to integrate a set of differential equations (specifically, the rates of resource loss (Equation (10.1)) and consumer loss (Equation (10.2))) numerically from 0 to T, and to use the predicted final densities of resource and consumer ($\hat{R}_T, \hat{N}_T$) in a nonlinear least-squares or maximum likelihood estimation procedure.

While modern computational methods solve the problem of fitting flexible models, we cannot achieve our goal of quantifying trait responses unless we couple them with response-surface experimental designs (Inouye 2001). As described above, the classic experimental designs evaluate resource depletion by consumers at a single level of resource and consumer, and few levels of predator density (present or absent, or present as a cue); thus they are not capable of resolving the dynamics of traits. Response-surface designs are more challenging logistically, but are necessary if we are to resolve the effects of flexible traits by fitting models.

Simulation example

Here, we show an example of estimating parameters based on the example we used in this chapter (Equations (10.1)–(10.4)). We used an (almost) minimal response-surface design, a factorial combination of three levels of each species (for details, see the caption of Fig. 10.2). We explored the effects of changing the number of replicates for each treatment combination from 1 (leading to 27 total trials, quite practical for some experimental systems) to 50 (1350 total trials, unrealistic for any approach not using automated systems to construct experimental microcosms). For perspective, McCoy *et al.* (2011) have implemented an experimental design with tadpoles and invertebrate predators that involved a total of 368 separate trials – in terms of logistical effort, the equivalent of about 14 replicates per trial in our simulated system. We used nonlinear least-squares to estimate the parameters (see Appendix for details).

Figure 10.2 shows that we can actually retrieve reasonable estimates of trait responses in a simple system. For the effects of trait response on foraging rate (α), we can recover estimated values within 10% of the true value most of the time if the number of replicates (n) is greater than 5–10 and the experimental duration (T) is 10 or greater. The results have little bias as long as $n > 5$. The effect on predation risk (β) is harder to quantify: the results were biased except for large amounts of replication ($n \geq 25$) and the estimates varied by about 50%

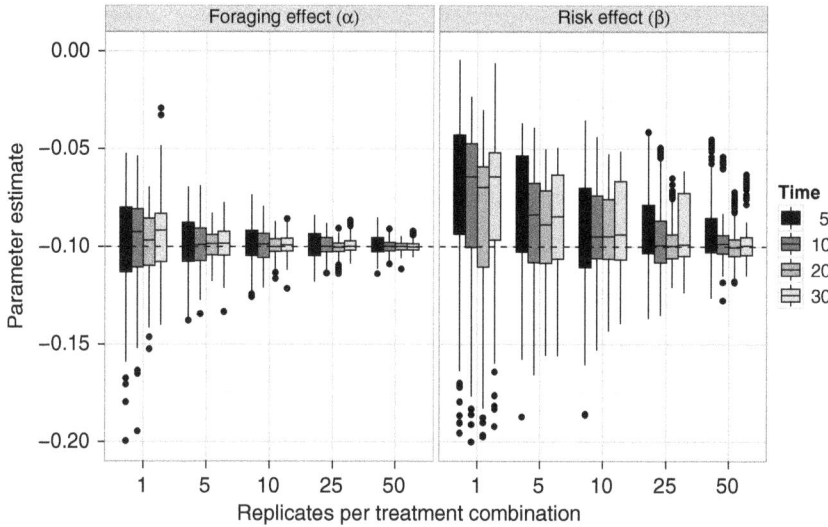

Figure 10.2 Estimated trait parameters for different experimental designs. The true parameters are $\{\alpha = -0.1, \beta = -0.1, a_N = 0.001, a_P = 0.01, h_N = h_P = 0.5\}$. The starting conditions are $R_0 = (50, 75, 100) \times N_0 = (30, 40, 50) \times P_0 = (0, 5, 10)$ (for a total of 27 treatment combinations). Some particularly low estimates (< -0.2) are not shown ($17/200$ cases for β with $n = 1$, $T = 5$ and $10/200$ for β with $n = 1$, $T = 10$; $21/7600$ other cases total).

around the true value. On the other hand, depending on the purpose of the study, it might be useful to know that adding an additional predator would modify consumer behaviour such that the predator attack rate decreased by approximately 5–15% (when β is small, we can interpret it as the approximate proportional change in attack rate per unit predator density). Furthermore, in all cases this experimental design correctly estimated that β was negative.

The estimates of the other parameters (a_N, a_P, h_N, h_P: not shown), especially the handling time parameters, were less well estimated by our design, but these uncertainties do not affect our conclusions about α and β. In order to get more reliable estimates of the baseline attack rates and handling times, one could augment the experimental design with additional trials of consumers on varying levels of resources (in the absence of predators) for α, and trials of fixed numbers of predators with varying numbers of consumers (for β).

More generally, one could use this framework to analyse the power of alternative designs. For example, we considered whether taking measurements at multiple time points within a single trial could decrease the necessity of multiple replicates – a highly desirable switch, because assessing population densities in an ongoing trial would often be easier (and require shorter overall experimental durations, or fewer experimental arenas) than setting up new trials. Somewhat surprisingly, we found that in an experimental setup with a single replicate at each starting density combination and a total duration of $T = 20$, taking samples as frequently as every two time steps did little to improve the estimates of α and β. Apparently, most of the information that can be gleaned from such a setup is already in the ending densities at $T = 20$; more information on the interim dynamics does not help.

One could similarly use these tools to evaluate the effects of using different ranges of starting conditions, or of using incomplete factorial designs (McCoy et al. 2011) to focus on particular parts of the $\{R,N,P\}$ space. Here we have evaluated the effect of replicating particular starting density combinations, but one could also allocate the same sampling effort to treatments spread evenly across the range of starting densities. In a series of simulations (not shown), we found that using evenly spaced designs gave a very slight increase in the precision with which we could estimate α, and had little effect on the estimates of β.

These conclusions depend on our assumptions about the system, particularly on the simple forms of C_α and C_β. There are (at least) two directions in which we can generalize this model. First, we could substitute more flexible basis functions such as splines for the exponential (or other simple) functional forms we have used to characterize the response of consumer traits. This approach would eliminate a fairly strong simplifying assumption, and might provide interesting knowledge on the shape of consumer responses – at the cost (as usual) of greater data requirements.

Second, we can make the consumers sensitive to more aspects of their environment. We might suppose that the densities of all three species affect foraging rates and predation risk of the foragers (i.e., $C_\alpha = \exp(\alpha_R R + \alpha_N N + \alpha_P P)$ and $C_\beta = \exp(\beta_R R + \beta_N N + \beta_P P)$). Or, as mentioned above, we might allow interactions in these terms: for example, consumers may downregulate their foraging less in the presence of predators with increasing conspecific density (a $-\gamma_{NP} NP$ term in C_α and/or C_β) or downregulate it more when resources are abundant (a $+\gamma_{RP} RP$ term).

Of course, in this way it is easy to construct a model that is too complex to estimate; the tools demonstrated above will be useful adjuncts to common sense in establishing whether a particular model can be successfully fitted to experimental data. One note of optimism is that recent studies have successfully used similar methods to characterize a variety of functional responses (McCoy and Bolker 2008; Kratina et al. 2009; Hauzy et al. 2010), including consumer-dependent functional responses which can be considered a specific example of flexible trait functional response (e.g., $C_\alpha = N^\alpha$).

Extrapolation and estimation

At this point, we have a parameterized model of flexible traits – but not what we initially wanted, a robust characterization of the effects of indirect interactions in a particular system. However, once we have parameterized a flexible-trait model such as Equations (10.1)–(10.4), we can also use it to estimate the strength of TMII for any state of the community $\{R,N,P\}$. Abrams (2008) proposed a general measure of TMII for a given community state based on quantifying the response of the per capita growth rate of the resource $((dR/dt)/R)$ to a change in the population mean value of the flexible trait C, multiplied by the sensitivity of the trait to a proportional change in the predator population:

$$\left(\frac{1}{P}\frac{\partial C}{\partial P}\right)\frac{\partial}{\partial C}\left(\frac{1}{R}\frac{dR}{dt}\right). \tag{10.5}$$

For our system described by Equations (10.1)–(10.4), treating consumer foraging rate C_α as the flexible trait, this expression reduces to

$$-\frac{\alpha a_N e^{\alpha P}N}{P(1 + a_N e^{\alpha P}h_N R)^2}; \tag{10.6}$$

this is an easily computed, if not straightforward, expression.

To demonstrate that one can use estimated trait responses from a real system to characterize the strength of TMII for any state of a community, we use the data of van Veen *et al.* (2005) from a one-predator–two-prey module involving two aphids and a specialist parasitoid. In this system, the parasitoid P consumes one of the consumer species (N_1) but does not consume the third (consumer) species N_2; N_1 and N_2 compete for the same resource. The functional response of the parasitoid on the prey species N_1 was characterized as

$$\frac{\alpha N_1}{1 + bN_1 + \omega N_2}. \tag{10.7}$$

In other words, although the competitor species (N_2) is not consumed by the parasitoid, it affects the predation of the parasitoid on the prey (N_1), which implies the existence of TMII between prey and competitor mediated by the parasitoid. According to Equations (10.5) and (10.7) (substituting N_2 for P, N_1 for R and assuming ωN_2 describes the flexible trait of the parasitoid affected by N_2), TMII increases as N_1 decreases and P increases (Fig. 10.3). In the conditions where the extinction probability of prey is largest, the positive TMII is also largest. van Veen *et al.* (2005) found that the prey are excluded in the absence of the competitor; the positive TMII is thought to be important for the persistence of the prey.

Another way to characterize the strength of TMII, one that is arguably more in line with the eventual goals of empirical and applied ecologists, is to describe the change in the equilibrium mean density of the resources that

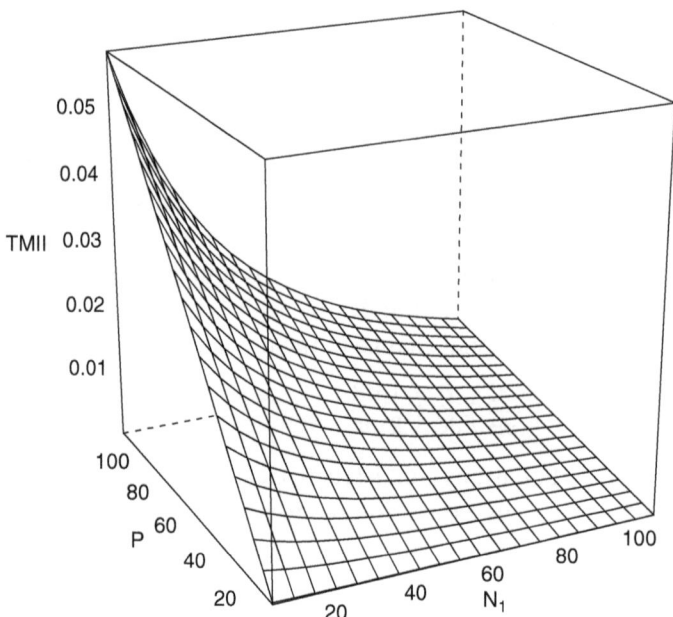

Figure 10.3 Response function of TMII based on Equations (10.5) and (10.7). $N_2 = 10$, $b = 0.233$, $\omega = 0.0434$ and $\alpha = 0.282$.

occurs due to trait- or density-mediated effects. In essence, we apply the now-standard decomposition of effects observed from threat predators and true predators (TMII effect $= R^*_{\text{threat}} - R^*_{\text{no}}$; DMII effect $= R^*_{\text{true}} - R^*_{\text{no}} - \text{TMII effect}$) to the estimated equilibrium rather than to the level of resources remaining after a short experiment.

We extend the short-term experimental system described by Equations (10.1)–(10.4) by adding a constant resource supply at rate g; a predator-independent per capita prey mortality rate μ; and a conversion efficiency b_N that translates resource uptake into prey reproduction:

$$\frac{dR}{dt} = g - \frac{a_N C_a R}{1 + a_N C_a h_N R} N \tag{10.8}$$

$$\frac{dN}{dt} = b_N \frac{a_N C_a R}{1 + a_N C_a h_N R} N - \mu N - \frac{a_P C_\beta N}{1 + a_P C_\beta h_P N} P. \tag{10.9}$$

Holding the predator population constant and assuming constant resource supply rather than a self-reproducing resource (i.e., exponential growth in the absence of prey) leads to an analytically tractable (albeit complicated), stable equilibrium. Fig. 10.4 shows the results. While even simple community models will often display much more complicated dynamics (Abrams *et al.* 2003), the point is that we can use parameterized models of the community to extrapolate

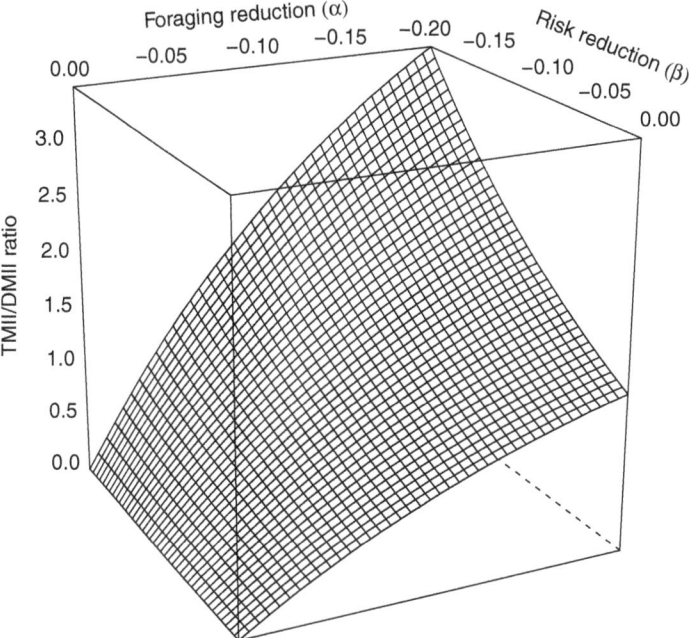

Figure 10.4 TMII/DMII ratio in equilibrium means as a function of foraging and risk reduction parameters (α, β). Parameters as in previous figures, with the addition of $\{g = 5, b_N = 1, \mu = 0.1\}$.

the effects of indirect interactions to the long term. In cases where analysis fails, we can predict relevant metrics of long-term dynamics (mean population size, amplitude or period of fluctuations, extinction risk, etc.) via simulation.

Conclusions

We have discussed problems associated with existing methods of quantifying indirect interactions. Our main contribution is to show how model-based analyses, in combination with response-surface designs, can actually overcome many of these difficulties. An inevitable trade-off of specifying a particular model is that our inferences are specific to the model we choose; however, as we have shown above and in the past (Okuyama and Bolker 2007), the apparent generality of index-based approaches is actually limited to particular experimental conditions.

Our discussion has focused primarily on estimating trait responses, rather than on TMII per se. However, once we have quantified a specific model, we can use it to derive the strength of TMII for any combination of population densities (Fig. 10.3). Furthermore, we can then use the parameterized model to examine how flexible traits will influence population dynamics. Thus, there is a direct connection between TMII and population dynamics.

One critical missing piece in our approach is the consideration of reproduction, which was introduced without comment in Equations (10.8) and
(10.9) (Okuyama and Bolker 2007; Abrams 2008; Preisser *et al.* 2009). Here,
we have focused on the effects of predation, mostly neglecting timescales
long enough for the reproduction of any of the species in the system. In the
past we have recommended that experimentalists should try to find systems
where they generate meaningful time-series data (e.g., experimental trials
can run long enough to include reproduction) (Bolker *et al.*, 2003). At that
time, we suggested that experimentalists could take data from short-term
(behavioural) studies such as those discussed in this chapter and extrapolate
them to community-dynamic timescales; however, we did not realize that
existing experimental designs and experimental analyses *will not allow such*
extrapolation: some approach such as we have suggested here would be
necessary to extrapolate to longer timescales. If it is feasible to collect
long-term experimental or observational data, it may be possible to parameterize models directly from time series data (e.g., the study by van Veen *et al.*
(2005) discussed above). However, for many other study systems (any with
prohibitively long life cycles or large habitat requirements, which includes
even the amphibians whose larval stages have been such important model
systems for TMIIs), we still do not have suggestions other than the stopgap
measures of trying to characterize the relationships between foraging success, size or condition, and resulting fecundity (Semlitsch *et al.* 1988; Taylor
et al. 1998).

Despite the difficulty of quantifying trait responses, and TMIIs, from laboratory or observational data, we are optimistic that model-based response-
surface approaches represent a step forward. Here and in the past, we have
pointed out problems in using the traditional approaches if one wants to do
more than establish the existence of TMIIs. By illustrating and advocating this
approach, we hope to go beyond complaining and provide empiricists with a
useful tool for expanding our understanding of community dynamics. In turn,
such specific information about the magnitudes and shapes of trait responses
will allow theoreticians to focus on the particular phenomena that are most
important to the dynamics of real communities.

Appendix

We used R versions 2.11.1 and 2.12.0 (R Development Core Team 2010) for all
computations and the ggplot (Wickham 2009) and lattice (Sarkar 2010) pack-
ages for all figures.

Simulation details

We implemented a simple stochastic simulation based on the Gillespie algorithm for continuous-time, stochastic systems. The probability per (small)

time interval dt that the consumer consumes a unit of resource, f_1, or that the predator consumes a consumer, f_2, are

$$f_1 = \frac{a_N e^{\alpha P} RN}{1 + a_N e^{\alpha P} h_N R} \tag{10.10}$$

$$f_2 = \frac{a_P e^{\beta P} NP}{1 + a_P e^{\beta P} h_P N}. \tag{10.11}$$

The following algorithm generates realizations from this system:

1. Initialize the number of each species and time: $R = R_0$, $N = N_0$, $P = P_0$ and $t = 0$.
2. Update time, $t = t + \Delta t$ where Δt is a (pseudo-)random number generated from the exponential distribution with rate parameter $f_1 + f_2$.
3. As long as $t < T$, determine which event occurs: Prob(event i) $= f_i/(f_1 + f_2)$.
4. If event 1, $R = R - 1$. If event 2, $N = N - 1$.
5. Repeat (2) to (4).

Estimation details

As discussed in the text, we used nonlinear least-squares to estimate the model parameters. Because the 'data' (simulation results) consist of numbers of surviving animals, sometimes fewer than 10, a maximum likelihood method might be preferable. Furthermore, in a real experimental setting it is likely that some form of experimental blocking would be necessary for logistical reasons (e.g., running each set of replicates during a different time period), which might require the estimation of models incorporating random effects for blocks. Such an analysis is possible, although computationally challenging, with software tools such as WinBUGS (Spiegelhalter et al. 2003) or AD Model Builder (Fournier 2012).

To generate nonlinear least-squares estimates of the parameters, we used an implementation of the Levenberg–Marquardt algorithm from the min-pack.lm package (Elzhov and Mullen 2009). For each set of starting densities represented in the data, we used the lsoda ordinary differential equation solver in the deSolve package (Soetaert et al. 2010) to find the ending densities of R and N based on Equations (10.1)–(10.4)). The gradient function was coded in C for computational efficiency; a typical combined simulation and estimation step (27 starting conditions × 5 replicates, $T = 5$) took 8 seconds on a laptop with a 1.66 GHz Intel Core 2 processor, mostly used by the estimation phase. These complexities (use of minpack.lm and gradients coded in C) were only necessary because we ran thousands of simulations; code that is intended only to be run for a few sets of data and could feasibly take several minutes to run could be much simpler. Code is available from the authors.

References

Abrams, P. A. (1992). Predators that benefit prey and prey that harm predators: unusual effects of interacting foraging adaptations. *American Naturalist*, **140**, 573–600.

Abrams, P. A. (1995) Implications of dynamically variable traits for identifying, classifying, and measuring direct and indirect effects in ecological communities. *American Naturalist*, **146**, 112–134.

Abrams, P. A. (2001) Modelling the adaptive dynamics of traits involved in inter- and intraspecific interactions: an assessment of three methods. *Ecology Letters*, **4**, 166–175.

Abrams, P. A. (2008) Measuring the impact of dynamic antipredator traits on predator–prey–resource interactions. *Ecology*, **89**, 1640–1649.

Abrams, P. A. (2010) Implications of flexible foraging for interspecific interactions: lessons from simple models. *Functional Ecology*, **24**, 7–17.

Abrams, P. A. and Ginzburg, L. R. (2000) The nature of predation: prey dependent, or ratio dependent or neither? *Trends in Ecology and Evolution*, **15**, 337–341.

Abrams, P. A., Brassil, C. E. and Holt, R. D. (2003) Dynamics and responses to mortality rates of competing predators undergoing predator–prey cycles. *Theoretical Population Biology*, **64**, 163–176.

Abrams, P. A., Menge, B. A., Mittelbach, G. G., Spiller, D. A. and Yodzis, P. (1996) The role of indirect effects in food webs. In G. A. Polis, and K. O. Winemiller, eds., *Food Webs: Integration of Pattern and Dynamics*. New York: Chapman and Hall, pp. 371–395.

Bolker, B. (2008) *Ecological Models and Data in R.* Princeton, NJ: Princeton University Press.

Bolker, B., Holyoak, M., Křivan, V., Rowe, L. and Schmitz, O. (2003) Connecting theoretical and empirical studies of trait-mediated interactions. *Ecology*, **84**, 1101–1114.

Caldwell, J. P. and Araújo, M. C. (1998) Cannibalistic interactions resulting from indiscriminate predatory behavior in tadpoles of poison frogs (Anura: Dendrobatidae). *Biotropica*, **30**, 92–103.

Caro, T. (2005) *Antipredator Defences in Birds and Mammals*. Chicago, IL: University of Chicago Press.

Cook, R. M. and Cockrell, B. J. (1978) Predator ingestion rate and its bearing on feeding time and the theory of optimal diets. *Journal of Animal Ecology*, **47**, 529–547.

Elzhov, T. V. and Mullen, K. M. (2009) minpack. lm: R interface to the Levenberg-Marquardt nonlinear least-squares algorithm found in MINPACK. R package version 1.1–4.

Fournier, D., Skaug, H. J., Ancheta, J. et al. (2012) AD model builder: using automatic differentiation for statistical inference of highly parameterized complex nonlinear models. *Optimization Methods and Software*, **27**, 233–249.

Fryxell, J. M. and Lundberg, P. (1998) *Individual Behavior and Community Dynamics*. London: Chapman and Hall.

Fussmann, G. F. and Blasius, B. (2005) Community response to enrichment is highly sensitive to model structure. *Biology Letters*, **1**, 9–12.

Fussmann, G. F., Weithoff, G. and Yoshida, T. (2005) A direct experimental test of resource vs. consumer dependence. *Ecology*, **86**, 2924–2930.

Fussmann, G. F., Weithoff, G. and Yoshida, T. (2007) A direct experimental test of resource vs. consumer dependence: reply. *Ecology*, **88**, 1603–1604.

Giller, P. S. (1980) The control of handling time and its effects on the foraging strategy of a heteropteran predator, *Notonecta*. *Journal of Animal Ecology*, **49**, 699–712.

Hall, S. R., Sivars-Becker, L., Becker, C. et al. (2007) Eating yourself sick: transmission of disease as a function of foraging ecology. *Ecology Letters*, **10**, 207–218.

Hassell, M. P. (1978) *The Dynamics of Arthropod Predator–Prey Systems*. Princeton, NJ: Princeton University Press.

Hauzy, C., Tully, T., Spataro, T., Paul, G. and Arditi, R. (2010) Spatial heterogeneity and functional response: an experiment in microcosms with varying obstacle densities. *Oecologia*, **163**, 625–636.

Holling, C. S. (1959) Some characteristics of simple types of predation and parasitism. *Canadian Entomologist*, **91**, 385–398.

Holt, R. D. (1997) Community modules. In A. C. Gange and V. K. Brown, eds., *Multitrophic Interactions in Terrestrial Systems*. Oxford: Blackwell Science, pp. 333–349.

Inouye, B. D. (2001) Response surface experimental designs for investigating interspecific competition. *Ecology*, **82**, 2696–2706.

Jensen, C. X. J., Jeschke, J. M. and Ginzburg, L. R. (2007) A direct experimental test of resource vs. consumer dependence: comment. *Ecology*, **88**, 1600–1602.

Jeschke, J. M., Kopp, M. and Tollrian, R. (2004) Consumer–food systems: why type I functional responses are exclusive to filter feeders. *Biological Reviews*, **79**, 337–349.

Jost, C. and Ellner, S. P. (2000) Testing for predator dependence in predator–prey dynamics: a non-parametric approach. *Proceedings of the Royal Society of London, Series B*, **267**, 1611–1620.

Juliano, S. A. and Williams, F. M. (1987) A comparison of methods for estimating the functional-response parameters of the random predator equation. *Journal of Animal Ecology*, **56**, 641–653.

Kondoh, M. (2003) Foraging adaptation and the relationship between food-web complexity and stability. *Science*, **299**, 1388–1391.

Kratina, P., Vos, M., Bateman, A. and Anholt, B. R. (2009) Functional responses modified by predator density. *Oecologia*, **159**, 425–433.

Křivan, V. (2007) The Lotka-Volterra predator–prey model with foraging-predation risk trade-offs. *American Naturalist*, **170**, 771–782.

Křivan, V. and Schmitz, O. J. (2004) Trait and density mediated indirect interactions in simple food webs. *Oikos*, **107**, 239–250.

Křivan, V. and Sirot, E. (2004) Do short-term behavioral response of consumers in tri-trophic food chains persist at population

time-scale? *Evolutionary Ecology Research*, **6**, 1063–1081.

Lima, S. L. and Dill, L. M. (1990) Behavioral decisions made under the risk of predation: a review and prospects. *Canadian Journal of Zoology*, **68**, 619–640.

Luttbeg, B., Rowe, L. and Mangel, M. (2003) Prey state and experimental design affect relative size of trait- and density-mediated indirect effects. *Ecology*, **84**, 1140–1150.

McCoy, M. W. (2007) Conspecific density determines the magnitude and character of predator-induced phenotype. *Oecologia*, **153**, 871–878.

McCoy, M. W. and Bolker, B. M. (2008) Trait-mediated interactions: influence of prey size, density and experience. *Journal of Animal Ecology*, **77**, 478–486.

MyCoy, M. W., Bolker, B. M., Warkentin, K. M. and Vonesh, J. R. (2011) Predicting predation through prey ontogeny using size-dependent functional response models. *American Naturalist*, **177**(6), 752–766.

Okuyama, T. (2006) *Maintenance of intraguild predation in jumping spiders*. PhD Thesis, University of Florida.

Okuyama, T. and Bolker, B. M. (2007) On quantitative measures of indirect interactions. *Ecology Letters*, **10**, 264–271.

Peacor, S. D. (2003) Phenotypic modifications to conspecific density: a new mechanism arising from predation risk assessment. *Oikos*, **100**, 409–415.

Peacor, S. D. and Werner, E. E. (2001) The contribution of trait-mediated indirect effects to the net effects of a predator. *Proceedings of the National Academy of Sciences of the United States of America*, **98**, 3904–3908.

Perdikis, D. C., Lykouressis, D. P. and Economou, L. P. (1990) The influence of temperature, photoperiod and plant type on the predation rate of *Macrolophus pygmaeus* on *Myzus persicae*. *BioControl*, **44**, 281–289.

Persons, M. H. and Rypstra, A. L. (2001) Wolf spiders show graded antipredator behavior in the presence of chemical cues from different sized predators. *Journal of Chemical Ecology*, **27**, 2493–2504.

Preisser, E. L., Bolnick, D. I. and Grabowski, J. H. (2009) Resource dynamics influence the strength of non-consumptive predator effects on prey. *Ecology Letters*, **12**, 315–323.

R Development Core Team (2010) R: A Language and Environment for Statistical Computing. Vienna, Austria: R Foundation for Statistical Computing.

Rogers, D. (1972) Random search and insect population models. *Journal of Animal Ecology*, **41**, 369–383.

Rosenthal, S. S. and Platts, B. E. (1990) Host specificity of *Aceria* (*Eriophyes*) *malherbe*, [*Acari: Eriophyidae*], a biological control agent for the weed, *Convolvulus arvensis* [*Convolvulaceae*]. *BioControl*, **35**, 459–463.

Rosenzweig, M. L. and MacArthur, R. H. (1963) Graphical representation and stability condition for predator–prey interactions. *American Naturalist*, **97**, 209–223.

Rudolf, V. H. W. (2007) Consequences of stage-structured predators: cannibalism, behavioral effects and trophic cascades. *Ecology*, **88**, 2991–3003.

Saha, N., Aditya, G., Bal, A. and Saha, G. K. (2007) Comparative study of functional response of common hemipteran bugs of east Calcutta wetlands, India. *International Review of Hydrobiology*, **92**, 242–257.

Samu, F. (1993) Wolf spider feeding strategies: optimality of prey consumption in *Pardosa hortensis*. *Oecologia*, **94**, 139–145.

Sanchez-Salazar, M. E., Griffiths, C. L. and Seed, R. (1987) The effect of size and temperature on the predation of cockles *Cerastoderma edule* (L.) by the shore crab *Carcinus maenas* (L.). *Journal of Experimental Marine Biology and Ecology*, **111**, 181–193.

Sarkar, D. (2010) lattice: Lattice Graphics. R package version 0.18–8.

Schenk, D., Bersier, L.-F. and Bacher, S. (2005) An experimental test of the nature of predation: neither prey- nor ratio-dependent. *Journal of Animal Ecology*, **74**, 86–91.

Schmitz, O. J. (2010) *Resolving Ecosystem Complexity*. Princeton, NJ: Princeton University Press.

Schmitz, O. J., Křivan, V. and Ovadia, O. (2004) Trophic cascades: the primacy of trait-mediated indirect interactions. *Ecology Letters*, **7**, 153–163.

Semlitsch, R. D., Scott, D. E. and Pechmann, J. H. K. (1988) Time and size at metamorphosis related to adult fitness in *Ambystoma talpoideum*. *Ecology*, **69**, 184–192.

Soetaert, K., Petzoldt, T. and Setzer, R. W. (2010) Solving Differential Equations in R: Package deSolve. *Journal of Statistical Software*, **33**, 1–25.

Spiegelhalter, D., Thomas, A., Best, N. and Lunn, D. (2003) *WinBUGS User Manual*. Version 1.4. Cambridge: BUGS.

Stephens, D. W. and Krebs, J. R. (1986) *Foraging Theory*. Princeton, NJ: Princeton University Press.

Taylor, B. W., Anderson, C. R. and Peckarsky, B. L. (1998) Effects of size at metamorphosis on stonefly fecundity, longevity, and reproductive success. *Oecologia*, **114**, 494–502.

Thomas, F., Renaud, F., de Meeus, T. and Poulin, R. (1998) Manipulation of host behaviour by parasites: ecosystem engineering in the intertidal zone? *Proceedings: Biological Sciences*, **265**, 1091–1096.

Trussell, G. C., Ewanchuk, P. J. and Matassa, C. M. (2006) Habitat effects on the relative importance of trait- and density-mediated indirect interactions. *Ecology Letters*, **9**, 1245–1252.

Turchin, P. (2003) *Complex Population Dynamics: A Theoretical/Empirical Synthesis*. Princeton, NJ: Princeton University Press.

van Veen, F. J., van Holland, P. D. and Godfray, H. C. J. (2005). Stable coexistence in insect communities due to density- and trait-mediated indirect effects. *Ecology*, **86**, 3182–3189.

Werner, E. E. and Peacor, S. D. (2003) A review of trait-mediated indirect interactions in ecological communities. *Ecology*, **84**, 1083–1100.

Wickham, H. (2009) *ggplot2: Elegant Graphics for Data Analysis*. New York: Springer.

PART II

Coevolution

Perspective: trait-mediated indirect interactions and the coevolutionary process

BENJAMIN J. RIDENHOUR

Department of Biological Sciences, University of Notre Dame

and

SCOTT L. NUISMER

Department of Biological Sciences, University of Idaho

Multispecific coevolution and the origins of coevolutionary trait-mediated indirect interactions

The notion of coevolution between interacting species can be traced at least as far back as Darwin's writings in the *Origin of Species*, where he outlines a process of coadaptation between a plant and its pollinator (Darwin 1859). Although Darwin's description of plant–pollinator coadaptation suggests a coevolutionary process, it was not until Ehrlich and Raven's (1964) study of butterfly and plant adaptive radiation that the term 'coevolution' was coined. Ehrlich and Raven used the term to signify the broad concept of any evolution resulting from biotic interactions. This broad definition of coevolution persisted until 1980.

Under the broad definition put forth by Ehrlich and Raven, it is clear that trait-mediated indirect interactions (TMIIs) are part of coevolutionary thinking. TMIIs occur when a third species affects the traits involved at the phenotypic interface of an interaction between two species (Wootton 1993; Abrams 1995; Brodie 2003); this is the definition we will use throughout this chapter for TMII. Thus, using Ehrlich and Raven's original interpretation of coevolution, TMIIs are a subset of coevolutionary interactions (assuming actual evolutionary change in trait values is occurring). TMIIs are often distinguished from density-mediated indirect interactions (DMIIs) where the third species alters the density of a species involved in another interaction and thereby alters the outcome of that interaction.

In 1980, however, Janzen formalized coevolution as reciprocal evolutionary change between pairs of interacting populations, arguing that previous work had used the term coevolution too loosely (Janzen 1980). Perhaps because of

Trait-Mediated Indirect Interactions: Ecological and Evolutionary Perspectives, eds. Takayuki Ohgushi, Oswald J. Schmitz and Robert D. Holt. Published by Cambridge University Press. © Cambridge University Press 2012.

Janzen's compelling arguments, or simply the difficulties of incorporating multiple species into coevolutionary studies, most subsequent research in coevolutionary biology focused on interactions between a single pair of species (Thompson 1994). Focusing on single pairs of species relies on the assumption that interactions with other community members do not alter the dynamics of coevolution between this focal pair of species. Thus, the bulk of existing coevolutionary studies implicitly assume TMIIs are not important. Although it is still true that the bulk of coevolutionary research considers only a single pair of species, there is an increasing realization that the broader community context in which pairwise interactions take place can have important consequences for the coevolutionary process (Stanton 2003; Strauss and Irwin 2004; Thompson 2005).

Two conceptual frameworks have focused our understanding of how additional species and community context shape the coevolutionary process. The first framework was developed in an effort to distinguish 'pairwise' from 'diffuse' coevolution, and thus to identify conditions under which coevolution could be understood by studying a single pair of interacting species in isolation (Hougen-Eitzman and Rausher 1994; Iwao and Rausher 1997; Inouye and Stinchcombe 2001; Strauss et al. 2005). The second framework, which has guided much recent research, is the geographic mosaic theory (Thompson 1994, 2005). This theory attempts to improve our understanding of coevolution by incorporating the ecological and spatial complexity of real species interactions. An important focus of this work has been identifying conditions under which the community context of pairwise interactions changes the intensity or trajectory of the coevolutionary process (Benkman et al. 2001; Thompson and Cunningham 2002; Berenbaum and Zangerl 2006).

Although neither of these conceptual frameworks explicitly invokes TMIIs, per se, such interactions are implicitly integrated into both. Our goals with this review are to: (1) illustrate how TMIIs can influence coevolution; (2) identify how TMIIs have already been implicitly integrated into existing paradigms of coevolution; and (3) evaluate the importance of TMIIs for the coevolutionary process.

Formalizing coevolutionary TMIIs

Throughout our review, we will adhere to Janzen's definition of coevolution as reciprocal evolutionary change between a pair of interacting species and will denote this as 'pairwise coevolution' to distinguish it from 'diffuse coevolution' following the convention within the coevolutionary literature. For pairwise coevolution to occur, two requirements must be met. First, there must be additive genetic variation for traits mediating the interface of the interaction (i.e., heritable variation for those traits that determine the fitness consequences of an interaction). Second, the fitness outcome of individual encounters must depend on the phenotypes of both interacting individuals such that selection is reciprocal. The extent to which these requirements are met when species exist within complex networks of interactions is a long-standing debate within coevolutionary biology, and is – to a large degree – determined by the prevalence of TMIIs (Hougen-Eitzman and Rausher 1994; Thompson 1994; Iwao and Rausher 1997; Thompson 1998; Leimu and Koricheva 2006; Morris *et al.* 2007).

We begin by illustrating how TMIIs can influence the coevolutionary process using simple communities of three species: two focal species, X and Y, and an additional species Z. Within this simple three species community there are two basic ways in which TMIIs can alter the course of coevolution, each of which we demonstrate by considering how interactions with species Z alter coevolution between X and Y. While Z could interact with either X or Y, or both X and Y to produce a TMII, for simplicity, we will assume Z interacts only with X.

One possible impact of TMIIs on coevolution arises when interactions between Z and X change the direction or magnitude of the response to coevolutionary selection in X either through direct selection on the trait mediating their interaction or through indirect selection acting on an additional genetically correlated trait (Box 11.1; Fig. 11.1b). This possibility has been long recognized and is essential to existing criteria for distinguishing pairwise and diffuse coevolution (Iwao and Rausher 1997). Another possible impact of TMIIs on coevolution arises when interactions between Z and X change the strength, direction, or form of coevolutionary selection between X and Y (Box 11.1; Fig. 11.1c). This possibility has been widely recognized by coevolutionary biologists since the early 1980s when books by Thompson (1982) and Futuyma and Slatkin (1983) began delving into the concepts and underpinnings of coevolution. In the following section, we expand upon the connections between existing coevolutionary theory and TMIIs, and illustrate how TMIIs can fundamentally change the dynamics and outcomes of the coevolutionary process.

Connecting TMIIs to existing coevolutionary theory

The introduction of TMIIs into coevolutionary biology can be traced back at least to 1982 when Thompson introduced similar ideas in his book, *Interaction and Coevolution*. Since that time, the concept has been of fundamental (although implicit) importance in efforts to distinguish between pairwise and diffuse coevolution and in the development of the geographic mosaic theory. Because these existing paradigms for coevolutionary research have not used the phrase TMII explicitly, however, this long history is underappreciated and opportunities for synthesis have been missed. In the next two sections, we illustrate connections between TMIIs and these

Box 11.1 Two types of TMIIs important to coevolution

We illustrate two specific ways in which TMIIs can influence the coevolutionary process using an example where reciprocal selection between Plant Species X and Herbivore Species Y is mediated by trait x in Species X (e.g., flowering phenology) and trait y in Species Y (e.g., emergence time). In the absence of the third species, traits x and y generate the pattern of reciprocal selection shown in Fig. 11.1.

In the presence of a third species, Z, this simple pairwise coevolutionary interaction can be significantly altered by TMIIs. Below, we consider two specific scenarios where a TMII imposed by Species Z alters the pattern of reciprocal selection between Species X and Species Y shown above.

Case 1: Indirect selection

Interactions between Species X and Species Z are mediated by traits $x's$ (e.g., toxic defensive compound) and z (e.g., detoxifying enzyme), respectively. If traits x and $x's$ in Species X are genetically correlated, indirect selection on trait x is generated, potentially altering the coevolutionary process (Fig. 11.1b). Craig *et al.* (Chapter 13) show the effects of indirect selection on interactions between *Solidago* (X) and *Eurostra* gall-flies (Y) are indirectly affected by the presence of the black-capped chickadee *Parus atricapillus* (Z).

Case 2: Interaction epistasis

The presence of Species Z, or interactions between Species Z and Species X or Y alters the strength or pattern of reciprocal selection (Fig. 11.3c). A classic example of this is given Chapter 15 where Benkman *et al.* describe the effects of squirrels (Z) on the coevolutionary interaction between lodgepole pine and Clark's nutcracker.

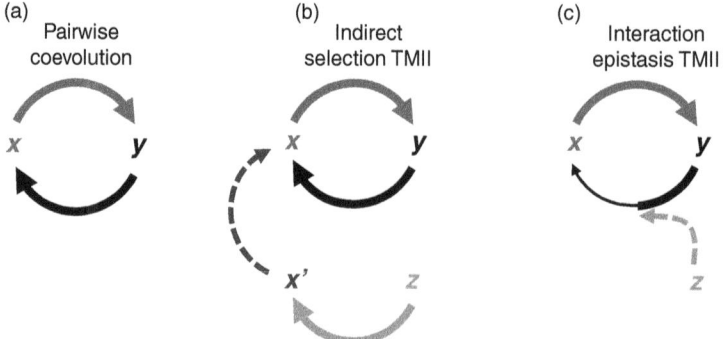

(a) Pairwise coevolution

(b) Indirect selection TMII

(c) Interaction epistasis TMII

Figure 11.1 Two types of TMIIs important to coevolution. See colour plate section.

existing coevolutionary paradigms; we then use studies developed within these paradigms to evaluate the prevalence of TMIIs within coevolutionary interactions.

TMIIs and 'pairwise' versus 'diffuse' coevolution

A central debate within coevolutionary biology has revolved around the extent to which coevolution is 'pairwise' versus 'diffuse' (Thompson 1994; Rausher 1996; Strauss *et al.* 2005; Leimu and Koricheva 2006; Morris *et al.* 2007). At its core, this debate boils down to evaluating the utility of a reductionist approach to studying coevolution. If coevolution is strictly 'pairwise' and occurs between a single pair of populations, we need not consider community context and can understand the coevolutionary process by studying pairs of species in isolation. In contrast, if coevolution is 'diffuse', then the full community context must be considered. Although the utility of the 'diffuse' coevolution concept has been justifiably questioned (Thompson 1994), we consider it here because we feel the presence or absence of TMIIs is one of the essential features which distinguish 'pairwise' and 'diffuse' coevolution.

The connection between TMIIs and the diffuse versus pairwise debate can be seen most clearly by considering conditions that distinguish diffuse coevolution from pairwise coevolution. These conditions have been formalized as a set of three criteria for pairwise coevolution by Rausher and colleagues (e.g., Hougen-Eitzman and Rausher 1994; Iwao and Rausher 1997; Inouye and Stinchcombe 2001). First, there must be no genetic correlation between the trait mediating the interaction of X with Y and the trait mediating the interaction of X with Z; in other words, there is no opportunity for *indirect selection* (Box 11.1; Fig. 11.2b). Second, the frequency or intensity of Y interacting with X cannot change in the presence or absence of Z (i.e., the total evolutionary change per generation is unaffected by Z). Third, the fitness consequence of Y interacting with X does not change in the presence or absence of Z (i.e., the per capita effect of an interaction on trait evolution is unchanged). The final two criteria can be combined into the single criterion that the presence or absence of species Z has no effect on the quantitative and qualitative pattern of (reciprocal) selection between X and Y; in other words there is no *interaction epistasis* (Box 11.1; Fig. 11.1c). Thus, for pairwise coevolution to occur, TMIIs affecting the coevolving traits of the focal species pair must be absent. With this realization, it becomes possible to gain insight into the frequency of TMIIs in coevolutionary interactions by considering the results of empirical studies attempting to distinguish between pairwise and diffuse coevolution.

Substantial empirical evidence exists for both indirect selection and interaction epistasis in putatively coevolving interactions. We begin by assessing evidence for TMIIs based on indirect selection, much of which comes from studies designed to evaluate the extent to which resistance to multiple enemies is genetically correlated (e.g., Fritz 1995; Stinchcombe and Rausher 2001). A recent meta-analysis of such studies (Leimu and Koricheva 2006) concluded that, within plants, genetic correlations often exist between traits conferring resistance to enemy species, suggesting that ample opportunity exists for selection imposed by one enemy to alter the trajectory of coevolution between the plant and the other enemy (e.g., Box 11.1; Fig. 11.1b). Thus, it appears that – at least for coevolutionary interactions between plants and their enemies – the raw material for TMIIs based on indirect selection is common. However, most of these studies fail to demonstrate conclusively coevolutionary TMIIs because selection, and the response to selection, is only rarely (if ever) evaluated within the interacting species. This is a critical step, because for TMIIs based on indirect selection to influence the coevolutionary process, evolutionary change in one enemy species must generate selection which is transmitted through the host genetic correlation to result in indirect evolutionary change in the other enemy species (e.g., Box 11.1; Fig. 11.1b). If this condition is met, TMIIs based on indirect selection can have important consequences for coevolution (Box 11.2; Fig. 11.2b).

As with TMIIs caused by indirect selection, evidence for TMIIs generated by interaction epistasis comes primarily from studies of interactions between plants, their natural enemies and mutualists. Many studies have investigated the potential for interaction epistasis by evaluating the fitness/performance of the host plant X in the presence of interacting species Y in isolation, interacting species Z in isolation, and interacting species Y and Z together (e.g., Strauss 1991; Pilson 1996; Wise and Sacchi 1996; Juenger and Bergelson 1998; Stinchcombe and Rausher 2002). If the fitness/performance effects of Y and Z on X are not additive or multiplicative, the results demonstrate interaction epistasis, suggesting diffuse coevolution and a potential role for TMIIs in the coevolutionary process (Box 11.1; Fig. 11.1c). A recent meta-analysis of such studies (Morris et al. 2007) revealed that interaction epistasis was very common across a wide range of studies, suggesting that TMIIs may be common and coevolution generally diffuse. Here too, however, current empirical evidence falls short of rigorously demonstrating TMIIs and implicating them as an essential force in coevolution. The reason is that interaction epistasis alone is insufficient for generating a TMII. Specifically, in addition to documenting non-additive fitness consequences of interactions, an important role for the traits of the interacting species on fitness would need to be demonstrated (e.g., Box 11.1; Fig. 11.1c). If this condition is met, TMIIs based on interaction epistasis can have important consequences for coevolution (Box 11.2; Fig. 11.2c). An important focus of future work on the role of TMIIs in the coevolutionary process will be to identify cases where interaction epistasis – which is demonstrably common – depends on the traits of the interacting species.

Box 11.2 How the two classes of TMIIs affect the local coevolutionary process

TMIIs can have important consequences for local coevolutionary dynamics. Here we illustrate this by showing how the two classes of TMIIs introduced in Box 11.1 can alter the dynamics of pairwise coevolution. Specifically, we consider the scenarios described in Box 11.1 with the added assumptions that the fitness of Species X individuals is decreased more by encountering Species Y individuals with a similar phenotype than by encountering Species Y individuals with a dissimilar phenotype. The opposite is assumed to be true for Species Y individuals. Making the additional assumption that additive genetic variance is present for traits x and y leads to the classical coevolutionary cycles shown in Fig. 11.2a.

In the presence of a third species, Z, this simple pairwise coevolutionary interaction can be significantly altered as we illustrate below for the two specific TMII scenarios introduced in Box 11.1. Z has no direct effect on the traits mediating the interaction between X and Y or the fitness consequence generated by specific X, Y pairs.

Case 1: Indirect selection

Interactions between Species X and Species Z are mediated by traits $x's$ (e.g., toxic defensive compound) and z (e.g., detoxifying enzyme), respectively. The fitness of species X individuals is increased by having larger values of trait x'. Making the additional assumption that additive genetic variance exists for trait x' and that this trait is genetically correlated with trait x significantly alters coevolutionary dynamics from those of the pairwise case shown in Fig. 11.2b.

Case 2: Interaction epistasis

The presence of Species Z, or interactions between Species Z and Species X or Y reduces the strength of selection imposed on Species X by Species Y. This type of TMII creates a qualitative shift in coevolutionary dynamics (Fig. 11.2c).

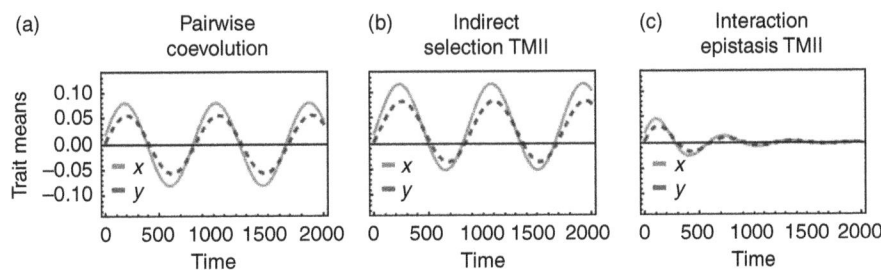

Figure 11.2 How the two classes of TMIIs affect the local coevolutionary process. See colour plate section.

TMIIs and the geographic mosaic theory of coevolution

The geographic mosaic theory of coevolution (Thompson 1994) attempts to integrate the complex spatial and ecological structure of real species interactions into coevolutionary theory. Three components are key: (1) selection mosaics; (2) intermingled coevolutionary hot and cold spots; and (3) trait remixing. Although no obvious connections between trait remixing – which is the combined action of gene flow, local extinction and genetic drift – and TMIIs exist, in many cases the causes of selection mosaics and coevolutionary hot and cold spots can be attributed to TMIIs. In fact, one of the empirical systems which largely motivated the early conceptual and mathematical development of the geographic mosaic theory did so precisely because of the opportunity it provides for TMIIs to generate a selection mosaic (Thompson and Pellmyr 1992). We are not suggesting that TMIIs cannot play a role in the process of trait remixing or the intermingling of hot and cold spots. For example, patterns of migration/gene flow could be altered as the result of interactions with a third species; however, we feel that the effects of TMIIs on the coevolutionary process are most easily understood by examining the concept of selection mosaics.

In the following paragraphs, we explore how TMIIs have been central to the development of the geographic mosaic theory, demonstrate how selection mosaics generated by TMIIs can result in novel coevolutionary dynamics and outcomes, and point to several well studied empirical systems which seem to provide evidence of TMIIs generating selection mosaics and intermingled coevolutionary hot and cold spots.

The ideas underlying the geographic mosaic theory can be traced back (at least) to Thompson's 1988 article, 'Variation in interspecific interactions' (Thompson 1988). In this article, Thompson proposed the idea of 'interaction norms' defined as $G \times G \times E$ interactions. In short, the idea was that embedding a specific coevolutionary interaction ($G \times G$) within different abiotic or biotic environments (E) could yield different outcomes to the coevolutionary process. Thus, those cases where interaction norms are generated by interactions between one of the coevolving species and a third species ($G \times G \times G$) are an example of a TMII for the member of the coevolving pair which does not interact with the third species. This early idea of interaction norms was later integrated into the geographic mosaic theory as one of the potential causes of selection mosaics (changes in the intensity or form of reciprocal selection across space) and coevolutionary hot and cold spots (areas in space where reciprocal selection is present or absent) (Thompson 1994).

In addition to playing an important role in the conceptual development of the geographic mosaic theory, TMIIs were the motivation for much of its early formal theoretical development. Specifically, early mathematical

models of the geographic mosaic theory were based on results from empirical studies of the interactions between the pollinating seed parasitic moth
Greya politella and one of its host plants *Lithophragma parviflorum* (Nuismer
et al. 1999; Gomulkiewicz *et al.* 2000; Nuismer *et al.* 2000). This interaction
is particularly intriguing because it has the potential to swing between
mutualism and antagonism depending on the local abundance of other
pollinator species (Thompson and Pellmyr 1992) and is thus a classical
example of a conditional mutualism (Cushman and Whitham 1989;
Bronstein 1994) and also – assuming the traits of the co-pollinator species
are relevant – of a TMII (Peacor and Werner 1997; Werner and Peacor 2003).
When local co-pollinator species are abundant, the host plant *L. parviflorum*
is not pollen limited and thus visitation by the moth *G. politella* does not
result in pollination of additional seeds but does result in the consumption
of some developing seeds by the moth's larvae (Thompson and Pellmyr
1992; Pellmyr *et al.* 1996). Under such conditions, the interaction is antagonistic. In contrast, when local co-pollinator density is low, plants may be
pollen limited and visitation by moths may result in the pollination of a
greater number of seeds than the moth's larvae ultimately consume. Under
such conditions, the interaction may be mutualistic. Thus, if co-pollinator
densities vary across space, some local interactions may be mutualistic
and others antagonistic, resulting in a selection mosaic (Thompson and
Cunningham 2002). Mathematical models of this scenario have demonstrated that, when combined with gene flow between locations, novel
coevolutionary dynamics and outcomes are produced (Nuismer *et al.* 1999,
2000). Thus, in this particular example, mathematical models have shown
that interactions between the host plant (*L. parviflorum*) and a third species
or suite of species (co-pollinators) can have indirect evolutionary consequences for the insect species (*G. politella*) and thus represent coevolutionary TMIIs.

Although early conceptual and theoretical development of the geographic
mosaic theory was based on selection mosaics caused by the TMIIs described
above, subsequent work has investigated how other types of selection mosaics
generated by more subtle TMIIs influence the coevolutionary process
(Nuismer 2006; Ridenhour and Nuismer 2007; Nuismer and Gandon 2008;
Gandon and Nuismer 2009). The general result of this work has been to show
that even moderate selection mosaics can have important consequences, as
we illustrate in Box 11.3 for the simple three species interaction introduced in
Boxes 11.1 and 11.2. Thus, we currently have a robust theoretical framework
demonstrating the importance of selection mosaics and coevolutionary hot
and cold spots. What remains much more elusive is a robust quantification of
the frequency with which TMIIs generate selection mosaics and intermingled
hot and cold spots within the natural world.

Box 11.3 TMIIs and the geographic mosaic theory

One of the central tenets of the geographic mosaic theory is the existence of selection mosaics. Here we illustrate how a TMII operating in a single habitat can generate a selection mosaic and significantly reshape coevolutionary trajectories.

Consider a scenario where Species X and Y, introduced in Boxes 11.1 and 11.2, exist in two physically separate habitats with only Species X individuals able to disperse between habitats. If the third species, Z, exists in only the second habitat and generates a TMII through interaction epistasis novel coevolutionary dynamics emerge in both habitats even though all parameters are identical to the case shown in Fig. 11.2a and c of Box 11.2.

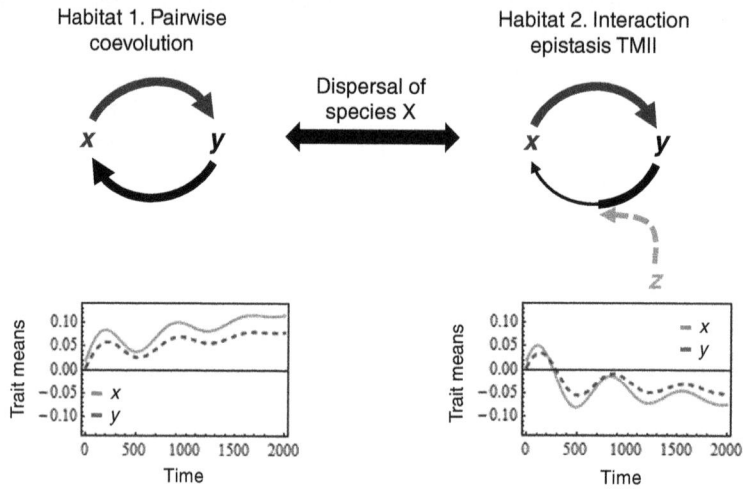

Figure 11.3 TMIIs and the geographic mosaic theory. See colour plate section.

Although we do not yet have a sufficient body of empirical research to draw firm conclusions regarding the overall frequency of selection mosaics generated by TMIIs in natural populations, we do have a handful of high quality studies suggesting TMIIs can indeed cause selection mosaics and intermingled hot and cold spots in a diverse range of organisms and types of interactions (Benkman *et al.* 2001; Thompson and Cunningham 2002; Berenbaum and Zangerl 2006). For instance, work by Craig Benkman and colleagues has shown that the coevolutionary interaction between crossbills and lodgepole pines can be modified by the presence or absence of squirrels (Benkman 1999; Benkman *et al.* 2001; Benkman *et al.* 2003). Specifically, in the absence of squirrels, lodgepole pine and crossbills appear to become locked in a coevolutionary arms race between beak and cone morphology. In the presence of squirrels, however, evidence appears to suggest that seed predation by squirrels effectively eliminates selection by crossbills rendering locations with squirrels coevolutionary cold spots. Thus, although no direct evolutionary interaction between squirrels and crossbills exists, squirrels alter the coevolutionary trajectory between crossbills and pine.

Another well-studied system which suggests that TMIIs generate selection mosaics is the interaction between wild parsnip and parsnip web worm (Berenbaum 1998; Zangerl and Berenbaum 2003; Berenbaum and Zangerl 2006). In this system, it has been argued that coevolution between the focal pair of species occurs with increased frequency when a third species – the alternative host plant cow parsnip – is absent. Apparently, the presence of this alternative host and its interactions with parsnip webworms causes these locations to be coevolutionary cold spots and results in a different coevolutionary outcome (mismatching of traits) than that found when the third species is absent (matching of traits) (Zangerl and Berenbaum 2003).

Conclusions and future directions

TMIIs have played an important role in the conceptual development of coevolutionary biology and are currently well integrated, albeit only implicitly, into two major paradigms of coevolutionary research. Specifically, the presence of TMIIs is one of the defining features of diffuse coevolution and a potential cause of selection mosaics and intermingled coevolutionary hot and cold spots. Because a significant body of theoretical work has demonstrated that the dynamics and outcome of coevolution differ between pairwise and diffuse scenarios (e.g., Gomulkiewicz et al. 2003; Nuismer and Doebeli 2004), and in the presence of selection mosaics or intermingled hot and cold spots (e.g., Nuismer et al. 1999; Gomulkiewicz et al. 2000; Ridenhour and Nuismer 2007) the potential importance of TMIIs for the coevolutionary process is substantial, although not formally recognized.

Although the conceptual and theoretical importance of TMIIs for the coevolutionary process is relatively easy to demonstrate, conclusive empirical evidence of their importance in natural populations is elusive. The primary reason for this is the sheer difficulty of demonstrating that the phenotype of a third species modifies the coevolutionary trajectory of a focal pair of interacting species in the field. Consequently, most available empirical support for coevolutionary TMIIs is indirect and comes from studies using experimentally manipulated populations (e.g., Strauss 1991; Juenger and Bergelson 1998) or field studies which demonstrate most – but almost never all – of the crucial ingredients of a TMII (e.g., Benkman et al. 2001). Perhaps not surprisingly, the same difficulties have confounded attempts to assess the prevalence and importance of the selection mosaics to which TMIIs contribute (Gomulkiewicz et al. 2007).

Several empirical studies presented in this book provide additional evidence for the importance of TMIIs in coevolutionary interactions. For instance, Craig, *et al.* (Chapter 15) show how TMIIs generate indirect selection that affects the coevolving interactions between *Solidago* and gall-forming flies of the genus *Eurosta* (Box 11.1; Fig. 11.1b). In Chapter 12, Lau explores how species invasions generate novel indirect interactions that have impacts on native communities. The coevolutionary interaction between the Clark's nutcracker and lodgepole pine is subject to fitness epistasis (Box 11.1; Fig. 11.1c) resulting from a TMII involving squirrels (Chapter 15). Finally, Irwin (Chapter 14) explores the role of multispecies mutualisms (i.e., those involving trait mediated interactions) on the dynamics of coevolution and community composition. Despite the fact that each of these chapters discusses the *potential* significance of TMIIs in particular systems, they also recognize that demonstrating the presence of such interactions and their consequences is a challenge that remains.

Because the prevalence of TMIIs has important consequences for the dynamics of multispecific coevolutionary interactions – in fact, defining the very scale at which we must study the coevolutionary process itself – developing new empirical and statistical approaches that can be used to identify coevolutionary TMIIs in natural populations is a pressing challenge. Of the statistical techniques that are currently available and could, in principle, be brought to bear on this problem, the selective source analysis techniques developed by Brodie and Ridenhour (2003) and Ridenhour (2005) seem to offer the best potential. Even these techniques, however, require a substantial amount of labour and are restricted to study systems with amenable natural history suggesting that, for the time being, the overall prevalence of TMIIs within natural communities and in coevolutionary systems is likely to remain a mystery.

References

Abrams, P. A. (1995) Implications of dynamically variable traits for identifying, classifying, and measuring direct and indirect effects in ecological communities. *American Naturalist,* **146**, 112–134.

Benkman, C. W. (1999) The selection mosaic and diversifying coevolution between crossbills and lodgepole pine. *American Naturalist,* **153**, S75–S91.

Benkman, C. W., Holimon, W. C. and Smith, J. W. (2001) The influence of a competitor on the geographic mosaic of coevolution between crossbills and lodgepole pine. *Evolution,* **55**, 282–294.

Benkman, C. W., Parchman, T. L., Favis, A. and Siepielski, A. M. (2003) Reciprocal selection causes a coevolutionary arms race between crossbills and lodgepole pine. *American Naturalist,* **162**, 182–194.

Berenbaum, M. R. (1998) Chemical phenotype matching between a plant and its insect herbivore. *Proceedings of the National Academy of Sciences of the United States of America,* **95**, 13743–13748.

Berenbaum, M. R. and Zangerl, A. R. (2006) Parsnip webworms and host plants at home and abroad: trophic complexity in a geographic mosaic. *Ecology*, **87**, 3070–3081.

Brodie, E. D. and Ridenhour, B. J. (2003) Reciprocal selection at the phenotypic interface of coevolution. *Integrative and Comparative Biology*, **43**, 408–418.

Bronstein, J. L. (1994) Conditional outcomes in mutualistic interactions. *Trends in Ecology and Evolution*, **9**, 214–217.

Cushman, J. H. and Whitham, T. G. (1989) Conditional mutualism in a membracid–ant association: temporal, age-specific, and density-dependent effects. *Ecology*, **70**, 1040.

Darwin, C. (1859) *On the Origin of Species by Means of Natural Selection*. London: J. Murray.

Ehrlich, P. R. and Raven, P. H. (1964) Butterflies and plants: a study in coevolution. *Evolution*, **18**, 586.

Fritz, R. S. (1995) Direct and indirect effects of plant genetic variation on enemy impact. *Ecological Entomology*, **20**, 18–26.

Futuyma, D. J. and Slatkin, M. (1983) *Coevolution*. Sunderland, MA: Sinauer Associates.

Gandon, S. and Nuismer, S. L. (2009) Interactions between genetic drift, gene flow, and selection mosaics drive parasite local adaptation. *American Naturalist*, **173**, 212–224.

Gomulkiewicz, R., Drown, D. M., Dybdahl, M. F. *et al.* (2007) Dos and don'ts of testing the geographic mosaic theory of coevolution. *Heredity*, **98**, 249–258.

Gomulkiewicz, R., Nuismer, S. L. and Thompson, J. N. (2003) Coevolution in variable mutualisms. *American Naturalist*, **162**, S80–S93.

Gomulkiewicz, R., Thompson, J. N., Holt, R. D., Nuismer, S. L. and Hochberg, M. E. (2000) Hot spots, cold spots, and the geographic mosaic theory of coevolution. *American Naturalist*, **156**, 156–174.

Hougen-Eitzman, D. and Rausher, M. D. (1994) Interactions between herbivorous insects and plant-insect coevolution. *American Naturalist*, **143**, 677–697.

Inouye, B. and Stinchcombe, J. R. (2001) Relationships between ecological interaction modifications and diffuse coevolution: similarities, differences, and causal links. *Oikos*, **95**, 353–360.

Iwao, K. and Rausher, M. D. (1997) Evolution of plant resistance to multiple herbivores: quantifying diffuse coevolution. *American Naturalist*, **149**, 316–335.

Janzen, D. H. (1980) When is it coevolution? *Evolution*, **34**, 611–612.

Juenger, T. and Bergelson, J. (1998) Pairwise versus diffuse natural selection and the multiple herbivores of scarlet gilia, *Ipomopsis aggregata*. *Evolution*, **52**, 1583–1592.

Leimu, R. and Koricheva, J. (2006) A meta-analysis of genetic correlations between plant resistances to multiple enemies. *American Naturalist*, **168**, E15–E37.

Morris, W. F., Hufbauer, R. A., Agrawal, A. A. *et al.* (2007) Direct and interactive effects of enemies and mutualists on plant performance: a meta-analysis. *Ecology*, **88**, 1021–1029.

Nuismer, S. L. (2006) Parasite local adaptation in a geographic mosaic. *Evolution*, **60**, 24–30.

Nuismer, S. L. and Doebeli, M. (2004) Genetic correlations and the coevolutionary dynamics of three-species systems. *Evolution*, **58**, 1165–1177.

Nuismer, S. L. and Gandon, S. (2008) Moving beyond common-garden and transplant designs: insight into the causes of local adaptation in species interactions. *American Naturalist*, **171**, 658–668.

Nuismer, S. L., Thompson, J. N. and Gomulkiewicz, R. (1999) Gene flow and geographically structured coevolution. *Proceedings of the Royal Society of London, Series B*, **266**, 605–609.

Nuismer, S. L., Thompson, J. N. and Gomulkiewicz, R. (2000) Coevolutionary clines across selection mosaics. *Evolution*, **54**, 1102–1115.

Peacor, S. D. and Werner, E. E. (1997) Trait-mediated indirect interactions in a simple aquatic food web. *Ecology*, **78**, 1146–1156.

Pellmyr, O., Thompson, J. N., Brown, J. M. and Harrison, R. G. (1996) Evolution of pollination and mutualism in the yucca moth lineage. *American Naturalist*, **148**, 827–847.

Pilson, D. (1996) Two herbivores and constraints on selection for resistance in *Brassica rapa*. *Evolution*, **50**, 1492–1500.

Rausher, M. (1996) Genetic analysis of coevolution between plants and their natural enemies. *Trends in Genetics*, **12**, 212–217.

Ridenhour, B. J. (2005) Identification of selective sources: partitioning selection based on interactions. *American Naturalist*, **166**, 12–25.

Ridenhour, B. J. and Nuismer, S. L. (2007) Polygenic traits and parasite local adaptation. *Evolution*, **61**, 368–376.

Stanton, M. L. (2003) Interacting guilds: moving beyond the pairwise perspective on mutualisms. *American Naturalist*, **162**, S10–S23.

Stinchcombe, J. R. and Rausher, M. D. (2001) Diffuse selection on resistance to deer herbivory in the ivyleaf morning glory, *Ipomoea hederacea*. *American Naturalist*, **158**, 376–388.

Stinchcombe, J. R. and Rausher, M. D. (2002) The evolution of tolerance to deer herbivory: Modifications caused by the abundance of insect herbivores. *Proceedings of the Royal Society of London, Series B*, **269**, 1241–1246.

Strauss, S. Y. (1991) Direct, indirect, and cumulative effects of three native herbivores on a shared host plant. *Ecology*, **72**, 543–558.

Strauss, S. Y. and Irwin, R. E. (2004) Ecological and evolutionary consequences of multispecies plant-animal interactions. *Annual Review of Ecology, Evolution, and Systematics*, **35**, 435–466.

Strauss, S. Y., Sahli, H. and Conner, J. K. (2005) Toward a more trait-centered approach to diffuse (co)evolution. *New Phytologist*, **165**, 81–89.

Thompson, J. N. (1982) *Interaction and Coevolution*. New York: Wiley.

Thompson, J. N. (1988) Variation in interspecific interactions. *Annual Review of Ecology and Systematics*, **19**, 65–87.

Thompson, J. N. (1994) *The Coevolutionary Process*. Chicago, IL: University of Chicago Press.

Thompson, J. N. (1998) The population biology of coevolution. *Researches on Population Ecology*, **40**, 159–166.

Thompson, J. N. (2005) *The Geographic Mosaic of Coevolution*. Chicago, IL: University of Chicago Press.

Thompson, J. N. and Cunningham, B. M. (2002) Geographic structure and dynamics of coevolutionary selection. *Nature*, **417**, 735–738.

Thompson, J. N. and Pellmyr, O. (1992) Mutualism with pollinating seed parasites amid co-pollinators: constraints on specialization. *Ecology*, **73**, 1780–1791.

Werner, E. E. and Peacor, S. D. (2003) A review of trait-mediated indirect interactions in ecological communities. *Ecology*, **84**, 1083–1100.

Wise, M. J. and Sacchi, C. F. (1996) Impact of two specialist insect herbivores on reproduction of horse nettle, *Solanum carolinense*. *Oecologia*, **108**, 328–337.

Wootton, J. T. (1993) Indirect effects and habitat use in an intertidal community – interaction chains and interaction modifications. *American Naturalist*, **141**, 71–89.

Zangerl, A. R. and Berenbaum, M. R. (2003) Phenotype matching in wild parsnip and parsnip webworms: causes and consequences. *Evolution*, **57**, 806–815.

CHAPTER TWELVE

Evolutionary indirect effects: examples from introduced plant and herbivore interactions

JENNIFER A. LAU

*W. K. Kellogg Biological Station and Department of Plant Biology,
Michigan State University*

Introduction

Biological invasions result in a wide range of novel ecological and evolutionary interactions with native community members. Invaders may be predators or competitors, with direct negative impacts on native taxa, and they may be prey or mutualists, which could have positive effects on native taxa. In addition to these initial direct effects, as invading species become integrated into native communities, they begin to interact both directly and indirectly with a multitude of native taxa in the invaded community. While the direct effects of biological invasions on natives are well appreciated, the indirect effects of biological invasions remain relatively unstudied (White *et al.* 2006), especially from an evolutionary perspective. However, indirect effects are likely common, can be strong, and may impact an even wider array of community members than the initial direct effects resulting from the invasion.

In this chapter, I first briefly discuss: (1) the evidence for evolutionary changes in both natives and invaders during the invasion process and (2) the many potential and documented indirect effects of invaders on natives. I then combine these two themes to address the main thesis: how evolutionary changes in both natives and invaders during the invasion process can alter interactions with additional community members (i.e. 'evolutionary indirect effects'). While specific tests for evolutionary indirect effects in the context of biological invasions are still rare, I present a series of examples linking evolutionary change and strong indirect effects to illustrate the possibility for important evolutionary indirect effects. I conclude by discussing open questions and experimental designs that could explicitly test the importance of evolutionary indirect effects to natural communities and the study of biological invasions.

Invasions lead to evolutionary change in both invaders and natives

Although the field of invasion biology was initially dominated by studies investigating the direct ecological effects of invaders, researchers have

Trait-Mediated Indirect Interactions: Ecological and Evolutionary Perspectives, eds. Takayuki Ohgushi, Oswald J. Schmitz and Robert D. Holt. Published by Cambridge University Press. © Cambridge University Press 2012.

begun to document evolutionary changes in both invaders and natives during the invasion process (for recent reviews see Lee 2002; Cox 2004; Lambrinos 2004; Strauss *et al.* 2006; Whitney and Gabler 2008). Evolutionary changes in invading species are expected because invaders occupy novel environments that may differ substantially from the habitats in which they evolved. As a result, different traits and genotypes may be favoured by natural selection in the invaded range than were favoured in the native range. These changes in trait values may influence how the invader interacts with native taxa in the invaded community. For example, the competitive dominance of some invaders may result from the post-invasion evolution of increased competitive ability (the EICA hypothesis, *sensu* Blossey and Notzold 1995). Selection for increased competitive ability in the invaded range occurs because of trade-offs between growth and defence (Herms and Mattson 1994). When invaders escape natural enemies, defence traits are no longer selected for and, instead, natural selection favours genotypes that allocate more resources to competition and growth traits than to defence (Blossey and Notzold 1995). EICA is one of the most tested evolutionary responses of invaders (reviewed in Bossdorf *et al.* 2005) and also is an example of an evolutionary indirect effect: the evolutionary change is a shift in allocation from defence to growth caused by (a lack of) natural enemies, and this increased growth may result in increased negative impacts on native competitors.

There are now many convincing examples of evolutionary changes in invaders, including phenological shifts (e.g., Wolfe *et al.* 2004; Dlugosch and Parker 2008; Allan and Pannell 2009), increased growth in the invaded range (Wolfe *et al.* 2004; Dlugosch and Parker 2008), decreased anti-herbivore defences (Siemann and Rogers 2003; Wolfe *et al.* 2004), decreased reliance on belowground mutualists (Seifert *et al.* 2009), altered morphology (Carroll *et al.* 2005) and changes in life history (reviewed in Barrett *et al.* 2008), such as increased selfing or clonal reproduction (Davis 2005; Xu *et al.* 2010). Such evolutionary changes may result from post-introduction evolution and adaptation to novel environmental conditions (e.g., Carroll *et al.* 2005) or from selection and/or genetic drift that occurs during the colonization phase (e.g., Bossdorf *et al.* 2008). Importantly for the thesis presented here: (1) evolutionary shifts in invaders may be common and (2) such genetically based changes in trait values may alter interactions with other species, including both mutualists (e.g., pollinators, dispersal agents, mycorrhizae, rhizobia) and antagonists (e.g., herbivores, competitors, predators).

Just as invaders may evolve during the invasion process, natives also may experience altered patterns of natural selection due to the strong ecological effects of invaders on native community members. In California grasslands, for example, an exotic plant (*Medicago polymorpha*) and an exotic herbivore (*Hypera brunneipennis*) interact to alter patterns of natural selection on the anti-herbivore defences (resistance and tolerance) of the native plant *Lotus wrangelianus* (Lau 2008). Given that both resistance and tolerance are genetically variable (Lau 2008), evolutionary changes in plant resistance and tolerance to herbivory are predicted to occur and could alter both the likelihood and fitness outcomes of the interaction between *Lotus* and *Hypera*. In addition, if these anti-herbivore defences are also effective against other herbivores that feed upon the native *Lotus*, evolutionary changes in *Lotus* defence traits could alter interactions with a suite of other insect herbivores in the community (e.g., *Bruchophagous* sp. seed predators, *Apion* sp. bud gallers, Lepidopteran folivores and aphids).

Several other studies also have documented evolutionary consequences of invasion for native community members. Common garden comparisons between invaded and uninvaded populations and/or studies of changes in trait values over time suggest that natives have responded to invasive species with evolutionary changes in traits such as gape size in gape-limited predators (Phillips and Shine 2004), resistance to novel alleochemicals (Callaway *et al.* 2005) and competitive ability (Leger 2008) (reviewed in Strauss *et al.* 2006). I am unaware, however, of any research explicitly linking these genetic changes in trait values to altered species interactions with other community members. Given recent studies documenting strong community and ecosystem-level responses to genetic variation in foundation species (reviewed in Hughes *et al.* 2008), such cascading effects may be likely.

Indirect ecological effects resulting from biological invasions

Exotic species can have a wide range of indirect effects on native community members (reviewed in White *et al.* 2006). These indirect effects result both from the numerical responses of interacting species and also from plastic or evolutionary changes in trait values that alter the per capita effects of one species on another (density-mediated indirect effects, DMIEs, driven by a numerical response or trait-mediated indirect effects, TMIEs). For example, an exotic herbivore may indirectly affect a native herbivore by consuming and reducing the abundance of a shared host plant (an indirect effect driven by a numerical response or change in abundance of the shared host plant). This same exotic herbivore could also indirectly affect a native herbivore if feeding by the exotic herbivore induces chemical defences in the shared host plant, which reduce the growth or fitness of the native herbivore (a trait-mediated indirect effect).

The importance of both DMIEs and TMIEs to ecological processes is well appreciated. Menge's (1995) seminal review suggests that on average 40% of changes in community structure resulting from experimental manipulations are driven by indirect effects. Menge's review focused largely on trophic interactions and on indirect effects driven by density responses; however, recent reviews illustrate that TMIEs also can have strong influences on populations, communities and ecosystems (Werner and Peacor 2003; Preisser *et al.* 2005). In fact, TMIEs are typically as strong or stronger than density-mediated indirect effects (DMIEs) (Werner and Peacor 2003; Preisser *et al.* 2005). In the three studies in Werner and Peacor's (2003) review that explicitly compared the magnitude of DMIEs versus TMIEs, TMIEs ranged from equal importance to twice as important as density-mediated effects. Similarly, a more formal meta-analysis showed that trait-mediated effects of predators had greater effects on prey demographics than the direct consumptive effects of predators and that cascading effects on prey resources were even stronger: density-mediated effects attenuated, but TMIEs on average explained 85% of the total predator effect on prey resource abundance (Preisser *et al.* 2005). With respect to plants, TMIEs can influence individual plant fitness (Goncalves-Souza *et al.* 2008), plant diversity and community composition (Wood *et al.* 2007; Gribben *et al.* 2009) and nutrient cycling (Stief and Holker 2006). Most of the empirical work on TMIEs has focused on plastic changes in trait values or behaviour. In this chapter, I limit discussion to evolutionary indirect effects.

Evolutionary indirect effects can result from two different processes (Fig. 12.1). The first mechanism focuses on the evolutionary outcomes of indirect ecological effects (i.e., a third species alters the outcome of interactions between two other species, potentially altering natural selection on one or both taxa (Abrams 2000; Abrams and Chen 2002; see Lau 2008 for an empirical example)). The second mechanism focuses on the ecological consequences that result from evolutionary changes in traits that affect species interactions (e.g., Abrams and Matsuda 1993; Matsuda *et al.* 1994; Abrams and Fung 2010). This type of evolutionary indirect effect occurs when selection imposed by one species causes genetic changes in a second intermediary species' traits that, in turn, influence interactions with other species in the community and is one type of trait-mediated indirect effect discussed in the preceding paragraph. These evolutionary indirect effects result because the trait has 'pleiotropic' effects and is relevant to interactions with multiple

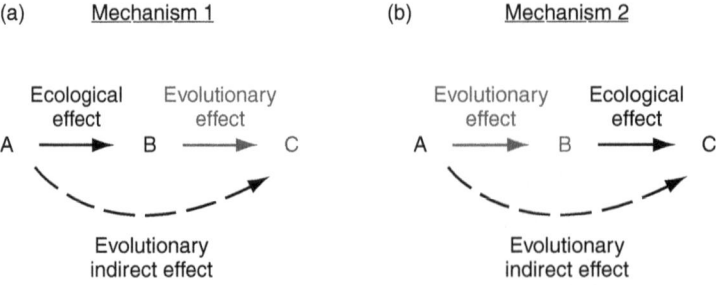

Figure 12.1 Mechanisms for evolutionary indirect effects. See colour plate section.

species. As described by Abrams (1995), 'a change in the population density of one species changes the mean trait value or trait distribution of a second species, which then affects the population growth rate and/or equilibrium density of a third species.' Although rarely investigated empirically, numerous theoretical studies, primarily focusing on the evolution of antipredator behaviour, have shown that this type of evolutionary indirect effect can alter the outcome of predator-mediated indirect interactions between prey (Abrams and Matsuda 1993), the outcome of intraguild predation (Abrams and Fung 2010) and trophic cascades (Abrams and Vos 2003) and community complexity and niche overlap (Matsuda et al. 1994). Many of these studies model adaptive antipredator behaviour; however, model outcomes are typically the same whether the trait change is a plastic or microevolutionary response (Abrams and Vos 2003).

This chapter primarily addresses the consequences of invasion-induced evolutionary changes in plant–herbivore interactions and, therefore, the second type of evolutionary indirect effect (Fig. 12.1b). In contrast to TMIEs that are due to plasticity, evolutionary indirect effects occur over multiple generations. As a result, evolutionary indirect effects may develop over longer timescales than most empirical experiments can detect and also may be less reversible than more labile, plastic or behavioural responses. Given the large impacts of TMIEs driven by plastic changes in trait values (Werner and Peacor 2003; Preisser et al. 2005), evolutionary changes in trait values of foundation species also may have large effects on communities and ecosystems (Holt 1994).

Putting it all together: examples illustrating potential evolutionary indirect effects

The study of introduced plants and herbivores may be a particularly tractable area for studying evolutionary indirect effects because: (1) escape from herbivores is thought to be a key factor promoting the success of plant invaders, (2) plant–herbivore interactions and the chemical, morphological and phenological traits that mediate these interactions are well studied and (3) plant defence guilds (i.e., the functional interdependence of plants within a community with respect to their herbivores, sensu Atsatt and Odowd 1976), herbivore suites (i.e., groups of insects that are attracted or repelled by similar plant traits, sensu Maddox and Root 1990) and large effects of plant defence expression on herbivore and predator community composition (see Utsumi 2011) are all common, suggesting that evolutionary changes in traits mediating plant–herbivore interactions will affect many community members. The following case studies illustrate the potential for strong evolutionary indirect effects resulting from plant or herbivore invasions and evolutionary changes in either invaders or native community members. Table 12.1 lists potential evolutionary indirect effects resulting from evolutionary responses of natives,

Table 12.1 *Evolutionary indirect effects of biological invasions resulting from the evolution of native plant or herbivore taxa in response to invasion. The table only includes examples where there is evidence documenting altered patterns of natural selection and/or evolutionary changes in native taxa in response to invasion and where effects of genetic changes in the native on other community members have been documented or can be hypothesized*

Indirect effect	Selective agent	Responding native taxa	Tertiary taxa	Comments	References
Herbivore–plant–herbivore	*Hypera brunneipennis* (invasive herbivore)	[a] *Lotus wrangelianus*, increased resistance to *Hypera*	[b] Increased abundance of *Aphidae* spp.	Reduced abundance of other insect herbivores could also result. Direction of effect depends on genetic correlations between resistance to different herbivores	[a]Lau 2008 [b]Lau 2012
Plant–plant–plant	*Bromus tectorum* (invasive plant)	[c] *Elymus multisetus*, increased competitive response	[?] Increased competitive effect on other natives?	Although effects on other natives have not yet been documented, *E. multisetus* genotypes from invaded sites had faster early growth compared to genotypes from uninvaded sites. Because size and growth rates are key determinants of competitive ability, increased competitive effects on co-occurring natives are possible.	[c]Leger 2008
Plant–herbivore–plant	*Koelreuteria elegans, Koelreuteria paniculata, Cardiospermum halicacabum* (invasive alternative host plants)	[d] *Jadera haematoloma,* altered proboscis length	[?] Reduced feeding on native host?	While host shifts in response to invasions of novel host plants have been documented, in many cases this has resulted in host race formation on introduced versus native hosts.[e] It is unknown whether attack rates on native hosts have been altered.	[d]Carroll and Boyd 1992 [e]Carroll et al. 1997

Table 12.2 *Evolutionary indirect effects of biological invasions resulting from the post-invasion evolution of invasive plant or herbivore taxa. The table only includes examples where there is evidence documenting altered patterns of natural selection and/or evolutionary changes in invasive taxa and where effects of genetic changes in the invader on other community members have been documented or can be hypothesized*

Indirect effect	Selective agent	Responding invasive taxa	Tertiary taxa	Comments	References
Herbivore–plant–herbivore	Escape from specialist natural enemies in invaded range	[a] *Lythrum salicaria*, size and competitive ability	? Reduced performance of native plant competitors	See discussion in main text for additional examples. Escape from enemies favours genotypes that allocate fewer resources to defence and more resources to growth, causing the evolution of increased competitive ability. Evolution of increased competitive ability in invasives could increase negative competitive effects on co-occurring natives.	[a]Blossey and Notzold 1995
Mutualist–plant–plant	Loss of mutualist (mycorrhizae) during invasion	[b] *Hypericum perforatum*, reduced dependence on mycorrhizae	? Increased competitive effect on natives (at least in high nutrient environments)	Studies on other taxa have demonstrated that plant populations that evolved reduced mycorrhizal dependence had increased growth in high nutrient environments. [c]This increased growth could translate into increased competitive effects on natives.	[b]Seifert et al. 2009 [c]Schultz et al. 2001

Table 12.2 (cont.)

Indirect effect	Selective agent	Responding invasive taxa	Tertiary taxa	Comments	References
			? Reduced (or increased) attractiveness to herbivores and/or herbivore performance	Association with mycorrhizae can both increase and decrease attractiveness to herbivores, depending on the mycorrhiza, plant and herbivore taxa studied.[d]	[d]Koricheva et al. 2009
			? Increased resources for pollinators	Mycorrhizae can increase flower number[e], nectar production[e], and pollen production[f], which could benefit consumers of floral resources.	[e]Gange and Smith 2005 [f]Poulton et al. 2002
Plant–plant–mycorrhizae	Competition with natives	[g,h] Alliaria petiolata, increased phytochemical production	[g,h] Phytochemicals negatively impact native competitors, because of negative effects on mycorrhizae	Interspecific competition selects for increased phytochemical production. In turn, increased production of allelopathic chemicals alters native plant–mycorrhizae interactions.	[g]Lankau 2010 [h]Lankau et al. 2009

and Table 12.2 provides examples of the types of evolutionary indirect effects that may result from evolutionary changes in the invader.

Evolution of increased competitive ability (EICA): a herbivore-plant-plant evolutionary indirect effect

There are two main components of an evolutionary indirect effect: (1) one species acts as a selective agent causing evolutionary change in a second species and (2) the resulting evolutionary response alters interactions with a third species (Fig. 12.1b). As discussed above, few studies in the biological invasions literature have investigated both components; however, the most convincing evidence for such effects comes from empirical tests of the EICA hypothesis. Although not explicitly stated in the original conception of EICA (Blossey and Notzold 1995), the EICA hypothesis involves three steps: (1) invasive plants escape from specialized natural enemies that are present in their native range, (2) the lack of enemies selectively favours invading genotypes that allocate fewer resources to defence and more resources to growth and competitive ability and (3) the evolution of increased competitive ability in invasive plants contributes to their competitive dominance over native plants. It is important to note, however, that EICA results from the removal of interacting species (specialist herbivores) rather than their presence.

Evidence for EICA is mixed (reviewed in Bossdorf et al. 2005). Several studies have documented cases where exotic plants and animals escape specialized natural enemies and receive less herbivory, disease or parasitism than natives (Torchin et al. 2003; Carpenter and Cappuccino 2005; Genton et al. 2005; Liu et al. 2007, reviewed in Mitchell and Power 2003 and Liu and Stiling 2006), and additional studies have demonstrated that this is especially true for the most invasive taxa (e.g., introduced plants characterized as 'invasive' experience less herbivory than introduced, non-invasive plants (Carpenter and Cappuccino 2005, see also Mitchell and Power 2003)). Invaders also accumulate new enemies in the invaded range, however, and the opposite pattern of increased damage to exotic taxa also has been observed (e.g., Agrawal and Kotanen 2003, reviewed in Colautti et al. 2004). Increased damage to exotics results when native generalist herbivores begin to feed on invaders that may be evolutionary naïve and lack defences against these novel herbivores (Parker et al. 2006a; Verhoeven et al. 2009).

Another suite of studies has shown that genotypes from invasive populations of a given plant taxa are larger in size (height, biomass, etc.) and, therefore, are likely more competitive than genotypes from comparable populations in the native range (see Bossdorf et al. 2005); however, several other studies have found little or no difference in size between populations in native versus invaded ranges (Genton et al. 2005; Williams et al. 2008). Although few studies investigate changes in both growth and defence, some studies do show a concordant reduction in defence associated with the increased size (Siemann and Rogers 2003). Other studies, however, show no difference (e.g., Genton *et al.* 2005) or even increased levels of defence in the invaded range (Cano *et al.* 2009), even when increases in size are observed (Ridenour *et al.* 2008).

Increases in an organism's size are expected to increase competitive ability (Goldberg and Werner 1983). However, one of the best studies to date found no differences between invasive and native populations in competitive effects, as measured by the effect of the invader on the growth of competitors, despite evidence for the evolution of increased size and vigour in the invasive range (He *et al.* 2009). This would be the key step in determining whether EICA is responsible for the invasiveness and negative impacts of exotic plant (or animal) taxa on native community members and would convincingly show the importance of this potential evolutionary indirect effect to determining outcomes and impacts of invasions.

Evolutionary loss of mutualism during invasion: a plant-mycorrhizae-herbivore (or pollinator or plant) evolutionary indirect effect

Just as invasive taxa may escape specialist natural enemies during the invasion process, invasives also might lose their mutualists. While mutualists are often believed to be more generalist than antagonists so that invasives may be able to easily acquire new mutualists in the invaded range, invaders engaged in specialized mutualism can be mutualist-limited (Richardson *et al.* 2000). There are several examples of outcrossing plant taxa, ranging from white clover to strangler figs, failing to become invasive until their coevolved pollinators also invaded (reviewed in Richardson *et al.* 2000). Likewise, a recent study (Parker *et al.* 2006b) suggests that a lack of compatible rhizobia (belowground resource mutualists that provide legumes with fixed nitrogen in exchange for photosynthate) may limit the range expansion of the invasive legume *Cytisus scoparius*.

The change in type or abundance of a mutualist during a biological invasion can alter patterns of natural selection. Seifert and co-authors (2009) have demonstrated that the invasive plant *Hypericum perforatum* is less likely to encounter compatible mycorrhizae (another belowground resource mutualist) in the invaded range. As a result, *H. perforatum* populations in the invaded range have evolved reduced dependence on mycorrhizae compared to populations in the native range (Seifert *et al.* 2009). Reduced mycorrhizal dependence may affect the spread of the invasive (e.g., it may be less able to invade low nutrient sites, but may be better able to invade high nutrient sites because it also escapes the costs of associating with mycorrhizae (Schultz *et al.* 2001)) and could also impact other organisms in the community (the evolutionary indirect effect). Colonization by mycorrhizae has been shown to influence plant susceptibility to herbivores and pathogens (Gange and West 1994; Gehring et al. 1997; Borowicz 2001), attractiveness to pollinators (Gange and Smith 2005; Wolfe et al. 2005) and competitive ability (Facelli et al. 2010). Reduced dependence on mycorrhizae, if accompanied by reduced colonization by mycorrhizae when they are present, could impact all of these interactions.

Similarly, when invaders leave behind specialized pollinators, selection may favour increased selfing because selfing may facilitate establishment when pollinators are rare (Barrett et al. 2008). Although rarely investigated empirically, selfers are overrepresented in both island flora (Barrett et al. 1996) and invasive taxa (Baker 1967; Rambuda and Johnson 2004). Results from empirical studies are mixed. Schueller (2004) and Colautti and coauthors (2010) do not detect evidence for the evolution of increased selfing in range expanding taxa, but other studies do find that populations at the edge of species ranges or in novel areas have evolved increased selfing capacity (e.g., Barrett et al. 1989). In these cases, the (lack of) coevolved pollinators likely contributes to the evolution of selfing (Barrett et al. 2008). The evolution of selfing has interesting implications for interactions with other community members including pathogens and herbivores. Increased selfing in invaders may result in increased (or in some cases, decreased) attack by natural enemies and could alter the outcome of the coevolutionary arms race between invasive plants and their pathogens and herbivores (Koslow and DeAngelis 2006; Johnson et al. 2009).

Evolution of increased anti-herbivore defences in natives: a herbivore–plant–herbivore evolutionary indirect effect

Evolutionary changes during the invasion process can occur in native as well as invading species (reviewed in Strauss et al. 2006). My prior work showed that an exotic annual plant (*Medicago*) and an exotic herbivore (*Hypera*) combined to alter the intensity of selection on the anti-herbivore defences of the native plant *Lotus wrangelianus* and that the strength of selection for increased resistance to *Hypera* was correlated with *Hypera* damage levels (Lau 2008). Two years of insecticide treatments that greatly reduced *Hypera* abundance and herbivory resulted in the evolution of reduced *Lotus* resistance: *Hypera* larvae reared on *Lotus* genotypes collected from insecticide-treated populations were three-fold larger than *Hypera* larvae reared on genotypes collected from no-insecticide control plots (Lau 2012). Given that resistance to multiple herbivores can be genetically correlated (Leimu and Koricheva 2006), the evolutionary responses to selection imposed by *Hypera* could influence the interaction between *Lotus* and other herbivores in the community and potentially could alter herbivore community composition. In fact, additional

experiments showed that *Lotus* genotypes collected from plots with historically high levels of *Hypera* damage actually supported higher abundances
of aphids (primarily *Acyrthosiphon pisum* and *Aphis craccivora*) in some environments than genotypes collected from plots that experienced less *Hypera* damage (either naturally or because of insecticide applications) (Lau 2012). This
evolutionary indirect effect on aphids was observed even though field insecticide treatments also were expected to reduce aphid densities, thereby potentially relaxing selection on resistance to aphids. In this case, however, *Hypera*
is by far the dominant herbivore in the system and may have been a stronger
selective agent than aphids. These findings suggest that invasions by exotic
herbivores, and the resulting strong selective impacts on native plants, may
cause evolutionary indirect effects that impact other members of the herbivore community.

Evolution of increased competitive ability in natives: a plant-plant-plant evolutionary indirect effect

Recent studies have shown that native taxa can evolve increased competitive-
ness against invaders (Callaway *et al.* 2005; Mealor and Hild 2007; Leger 2008).
In one example, genotypes of the native perennial bunch grass *Elymus multisetus* collected from habitats invaded by the annual invasive grass *Bromus
tectorum* had greater fitness relative to genotypes from uninvaded areas
when grown in competition with *B. tectorum* (Leger 2008). This increased
competitive ability appears to result from increased early growth rates.
Although not explicitly demonstrated in the *E. multisetus–B. tectorum* system,
this evolutionary change in early growth could alter competitive effects of *E.
multisetus* on other native competitors. Given that competition is often a
strong selective agent favouring increased size (Thomas and Bazzaz 1993;
Miller 1995; Dudley and Schmitt 1996; Donohue *et al.* 2000; Dorn *et al.* 2000),
this type of evolutionary indirect effect resulting from plant invasions may be
common.

Future directions

In this chapter I have highlighted some of the most convincing examples of
evolutionary indirect effects resulting from biological invasions and their
impact on surrounding community members. Few examples experimentally
assessed all parts of the process driving evolutionary indirect effects. A conclusive study would show that: (1) one species is a strong selective agent on a
second interacting species, (2) that evolutionary change has occurred in the
second species and (3) that the evolutionary changes in the second species
alter interactions with a third species. No published example to date has met
these three criteria. For example, in studies investigating EICA, even though
evolutionary changes in plant size (often a close proxy for competitive ability)

have been observed in some cases, to my knowledge, no study has been able convincingly to attribute the evolution of increased size to release from natural enemies, and no study has yet demonstrated that these evolutionary responses in the invader actually change the magnitude of competitive impacts on co-occurring native plants.

Although such experiments would be challenging, one potential way to investigate experimentally evolutionary indirect effects is through 'controlled natural selection' experiments (*sensu* Conner 2003). In these types of studies, replicated populations are grown (and evolve) under different environmental conditions (e.g., the presence or absence of an invasive herbivore) for multiple generations. Observed differences in population means between herbivore present/absent treatments provide evidence that the invasive herbivore is an agent of selection and that evolutionary responses have occurred. One could then use these evolved populations in a second set of experiments measuring the strength or outcomes of interactions with other community members. An evolutionary indirect effect could be inferred if the populations evolved in the presence versus absence of the invasive herbivore differ in how they interact with other community members.

While rigorously testing for evolutionary indirect effects in the context of biological invasions is the obvious next step, additional unanswered questions concerning the long-term consequences of evolutionary indirect effects abound: How do evolutionary indirect effects influence the long-term impacts of invasive species on native community members? What type/proportion of community members will be influenced by evolutionary indirect effects? And finally, how do evolutionary indirect effects alter the consequences of invasive species control or eradication? I discuss each of these questions below.

How do evolutionary indirect effects influence the long-term impacts of invasive species on native community members?

The influence of evolutionary indirect effects on long-term impacts of biological invasions likely depends on the type of selection and whether it is the native or invader that exhibits the evolutionary response. For example, in the EICA and loss of mutualist examples described above, the evolutionary response of the invader increased its competitive dominance over natives in at least some environments. Such evolutionary indirect effects may exacerbate the effects of invasion on native community members. In contrast, evolutionary responses in native species could reduce the immediate impacts of the invasion. For example, the evolution of increased resistance in the native *Lotus* described above could reduce herbivore growth and fitness. If these effects translate into impacts on herbivore population growth rates, then the abundance of the invading herbivore (and its associated impacts on the native *Lotus*) may be reduced. Still, long-term consequences of the invasion are likely to persist. The effects of the evolution of increased anti-herbivore defences on other members of the herbivore community may remain (discussed in following section), and strong selection will reduce genetic variation of the affected population (Fisher 1930). Biological invasions could decrease genetic diversity of native populations by decreasing population sizes and thereby increasing genetic drift, via their role as strong agents of natural selection, and/or because they may homogenize selection across communities (Benkman *et al.* 2008). This reduced genetic diversity may limit the ability of natives to respond to future environmental challenges. In addition, although the consequences of this reduced genetic variation in natives for other community members are unknown, several recent studies have shown that genetic diversity can influence a wide range of community and ecosystem properties, including herbivore abundance and diversity (Crutsinger *et al.* 2006; Johnson *et al.* 2006), invasion by additional plant taxa (Crutsinger *et al.* 2008) and ecosystem functions such as primary productivity (Crutsinger *et al.* 2006) and decomposition and nutrient flux (Schweitzer *et al.* 2005) (reviewed in Hughes *et al.* 2008).

What proportion of community members will be influenced by evolutionary indirect effects?

The number of species affected by evolutionary indirect effects resulting from biological invasions will no doubt be influenced by community properties, including connectance and species richness, as well as the genetic correlations between traits mediating different types of species interactions. In highly diverse and tightly integrated communities, any given taxon may interact with dozens of other species. For example, Richard Root's classic studies on goldenrods showed that a single host plant species can be attacked by over 138 herbivore species, although very few of these herbivores are abundant (<5%) and the abundance and composition of the herbivore community varies over space and time (Root and Cappuccino 1992). While changes in any given plant resistance trait are unlikely to impact all of these taxa, other studies suggest that goldenrod resistance to many herbivore taxa is genetically correlated (Maddox and Root 1990). Research on 17 herbivore species on goldenrod detected 'suites' of herbivore taxa that showed similar preference and avoidance behaviours for different goldenrod genotypes (Maddox and Root 1990). Numerous positive correlations between resistances to different herbivore taxa were detected, implying that strong selection imposed by any one herbivore for increased defences will likely impact other herbivores in the same suite. In this study, one-quarter of all possible genetic correlations between resistances to different herbivore taxa were statistically significant. If the subset of herbivores studied by Maddox and Root is representative of the entire goldenrod herbivore community, then we might predict changes in attack rates of 34 herbivore species (0.25×138 species in the community) if an exotic herbivore entered the community and altered the evolution of anti-herbivore resistance traits. And this thought experiment only considers the herbivores! The number of potentially affected species increases when one considers that changes in plant quality or resistance that impact herbivore abundance or growth also may impact higher trophic levels (predators and parasitoids) (Dickson and Whitham 1996; Johnson 2008; Utsumi *et al.* 2009) and that resistance-linked changes in litter quality may affect belowground detritivore and microbial communities (Wardle *et al.* 2004; Schweitzer *et al.* 2008). While the goldenrod example may be extreme, examples to date suggest the potential for evolutionary changes in key plant defence traits could impact many community members across different trophic levels.

How do evolutionary indirect effects alter the consequences of invasive species control or eradication?

Complete eradication of any invasive species is rare, and complete control is also infrequently achieved. Even in areas where control is achieved, however, natives do not always fully recover. Recent studies point to key changes in the soil microbial community that inhibit native species establishment and facilitate reinvasion by exotic species (e.g., Vogelsang and Bever 2009). However, when native community members evolve in ways that influence interactions with other species and/or change the abiotic environment, they may also alter their selective environment (i.e. 'niche construction' *sensu* Laland *et al.* 1999; Odling-Smee *et al.* 2003; see also Post and Palkovacs 2009). As a result, the community may not return to its initial state following the removal of the invader (the initial selective agent) because of residual differences between the pre- and post-invasion phenotypes of surviving native taxa. Because the evolutionary changes resulting from invasion might alter the selective environment, longer term effects may result that are not easily reversed by the removal of the initial selective agent. It is unknown to what extent these genetic changes influence the recovery of communities following invasive species control or the response of these communities to further anthropogenic stressors including additional invasions.

Conclusions

Altered species interactions are an inevitable consequence of biological invasions. Novel interactions automatically result when a new predator, prey, competitor or mutualist enters a community (direct effects) and also occur when the invader alters interactions between native community members (indirect effects). Although several studies have demonstrated how the novel species interactions that result from biological invasions alter patterns of natural selection and even cause evolutionary changes in invaders or natives (reviewed in Lee 2002; Cox 2004; Lambrinos 2004; Strauss *et al.* 2006; Whitney and Gabler 2008), few studies document both components of an evolutionary indirect effect resulting from a biological invasion: (1) evolutionary changes in response to invasion and (2) altered interactions with other community members as a result of the evolutionary change. However, given the potential evolutionary effects of invasion, the strength of indirect effects and the importance of species interactions to population, community and ecosystem process, evolutionary indirect effects are no doubt important to understanding and predicting the long-term, community-wide impacts of biological invasions.

Reciprocally, just as biological invasions have been used to advance understanding of basic topics in ecology and evolution, such as community assembly, niches and species distributions, and rapid adaptation (see Sax *et al.* 2007), the invasion of exotic plants and herbivores may be ideal systems for studying evolutionary indirect effects. Specifically, biological invasions provide an excellent forum for investigating how novel species interactions can influence patterns of natural selection and how evolutionary changes in ecologically relevant traits influence the outcomes of species interactions. Comparisons between native genotypes collected from naturally invaded and uninvaded areas combined with experimental manipulations (e.g., invasive species removals) and comparisons between invasives in their native versus introduced ranges can be used to investigate evolutionary indirect effects. The typically patchy nature of ongoing invasions facilitates comparisons between native communities that differ in the presence/absence of one or a few exotic species at much greater spatial scales than can be conducted in more typical manipulative plot-based experiments. Native genotypes could be collected from invaded versus uninvaded areas and used in experiments to investigate how evolutionary responses to interactions with the invader alter the outcome of interactions with other community members. Similarly, genotypes of invaders collected from native versus invaded ranges could be used in similar experiments that focus on how evolutionary changes in the invader alter species interactions. Invasive plants and herbivores, in particular, may be especially well-suited for such experiments given that their presence can be easily manipulated by hand-removal (plants) or chemical removal (both plants and herbivores), their rapid generation times and the well-documented importance of community composition and structure to the ecological and evolutionary outcomes of plant–herbivore interactions (reviewed in Rausher 1996; Agrawal *et al.* 2006; Poelman *et al.* 2008).

The ideas and examples presented here highlight the advances that can be gained by linking the fields of evolutionary ecology and invasion biology. While much effort in traditional evolutionary ecology has focused on studying the evolutionary consequences of selection, much less work has investigated the ecological consequences of evolutionary change (reviewed in Pelletier *et al.* 2009 and citations therein). This burgeoning field of eco-evolutionary dynamics may have much relevance to invasion biology. Even though the importance of evolutionary indirect effects to outcomes of invasion remains untested, the existing data suggest that these effects may be important to understanding and predicting the long-term outcomes of invasion.

Acknowledgements

I thank R. Holt, D. Moeller, E. Preisser, S. Strauss and one anonymous reviewer for their careful review of the ideas presented here. Research presented in this chapter was performed at the University of California Natural Reserve System's Donald and Sylvia McLaughlin Reserve and was funded by a National Science Foundation Dissertation Improvement Grant IBN-0206601 to S. Y. Strauss and J. A. Lau. Partial support during the writing of this chapter came from National Science Foundation grant DEB-0918963 to J.A.L.

Glossary

Eco-evolutionary feedback   Evolutionary (genetic) changes that alter ecological processes (e.g., population dynamics, community assembly, and nutrient cycling) (see Pelletier *et al.* 2009).

Evolution of increased competitive ability (EICA)   A hypothesis to explain the competitive dominance and success of invading plant species (Blossey and Notzold 1995). Invasive plants escape specialist natural enemies, resulting in the evolution of reduced resource allocation to defence and increased resource allocation to growth and competitive ability.

Herbivore suite   Groups of herbivore taxa that are attracted or repelled by similar plant traits (see Maddox and Root 1990).

Indirect effect   Indirect effects result when one species or resource alters the likelihood or intensity of interaction between two other community members. Examples of indirect effects include: apparent competition, keystone predation and indirect defence (see Menge 1995).

Density-mediated indirect effect   Indirect effects mediated through changes in the abundance of intermediary species. For example, herbivore species A reduces the abundance of host plant B, decreasing the food supply of herbivore species C, which feeds on the same host.

Trait-mediated indirect effect   Indirect effects mediated through changes in traits or behaviours of intermediary species. For example, herbivore species A feeds on host plant B, inducing host plant B to increase production of defensive chemicals, which then negatively impacts herbivore species C.

Evolutionary indirect effect   A special case of trait-mediated indirect effect: indirect effects mediated through evolutionary changes in traits or behaviours of intermediary species. For example, herbivore species A feeds on host plant B exerting selection for and causing the evolution of increased resistance, which then nega-tively impacts herbivore species C. An alternative definition is: evolutionary changes resulting from indirect interactions (see Abrams 2000; Abrams and Chen 2002). For example, plant species A increases herbivory on plant species B through apparent competition, causing selection for increased resistance in plant B.

Plant defence guild   A concept emphasizing the interconnectedness of plants within a community with regards to their herbivores. Neighbouring plants may increase or decrease herbivory on a focal plant through a variety of mechanisms (see Atsatt and O'Dowd 1976).

References

Abrams, P. A. (1995) Implications of dynamically variable traits for identifying, classifying, and measuring direct and indirect effects in ecological communities. *American Naturalist*, **146**, 112–134.

Abrams, P. A. (2000) Character shifts of prey species that share predators. *American Naturalist*, **156**, S45–S61.

Abrams, P. A. and Chen, X. (2002) The effect of competition between prey species on the evolution of their vulnerabilities to a shared predator. *Evolutionary Ecology Research*, **4**, 897–909.

Abrams, P. A. and Fung, S. R. (2010) The impact of adaptive defence on top-down and bottom-up effects in systems with intraguild predation. *Evolutionary Ecology Research*, **12**, 307–325.

Abrams, P. A. and Matsuda, H. (1993) Effects of adaptive predatory and anti-predator behaviour in a two-prey-one-predator system. *Evolutionary Ecology*, **7**, 312–326.

Abrams, P. A. and Vos, M. (2003) Adaptation, density dependence and the responses of trophic level abundances to mortality. *Evolutionary Ecology Research*, **5**, 1113–1132.

Agrawal, A. A. and Kotanen, P. M. (2003) Herbivores and the success of exotic plants: a phylogenetically controlled experiment. *Ecology Letters*, **6**, 712–715.

Agrawal, A. A., Lau, J. A. and Hambäck, P. A. (2006) Community heterogeneity and the evolution of interactions between plants and insect herbivores. *Quarterly Review of Biology*, **81**, 349–376.

Allan, E. and Pannell, J. R. (2009) Rapid divergence in physiological and life-history traits between northern and southern populations of the British introduced neo-species, *Senecio squalidus*. *Oikos*, **118**, 1053–1061.

Atsatt, P. R. and O'Dowd, D. J. (1976) Plant defense guilds. *Science*, **193**, 24–29.

Baker, H. G. (1967) Support for Baker's Law as a rule. *Evolution*, **21**, 853–856.

Barrett, S. C. H., Colautti, R. I. and Eckert, C. G. (2008) Plant reproductive systems and evolution during biological invasion. *Molecular Ecology*, **17**, 373–383.

Barrett, S. C. H., Emerson, B. and Mallet, J. (1996) The reproductive biology and genetics of island plants (and discussion). *Philosophical Transactions of the Royal Society of London, Series B*, **351**, 725–733.

Barrett, S. C. H., Morgan, M. T. and Husband, B. C. (1989) The dissolution of a complex genetic polymorphism: the evolution of self-fertilization in tristylous *Eichhornia paniculata* (Pontederiaceae). *Evolution*, **43**, 1398–1416.

Benkman, C. W., Siepielski, A. M. and Parchman, T. L. (2008) The local introduction of strongly interacting species and the loss of geographic variation in species and species interactions. *Molecular Ecology*, **17**, 395–404.

Blossey, B. and Notzold, R. (1995) Evolution of increased competitive ability in invasive nonindigenous plants: a hypothesis. *Journal of Ecology*, **83**, 887–889.

Borowicz, V. A. (2001) Do arbuscular mycorrhizal fungi alter plant-pathogen relations? *Ecology*, **82**, 3057–3068.

Bossdorf, O., Auge, H., Lafuma, L. *et al.* (2005) Phenotypic and genetic differentiation between native and introduced plant populations. *Oecologia*, **144**, 1–11.

Bossdorf, O., Lipowsky, A. and Prati, D. (2008) Selection of preadapted populations allowed *Senecio inaequidens* to invade Central Europe. *Diversity and Distributions*, **14**, 676–685.

Callaway, R. M., Ridenour, W. M., Laboski, T., Weir, T. and Vivanco, J. M. (2005) Natural selection for resistance to the allelopathic effects of invasive plants. *Journal of Ecology*, **93**, 576–583.

Cano, L., Escarre, J., Vrieling, K. and Sans, F. X. (2009) Palatability to a generalist herbivore, defence and growth of invasive and native *Senecio* species: testing the evolution of increased competitive ability hypothesis. *Oecologia*, **159**, 95–106.

Carpenter, D. and Cappuccino, N. (2005) Herbivory, time since introduction and the invasiveness of exotic plants. *Journal of Ecology*, **93**, 315–321.

Carroll, S. P. and Boyd, C. (1992) Host race radiation in the soapberry bug: natural history with the history. *Evolution*, **46**, 1052–1069.

Carroll, S. P., Dingle, H. and Klassen, S. P. (1997) Genetic differentiation of fitness-associated traits among rapidly evolving populations of the soapberry bug. *Evolution*, **51**, 1182–1188.

Carroll, S. P., Loye, J. E., Dingle, H. *et al.* (2005) And the beak shall inherit: evolution in response to invasion. *Ecology Letters*, **8**, 944–951.

Colautti, R. I., Ricciardi, A., Grigorovich, I. A. and MacIsaac, H. J. (2004) Is invasion success explained by the enemy release hypothesis? *Ecology Letters*, **7**, 721–733.

Colautti, R. I., White, N. A. and Barrett, S. C. H. (2010) Variation of self-incompatibility within invasive populations of purple loosestrife (*Lythrum salicaria* L.) from eastern North America. *International Journal of Plant Sciences*, **171**, 158–166.

Conner, J. K. (2003) Artificial selection: a powerful tool for ecologists. *Ecology*, **84**, 1650–1660.

Cox, G. W. (2004) *Alien Species and Evolution: The Evolutionary Ecology of Exotic Plants, Animals, Microbes, and Interacting Native Species*. Washington DC: Island Press.

Crutsinger, G. M., Collins, M. D., Fordyce, J. A. *et al.* (2006) Plant genotypic diversity predicts community structure and governs an ecosystem process. *Science*, **313**, 966–968.

Crutsinger, G. M., Souza, L. and Sanders, N. J. (2008) Intraspecific diversity and dominant genotypes resist plant invasions. *Ecology Letters*, **11**, 16–23.

Davis, H. G. (2005) *r*-Selected traits in an invasive population. *Evolutionary Ecology*, **19**, 255–274.

Dickson, L. L. and Whitham, T. G. (1996) Genetically based plant resistance traits affect arthropods, fungi, and birds. *Oecologia*, **106**, 400–406.

Dlugosch, K. M. and Parker, I. M. (2008) Invading populations of an ornamental shrub show rapid life history evolution despite genetic bottlenecks. *Ecology Letters*, **11**, 701–709.

Donohue, K., Messiqua, D., Pyle, E. H., Heschel, M. S. and Schmitt, J. (2000) Evidence of adaptive divergence in plasticity: density- and site-dependent selection on shade-avoidance responses in *Impatiens capensis*. *Evolution*, **54**, 1956–1968.

Dorn, L. A., Pyle, E. H. and Schmitt, J. (2000) Plasticity to light cues and resources in *Arabidopsis thaliana*: testing for adaptive value and costs. *Evolution*, **54**, 1982–1994.

Dudley, S. A. and Schmitt, J. (1996) Testing the adaptive plasticity hypothesis: density-dependent selection on manipulated stem length in *Impatiens capensis*. *American Naturalist*, **147**, 445–465.

Facelli, E., Smith, S. E., Facelli, J. M., Christophersen, H. M. and Smith, F. A. (2010) Underground friends or enemies: model plants help to unravel direct and indirect effects of arbuscular mycorrhizal fungi on plant competition. *New Phytologist*, **185**, 1050–1061.

Fisher, R. A. (1930) *The Genetical Theory of Natural Selection*. Oxford: Clarendon Press.

Gange, A. C. and Smith, A. K. (2005) Arbuscular mycorrhizal fungi influence visitation rates of pollinating insects. *Ecological Entomology*, **30**, 600–606.

Gange, A. C. and West, H. M. (1994) Interactions between arbuscular mycorrhizal fungi and foliar-feeding insects in *Plantago lanceolata* L. *New Phytologist*, **128**, 79–87.

Gehring, C. A., Cobb, N. S. and Whitman, T. G. (1997) Three-way interactions among ectomycorrhizal mutualists, scale insects, and resistant and susceptible pinyon pines. *American Naturalist*, **149**, 824–841.

Genton, B. J., Kotanen, P. M., Cheptou, P. O., Adolphe, C. and Shykoff, J. A. (2005) Enemy release but no evolutionary loss of defence in a plant invasion: an inter-continental reciprocal transplant experiment. *Oecologia*, **146**, 404–414.

Goldberg, D. E. and Werner, P. A. (1983) Equivalence of competitors in plant communities – A null hypothesis and field experimental approach. *American Journal of Botany*, **70**, 1098–1104.

Goncalves-Souza, T., Omena, P. M., Souza, J. C. and Romero, G. Q. (2008) Trait-mediated effects on flowers: artificial spiders deceive pollinators and decrease plant fitness. *Ecology*, **89**, 2407–2413.

Gribben, P. E., Byers, J. E., Clements, M. *et al.* (2009) Behavioural interactions between ecosystem engineers control community species richness. *Ecology Letters*, **12**, 1127–1136.

He, W., Feng, Y., Ridenour, W. M., *et al.* (2009) Novel weapons and invasion: biogeographic differences in the competitive effects of *Centaurea maculosa* and its root exudate (±)-catechin. *Oecologia*, **159**, 803–815.

Herms, D. A. and Mattson, W. J. (1992) The dilemma of plants: to grow or defend. *Quarterly Review of Biology*, **67**, 283–335.

Holt, R. D. (1994) Linking species and ecosystems: where's Darwin? In C. G. Jones and J. H. Lawton, eds., *Linking Species and Ecosystems*. New York: Chapman and Hall, pp. 273–279.

Hughes, A. R., Inouye, B. D., Johnson, M. T. J., Underwood, N. and Vellend, M. (2008) Ecological consequences of genetic diversity. *Ecology Letters*, **11**, 609–623.

Johnson, M. T. J. (2008) Bottom-up effects of plant genotype on aphids, ants, and predators. *Ecology*, **89**, 145–154.

Johnson, M. T. J., Lajeunesse, M. J. and Agrawal, A. A. (2006) Additive and interactive effects of plant genotypic diversity on arthropod communities and plant fitness. *Ecology Letters*, **9**, 24–34.

Johnson, M. T. J., Smith, S. D. and Rausher, M. D. (2009) Plant sex and the evolution of plant defenses against herbivores. *Proceedings of the National Academy of Sciences of the United States of America*, **106**, 18079–18084.

Koricheva, J., Gange, A. C. and Jones, T. (2009) Effects of mycorrhizal fungi on insect herbivores: a meta-analysis. *Ecology*, **90**, 2088–2097.

Koslow, J. M. and DeAngelis, D. L. (2006) Host mating system and the prevalence of disease in a plant population. *Proceedings of the Royal Society of London, Series B*, **273**, 1825–1831.

Laland, K. N., Odling-Smee, F. J. and Feldman, M. W. (1999) Evolutionary consequences of niche construction and their implications for ecology. *Proceedings of the National Academy of Sciences of the United States of America*, **96**, 10242–10247.

Lankau, R. (2010) Soil microbial communities alter allelopathic competition between *Alliaria petiolata* and a native species. *Biological Invasions*, **12**, 2059–2068.

Lankau, R. A., Nuzzo, V., Spyreas, G. and Davis, A. S. (2009) Evolutionary limits ameliorate the negative impact of an invasive plant. *Proceedings of the National*

Academy of Sciences of the United States of America, **106**, 15362–15367.

Lambrinos, J. G. (2004) How interactions between ecology and evolution influence contemporary invasion dynamics. *Ecology*, **85**, 2061–2070.

Lau, J. A. (2008) Beyond the ecological: biological invasions alter natural selection on a native plant species. *Ecology*, **89**, 1023–1031.

Lau, J. A. (2012) Evolutionary indirect effects of biological invasions. *Oecologia*, doi: 10.1007/s00442-012-2288-x.

Lee, C. E. (2002) Selection and physiological evolution during biological invasion events. *Integrative and Comparative Biology*, **42**, 1264–1264.

Leger, E. A. (2008) The adaptive value of remnant native plants in invaded communities: an example from the Great Basin. *Ecological Applications*, **18**, 1226–1235.

Leimu, R. and Koricheva, J. (2006) A meta-analysis of genetic correlations between plant resistances to multiple enemies. *American Naturalist*, **168**, E15–E37.

Liu, H. and Stiling, P. (2006) Testing the enemy release hypothesis: a review and meta-analysis. *Biological Invasions*, **8**, 1535–1545.

Liu, H., Stiling, P. and Pemberton, R. W. (2007) Does enemy release matter for invasive plants? Evidence from a comparison of insect herbivore damage among invasive, non-invasive and native congeners. *Biological Invasions*, **9**, 773–781.

Maddox, G. D. and Root, R. B. (1990) Structure of the encounter between goldenrod (*Solidago altissima*) and its diverse insect fauna. *Ecology*, **71**, 2115–2124.

Matsuda, H., Hori, M. and Abrams, P. A. (1994) Effects of predator-specific defence on community complexity. *Evolutionary Ecology*, **8**, 628–638.

Mealor, B. A. and Hild, A. L. (2007) Post-invasion evolution of native plant populations: a test of biological resilience. *Oikos*, **116**, 1493–1500.

Menge, B. A. (1995) Indirect effects in marine rocky intertidal interaction webs – Patterns and importance. *Ecological Monographs*, **65**, 21–74.

Miller, T. E. (1995) Evolution of *Brassica rapa* L (Cruciferae) populations in intra- and interspecific competition. *Evolution*, **49**, 1125–1133.

Mitchell, C. E. and Power, A. G. (2003) Release of invasive plants from fungal and viral pathogens. *Nature*, **421**, 625–627.

Odling-Smee, F. J., Laland, K. N. and Feldman, M. W. (2003) *Niche Construction: The Neglected Process in Evolution*. Princeton, NJ: Princeton University Press.

Parker, J. D., Burkepile, D. E. and Hay, M. E. (2006a) Opposing effects of native and exotic herbivores on plant invasions. *Science*, **311**, 1459–1461.

Parker, M. A., Malek, W. and Parker, I. M. (2006b) Growth of an invasive legume is symbiont limited in newly occupied habitats. *Diversity and Distributions*, **12**, 563–571.

Pelletier, F., Garant, D. and Hendry, A. P. (2009) Eco-evolutionary dynamics. *Philosophical Transactions of the Royal Society of London, Series B*, **364**, 1483–1489.

Phillips, B. L. and Shine, R. (2004) Adapting to an invasive species: toxic cane toads induce morphological change in Australian snakes. *Proceedings of the National Academy of Sciences of the United States of America*, **101**, 17150–17155.

Poelman, E. H., van Loon, J. J. A. and Dicke, M. (2008) Consequences of variation in plant defense for biodiversity at higher trophic levels. *Trends in Plant Science*, **13**, 534–541.

Post, D. M. and Palkovacs, E. P. (2009) Eco-evolutionary feedbacks in community and ecosystem ecology: interactions between the ecological theatre and the evolutionary play. *Philosophical Transactions of the Royal Society of London, Series B*, **364**, 1629–1640.

Poulton, J. L., Bryla, D., Koide, R. T. and Stephenson, A. G. (2002) Mycorrhizal infection and high soil phosphorus improve vegetative growth and the female and male functions in tomato. *New Phytologist*, **154**, 255–264.

Preisser, E. L., Bolnick D. I., and Benard, M. F. (2005) Scared to death? The effects of intimidation and consumption in predator–prey interactions. *Ecology*, **86**, 501–509.

Rambuda, T. D. and Johnson, S. D. (2004) Breeding systems of invasive alien plants in South Africa: does Baker's rule apply? *Diversity and Distributions*, **10**, 409–416.

Rausher, M. D. (1996) Genetic analysis of coevolution between plants and their natural enemies. *Trends in Genetics*, **12**, 212–217.

Richardson, D. M., Allsopp, N., D'Antonio, C. M., Milton, S. J. and Rejmanek, M. (2000) Plant invasions: the role of mutualisms. *Biological Reviews*, **75**, 65–93.

Ridenour, W. M., Vivanco, J. M., Feng, Y. L., Horiuchi, J. and Callaway, R. M. (2008) No evidence for trade-offs: *Centaurea* plants from America are better competitors and defenders. *Ecological Monographs*, **78**, 369–386.

Root, R. B. and Cappuccino, N. (1992) Patterns in population change and the organization of the insect community associated with goldenrod. *Ecological Monographs*, **62**, 393–420.

Sax, D. F., Stachowicz, J. J., Brown, J. H. *et al.* (2007) Ecological and evolutionary insights from species invasions. *Trends in Ecology and Evolution*, **22**, 465–471.

Schueller, S. K. (2004) Self-pollination in island and mainland populations of the introduced hummingbird-pollinated plant, *Nicotiana glauca* (Solanaceae). *American Journal of Botany*, **91**, 672–681.

Schultz, P. A., Miller, R. M., Jastrow, J. D., Rivetta, C. V. and Bever, J. D. (2001) Evidence of a mycorrhizal mechanism for the adaptation of *Andropogon gerardii* (Poaceae) to high- and low-nutrient prairies. *American Journal of Botany*, **88**, 1650–1656.

Schweitzer, J. A., Bailey, J. K., Hart, S. C. and Whitham, T. G. (2005) Nonadditive effects of mixing cottonwood genotypes on litter decomposition and nutrient dynamics. *Ecology*, **86**, 2834–2840.

Schweitzer, J. A., Madritch, M. D., Bailey, J. K. *et al.* (2008) From genes to ecosystems: the genetic basis of condensed tannins and their role in nutrient regulation in a *Populus* model system. *Ecosystems*, **11**, 1005–1020.

Seifert, E. K., Bever, J. D. and Maron, J. L. (2009) Evidence for the evolution of reduced mycorrhizal dependence during plant invasion. *Ecology*, **90**, 1055–1062.

Siemann, E. and Rogers, W. E. (2003) Increased competitive ability of an invasive tree may be limited by an invasive beetle. *Ecological Applications*, **13**, 1503–1507.

Stief, P. and Holker, F. (2006) Trait-mediated indirect effects of predatory fish on microbial mineralization in aquatic sediments. *Ecology*, **87**, 3152–3159.

Strauss, S. Y., Lau, J. A. and Carroll, S. P. (2006) Evolutionary responses of natives to introduced species: what do introductions tell us about natural communities? *Ecology Letters*, **9**, 354–371.

Thomas, S. C. and Bazzaz, F. A. (1993) The genetic component in plant size hierarchies: norms of reaction to density in a *Polygonum* species. *Ecological Monographs*, **63**, 231–249.

Torchin, M. E., Lafferty, K. D., Dobson, A. P., McKenzie, V. J. and Kuris, A. M. (2003) Introduced species and their missing parasites. *Nature*, **421**, 628–630.

Utsumi, S. (2011) Eco-evolutionary dynamics in herbivorous insect communities mediated by induced plant responses. *Population Ecology*, **53**, 23–34.

Utsumi, S., Nakamura, M. and Ohgushi, T. (2009) Community consequences of herbivore-induced bottom-up trophic cascades: the importance of resource heterogeneity. *Journal of Animal Ecology*, **78**, 953–963.

Verhoeven, K. J. F., Biere, A., Harvey, J. A. and van der Putten, W. H. (2009) Plant invaders and their novel natural enemies: who is naïve? *Ecology Letters*, **12**, 107–117.

Vogelsang, K. M. and Bever, J. D. (2009) Mycorrhizal densities decline in association with nonnative plants and contribute to plant invasion. *Ecology*, **90**, 399–407.

Wardle, D. A., Bardgett, R. D., Klironomos, J. N. et al. (2004) Ecological linkages between aboveground and belowground biota. *Science*, **304**, 1629–1633.

Werner, E. E. and Peacor, S. D. (2003) A review of trait-mediated indirect interactions in ecological communities. *Ecology*, **84**, 1083–1100.

White, M., Wilson, J. C. and Clarke, A. R. (2006) Biotic indirect effects: a neglected concept in invasion biology. *Diversity and Distributions*, **12**, 443–455.

Whitney, K. D. and Gabler, C. A. (2008) Rapid evolution in introduced species, 'invasive traits' and recipient communities: challenges for predicting invasive potential. *Diversity and Distributions*, **14**, 569–580.

Williams, J. L., Auge, H. and Maron, J. L. (2008) Different gardens, different results: native and introduced populations exhibit contrasting phenotypes across common gardens. *Oecologia*, **157**, 239–248.

Wolfe, B. E., Husband, B. C. and Klironomos, J. N. (2005) Effects of a belowground mutualism on an aboveground mutualism. *Ecology Letters*, **8**, 218–223.

Wolfe, L. M., Elzinga, J. A. and Biere, A. (2004) Increased susceptibility to enemies following introduction in the invasive plant *Silene latifolia*. *Ecology Letters*, **7**, 813–820.

Wood, C. L., Byers, J. E., Cottingham, K. L. et al. (2007) Parasites alter community structure. *Proceedings of the National Academy of Sciences of the United States of America*, **104**, 9335–9339.

Xu, C. Y., Julien, M. H., Fatemi, M. et al. (2010) Phenotypic divergence during the invasion of *Phyla canescens* in Australia and France: evidence for selection-driven evolution. *Ecology Letters*, **13**, 32–44.

CHAPTER THIRTEEN

Indirect evolutionary interactions in a multitrophic system

TIMOTHY P. CRAIG and JOANNE K. ITAMI
Department of Biology, University of Minnesota-Duluth

MICHAEL DIXON
US Fish and Wildlife Service

and

TERRY R. HAMS
Golder Associates, Saskatoon

Indirect evolutionary effects

Species interact both directly and indirectly through interactions mediated by a third species. Recently, it has been established that these indirect interactions have important effects on the structure of ecological communities and that they can produce more complex interaction webs than the food webs of direct trophic interactions (Ohgushi 2005, 2007). In this chapter we address the question: can indirectly interacting species separated in space or time influence each other's evolution? These indirect interactions can evolve if there is a chain reaction where one species exerts selection via a second species on a heritable trait of a third species (Fig. 13.1). If indirect interactions can evolve they may play an important role in community evolution. Measuring indirect interactions is more difficult than measuring direct evolutionary interactions. We will suggest strategies for measuring indirect evolutionary interactions using examples from the community centred on the gall-inducing fly *Eurosta solidaginis*, its host plants and natural enemies.

Goldenrod–herbivore–natural enemy interaction

Eurosta has formed three partially reproductively isolated populations on *Solidago altissima altissima*, *S. a. gilvocanescens and S. gigantea* which we refer to as the forest *altissima*, prairie *altissima* and forest *gigantea* host races, respectively. The forest *altissima* and forest *gigantea* host races occur sympatrically in the forest biome of North America (Waring *et al.* 1990; Craig *et al.* 1993, 1997, 2001, 2007a, b; Brown *et al.* 1996; Itami *et al.* 1998; Stireman *et al.* 2005; Horner *et al.* 2008) and the host races on *S. altissima gilvocanescens* and *S. gigantea* occur sympatrically in the prairie (Craig *et al.* 2007a; Craig and Itami 2011). Along the

Trait-Mediated Indirect Interactions: Ecological and Evolutionary Perspectives, eds. Takayuki Ohgushi, Oswald J. Schmitz and Robert D. Holt. Published by Cambridge University Press. © Cambridge University Press 2012.

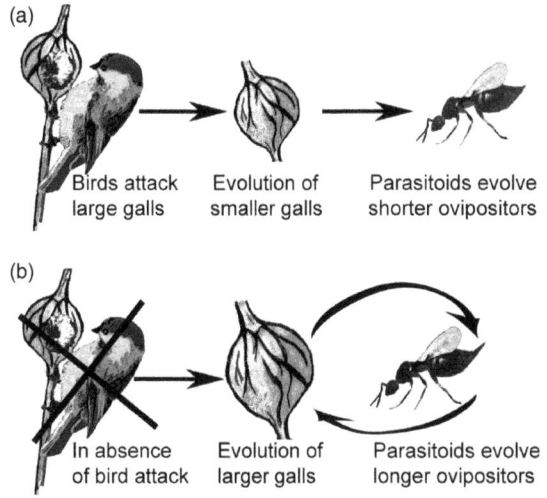

(a) Birds attack large galls → Evolution of smaller galls → Parasitoids evolve shorter ovipositors

(b) In absence of bird attack → Evolution of larger galls → Parasitoids evolve longer ovipositors

Figure 13.1 Indirect evolutionary interactions of three species: Species 1, black-capped chickadees, *Parus atricapillus*; Species 2, gall-inducing fly; *Eurosta solidaginis*; Species 3, a parasitoid, *Eurytoma gigantea*. The presence or absence of birds indirectly determines the ovipositor length. (a) Black-capped chickadees exert selection for smaller gall diameter. Smaller gall diameter evolves. Smaller gall selects for shorter ovipositor length. (b) In the absence of birds there is no selection against large galls. Parasitoids exert selection for larger galls size which in a positive feedback loop leads to longer ovipositor lengths.

forest–prairie biome border there is geographic mosaic in the distribution of *S. a. altissima* and *S. a. gilvocanescens* growing in close proximity to each other.

Eurosta emerge in early summer and mate and oviposit on the bud of the host plant (see Abrahamson and Weis 1997 for life history description). The larvae burrow into the stem of the host plant and induce the formation of a gall in which the larva develops, subsequently it is attacked by natural enemies and surviving larvae overwinter in the gall. The *Eurosta* larvae are attacked by a parasitoid, *Eurytoma gigantea*, which oviposits through the gall wall and into the *Eurosta* larvae. *Eurosta* also suffer mortality from an inquiline beetle *Mordellistena convicta* that oviposits on the surface of the gall and the larvae tunnel through the gall, feeding on gall tissue and *Eurosta* larvae. During the autumn and winter *Eurosta* larvae are attacked by two birds: downy woodpeckers, *Poecile pubescens*, and black-capped chickadees, *Parus atricapillus*.

Many indirect interactions could potentially occur among the *Solidago* species, the *Eurosta* host races and their natural enemies, but we will consider only three potential interactions in this chapter. First, that *Eurosta* mediate an indirect interaction between the natural enemies that attack it. Second, that *Solidago* species mediate indirect interactions between *Eurosta* host races. Third, that *Eurosta* host races mediate indirect interactions between the *Solidago* species.

Measuring indirect interactions
The measurement of selection in direct interactions has become common through the use of phenotypic selection analysis introduced by Lande and Arnold (1983) where selection on a trait is measured by regressing variation in that trait on mortality caused by another species. However, phenotypic selection analysis cannot be used in the analysis of indirect interactions

because species 1 does not directly induce any mortality in species 3 (Fig. 13.1). As alternatives to phenotypic selection analysis we suggest four complementary approaches to testing the indirect interaction hypothesis. First, structured equation models can be used to explore connections among pairs of direct interactions to determine whether when these interactions are coupled they have the potential to produce the evolution of indirect interactions. Second, the assumptions of the indirect selection hypothesis can be tested to determine if the evolution of an indirect interaction is possible. Third, geographical variation in the interactions of three species could be used to measure the impact of indirect interactions in producing geographic variation in species. Fourth, direct experimental tests of the hypothesis could be done by experimental exclusion or alteration of species in the interaction.

Structured equation modelling

Indirect evolutionary selection could be modelled by using structured equation modelling (SEM) to link two different structured equation models developed for two species hypothesized to have indirectly evolving traits. For example, the hypothesis that bird predation indirectly exerts selection for shorter *Eurytoma* ovipositors by selecting for decreased gall size could be analysed by linking together a structured SEM for selection on *Eurosta* gall size developed by Weis and Kapelinski (1994), and a SEM for *Eurytoma* ovipositor length. A hypothetical model is shown in Fig. 13.2 that includes gall size as well as other factors that exert selection on the evolution of *Eurytoma* ovipositors. For example, since longer ovipositor size is linked to larger body size (Craig *et al.* 2007a) constraints on body size such as vulnerability to predators and foraging ability will form an

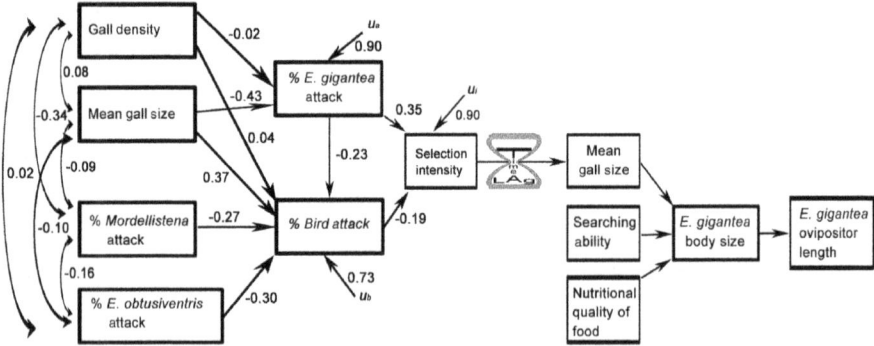

Figure 13.2 A structured equation model developed to test hypotheses about indirect evolutionary selection between two species. The first part of the model was developed to explore selective forces acting on the evolution of gall size in *Eurosta solidaginis* by Weis and Kapelenski (1994). The second part of the model shows the hypothetical forces selecting for ovipositor length in *Eurytoma gigantea*. The models operate separately in each generation, but because they produce changes in traits with a time lag they could be joined together in a simulation model where traits evolved through time.

indirect path influencing the evolution of ovipositor length. There will be a lag between the time when bird predation exerts selection on gall size, and the time when heritable changes in gall size exert selection on ovipositor length. A multi-generation simulation model where selection on gall size leads to changes of gall size linked to evolution of ovipositor length could be used to study the indirect evolutionary effects as the interaction of species one is varied.

Testing assumptions of indirect selection

For an indirect interaction to occur there must be a chain reaction where selection on one species results in the evolution of a trait in a second species that causes evolution of traits in a third species. For evolution to occur for each link in the chain there must be selection, and genetic variation in the trait under selection allowing a response to selection. Therefore, the following criteria must be met for indirect evolutionary interaction to occur (Fig. 13.1). (1) Species one exerts selection on a trait of species two. (2) There is genetic variation in the trait under selection in species two. (3) Species two exerts selection on a trait in species three. (4) There is heritable variation in species three. If species one alters the phenotype of species two within a generation then assumption (2) is not necessary.

Eurosta mediates indirect interactions among its natural enemies

The hypothesis that *E. solidaginis* mediates the indirect evolutionary interactions among its natural enemies has been strongly supported. Black-capped chicka-dees and downy woodpeckers exert selection on *Eurytoma gigantea* ovipositor length via selection on gall size in *Eurosta solidaginis*. All four assumptions for indirect selection are met for these interactions. First, birds exert selection on *Eurosta solidaginis* for smaller gall size by selectively preying on *Eurosta* larvae in large galls (Weis and Abrahamson 1986; Craig *et al.* 2007a). In contrast, *Mordellistena convicta* exerts selection on *Eurosta solidaginis* for increased gall size by causing higher mortality of *Eurosta* larvae in smaller galls (Craig *et al.* 2007a). Second, gall size is a heritable trait of *Eurosta solidaginis* (Weis and Abrahamson 1986; Craig and Itami 2011). Third, gall size exerts selection on *Eurytoma gigantea*. *Eurytoma* is limited by its ovipositor length in the size of galls it can attack (Weis *et al.* 1985) so that smaller gall size will select for shorter ovipositors. The fourth assumption is that there is genetic variation in ovipositor length in *Eurytoma gigantea*. Parasitoid populations in the prairie have significantly larger oviposi-tors than those in the forest (Craig *et al.* 2007a), and new analysis indicates that it is a genetically based trait (Craig *et al.* unpublished data).

 Eurytoma gigantea exerts selection on *E. solidaginis* for larger gall size and as result it potentially exerts selection on *M. convicta* traits. As stated above, the assumptions of selection on *E. solidaginis* gall size by *E. gigantea*, and of genetic variation in *E. solidaginis* gall size are supported. The third assumption that gall size exerts selection on *M. convicta* is also met because larger gall size increases

M. convicta fitness: larger galls produce larger beetles (Dixon *et al.* 2009). As a result *M. convicta* preference should be under selection by the parasitoid, the beetle should be selected to shift the range of gall sizes it prefers for oviposition upwards to increase larval performance. We have not tested the fourth assumption that there is genetic variation in beetle oviposition preference among gall sizes. Dixon *et al.* (2009) reported that both forest *M. convicta* populations, where galls are small, and prairie *M. convicta* populations, where galls are large, both preferred large galls. This suggests that gall size preference is a phenotypically plastic trait that shifts as the distribution of gall sizes changes, and this would not support the indirect evolution hypothesis.

Eurytoma gigantea and *M. convicta* could exert indirect selection on the bird's behaviour via selection on gall size. As stated previously the first two assumptions are met because: *Eurytoma* and *M. convicta* exert selection on *Eurosta solidaginis* for larger gall size, and gall size is a heritable trait of *Eurosta solidaginis*. The third assumption is also met, as there is selection for birds to shift their gall size preference. Attacking large galls leads to greater success because large galls are more likely to contain larvae (Weis and Abrahamson 1986; Craig *et al.* 2007a). As larger gall sizes evolve then to increase foraging efficiency birds should alter their preferences for the ranges of gall size attacked. We have not tested the hypothesis that gall size preference is a heritable trait in birds.

Web of indirect interactions among *Solidago* species and *Eurosta* host races

The *Eurosta* host races and their *Solidago* host plants are connected in a web of indirect interactions (Craig 2007) where there is the potential for each host race to influence indirectly the fitness of the other host race via the host plants, and each plant species to influence the other host plant species via the host races (Fig. 13.3). If each species in Fig. 13.3 exerts selection on the next one in the chain and there is genetic variation that permits a response to selection then indirect evolutionary interaction chains will be formed.

1. ***Eurosta* exert selection on *Solidago*.** *Eurosta* galls have a complex impact on *Solidago* fitness (summarized in Abrahamson and Weis 1997) including the reduction of the production of ramets and inflorescences (Hartnett and Abrahamson 1979), and as a result *Eurosta* have the potential to decrease the lifetime fitness of *Solidago* genets and exert selection for a change in traits that influence vulnerability to *Eurosta* (Abrahamson and Weis 1997).

2. **There is genetic variation among and within *Solidago* species on traits under selection by the host races.** There is genetic variation among *Solidago* genotypes that influences *Eurosta* fitness. *Solidago* genotypes vary in how preferred they are for oviposition by *Eurosta*, and how suitable they are for larval survival by the host races (Craig *et al.* 1999, 2000; Cronin *et al.* 2001; Anderson and Craig unpublished data). Genotypic variation in

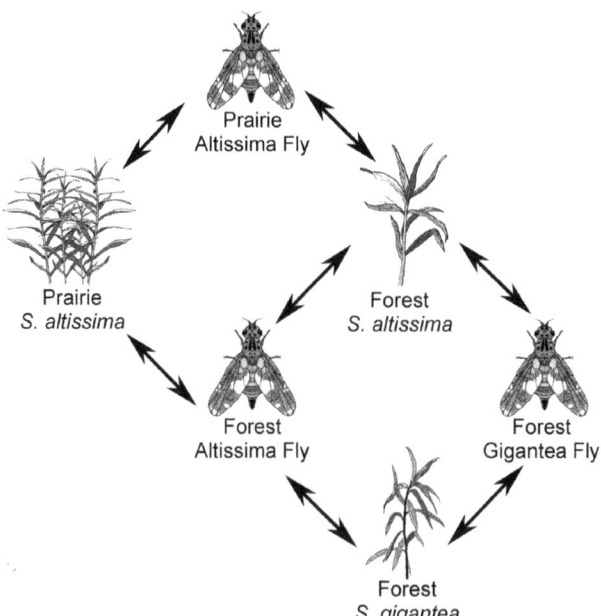

Figure 13.3 Indirect interactions among *Solidago* host plants and their associated *E. solidaginis* host races: the forest *Solidago a. altissima* and its associated forest *altissima* fly host race, the prairie *Solidago a. gilvocanescens* and its associated prairie *altissima* fly host race, the forest *Solidago gigantea* and its associated forest *gigantea* fly host race. Arrows represent selection of one organism by another, and the arrows are double headed because selection could be reciprocal. One example of chain of indirect interactions would be the forest *altissima* fly exerting selection on *S. a. altissima* which would exert selection on the forest *gigantea* race resulting in indirect selection of the forest *altissima* race on the forest *gigantea* race. The forest *gigantea* race could then exert selection on forest *S. gigantea* resulting indirect selection of *S. a. altissima* on *S. gigantea*.

plant resistance involves oviposition cues or deterrents, chemical defences against the larvae, the necrotic reaction that kills larvae and increased susceptibility to natural enemies (Abrahamson and Weis 1997).

3. **Solidago exerts selection on the host races**. *Solidago* host plants exert strong selection on *Eurosta* traits because larvae complete development in host plant tissue so that the plant exerts strong selection on their physiological adaptation and oviposition preference. Each host race is under strong selection to prefer their natal *Solidago* species for oviposition and mating sites and to avoid other *Solidago* species and subspecies. The forest *altissima* and forest *gigantea* host races have lower fitness on the alternate sympatric host species (Craig *et al.* 1997), as do the prairie and forest *altissima* host races (Craig and Itami 2011). Mating takes place on the host plant (Craig *et al.* 1993) and non-assortative mating results in low survival of hybrid offspring both between the forest *gigantea* and forest *altissima* host races (Craig *et al.* 1997) and between the prairie and forest

altissima races (Craig and Itami 2011). There also is high intraspecific variation among plant genotypes in larval mortality due to host plant effects that exerts selection for preference among plant genotypes within a *Solidago* species or subspecies (Craig *et al.* 1999, 2000; Cronin *et al.* 2001).

4. **There is genetic variation among *Eurosta* host races for traits under selection by the host plant**. In choice experiments individual forest *gigantea* and forest *altissima* flies (Craig *et al.* 1993, 2001) and individual prairie and forest *altissima* flies (Craig and Itami 2011) showed significant variation in their acceptance of the alternate host species. This indicates that there is genetic variation among the *Eurosta* host races in their response to oviposition cues and deterrents from the alternate host plant permitting the further evolution of host preference.

Since these four criteria are met it indicates that there is the potential for a host race to exert selection indirectly on another host race via selection on a host plant species. For example, if a host plant compound is an oviposition stimulant for the forest *altissima* host race then there will be selection on *S. a. altissima* to reduce or alter this compound to avoid attack. If the same compound is an oviposition deterrent for the *gigantea* host race then as this compound is changed in *S. a. altissima* there will be selection for a change in the ability to detect or respond to this compound by the *gigantea* host race. We do not know if there are such compounds in this interaction, but the same compound can act as an attractant to one herbivore and as deterrent to a closely related species (Honda *et al.* 2011).

Meeting the four criteria also creates the potential for indirect selection by one host plant species on traits of another host species via the host races. The flies encounter both their natal and alternative host plants when searching for mating and oviposition sites so a change in any host species could influence the evolution of all host races in an area. Using the same example of host plant cues we can hypothesize how *Eurosta* host races could indirectly mediate the interaction among *Solidago* species. The original host in the forest region was probably *S. a. altissima* (Brown *et al.* 1996) and so initially there would have been selection by *Eurosta* on *S. gigantea* to diverge in its oviposition cues from those of *S. a. altissima* to avoid attack. When the forest *gigantea* host race evolved it would have exerted selection on *S. gigantea* oviposition attractants and deterrents that would have indirectly exerted selection on the forest *altissima* host race. These interactions are connected in a feedback loop and so the indirect selection could continue indefinitely (Fig. 13.3). Since there are also parapatric host races in the prairie that interact with the forest host races this indirect interaction web could expand to include additional host plant species or subspecies and host races. The same kind of indirect interactions could evolve between distinct insect species as well as host races, and in the interaction of more distantly related plants than the *Solidago* species and subspecies.

Using geographic variation to test the indirect
selection hypothesis

Until time machines are made widely available for evolutionary studies, documenting the evolutionary changes in an indirect interaction webs through time will be difficult. Fortunately, geographic variation in the interactions among species can be used as a proxy for time travel as natural experiments to test the indirect selection hypothesis. Thompson (2005) in his geographic mosaic theory of coevolution suggested that coevolution can be detected by comparing traits of species in coevolutionary hotspots where species strongly interact with coevolutionary coldspots where species weakly interact.

Indirect evolutionary selection can also be detected by comparing traits of species in indirect evolution of hotspots and coldspots. The indirect selection that birds and mordellid beetles exert on *Eurytoma* ovipositors length via *Eurosta* gall size illustrates this. The indirect selection hypothesis predicts that *Eurytoma* ovipositor length will decrease in habitats with high bird predation which selects for smaller gall size (Fig. 13.1). Conversely, it predicts that *Eurytoma* ovipositor length will increase in habitats with high *M. convicta* density which will select for larger gall size. In regions without bird predation reciprocal selection would form a positive feedback loop with increasing gall size and increasing ovipositor length (Fig. 13.1). The geographic patterns support these predictions (Craig 2007; Craig *et al.* 2007a). *Eurytoma* ovipositors are longer in the prairie where gall size is large: the prairie is a coldspot for the bird–*Eurosta* interaction and a hotspot for the *E. gigantea*–*M. convicta*–*Eurosta* interaction. In contrast, *Eurytoma* ovipositors are shorter in the forest where gall size is smaller: the forest is a hotspot for the bird–*Eurosta* interaction, and a coldspot for the *M. convicta*–*Eurosta* interaction.

While the geographic pattern of ovipositor length is consistent with the indirect evolution hypothesis two points are not sufficient to calculate the shape of the interaction curve between mortality and ovipositor length necessary to conduct phenotypic selection analysis. We propose that the mean of a character from a range of sites be regressed on the mean of the mortality due to another species to measure the phenotypic selection curve. This would be analogous to calculating the mortality rates for different phenotypes within a population due to another organism. For example, parasitoid ovipositor lengths from a range of sites where bird densities varied could be regressed on the proportion of *Eurosta* mortality due to bird predation. The same could be done with *M. convicta* densities and parasitoid ovipositor length. A multivariate regression analysis using both of these variables would provide a better picture of the selection surface since mortality factors may interact. For example, birds prefer large galls, but they avoid galls with *M. convicta* which are predominately found in large galls (Hams and Craig unpublished data). Such a multivariate approach would provide an understanding of indirect selection not possible from analysis of selection in single populations.

There are also other examples of geographic hotspots and coldspots in the interactions of *Solidago* and the host races that could produce tests of the indirect evolutionary interaction hypothesis. *Solidago altissima* was introduced to Japan about 150 years ago, but *E. solidaginis* has not been introduced. Preliminary data indicate that the population in Japan originated from *S. a. altissima* (Craig unpublished data; Ando unpublished data). In the absence of *Eurosta* in Japan, selection for *S. altissima* resistance would be relaxed which we would predict would lead to lower resistance to the forest *altissima* host race. The indirect interaction hypothesis would also predict that this would alter its susceptibility to the prairie *altissima* race. Contrary to these predictions the Japanese *S. altissima* population was more resistant to forest *altissima* race flies than Minnesota populations of *S. a. altissima*, and it was as resistant to prairie *altissima* host races as Minnesota populations (Ando *et al.* unpublished data). This may indicate that Japanese *S. altissima* have evolved to adapt to their new environment, and that these changes have coincidently produced resistance to both host races. It also shows that indirect interactions may be influenced by a plethora of forces counteracting or obscuring their effects.

Reciprocal transplant experiments

Complete reciprocal transplant experiments, where the fitness of all geographically variable members of all interacting species (plants, flies and parasitoids) are grown in all possible combinations at each site, can measure the selection that each species exerts on the other members of the interaction (Nuismer and Gandon 2008). Such experiments can test the prediction of the indirect evolution hypothesis that populations from hotspots and coldspots in the interaction of natural enemies will have higher fitness in their local environments compared to foreign environments. For example, *Eurytoma* from the forest, a 'hotspot' for interaction of birds and parasitoids, will be predicted to have low fitness on *Eurosta* from the prairie, an interaction 'coldspot' where parasitoids are common but birds are rare (Fig. 13.1). Forest *Eurytoma* with short ovipositors that evolved where gall size is small are predicted to have low fitness when attacking galls from the prairie where galls are large because they will not be able to reach many of the larvae. Prairie *Eurytoma* in the forest may have reduced fitness for other reasons. Prairie *Eurytoma* have ovipositors long enough to reach all of the larvae in forest galls; however, we hypothesize that there are trade-offs that lower the fitness of individuals with long ovipositors and large body size where they are not needed. For example, smaller parasitoids could survive on smaller hosts which are found in smaller galls (Weis *et al.* 1989).

Geographic variation within the parapatric forest *altissima* and prairie *altissima* host races can also be used to measure indirect selection in *Solidago* and *Eurosta*. Hotspots would be near the biome border where both host plants and both host races interact, and coldspots would occur away from the biome border where only one host plant and one host race interact.

The indirect selection hypothesis would predict that this would result in differentiation between population hotspots and coldspots within each *Solidago* subspecies and host race. Due to similarities in indirect selection, populations from hotspots and coldspots would have greater similarities to each other than would be predicted on the basis of distance alone. Populations closer together would have similarity in traits due to gene flow, but the indirect selection hypothesis would predict that there would be higher levels of similarity among different populations in coldspots and among different hotspots due similar selection pressures.

Selection experiments

Selection experiments could test the indirect selection hypothesis by comparing the evolution of traits in species two and three (Fig. 13.1) in the presence and absence of species one. Reciprocal transplant experiments, such as those outlined above, can function as indirect selection experiments if they are maintained for a number of generations. In the example above, the indirect selection hypothesis predicts that the forest *Eurytoma* would evolve a longer ovipositor in the absence of birds. Larger gall size would evolve first followed by evolution of longer ovipositors. Rapid evolution in directly selected traits in reciprocal transplant experiments has been documented in several studies such as the classic studies of colour evolution in guppies in the presence and absence of predators (Endler 1980). Such experiments are feasible with organisms with rapid generation times, but would be impractical with the univoltine life histories of species in the *Eurosta* system. However, *Solidago* herbivores such as aphids with multivoltine life cycles could provide more suitable opportunities for selection experiments.

Conclusions

Our studies indicate that the *Solidago*–*Eurosta*–natural enemy community is evolving due to indirect evolutionary interactions, indicating that evolution of this component community is due to both direct and indirect interactions. A fundamental question in biology is whether there is community evolution. Do communities evolve as a result of the pairwise direct interaction of species or are there genetically integrated webs that link genetic variation in one species directly and indirectly to other organisms separated in space and time? Whitham *et al.* (2006) have proposed that the effects of individual genes of large effect in foundation species can propagate through an ecological community exerting influence on other species and producing community evolution. For example, they describe how genetically based variation in tannin levels of cottonwood trees can influence the arthropod community feeding on the trees, and the community decomposers that feed on the leaves. Many of these are direct effects but they could also produce indirect evolutionary effects such as we have discussed. An important future direction for research is to trace the effects of genes that could produce indirect evolutionary interactions and produce community evolution.

Do the indirect evolutionary interactions we have described extend from the component community to influence the larger ecological community evolution? Looking at a specific indirect evolutionary interaction web allows us to examine the potential for community evolution. We have discussed how bird preference for larvae in large galls exerts selection on *E. solidaginis* gall size and indirectly exerts selection on *E. gigantea* ovipositor length. If other prey exerted selection on black-capped chickadee and downy woodpecker foraging behaviour this could lengthen the indirect evolutionary interaction chain exerting selection on *E. gigantea* ovipositor length. For example, if black-capped chickadees were selected to prefer large galls as the result of selection for preference of large galls of other species then the indirect interaction chains would be four species in length with three links: other gall-forming insects exerting selection on black-capped chickadee gall preference which exerts selection on *Eurosta* gall size in turn exerting selection on *E. gigantea* ovipositor lengths. However, if any one of the criteria for the existence of the indirect interaction chains is not met then the indirect evolutionary forces will not propagate through the system. For example, if there is no heritable variation in black-capped chickadee gall size preference then the entire hypothesized extended web of potential indirect interactions pathways disappears. If the criteria for this link and for further links lengthening the chains are met then the web could increase exponentially as indirect interaction webs are even more complex than trophic webs, and they can contain a wide variety of kinds of linkages (Ohgushi 2007).

Measuring selection and community evolution in an extensive indirect interaction web is a daunting task. As the web expands the evolution of each species would potentially be influenced by many species linked in complex and variable ways with many similar and opposing forces exerting selection on a potentially large number of traits with variable time lags. The problems of accurately measuring selection and evolution are indicated by the history of coevolutionary studies with the difficulties of measuring reciprocal evolution between even two interacting species leading many to dismiss the importance of coevolution for a time (Thompson 1994, 2005). However, just because forces are complex and difficult to measure does not mean that they are unimportant, and the development of new techniques has led to rapid progress in the study of coevolution (Thompson 2005). This indicates that techniques that will allow understanding of selection in more intricate interactions are possible. Since studies of community genetics are reshaping our views of how communities evolve (Whitham *et al.* 2006), it is important that techniques permitting a quantitative understanding of indirect evolutionary interactions be developed. Our studies indicate that an understanding of indirect evolutionary interactions can be initiated by studying organisms with relatively tight host specificity and strong interactions such as the *Eurosta* interaction web. This and other herbivore–host plant–enemy interaction webs that are easily manipulated can be used as model systems for understanding indirect evolutionary interaction webs. In such systems a thorough understanding of each pairwise interaction can be obtained, and gradually each individual potential indirect link in a web can be tested to indicate the extent and limits of indirect interaction webs.

Acknowledgements

We thank R. D. Holt for his valuable comments on an earlier draft of this chapter. This study was partly supported by Ministry of Education, Culture, Sports, Science and Technology, by the JSPS Core-to-core program, a grant from NSF DEB 0949280 to T. P. Craig and J. K. Itami.

References

Abrahamson, W. G. and Weis, A. E. (1997) *Evolutionary Ecology Across Three Trophic Levels: Goldenrods, Gallmakers, and Natural Enemies*. Princeton, NJ: Princeton University Press.

Brown, J. M., Abrahamson, W. G. and Way, P. A. (1996) Mitochondrial DNA phylogeography of host races of the goldenrod ball gallmaker, *Eurosta solidaginis* (Diptera: Tephritidae). *Evolution*, **50**, 777–786.

Craig, T. P. (2007) Evolution of plant-mediated interactions among natural enemies. In T. Ohgushi, T. P. Craig and P. W. Price, eds., *Ecological Communities: Plant Mediation in Indirect Interaction Webs*. Cambridge: Cambridge University Press, pp. 331–353.

Craig, T. P. and Itami, J. K. (2011) Divergence of *Eurosta solidaginis* in response to host plant variation and natural enemies. *Evolution*, **65**, 802–17.

Craig, T. P., Horner, J. D. and Itami, J. K. (1997) Hybridization studies on the host races of *Eurosta solidaginis*: implications for sympatric speciation. *Evolution*, **51**, 1552–1560.

Craig, T. P., Horner J. D. and Itami, J. K. (2001) Genetics, experience and host–plant preference in *Eurosta solidaginis*: implications for host shifts and speciation. *Evolution*, **55**, 773–782.

Craig, T. P., Itami, J. K., Abrahamson, W. G. and Horner, J. D. (1993) Behavioral evidence for host-race formation in *Eurosta solidaginis*, *Evolution*, **47**, 1696–1710.

Craig, T. P., Itami, J. K., Abrahamson, W. G. and Horner, J. D. (1999) Oviposition preference and offspring performance of *Eurosta solidaginis* on genotypes of *Solidago altissima*. *Oikos*, **86**, 119–126.

Craig, T. P., Itami, J. K. and Craig, J. V. (2007a) Host plant genotype influences survival of hybrids between *Eurosta solidaginis* host races. *Evolution*, **61**, 2607–2613.

Craig, T. P., Itami, J. K. and Horner, J. D. (2007b) Geographic variation in the evolution and coevolution of a tritrophic interaction. *Evolution*, **61**, 1137–1152.

Craig, T. P., Itami, J. K., Schantz, C. *et al.* (2000) The influence of host plant variation and intraspecific competition on oviposition preference in the host races of *Eurosta solidaginis*. *Ecological Entomology*, **25**, 7–18.

Cronin, J.T., Abrahamson, W.G. and Craig, T.P. (2001) Temporal variation in herbivore host-plant preference and performance: constraints on host-plant adaptation. *Oikos*, **93**:312–320.

Dixon M.D., Craig, T.P. and Itami, J.K. (2009) The geographic mosaic of coevolution and the natural enemies of *Eurosta solidaginis*. *Evolutionary Ecology Research* **11**, 871–887.

Endler, J.A. (1980) Natural selection on color patterns in *Poecilia reticulata*. *Evolution*, **34**, 76–91.

Hartnett, D.C. and Abrahamson, W.G. (1979) The effects of stem gall insects on life history patterns in *Solidago canadensis*. *Ecology*, **60**, 910–917.

Honda, K., Omura, H., Chachin, M., Kawano, S. and Inoue, T. (2011) Synergistic or antagonistic modulation of oviposition response of two swallowtail butterflies, *Papilio maackii*, and *P. protnor*, to *Phellodendron amurense* by its constitutive prenylated flavonoid, phellamurin. *Journal of Chemical Ecology*, **37**, 575–581.

Itami, J.K., Craig, T.P. and Horner, J.D. (1998) Factors affecting gene flow between the host races of *Eurosta solidaginis*. In S. Mopper and S. Strauss, eds., *Genetic Structure and Local Adaptation in Natural Insect Populations: Effects of Ecology, Life History, Behavior*. New York: Chapman and Hall, pp. 375–404.

Lande, R. and Arnold, S.J. (1983) The measurement of selection on correlated characters. *Evolution*, **37**, 1210–1226.

Nuismer, S.L. and Gandon, S. (2008) Moving beyond common-garden and transplant designs: insight into the causes of local adaptation in species interactions. *American Naturalist*, **171**, 658–668.

Ohgushi, T. (2005) Indirect interaction webs: herbivore-induced effects through change in plants. *Annual Review of Ecology, Evolution, and Systematics*, **36**, 81–105.

Ohgushi, T. (2007) Nontrophic, indirect interaction webs of herbivorous insects. In T. Ohgushi, T.P. Craig and P.W. Price, eds., *Ecological Communities: Plant Mediation in Indirect Interaction Webs*. Cambridge: Cambridge University Press, pp. 221–245.

Stireman, J.O. III, Nason, J.K. and Heard, S.B. (2005) Host-associated genetic differentiation in phytophagous insects: general phenomenon or isolated exceptions? Evidence from a goldenrod-insect community, *Evolution* **59**, 2573–2587.

Thompson, J.N. (1994) *The Coevolutionary Process*. Chicago, IL: University of Chicago Press.

Thompson, J.N. (2005) *The Geographic Mosaic of Coevolution*, Chicago, IL: University of Chicago Press.

Waring, G.L., Abrahamson, W.G. and Howard, D.J. (1990) Genetic differentiation among host-associated populations of the gallmaker *Eurosta solidaginis* (Diptera: Tephritidae). *Evolution*, **44**, 1648–1655.

Weis, A.E. and Abrahamson, W.G. (1986) Evolution of host plant manipulation by gall makers: Ecological and genetic factors in the *Solidago-Eurosta* interaction. *American Naturalist*, **127**, 681–695.

Weis, A.E. and Kapelinski, A.D. (1994) Variable selection on *Eurosta's* gall size, II, A path analysis of the ecological factors behind selection. *Evolution*, **48**, 734–745.

Weis, A.E., Abrahamson, W.G. and McCrea, K.D. (1985) Host gall size and oviposition success by the parasitoid *Eurytoma gigantea*. *Ecological Entomology*, **10**, 341–348.

Weis, A.E., McCrea, K.D. and Abrahamson, W.G. (1989) Can there be an escalating arms race without coevolution? Implications from a host–parasitoid simulation. *Evolutionary Ecology*, **3**, 361–370.

Whitham, T.G., Bailey, J.K., Schweitzer, J.A. *et al.* (2006) A framework for community and ecosystem genetics: from genes to ecosystems. *Nature Reviews Genetics*, **7**, 510–523.

CHAPTER FOURTEEN

The role of trait-mediated indirect interactions for multispecies plant–animal mutualisms

REBECCA E. IRWIN

Biology Department, Dartmouth College

Introduction

Organisms experience myriad interactions with both antagonists and mutualists. There is widespread recognition that these multispecies interactions are not independent and that community membership can have important consequences for host fitness as well as patterns of natural selection (Strauss and Irwin 2004; Morris *et al.* 2007). For example, herbivore feeding can influence plant interactions with other herbivores or mutualists (e.g., pollinators), which can have subsequent effects on host plant fitness (Karban and Baldwin 1997; Mothershead and Marquis 2000; Strauss *et al.* 2001). Moreover, traits involved in these multispecies interactions can represent an adaptive compromise due to host interactions with antagonists and mutualists (Galen and Cuba 2001). While the effects of antagonist–antagonist and antagonist–mutualist interactions on hosts have received attention from both theoreticians and empiricists, the ecological and evolutionary consequences of host interactions with multiple mutualists have received less study (Hoeksema and Bruna 2000). That the study of multispecies mutualisms has lagged behind other suites of multispecies interactions is not surprising, given that mutualisms in general receive less study than competition and predation (Bronstein 1994). However, because many species interact with multiple mutualists either simultaneously or sequentially throughout their lifetimes (Janzen 1985) and mutualisms can have powerful impacts on host fitness and evolution (Bronstein *et al.* 2006), it is germane to ask how these multispecies mutualisms affect the ecology and evolution of their shared hosts (also see Stanton 2003).

Here I review evidence that host interactions with multiple mutualists may not be independent and the consequences of these multispecies mutualisms for host ecology and evolution. I focus on plant interactions with mutualists because plants make up a basal resource upon which other trophic interactions are structured; thus, it is important to understand how multispecies mutualisms affect plants. Moreover, notwithstanding notable examples from

Trait-Mediated Indirect Interactions: Ecological and Evolutionary Perspectives, eds. Takayuki Ohgushi, Oswald J. Schmitz and Robert D. Holt. Published by Cambridge University Press. © Cambridge University Press 2012.

other systems (e.g., Stachowicz and Whitlatch 2005), many of the examples of the effects of multispecies mutualisms are with plant hosts. It is important to note, however, that the general mechanisms that I outline for how multi-species mutualisms potentially affect host fitness can also be applied to other multispecies mutualisms. Throughout this chapter, when I refer to hosts shared by multiple mutualists, the host is the plant. First, I describe some of the mechanisms or pathways by which plant interactions with multiple mutualists may shape plant fitness. These pathways share similarities with the general mechanisms by which plants are affected by other types of multi-species interactions, such as those involving antagonists. Second, I review studies that have documented the consequences of multispecies mutualisms for plants. Third, I present unpublished data from a case study assessing whether plant interactions with one mutualist (pollinators) have the potential to alter subsequent plant interactions with another mutualist (seed dispersers), and discuss the potential ecological and evolutionary consequences of the interactions. I conclude with unanswered areas in need of further work.

Mechanisms by which multiple mutualists affect hosts

There are a number of mechanisms by which interactions with multiple mutu-alists may affect host fitness. First, trait-mediated indirect interactions (TMIIs) can occur when interactions with one mutualist cause a change in host mor-phology, physiology, biochemistry or behaviour (hereafter referred to simply as host traits) that affects host interactions with another mutualist (Abrams et al. 1996). For example, plant interactions with mycorrhizal fungi can influence flowering traits which alter subsequent plant–pollinator interactions (Wolfe et al. 2005). Second, density-mediated indirect interactions (DMIIs) can occur when one mutualist alters the population density of community members, which affects subsequent host–mutualist interactions (Abrams et al. 1996). For example, fruit and seed interactions with seed dispersers can affect seedling and plant density (Kalisz et al. 1999), and changes in plant and flower density can affect subsequent plant–pollinator interactions (Waites and Ågren 2004). TMIIs and DMIIs are not mutually exclusive and can occur simultaneously (e.g., Peacor and Werner 2001). They often involve host interactions with mutualists of different guilds who offer different benefits or rewards (i.e., pollen transfer versus nutrient acquisition), and it is primarily these forms of multispecies mutualisms that I review. In this chapter, I primarily focus on multispecies mutualisms mediated by TMIIs and DMIIs, and I do not extensively review exploitation competition between mutualists of the same guild (Palmer et al. 2003; Stanton 2003), long-lived hosts who interact with multiple mutualists of the same guild throughout their ontogeny (Palmer et al. 2010), mutualism net-works (Bascompte and Jordano 2007), direct interference competition between mutualists of the same or different guilds (Altshuler 1999; Ness 2006) or other types of interactions among mutualists that do not occur via TMIIs or DMIIs.

While at first glance one might envision that associating with multiple mutualists may be more beneficial to host fitness then associating with only one mutualist, that need not always be the case. The net outcome of these multispecies mutualisms may be positive or negative for host fitness, depending on the combined costs and benefits of all mutualistic interactions (Bronstein 2001). Positive effects could occur if a change in trait value (or density) caused by one mutualist increases the attraction of other mutualists to the host or if the change in trait value increases the beneficial effect that a subsequent mutualist has on the host. Because some mutualisms result in changes in plant physiology or morphology via increased nutrient uptake, they can increase resources required to attract or reward other mutualists. Alternatively, associating with more mutualists is not always beneficial for host fitness if changes in traits (or density) reduce host attractiveness to or efficiency of other mutualists that are essential for host growth, reproduction or survival. Associating with mutualist partners typically exacts a cost on host resources, but the benefits outweigh the costs (Bronstein 2001). However, if the costs of one mutualist affect trait expression and host attraction of other mutualists, or the ability of the host to reward the other mutualist (Hoeksema and Bruna 2000), the net outcome on host fitness may be negative. For example, mycorrhizal fungi can reduce the production of extra-floral nectaries used to attract ants that protect plants from herbivores (Laird and Addicott 2007), presumably due to carbon costs associated with maintaining the mycorrhizal mutualism. If the benefits of mycorrhizae are outweighed by the costs of reduced extra-floral nectaries (EFNs) and increased herbivory, then the net result of the interactions for host fitness may be negative.

One trend emerging from studies of multispecies plant–animal mutualisms, as with studies of other multispecies plant–animal interactions (Strauss and Irwin 2004), is that TMIIs are common (see *Multispecies mutualisms: trait-mediated indirect effects* below). Because plants can plastically respond to species interactions, there are a suite of physiological and biochemical mechanisms by which interactions with one mutualist can alter traits important to interactions with another mutualist. Moreover, because plants often must 'advertise' to attract mutualist partners and/or provide a reward, mutualisms that alter attractive or reward traits are particularly likely to alter subsequent mutualisms. Although TMIIs are emergent properties of species assemblages and difficult to predict in isolation (Wootton 1993), taking a trait-based approach and focusing on the plant traits or rewards that plant–mutualist interactions modify may provide initial predictive insight into how subsequent plant–mutualist interactions may be affected. Given that indirect species interactions are now considered key to understanding the dynamics of natural communities (McCann 2000), it is essential to redirect the study of mutualisms from a pairwise to a multispecies context.

Multispecies mutualisms: trait-mediated indirect effects

Here I review evidence that interactions with one mutualist may alter subsequent host–mutualist interactions (Table 14.1), including the traits involved and the consequences for the host plant. I categorize the studies based on the types of mutualisms interacting, adopting terminology of nutritional, protection and transport mutualisms (Boucher 1982). Nutritional (or energetic) mutualisms involve a transfer of energy or nutrients from one host to another in return for a reward. Protection mutualisms involve protection of the host from environmental variation or natural enemies in return for a reward, typically food or shelter. Transport mutualisms involve the movement of gametes or individuals in return for a reward. Each of these mutualisms typically involves a cost, such as the provision of food or photosynthate in

Table 14.1 *Examples of multiple mutualists interacting with the same host plants, and the potential trait-based mechanisms involved*

Types of mutualisms	Interactors	Trait change	Effect on second mutualism	Reference
Nutritional → transport	AMF → pollinators	Inflorescence size	Positive	Wolfe *et al.* 2005
Nutritional → protection	AMF → natural enemies (parasitoids)	Presumed plant size or plant volatiles	Negative, but variable	Gange *et al.* 2003
	AMF → natural enemies (ants)	EFNs	Not measured but presumed negative	Laird and Addicott 2007
Nutritional → nutritional	Endophytic fungi → AMF	Not measured	Negative	Mack and Rudgers 2008
	AMF → rhizobia	Not measured	Positive	Ferrari and Wall 2008
Protection → transport	Natural enemies (ants) → pollinators	Flower number	Positive	Schemske and Horvitz 1988
Protection → protection	Natural enemies → natural enemies	VOCs enhanced the production of EFNs	Presumed positive	Kost and Heil 2008
Transport → transport	Pollination → seed dispersal	Elaiosome size and density	Positive	Irwin (see case study)

Notes: AMF refers to arbuscular mycorrhizal fungi; EFN refers to extra-floral nectary; VOC refers to volatile organic compound.

return for a service (Bronstein 2001). I focus on mutualisms in which the benefits of the interaction outweigh the costs for host fitness, in at least a pairwise perspective, and do not cover abiotic environmental conditions which shift the interactions from mutualistic to parasitic. Categorizing mutualisms as nutritional, protection or transport provides natural trait-, or reward-, mediated routes of indirect interactions between very different mutualisms. However, it is important to note that mutualisms can also be categorized in other ways, such as consumer–resource interactions (Holland *et al.* 2005). Doing so was beyond the scope of this chapter but could yield additional mechanistic insight into how mutualisms interact.

Nutritional-transport mutualisms

By altering plant resource status, nutritional mutualists can modify plant interactions with other mutualists via the interface of plant traits. For example, mycorrhizal fungi are associated with the roots of many vascular plant species. In general, the fungi enhance plant uptake of inorganic nutrients (in particular phosphorous for arbuscular-mycorrhizal fungi (AMF)) in return for plant photosynthate (Allen *et al.* 1991; Smith and Read 1997). Because mycorrhizae alter the physiology and morphology of their host plants by affecting resource uptake and allocation, they have the potential to affect plant traits that can plastically respond to resource availability, such as floral traits, which can affect subsequent plant–pollinator transport mutualisms and plant reproduction (Wolfe *et al.* 2005).

The effects of AMF on pollination have been measured in studies manipulating AMF on individual plants and in field plots. At the individual plant level, the addition of mycorrhizae to the soil of sterile potted plants can increase pollinator visitation (Gange and Smith 2005; Wolfe *et al.* 2005), presumably due to increased floral attractive and reward traits, including inflorescence height, flower number, flower size, pollen production and nectar production and quality (e.g., Lau *et al.* 1995; Koide 2000; Poulton *et al.* 2002; Gange *et al.* 2005; Gange and Smith 2005; Varga and Kytöviita 2010). If plants are pollen-limited for seed set, increased pollinator visitation and pollen deposition due to AMF can result in increased seed production. For example, inoculation of the roots of fireweed, *Chamerion angustifolium* (Onagraceae), with AMF results in increased per plant pollinator visitation by bees and increased per cent seed set (Wolfe *et al.* 2005). The authors speculate that because AMF inoculation favoured shoot and inflorescence growth, the larger flowering inflorescence was more conspicuous to pollinators. The beneficial effects of AMF on pollination are not universal, however. Varga and Kytöviita (2010) found that AMF inoculation of *Geranium sylvaticum* (Geraniaceae) by *Glomus hoi* increased flower size and pollen production but did not affect flower visitation by bumble bees and reduced flower visitation by other nectar-feeding Hymenoptera. One mechanism that may be driving this result is that if AMF consume a significant portion of plant photosynthate, then less carbon may be available for nectar production (Laird and Addicott 2007).

One study has manipulated AMF in wild-growing plants and measured pollinator response. Cahill *et al.* (2008) manipulated AMF in field plots for 3 years using fungicide and found that AMF suppression caused a shift in the pollinator community from large-bodied to small-bodied bees and flies and a reduction in the average per stem pollinator visitation rate per plot. However, the mechanism by which AMF affected pollinator visitation was unexpected. Rather than finding direct effects of AMF on floral traits, the authors speculate that their result was driven by the indirect effects of AMF on the flowering plant assemblage and competitive interactions among neighbouring species, affecting plot-level floral display and subsequent pollinator behaviour. Whether or not the changes in pollinator assemblages affected plant fitness was not reported.

Two caveats about the studies to date on the effects of AMF on pollination and plant fitness are that, first, no studies have fully teased apart the direct effects of AMF on plant fitness versus the indirect effects mediated through changes in pollination. To do so would require experiments in which both AMF and pollination are manipulated in a factorial design. Second, studies have only provided correlative evidence of the trait-based mechanisms by which AMF affect pollinator visitation and pollination. Studies manipulating AMF, pollination and traits would provide additional insight into the mechanisms driving pollinator response. These two caveats, namely the lack of factorial experiments and manipulations of plant traits, plague the majority of studies described in this chapter.

I am not aware of studies that have manipulated nutritional mutualists and measured subsequent effects on seed dispersal transport mutualisms. However, given that nutritional mutualists can enhance nutrient uptake in plants, which could affect fruit quality or quantity, and given that frugivores can respond to the quality and abundance of available fruits (Jordano 1987; Sallabanks 1993; Russo 2003), the potential exists for nutritional mutualisms to alter seed dispersal via TMIIs and DMIIs.

All of the examples thus far have focused on how nutritional mutualisms affect transport mutualisms. However, the interaction could occur in the opposite direction, with transport mutualisms affecting nutritional mutualisms. For example, because flower pollination typically results in photosynthate supplied to the root zone being redirected towards the production of seeds, the amount of carbon a plant provides to mycorrhizae may decline, which could affect root–mycorrhizal associations (Zitzer *et al.* 1996; Heinemeyer *et al.* 2004).

Nutritional–protection mutualisms

Nutritional mutualisms may also affect protection mutualisms. Studies have documented that AMF can affect the attraction of natural enemies of hosts' herbivores as well as herbivore parasitism rates. For example, Guerrieri *et al.* (2004) found that tomato plants (*Lycopersicon esculentum*, Solanaceae) inoculated with AMF attracted significantly more parasitic wasps than control plants, but not more parasitoids. Conversely, Gange *et al.* (2003) found that AMF infection of *Leucanthemum vulgare* (Asteraceae) reduced parasitism of its herbivores, depending on the fungal associate. Although the mechanisms by which AMF impact the attraction of natural enemies are not well explored, one hypothesis is that mycorrhizae, which are known to affect plant secondary metabolism, alter the production of volatile organic compounds (VOCs) emitted by plants to signal to the natural enemies of herbivores (Leitner *et al.* 2010). Evidence is accumulating that AMF can affect VOC emissions (Rapparini *et al.* 2008; Fontana *et al.* 2009), but often by *decreasing* the emissions of certain VOCs, either because AMF deplete plant carbon supply that would be used for VOC emission or AMF infection modulates plant defence signalling pathways and subsequent VOC production (García-Garrido and Ocampo 2002). Thus, the nature of the interactions among nutritional and protection mutualisms for host fitness and the trait-based mechanisms involved remain unclear.

The effects of nutritional mutualists on VOC emission are not limited to mycorrhizal fungi. Endophytic fungi, although sometimes categorized as protection mutualists via the production of alkaloids that act in plant defence, can also act as nutritional mutualists by aiding plants in nutrient acquisition in return for a carbon reward (Rahman and Saiga 2005). Endophytic fungi can affect VOC emission (Jallow *et al.* 2008). For example, Jallow *et al.* (2008) found that fungal infection of tomato by the endophyte *Acremonium strictum* reduced the emissions of most volatile terpenes and sesquiterpenes while increasing the production of one, *trans-β*-caryophyllene, which is reported as a cue for entomopathogenic nematodes locating herbivorous larval prey (Rasmann *et al.* 2005).

Nutritional mutualisms may also alter protection mutualisms via changes in extra-floral nectaries (EFNs) used to attract ant predators of herbivores, but the combined outcome of these mutualisms may not benefit host fitness. For example, *Vicia faba* (Fabaceae) inoculated with AMF in growth chambers constructed significantly fewer EFNs (Laird and Addicott 2007). A reduction in nectar reward size or quality in other systems can lead to reduced plant attendance by ant bodyguards (Ness 2003; Rudgers 2004), and reduced plant fitness due to increased levels of herbivory (Ness *et al.* 2006). The authors speculate that the reduction in EFNs in the presence of AMF was caused by carbon costs associated with maintaining the AMF.

Nutritional–nutritional mutualisms

Host plants may also need to balance the costs and benefits of interacting with concurrent nutritional mutualists, such as mycorrhizal fungi (arbuscular mycorrhizal or ectomycorrhizal), endophytic fungi and nitrogen (N)-fixing bacteria. Studies report both antagonisms and synergisms among multiple nutritional mutualists (reviewed in Larimer *et al.* 2010). Here I limit discussion to interactions between AMF and endophytes and between AMF and N-fixing bacteria. Considering AMF and endophytic fungi, because of the spatial dis-association between where these two nutritional mutualists interact with the plant (typically roots versus leaves, respectively), understanding how the mutualists affect the shared plant interface is key to understanding mechanis-tically how they affect each other and the outcome for the host plant. In this case, having one mutualist typically does not increase the likelihood of inter-actions with or benefits derived from the other. Endophyte infection of plant tissue generally reduces AMF colonization (e.g., Guo *et al.* 1992; Mack and Rudgers 2008), although there are exceptions (Novas *et al.* 2005). Although the mechanism remains unknown, endophytes may reduce AMF colonization via TMIIs, such as chemical inhibition or increased plant nutrition thus reduc-ing the need for AMF, although exploitation competition for host-plant carbon cannot be ruled out. Because endophytes typically inhabit shoots, they may receive spatial priority in carbon allocation over mycorrhizae inhabiting roots, thus decreasing AMF root colonization. Interestingly, however, the effect of the two nutritional mutualists on one another can be asymmetric, with AMF having no effect on endophyte colonization (Mack and Rudgers 2008). The combined plant-fitness consequences of having both endophytes and AMF are variable, ranging from negative to mixed to neutral, depending on the study as well as the plant and fungal cultivars used (reviewed in Larimer *et al.* 2010).

In contrast, AMF and N-fixing bacteria commonly interact in a synergistic fashion (e.g., Weber *et al.* 2005; Yamanaka *et al.* 2005), with increased mycorrhizal colonization and nodule activity when plants are co-infected (reviewed in Larimer *et al.* 2010). These co-infections with both mutualists can result in additive as well as synergistic effects on host plants (Pacovsky *et al.* 1986; Ferrari and Wall 2008). For example, co-inoculations of black locust (*Robinia pseudoacacia*) with mycorrhizae and rhizobia resulted in higher plant shoot bio-mass, nodule biomass and N-fixation compared to inoculation with only one of the mutualists (Ferrari and Wall 2008). Co-inoculation may help plants meet both nitrogen and phosphorous demands in nutrient-limited environments.

Protection–transport mutualisms

Although the direct effects of protection mutualisms on transport mutual-isms have been well documented (such as ant bodyguards attracted to EFNs reducing pollinator visits to flowers or seed disperser visits to fruits, Altshuler 1999; Ness 2006), the indirect effects among these mutualisms via TMIIs or

DMIIs have received limited study. However, given well-known effects of herbivore damage on pollination via changes in floral and nectar traits (e.g., Strauss 1997; Mothershead and Marquis 2000), protection mutualisms that reduce herbivory on plants may increase pollination relative to unprotected plants, assuming the bodyguards do not directly interfere with pollinators. Thus, the benefits to plant fitness of protection mutualisms may not only be due to a reduction in herbivory but also due to an increase in pollination (Schemske and Horvitz 1988). In a similar vein, protection mutualisms may also increase seed dispersal through TMIIs or DMIIs if protection results in increased fruit quality (i.e., fruit size, sugar content, etc.) or density, and many frugivorous seed dispersers are known to respond to the quality or quantity of fruits (see references in *Nutitional–transport mutualisms*).

Protection-protection mutualisms

Many plants have multiple lines of defence against herbivores, including multiple traits that confer protection mutualisms, such as EFNs that attract ant bodyguards and VOCs that attract flying predators and parasitoids of herbivores (Arimura *et al.* 2005; Heil 2008). Just like other multispecies mutualisms, protection mutualisms may also interact, and TMIIs may play a role. For example, protection from herbivory by ant bodyguards patrolling EFNs may reduce herbivore-induced VOCs and subsequent protection mutualism via VOC emission. Or, VOC-enhanced protection could increase ant-patrolling bodyguards via increased plant quality and EFN secretion.

One exemplary study that has examined the defensive role of two simultaneous protection mutualisms is by Kost and Heil (2008) who studied the simultaneous impacts of two indirect plant defences, EFNs and VOCs, in a field study of lima bean (*Phaseolus lunatus*, Fabaceae). They found that application of a VOC mixture to plants induced secretion of extra-floral nectar (also see Choh *et al.* 2006; Heil and Kost 2006; Kost and Heil 2006), and other studies suggest that EFNs and VOCs share portions of the same biochemical signalling pathway (Heil *et al.* 2001). However, under the environmental conditions of their study, Kost and Heil (2008) found that the presence of extra-floral nectar was more important for plant defence against herbivores than was the emission of VOCs, and having both EFNs and VOCs did not benefit plant fitness beyond having only EFNs. How both lines of defence interact when ants are not the most common natural enemies, as was the case in Kost and Heil (2008), deserves further investigation.

Transport-transport mutualisms

Patterns of host use by one transport mutualist can alter patterns of host-use by others. A number of examples come from interactions among pollinators, in which interactions with one pollinator modify traits important to subsequent floral attraction of other pollinators. In some cases, visitation by one pollinator can make flowers less attractive to other pollinators. For example, many insect pollinators leave behind scent marks on flowers or cause floral colour changes, making flowers less attractive to subsequent pollinators (Gori 1983; Stout and Goulson 2001). Alternatively, visitation by one pollinator can encourage visitation by others, especially in cases where visitation stimulates nectar production (Gill 1988). In a similar vein, multiple seed dispersers can interact through the interface of fruit and seed traits. Diplochory is seed dispersal by a sequence of two or more phases or dispersers (Vander Wall and Longland 2004). For example, seeds of the Malagasy tree *Commiphora guillanumini* (Burseraceae) experience primary dispersal by parrots and secondary dispersal by ants (Böhning-Gaese *et al.* 1999). Primary dispersal could affect the likelihood of secondary dispersal, for example by removal of the fruit and deposition of seeds in faeces.

Pollination may also affect the likelihood or intensity of seed-dispersal transport mutualisms, although to my knowledge, the experimental links have not been assessed. Given that pollination and seed dispersal are temporally separated, their interaction may be linked via pollination-mediated changes in fruit quality or density, with subsequent effects on the abundance or activity of seed dispersers. For example, pollination can affect the quantity and quality of fleshy fruits produced (Gonzalez *et al.* 1998; Dimou *et al.* 2008), and seed dispersers can respond to fruit quantity and quality, providing suggestive evidence that pollination and seed dispersal transport mutualisms may not be independent (also see Strauss and Irwin 2004). The potential also exists for seed dispersal to affect pollination via changes in the density or quality of the subsequent plants produced by the dispersed seeds.

Case study: interactions between pollination and seed dispersal mutualisms

Although evidence is mounting that mutualisms interact, we remain relatively unaware of how pollination mutualisms alter the likelihood and intensity of seed dispersal mutualisms (see *Transport–transport mutualisms* above), and the ecological and evolutionary consequences of their interactions. Pollination and seed dispersal have been exemplar mutualisms in our understanding of coevolutionary relationships between plants and animals and the ecological and evolutionary consequences of mutualisms (Bronstein *et al.* 2006). However, understanding if and how these mutualisms interact is key to understanding host plant fitness, patterns of spatial relatedness, and evolutionary response to selection on floral traits. In particular, evolutionary response to pollinator-mediated selection on floral traits may be masked or magnified depending on if and how pollination alters patterns of seed dispersal. Here I start to fill this gap by examining how pollination quality and quantity alter the potential for seed dispersal through TMIIs and DMIIs.

Methods

I studied the long-lived, perennial, understory herb *Trillium erectum* (red trillium, Trilliaceae), common in northeastern forests in North America, in Grafton County, New Hampshire, USA. Plants have a single stem emerging from a tuber-like rhizome (Davis 1981). Each stem produces a single, red flower with three petals; flowers are nectarless and produce a fetid odour attractive to primarily dipteran pollinators (Irwin 2000). *Trillium erectum* has a mixed mating system (Broyles *et al.* 1997; Sage *et al.* 2001). Pollen limitation of fruit and seed set is variable among years and sites (Irwin 2000). The seeds of *T. erectum* produce elaiosomes (lipid-rich appendages) that are attractive to ant (and other less common) seed dispersers (Gunther and Lanza 1989).

I predicted that pollination could alter the abundance or activity of seed dispersers through both TMIIs and DMIIs. Pollination could alter seed dispersal via TMIIs if pollen quality or quantity alters elaiosome traits that affect the attractiveness of seeds to seed-dispersing ants. In addition, pollination could alter seed dispersal through DMIIs if pollen quality or quantity affects the number of seeds produced, and if seed number affects the attractiveness of seeds to seed-dispersing ants. These two hypotheses are not mutually exclusive. I tested these hypotheses using hand-pollination experiments, measurements of seed number and traits, estimates of seed removal and information from the primary literature.

First, I used a hand-pollination experiment to ask whether pollen quality (selfed versus outcrossed pollination) or quantity (supplemental versus open pollination) affected seed number and elaiosome traits (size). In each of three populations in 2005 and 2006, I chose 45 plants and randomly assigned them to one of three treatments: (1) self pollination, (2) supplemental hand pollination and (3) open pollination. I used different plants in 2005 and 2006. Plants in the self-pollination treatment were bagged with bridal veil. Upon flowering and anther dehiscence, I hand pollinated flowers with self-pollen. In a prior study, I found that bagging flowers ensures that no outcrossed pollen is deposited onto stigmas (Irwin 2000). Bagging flowers in the self-pollination treatment had no direct effect, per se, on fruit or seed set or elaiosome size (data not shown). In the supplemental hand-pollination treatment, flowers were accessible to pollinators and I provided supplemental pollen to receptive stigmas using a camel-hair paintbrush from a mixture of pollen of at least five *T. erectum* growing within 30 m of the treatment plants. In the open-pollination treatment, flowers were accessible to pollinators and handled to control for flower handling in the self-pollination and supplemental-pollination treatments.

In each year, pollination treatments were performed in May, and whether or not flowers made seed-bearing fruits was recorded in July. If a plant made a fruit, I collected it and counted the number of seeds. In addition, in 2005 I recorded the weight of each seed and elaiosome as an estimate of elaiosome quality to the nearest 0.01 mg. I estimated elaiosome quality as absolute elaiosome weight and relative elaiosome weight (elaiosome divided by seed weight). Relative elaiosome weight provides an estimate of the net reward for the dispersing agent (i.e., reward size relative to the size of the seed that must be moved, Gunther and Lanza 1989). Testing whether pollination affected seed dispersal through changes in elaiosome lipid chemistry was beyond the scope of this study, but could be addressed in future research.

To test whether pollination has the potential to affect seed dispersal through DMIIs, I tested whether pollen quality or quantity affected seed number using an analysis of variance (ANOVA) with pollination treatment (self, supplemental pollination and open pollination), site, year and their interactions. Non-significant interactions were removed from the final analysis (here and below). I tested whether pollination treatment, site, year and their interactions affected the probability of fruit set using nominal logistic regression. To evaluate whether pollination has the potential to affect seed dispersal through TMIIs, I tested whether pollination quality or quantity affected absolute and relative elaiosome weight using ANOVAs with pollination treatment, site and their interaction.

I then asked whether variation in elaiosome weight and seed density affected seed dispersal. To understand whether ant seed dispersers respond to variation in elaiosome weight, I used information from the literature on prior studies with *Trillium*. To understand whether ant seed dispersers responded to variation in seed density, I used a field experiment. I used six seed-density treatments, representing a range of seed densities found in the pollination treatments (see *Results and discussion*). I used seed treatments of 2, 5, 10, 20, 30 and 60 seeds. I conducted this study using two trials on 22 July and 29 July 2009, with 10 replicates per seed density per trial. I collected seeds from one site used in the pollination study, mixed the seeds, counted the seed treatments and placed them back in the same site approximately 1 m apart along transects, with treatment assigned to transect location at random. Transects were run through areas with fruiting *T. erectum* so that ant seed dispersers would be present in the area. The seeds were placed inside wire cages to deter rodent and other seed predators, but the cage holes were large enough to allow ants to access and remove seeds. Ants were observed foraging in the cages in preliminary trials in 2007, so I assumed that all seeds were removed by ant seed dispersers in these experimental trials although one weakness of this study is that I did not observe ant seed dispersers directly. Seeds were placed in the field for at least 7 hours, and the number of seeds removed was recorded. I assumed that any seeds removed were dispersed; following seed fate post-dispersal was beyond the scope of this study but should be addressed in future work. I first asked whether seed density affected the probability of at least one seed being dispersed using logistic regression with seed density, trial and their interaction as factors. This analysis assesses the attractiveness of the seed treatments to ant dispersers. I then asked whether seed density affected the number and proportion of seeds removed (square-root and arcsine-square root transformed) using analyses of covariance (ANCOVAs) with seed density, trial and their interaction as factors. In the analyses, seed density was considered a continuous variable.

Results and discussion

I found that both pollination quality and quantity affected seed production ($F_{2,213} = 38.01$, $P < 0.0001$). Plants in the self-pollination treatment produced 10-times fewer seeds than plants in the open-pollination treatment. In addition, plants in the supplemental-pollination treatment produced over 30% more seeds then plants in the open-pollination treatment (Fig. 14.1). Seed set varied between years ($F_{1,213} = 19.27$, $P < 0.0001$), and I found a significant treatment × year interaction ($F_{2,213} = 4.00$, $P = 0.02$), primarily driven by a stronger effect of open and supplemental pollination on seed set in 2005 than 2006, but no difference in seed set in the self-pollination treatment between the two years. I also found significant site × year ($F_{2,213} = 6.12$, $P = 0.003$) and treatment × site × year interactions ($F_{4,213} = 2.85$, $P = 0.03$). For probability of fruit set, I found that pollination treatment significantly

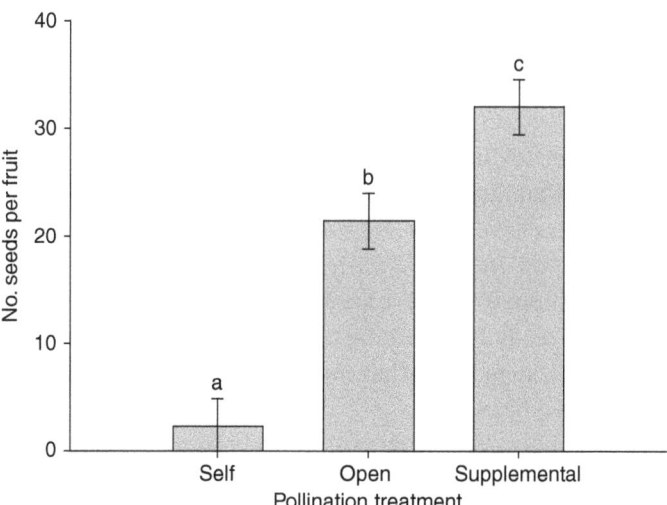

Figure 14.1 Pollination quality and quantity significantly affect seed set of *Trillium erectum*. Bars are means + SE. Different lower-case letters represent statistically significant differences among pollination treatments.

affected fruit set ($\chi^2_2 = 80.87$, P < 0.0001), but only pollination quality and not quantity was important. Only 33% of self-pollinated flowers produced fruits whereas 89% of open- and 91% of hand-pollinated flowers produced fruit. These results match prior studies showing that, first, *T. erectum* are weakly self-compatible, with self-pollination producing significantly fewer seeds then cross-pollination (Irwin 2001; Sage *et al.* 2001), and second, *T. erectum* can be pollen-limited for seed set (Irwin 2000). Taken together, these results suggest that pollination quality and quantity can affect the number of seeds available to seed dispersers. If seed dispersers respond to differences in seed number, then pollination may alter the likelihood or intensity of the seed dispersal via DMIIs.

I found that pollination also affected traits important to seed dispersal, namely absolute and relative elaiosome weight. Pollination treatment significantly affected absolute elaiosome weight ($F_{2,56} = 4.32$, P = 0.02) and had a marginal effect on relative elaiosome weight ($F_{2,56} = 2.77$, P = 0.07). Seeds from the self-pollination treatment produced the smallest elaiosomes whereas seeds from the open-pollination treatment produced the largest (Fig. 14.2a). Interestingly, hand-pollinated flowers produced intermediate weight elaiosomes even though they produced the most seeds. Mechanistic studies are needed to understand the physiological processes that drove these results. Relative elaiosome weight showed a similar trend to absolute weight (Fig. 14.2b). Prior studies have shown that elaiosome size can affect the probability of seed dispersal, and plants with seeds with larger elaiosomes typically have more seeds dispersed by ants (Gunther and Lanza 1989). Dispersers are presumably attracted to larger elaiosomes because they provide higher food reward, and if there is a cost of transport of heavier seeds, elaiosomes with the highest relative weight would provide the largest net benefit for the disperser.

Changes in seed density due to differences in pollination can also affect seed dispersal probability. The probability that at least one seed will be removed by a seed disperser increased with seed density ($\chi^2_1 = 23.81$, P < 0.0001). In addition, the number of seeds removed increased with seed density ($F_{1,97} = 72.82$, P < 0.0001), in part because there were more seeds available for seed dispersal. However, the proportion of seeds removed did not vary as a function of seed density ($F_{1,97} = 0.35$, P = 0.56), suggesting that increased seed density may not benefit dispersal on a per seed basis.

Taken together, these results suggest that pollination mutualisms have the potential to alter seed dispersal mutualisms through both TMIIs and DMIIs. The two mechanisms are not mutually exclusive and may combine in additive or non-additive ways. In this case, increased pollination quality (comparing selfed versus open pollination) can benefit seed dispersal through both trait- and density-mediated effects. However, the benefits of increased pollination

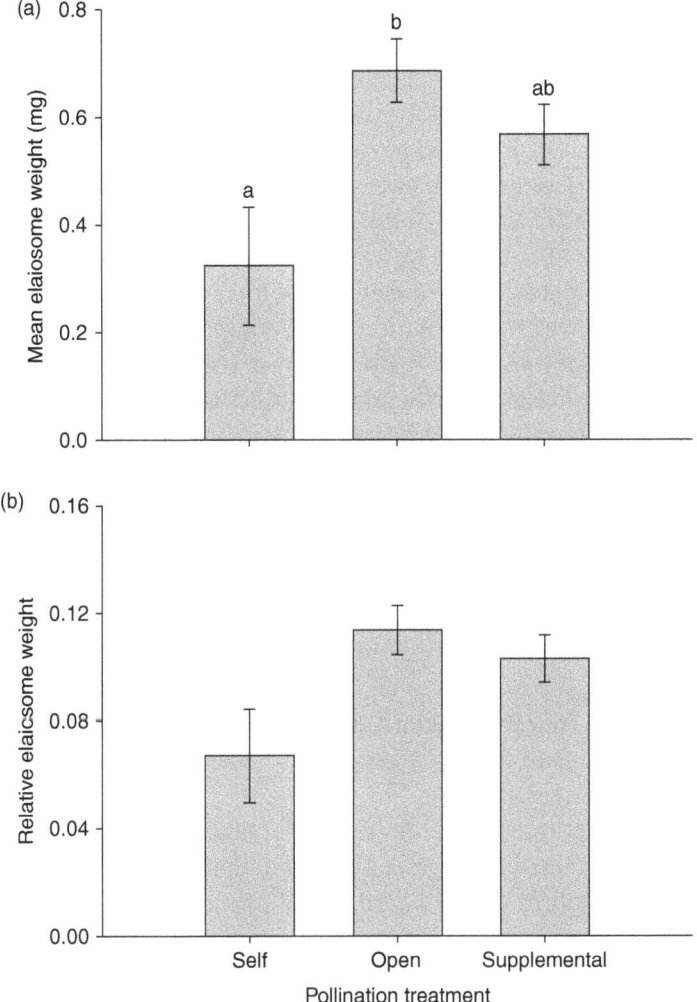

Figure 14.2 Pollination significantly affects (a) absolute and (b) relative elaiosome size of *Trillium erectum*. Bars and letters as in Figure 14.1.

quantity (comparing open versus hand pollination) are less clear. From a density-mediated perspective, increased pollen deposition increases seed number and increased seed number increases the likelihood that at least one seed will be dispersed. However, from a trait-mediated perspective, hand-pollinated seeds had slightly smaller elaiosomes than open-pollinated seeds, suggesting that the net benefits of pollen quantity may be reduced in hand-pollinated plants if seeds with smaller elaiosomes are dispersed less. Experiments are needed that manipulate seed number and elaiosome size in a factorial design to assess the relative importance of density- and trait-mediated effects. Nonetheless, given the beneficial effects of pollination and seed dispersal on plant reproduction and demography, interactions between these mutualisms may have important ramifications for host plants.

The implications of these multispecies mutualisms for patterns of evolution on floral traits generate some interesting hypotheses. A prior study in Vermont, USA, populations of *T. erectum* identified positive phenotypic selection on the size of morphological characters, including petal length and width, but not the shape of characters (Irwin 2000). The mechanisms driving these patterns are unknown but could include selection driven by the biotic (i.e., pollinators) and/or abiotic environment. If pollinators are important agents of selection on flower size, evolutionary response to selection could be magnified if flowers favoured by pollinators produce seed with the highest probability of germination and survivorship because they are most likely dispersed by ants through trait- or density-mediated effects. However, given that *T. erectum* exhibits variation in floral traits, other factors likely constrain or oppose consistent selection for larger size. These hypotheses warrant further investigation.

Conclusions

While our current framework of trait-mediated and density-mediated indirect interactions can be used successfully to describe multispecies mutualisms, the challenge remains in collecting empirical data to assess the relevance of these multispecies mutualisms for host ecology and evolution. Here I list two areas in need of further research.

First, I organized the literature based on the types of mutualists interacting, using the categories of nutritional, protection, and transport mutualisms (Boucher 1982). This literature review highlighted known examples as well as posited potential interactions that currently remain unstudied. As these gaps in knowledge are filled, synthetic studies are needed that will address whether certain types of mutualisms are more likely to interact than others and whether they yield small versus large and positive versus negative fitness outcomes for the host. For example, do nutritional mutualisms have larger effects on other mutualisms than protection or transport mutualisms? If so, why? Are some types of interacting mutualisms more or less likely to occur via TMIIs versus DMIIs? Does categorizing mutualisms by the rewards exchanged provide predictive insight into the outcome of these multispecies assemblages? Answers to these questions may strengthen the conceptual foundation underlying these complex species interactions.

Second, while the ecological consequences of some multispecies mutualisms have been assessed, their evolutionary implications are poorly understood. Their lack of study represents a critical gap in our understanding of how mutualists in concert shape host evolution. For multispecies mutualisms to affect patterns of natural selection on host plants through TMIIs or DMIIs, not only do the multispecies interactions need to affect host fitness but the relationship between traits and fitness must vary in a pairwise compared to multispecies context (Strauss *et al.* 2005). In other words, the patterns of selection generated by pairwise plant–mutualist interactions must be modified by the presence or abundance of another mutualist. Investigations of other forms of multispecies plant–animal interactions have documented the importance of community membership in altering patterns of natural selection, including through TMIIs (Irwin 2006). Given the commonality of TMIIs in multispecies mutualisms and their ecological impacts on host fitness, their study may be critical to our understanding of how multispecies mutualisms shape evolutionary dynamics.

Acknowledgements

I thank L. Rolfe, J. Stevens and M. Hamilton for help in the field and J. Evans, Z. Gezon, J. Manson, L. Richardson, R. Schaeffer and two anonymous reviewers for comments on the manuscript. Research was funded by the National Science Foundation (DEB-0455348).

References

Abrams, P. A., Menge, B. A., Mittelbach, G. G., Spiller, D. A. and Yodzis, P. (1996) The role of indirect effects in food webs. In G. A. Polis and K. O. Winemiller, eds., *Food Webs: Integration of Patterns and Dynamics*. New York: Chapman and Hall, pp. 371–395.

Allen, M. F., Birks, H. J. B. and Weins, J. A. (1991) *The Ecology of Mycorrhizae*. Cambridge: Cambridge University Press.

Altshuler, D. L. (1999) Novel interactions of non-pollinating ants with pollinators and fruit consumers in a tropical forest. *Oecologia*, **119**, 600–606.

Arimura, G., Kost, C. and Boland, W. (2005) Herbivore-induced, indirect plant defences. *Biochemica et Biophysica Acta*, **1734**, 91–111.

Bascompte, J. and Jordano, P. (2007) Plant-animal mutualistic networks: the architecture of biodiversity. *Annual Review of Ecology, Evolution, and Systematics*, **38**, 567–593.

Böhning-Gaese, K., Gaese, B. H. and Rabemanantsoa, S. B. (1999) Importance of primary and secondary seed dispersal in the Malagasy tree *Commiphora guillaumini*. *Ecology*, **80**, 821–832.

Boucher, D. H. (1982) The ecology of mutualism. *Annual Review of Ecology and Systematics*, **13**, 315–347.

Bronstein, J. L. (1994) Our current understanding of mutualism. *Quarterly Review of Biology*, **69**, 31–51.

Bronstein, J. L. (2001) The costs of mutualism. *American Zoologist*, **41**, 825–839.

Bronstein, J. L., Alarcon, R. and Geber, M. (2006) The evolution of plant-insect interactions. *New Phytologist*, **172**, 412–428.

Broyles, S. B., Sherman-Broyles, S. L. and Rogati, P. (1997) Evidence of outcrossing in *Trillium erectum* and *Trillium grandiflorum* (Liliaceae). *Journal of Heredity*, **88**, 325–329.

Cahill, J. F., Elle, E., Smith, G. R. and Shore, B. Y. (2008) Disruption of a belowground mutualism alters interactions between plants and their floral visitors. *Ecology*, **89**, 1791–1801.

Choh, Y., Kugimiya, S. and Takabayashi, J. (2006) Induced production of extrafloral nectar in intact lima bean plants in response to volatiles from spider mite-infested conspecific plants as a possible indirect defense against spider mites. *Oecologia*, **147**, 455–460.

Davis, M. A. (1981) The effect of pollinators, predators, and energy constraints on the floral ecology and evolution of *Trillium erectum*. *Oecologia*, **48**, 400–406.

Dimou, M., Taraza, S., Thrasyvoulou, A. and Vasilakakis, M. (2008) Effect of bumble bee pollination on greenhouse strawberry production. *Journal of Apicultural Research*, **47**, 99–101.

Ferrari, A. E. and Wall, L. G. (2008) Coinoculation of black locust with rhizobium and *Glomus* on a desurfaced soil. *Soil Science*, **173**, 195–202.

Fontana, A., Reichelt, M., Hempel, S., Gershenzon, J. and Unsicker, S. B. (2009) The effects of arbuscular mycorrhizal fungi on direct and indirect defense metabolites of *Plantago lanceolata* L. *Journal of Chemical Ecology*, **35**, 833–843.

Galen, C. and Cuba, J. (2001) Down the tube: pollinators, predators, and the evolution of flower shape in the alpine skypilot, *Polemonium viscosum*. *Evolution*, **55**, 1963–1971.

Gange, A. C. and Smith, A. K. (2005) Arbuscular mycorrhizal fungi influence visitation rates of pollinating insects. *Ecological Entomology*, **30**, 600–606.

Gange, A. C., Brown, V. K. and Alpin, D. M. (2003) Multitrophic links between arbuscular mycorrhizal fungi and insect parasitoids. *Ecology Letters*, **6**, 1051–1055.

Gange, A. C., Brown, V. K. and Aplin, D. M. (2005) Ecological specificity of arbuscular mycorrhizae: evidence from foliar- and seed-feeding insects. *Ecology*, **86**, 603–611.

García-Garrido, J. M. and Ocampo, J. A. (2002) Regulation of the plant defence response in arbuscular mycorrhizal symbiosis. *Journal of Experimental Biology*, **53**, 1377–1386.

Gill, F. B. (1988) Effects of nectar removal on nectar accumulation in flowers of *Heliconia*

imbricata (Heliconianaceae). *Biotropica*, **20**, 168–171.

Gonzalez, M. V., Coque, M. and Herrero, M. (1998) Influence of pollination systems on fruit set and fruit quality in kiwifruit (*Actinidia deliciosa*). *Annals of Applied Biology*, **132**, 349–355.

Gori, D. F. (1983) Post-pollination phenomena and adaptive floral changes. In C. E. Jones and R. J. Little, eds., *Handbook of Experimental Pollination Biology*. New York: Van Nostrand Reinhold, pp. 31–49.

Guerrieri, E., Lingua, G., Digilio, M. C., Massa, N. and Berta, G. (2004) Do interactions between plant roots and the rhizosphere affect parasitoid behavior. *Ecological Entomology*, **29**, 753–756.

Gunther, R. W. and Lanza, J. (1989) Variation in attractiveness of *Trillium* diaspores to a seed-dispersing ant. *American Midland Naturalist*, **122**, 321–328.

Guo, B. Z., Hendrix, J. W., An, Z.-Q. and Ferriss, R. S. (1992) Role of *Acremonium* endophyte of fescue on inhibition of colonization and reproduction of mycorrhizal fungi. *Mycologia*, **84**, 882–885.

Heil, M. (2008) Indirect defence via tritrophic interactions. *New Phytologist*, **178**, 41–61.

Heil, M. and Kost, C. (2006) Priming of indirect defenses. *Ecology Letters*, **9**, 813–817.

Heil, M., Koch, T., Hilpert, A., Fiala, B., Boland, W. and Linsenmair, K. E. (2001) Extrafloral nectar production of the ant-associated plant, *Macaranga tanarius*, is an induced, indirect, defensive response elicited by jasmonic acid. *Proceedings of the National Academy of Sciences of the United States of America*, **98**, 1083–1088.

Heinemeyer, A., Ridgway, K. P., Edwards, E. J., Benham, D. G., Young, J. P. W. and Fitter, A. H. (2004) Impact of soil warming and shading on colonization and community structure of arbuscular mycorrhizal fungi in roots of a native grassland community. *Global Change Biology*, **10**, 52–64.

Hoeksema, J. D. and Bruna, E. M. (2000) Pursuing the big questions about interspecific mutualism: a review of theoretical approaches. *Oecologia*, **125**, 321–330.

Holland, J. N., Ness, J. H., Boyle, A. L. and Bronstein, J. L. (2005) Mutualisms as consumer–resource interactions. In P. Barbosa and I. Castellanos, eds., *Ecology of Predator–Prey Interactions*. New York: Oxford University Press, pp. 17–33.

Irwin, R. E. (2000) Morphological variation and female reproductive success in two sympatric *Trillium* species: evidence for phenotypic selection in *Trillium erectum* and *Trillium grandiflorum* (Liliaceae). *American Journal of Botany*, **87**, 205–214.

Irwin, R. E. (2001) Field and allozyme studies investigating optimal mating success in two sympatric spring-ephemeral plants, *Trillium erectum* and *T. grandiflorum*. *Heredity*, **87**, 178–189.

Irwin, R. E. (2006) Consequences of direct versus indirect species interactions to selection on traits: pollination and nectar robbing in *Ipomopsis aggregata*. *American Naturalist*, **167**, 315–328.

Jallow, M. F. A., Dugassa-Gobena, D. and Vidal, S. (2008) Influence of an endophytic fungus on host plant selection by a polyphagous moth via volatile spectrum changes. *Arthropod–Plant Interactions*, **2**, 53–62.

Janzen, D. H. (1985) The natural history of mutualisms. In D. H. Boucher, ed., *The Biology of Mutualism*. New York: Oxford University Press, pp. 40–99.

Jordano, P. (1987) Avian fruit removal: effects of fruit variation, crop size, and insect damage. *Ecology*, **68**, 1711–1723.

Kalisz, S., Hanzawa, F. M., Tonsor, S. J., Thiede, D. A. and Voigt, S. (1999) Ant-mediated seed dispersal alters pattern of relatedness in a population of *Trillium grandiflorum*. *Ecology*, **80**, 2620–2634.

Karban, R. and Baldwin, I. T. (1997) *Induced Responses to Herbivory*. Chicago, IL: University of Chicago Press.

Koide, R. T. (2000) Mycorrhizal symbiosis and plant reproduction. In Y. Kapulnik and D. D. Douds, eds., *Arbuscular Mycorrhizas:*

Physiology and Function. Dodrecht, The Netherlands: Kluwer Academic, pp. 19–46.

Kost, C. and Heil, M. (2006) Herbivore-induced plant volatiles induce an indirect defense in neighbouring plants. *Journal of Ecology*, **94**, 619–628.

Kost, C. and Heil, M. (2008) The defensive role of volatile emission and extrafloral nectar secretion for lima bean in nature. *Journal of Chemical Ecology*, **34**, 2–13.

Laird, R. A. and Addicott, J. F. (2007) Arbuscular mycorrhizal fungi reduce the construction of extrafloral nectaries in *Vicia faba*. *Oecologia*, **152**, 541–551.

Larimer, A. L., Bever, J. D. and Clay, K. (2010) The interactive effects of plant microbial symbionts: a review and meta-analysis. *Symbiosis*, **51**, 139–148.

Lau, T. C., Lu, X., Koide, R. T. and Stephenson, A. G. (1995) Effects of soil fertility and mycorrhizal infection on pollen production and pollen grain-size of *Cucurbita pepo* (Cucurbitaceae). *Plant Cell and Environment*, **18**, 169–177.

Leitner, M., Roland, K., Hause, B., Boland, W. and Mithöfer, A. (2010) Does mycorrhization influence herbivore-induced volatile emission in *Medicago truncatula*? *Mycorrhiza*, **20**, 89–101.

Mack, K. M. L. and Rudgers, J. A. (2008) Balancing multiple mutualists: asymmetric interactions among plants, arbuscular mycorrhizal fungi, and fungal endophytes. *Oikos*, **117**, 310–320.

McCann, K. S. (2000) The diversity-stability debate. *Nature*, **405**, 228–233.

Morris, W. F., Hufbauer, R. A., Agrawal, A. A. *et al.* (2007) Direct and indirect interactive effects of enemies and mutualists on plant performance: a meta-analysis. *Ecology*, **88**, 1021–1029.

Mothershead, K. and Marquis, R. J. (2000) Fitness impacts of herbivory through indirect effects on plant-pollinator interactions in *Oenothera macrocarpa*. *Ecology*, **81**, 30–40.

Ness, J. H. (2003) *Catalpa bignonioides* alters extrafloral nectar production after herbivory and attracts ant bodyguards. *Oecologia*, **134**, 210–218.

Ness, J. H. (2006) A mutualism's indirect costs: the most aggressive plant bodyguards also deter pollinators. *Oikos*, **113**, 506–514.

Ness, J. H., Morris, W. F. and Bronstein, J. L. (2006) Integrating quality and quantity of mutualistic service to contrast ant species protecting *Ferocactus wislizeni*. *Ecology*, **87**, 912–921.

Novas, M. V., Cabral, D. and Godeas, A. M. (2005) Interaction between grass endophytes and mycorrhizas in *Bromus setifolius* from Patagonia, Argentina. *Symbiosis*, **40**, 23–30.

Pacovsky, R. S., Fuller, G., Stafford, A. E. and Paul, E. A. (1986) Nutrient and growth interactions in soybean colonized with *Glomus fasciculatum* and *Rhizobium japonicum*. *Plant and Soil*, **92**, 37–45.

Palmer, T. M., Doak, D. F., Stanton, M. L. et al. (2010) Synergy of multiple partners, including freeloaders, increases host fitness in a multispecies mutualism. *Proceedings of the National Academy of Sciences of the United States of America*, **107**, 17234–17239.

Palmer, T. M., Stanton, M. L. and Young, T. P. (2003) Competition and coexistence: exploring mechanisms that restrict and maintain diversity within mutualist guilds. *American Naturalist*, **162**, S63–S79.

Peacor, S. D. and Werner, E. E. (2001) Contribution of trait-mediated indirect effects to the net effects of a predator. *Proceedings of the National Academy of Sciences of the United States of America*, **98**, 3904–3908.

Poulton, J. L., Bryla, D. R., Koide, R. T. and Stephenson, A. G. (2002) Mycorrhizal infection and high soil phosphorus improve vegetative growth and the female and male functions in tomato. *New Phytologist*, **154**, 255–264.

Rahman, M. H. and Saiga, S. (2005) Endophytic fungi (*Neotyphodium coenophialum*) affect the growth and mineral uptake, transport and efficiency ratios of tall fescue (*Festuca arundinacea*). *Plant and Soil*, **272**, 163–171.

Rapparini, F., Llusià, J. and Penuelas, J. (2008) Effect of arbuscular mycorrhizal (AM) colonization on terpene emission and content of *Artemisia annua* L. *Plant Biology*, **10**, 108–122.

Rasmann, S., Köllner, T. G., Degenhardt, J. et al. (2005) Recruitment of endomopathogenic nematodes by insect-damaged maize roots. *Nature*, **434**, 732–737.

Rudgers, J. A. (2004) Enemies of herbivores can shape plant traits: selection in a facultative ant-plant mutualism. *Ecology*, **85**, 192–205.

Russo, S. E. (2003) Responses of dispersal agents to tree and fruit traits in *Virola calophylla* (Myristicaceae): implications for selection. *Oecologia*, **136**, 80–87.

Sage, T. L., Griffin, S. R., Pontieri, V. et al. (2001) Stigmatic self-incompatibility and mating patterns in *Trillium grandiflorum* and *Trillium erectum* (Melanthiaceae). *Annals of Botany*, **88**, 829–841.

Sallabanks, R. (1993) Hierarchical mechanisms of fruit selection by an avian frugivore. *Ecology*, **74**, 1326–1336.

Schemske, D. W. and Horvitz, C. C. (1988) Plant-animal interactions and fruit production in a neotropical herb: a path analysis. *Ecology*, **69**, 1128–1137.

Smith, S. E. and Read, D. J. (1997) *Mycorrhizal Symbiosis*. San Diego, CA: Academic Press.

Stachowicz, J. J. and Whitlatch, R. B. (2005) Multiple mutualists provide complementary benefits to their seaweed host. *Ecology*, **86**, 2418–2427.

Stanton, M. L. (2003) Interacting guilds: moving beyond the pairwise perspective on mutualisms. *American Naturalist*, **162**, S10–S23.

Stout, J. C. and Goulson, D. (2001) The use of conspecific and interspecific scent marks by foraging bumblebees and honeybees. *Animal Behavior*, **62**, 183–189.

Strauss, S. Y. (1997) Floral characters link herbivores, pollinators, and plant fitness. *Ecology*, **78**, 1640–1645.

Strauss, S. Y. and Irwin, R. E. (2004) Ecological and evolutionary consequences of multispecies plant-animal interactions. *Annual Review of Ecology, Evolution, and Systematics*, **35**, 435–466.

Strauss, S. Y., Conner, J. K. and Lehtilä, K. P. (2001) Effects of foliar herbivory by insects on the fitness of *Raphanus raphanistrum*: damage can increase male fitness. *American Naturalist*, **158**, 496–504.

Strauss, S. Y., Sahli, H. and Conner, J. K. (2005) Toward a more trait-centered approach to diffuse (co)evolution. *New Phytologist*, **165**, 81–90.

Vander Wall, S. B. and Longland, W. S. (2004) Diplochory: are two seed dispersers better than one? *Trends in Ecology and Evolution*, **19**, 155–161.

Varga, S. and Kytöviita, M.-M. (2010) Gender dimorphism and mycorrhizal symbiosis affect floral visitors and reproductive output in *Geranium sylvaticum*. *Functional Ecology*, **24**, 750–758.

Waites, A. R. and Ågren, J. (2004) Pollinator visitation, stigmatic pollen loads and among-population variation in seed set in *Lythrum salicaria*. *Journal of Ecology*, **92**, 512–526.

Weber, J., Ducousso, M., Tham, F. Y. *et al.* (2005) Co-inoculation of *Acacia mangium* with *Glomus intraradices* and *Bradyrhizobium* sp. in aeroponic culture. *Biology and Fertility of Soils*, **41**, 233–239.

Wolfe, B. E., Husband, B. C. and Klironomos, J. N. (2005) Effects of belowground mutualism on an aboveground mutualism. *Ecology Letters*, **8**, 218–223.

Wootton, J. T. (1993) Indirect effects and habitat use in an intertidal community: interaction chains and interaction modifications. *American Naturalist*, **141**, 71–89.

Yamanaka, T., Akama, A., Li, C.-Y. and Okabe, H. (2005) Growth, nitrogen fixation and mineral acquisition of *Alnus sieboldiana* after inoculation of *Frankia* together with *Gigaspora margarita* and *Pseudomonas putida*. *Journal of Forest Research*, **10**, 21–26.

Zitzer, S. F., Archer, S. R. and Boutton, T. W. (1996) Spatial variability in the potential for symbiotic N_2 fixation by woody plants in a subtropical savanna ecosystem. *Journal of Applied Ecology*, **33**, 1125–1136.

CHAPTER FIFTEEN

Consequences of trait evolution in a multispecies system

CRAIG W. BENKMAN

Department of Zoology and Physiology, University of Wyoming

ADAM M. SIEPIELSKI

Department of Biology, University of San Diego

and

JULIE W. SMITH

Department of Biology, Pacific Lutheran University

Introduction

Ecologists interested in how traits mediating species interactions evolve have increasingly recognized that trait evolution is a consequence of multiple interacting species (Miller and Travis 1996; Strauss and Ambruster 1997; Thompson 1999; Strauss and Irwin 2004; Haloin and Strauss 2008). Species may alter the evolution of traits affecting multiple species through a number of direct or indirect pathways; examples include exerting conflicting selection pressures on the same traits (Siepielski and Benkman 2007a; Manzaneda *et al.* 2009) and predators indirectly altering the strength of species interactions (Werner and Peacor 2003). Although our understanding of the community context of species interactions has sharpened in the past decade (e.g., Strauss and Irwin 2004; Bascompte and Jordano 2007; Johnson and Stinchcombe 2007), particularly with regards to spatial dynamics (e.g., Thompson 2005; Urban *et al.* 2008), a number of outstanding questions remain. For example, to what extent do interactions evolve because of adaptive evolution of traits mediating other species interactions? Similarly, how does the loss of an interacting species (e.g., relaxed selection) affect the evolution of other interactions? Finally, when and to what extent is variation in community and ecosystem patterns and processes influenced by adaptive evolution of traits mediating species interactions? Answers to these questions have important implications for our understanding of major topics in evolutionary biology including the evolutionary outcome of selection in multispecies interactions, the geographic mosaic of coevolution and even patterns of adaptive radiation.

Seeds represent an important life history character for plants and they are a shared resource for many species, including multiple seed dispersers (Herrera 2002) and seed predators (Hulme and Benkman 2002). Moreover, the production

Trait-Mediated Indirect Interactions: Ecological and Evolutionary Perspectives, eds. Takayuki Ohgushi, Oswald J. Schmitz and Robert D. Holt. Published by Cambridge University Press. © Cambridge University Press 2012.

of seeds is often dependent on successful pollination by animal pollinators (Pellmyr 2002), adding further to interaction complexity. The result is that adaptive evolution of traits (i.e., those related to seed production, protection and dispersal) that are important for one interaction is very likely to have consequences for other species interactions, and in some cases may be critical in driving the evolution of other interactions (e.g., Armbruster *et al.* 1997, 2009; Siepielski and Benkman 2004, 2008b; see below for examples). While increasing attention has focused on the ecological and evolutionary consequences of these kinds of multispecies interactions, less appreciated is that adaptive evolution may also have important consequences at the community and ecosystem level (Holt 1994; Bailey *et al.* 2004, 2009; Benkman and Siepielski 2004; Siepielski and Benkman 2008a; Palkovacs *et al.* 2009; Harmon *et al.* 2009; Bassar *et al.* 2010). For example, many plant populations are seed limited (Clark *et al.* 2007), so that trait evolution affecting the number of seeds dispersed can also affect plant ecology in a number of ways (e.g., regeneration, colonization, spatial structure; see Hulme and Benkman 2002). These effects may further extend to the ecosystem level, particularly in cases where the plant is the dominant or foundation species in a community, such as lodgepole pine (*Pinus contorta latifolia*) in upland forests (e.g., Benkman and Siepielski 2004) and cottonwood (*Populus* spp.) in riparian habitats (Whitham *et al.* 2006; Bailey *et al.* 2009) in the Rocky Mountains of western North America.

Here we focus on the direct and indirect effects of variation in the reproductive traits of conifers (seeds and cones) that are evolving in response to species interactions. Conifers are wind pollinated, but many animal species consume conifer seeds and some species are also potential seed dispersers (Smith and Balda 1979; Tomback and Linhart 1990; Vander Wall 2008). Evolutionary responses to phenotypic selection are expected because cone and seed traits have high heritabilities (Khalil 1984; Matziris 1998), and we have no evidence of phenotypic plasticity in cone and seed traits in response to species interactions (Benkman *et al.* 2001). Our general approach involves measuring phenotypic selection and using broad-scale geographic comparisons of ecologically important traits among communities with and without different sets of interacting species. This enables us to gain insight into how changes in community structure affect trait evolution.

The ecological and evolutionary effects of indirect interactions: an example

Just as indirect interactions can be either density mediated (DMII), trait mediated (TMII) or both (Werner and Peacor 2003), the effects of indirect interactions on other species can pertain to the alteration of their densities or to their evolution or both. For example, selection exerted by granivorous pine squirrels (*Tamiasciurus* spp.) has favoured the evolution of thicker cone scales in

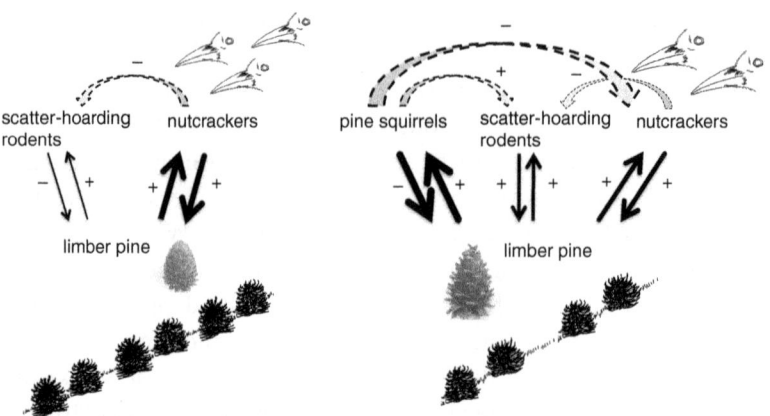

Figure 15.1 Where granivorous pine squirrels (*Tamiasciurus* spp.) are absent (left side), the mutualistic direct interaction between Clark's nutcrackers (*Nucifraga columbiana*) and limber pine (*Pinus flexilis*) is strong (solid arrows, where thickness indicates strength of interaction) and cones evolve so that nutcrackers can harvest and disperse seeds efficiently. Because relatively few seeds fall to the ground, scatter-hoarding rodents are less likely to be satiated and more likely to act as seed predators (a negative trait- and density-mediated indirect effect as shown by the dashed arrow). Where pine squirrels are present (right side), the direct interaction between pine squirrels and limber pine is the dominant direct interaction that favours the evolution of increased seed defences (e.g., larger scales) and cone removal in particular results in a decrease in seed availability to nutcrackers. The increase in cone armature favours an increase in nutcracker bill size, but results in fewer seeds harvested and dispersed by fewer nutcrackers (indicated by the size and number of nutcracker heads), and more seeds falling to the ground so that scatter-hoarding rodents are more likely to act as seed dispersers (a negative trait-mediated indirect effect for nutcrackers, but a positive one for scatter-hoarding rodents). An important trait- and density-mediated indirect effect of this interaction between pine squirrels and limber pine is the lower density of limber pine in areas with than without pine squirrels.

limber pine (*P. flexilis*), and as a result the Clark's nutcracker (*Nucifraga colum-biana*), a mutualistic seed disperser, has apparently evolved larger bills in regions with pine squirrels than in regions without them (Fig. 15.1; Siepielski and Benkman 2007a). In addition, the densities of nutcrackers are lower per cone produced in regions with pine squirrels than in regions without them, apparently in large part because cone removal by pine squirrels reduces seed availability (Fig. 15.1; Siepielski and Benkman 2007a; see below). The evolution of nutcracker bill size, therefore, is the result of a TMII, while the decrease in nutcrackers is caused mostly by DMIIs (i.e., exploitative competition).

Seed predation and selection by pine squirrels, which has favoured an increased investment in cones (i.e., seed defence) over seeds and has caused a reduction in nutcracker seed harvesting rates, have reduced the number of seeds dispersed and thus led to lower densities of limber pine in the presence than in the absence of pine squirrels (Fig. 15.1). These effects on limber pine density will have large impacts on community and ecosystem characteristics, as limber pine is a foundation species in the Great Basin and Rocky Mountain region (Siepielski and Benkman 2008a). Community and ecosystem properties that are likely altered by variation in density and biomass of limber pine include: habitat for a wide range of species (Tomback and Kendall 2001); the amount of influxes of energy and nutrients (Barbour *et al.* 1987) and the extent to which the trees serve as carbon sinks (Reich *et al.* 2006); regulation of runoff and erosion from snowmelt and stream flow (Tomback *et al.* 2001); and facilitation of succession (Baumeister and Callaway 2006). Thus, the pine squirrel–limber pine interaction has profound ecological and evolutionary consequences not only because of seed predation and selection exerted by pine squirrels, but also because of the trait- and density-mediated indirect effects (DMIEs) arising from this pairwise interaction. Because such indirect interactions appear to arise commonly from the pairwise interactions between conifers and tree squirrels (*Sciurus* spp. and especially pine squirrels *Tamiasciurus* spp.; Benkman *et al.* 2008, 2010), we will focus on this system. First, we consider factors that influence the type and strength of indirect interactions. Then we will discuss the consequences of these interactions.

Causes of variation in DMIIs and TMIIs

One factor that influences the importance of DMIIs is the timing of seed and cone exploitation (Fig. 15.2a). Pine squirrels harvest vast numbers of cones soon after the seeds mature but before the cones open in early autumn. Thus,

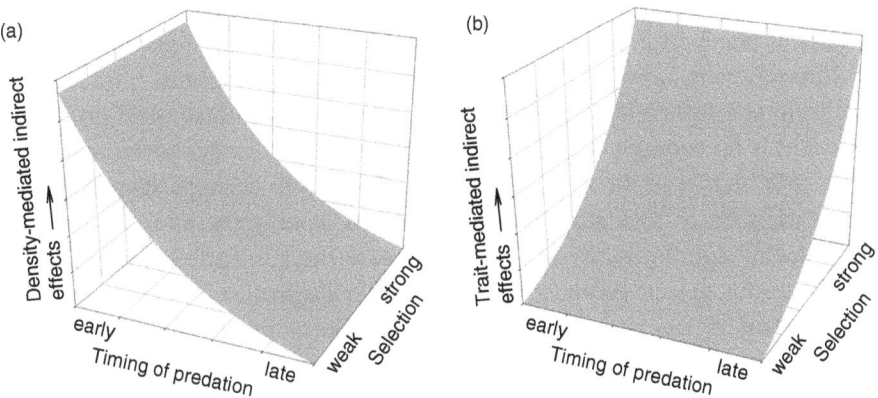

Figure 15.2 While the strength of density-mediated indirect effects is mostly determined by the strength of competition, which is determined by the intensity of seed predation and especially its timing in the conifer seed system we describe (a), the strength of trait-mediated effects depend mostly on the strength of selection (b).

pine squirrels potentially have strong DMIEs on species that harvest seeds mostly later in the season (i.e., preemptive competition). However, pine squirrels will have little if any density-mediated effects on those species that utilize seeds and cones earlier (Smith and Balda 1979). In contrast, pine squirrels can have strong trait-mediated indirect effects (TMIEs) on other species regardless of when they forage (Fig. 15.2b).

An example of a TMII where there is no opportunity for density-mediated effects concerns the lodgepole pine cone borer moth (*Eucosma recissoriana*). This moth pupates out of the cones before pine squirrels begin harvesting many cones, and is the only insect that feeds on developing seeds and cones of Rocky Mountain lodgepole pine. Even though developing cones and seeds are plentiful during most years, this moth consumes < 0.1% of the seeds in many areas of the Rocky Mountains (Miller 1986). This low percentage cannot be the result of seed depletion by pine squirrels, but instead is probably related to the evolution of enhanced seed defences in response to selection exerted by pine squirrels (Siepielski and Benkman 2004); with the exception of true firs (*Abies*), defences against conifer seed predators are mostly physical rather than chemical (Smith 1970; Benkman *et al.* 2010). Moths prefer cones with more seeds, and moths are uncommon presumably because strong selection exerted by pine squirrels has led to the evolution of cones with few seeds (seeds comprise only about 1% of the total cone mass; Smith 1970; Benkman 1999). In ranges east and west of the Rocky Mountains, where pine squirrels have been absent for about 12 000 years, cones have lost defences directed toward pine squirrels (the ratio of cone mass to seed mass has decreased by 64% from the values in areas with pine squirrels; Benkman 1999). Here seed predation by moths is over twice as high as in ranges with pine squirrels (Siepielski and Benkman 2004). The low seed predation rates by moths in the presence of pine squirrels are therefore likely the result of a TMII. Cone evolution in response to selection exerted by pine squirrels, however, is not the only factor affecting seed predation by moths. In one squirrel-less range (the Little Rocky Mountains), seed predation by moths is exceedingly high (20 times higher than in ranges with pine squirrels). This high level of seed predation by moths is perhaps related to the limestone soils in the Little Rocky Mountains. Poor soils, like those in the Little Rocky Mountains, are known to increase the susceptibility of plants to phytophagous insects (Waring and Cobb 1992). Regardless of why moths are so abundant in the Little Rocky Mountains, moths exert selection on cone structure in a similar manner to that exerted by pine squirrels. Consequently with the increase in selection exerted by moths, the cones in the Little Rocky Mountains are more similar to cones in regions with pine squirrels than are cones from other squirrel-less ranges (Siepielski and Benkman 2004).

Those species that harvest seeds later in the year are more likely to experience DMIEs (Fig. 15.2a). For example, the densities of nutcrackers in limber pine woodlands (comparing five mountain ranges with pine squirrels and six ranges without) appear to be more adversely affected by the presence of pine squirrels than by the increase in cone defences as a result of selection exerted by pine squirrels (multiple regression: $P = 0.059$ for the effect of the presence and absence of pine squirrels, and $P = 0.17$ for the effect of the first principal component [PC1] of ten cone traits; see Siepielski and Benkman 2007a). An additional contributing factor for why we did not find much evidence for a trait-mediated indirect effect on nutcracker abundance is that nutcrackers harvest a large fraction of seeds from limber pine cones after they have opened, which is when variation in cone structure has little impact on foraging rates (Siepielski and Benkman 2007a). On the other hand, variation in cone structure has a large impact on seed accessibility when cones are closed. This may explain why the abundances of hairy woodpeckers (*Picoides villosus*) (Benkman *et al.*, in press), which forage on seeds in closed Rocky Mountain lodgepole pine cones, appear to have been negatively impacted by the evolution of cone traits resulting from selection exerted by pine squirrels. The ratio of seed mass to cone mass in lodgepole pine, which is a trait under similar selection by both pine squirrels and hairy woodpeckers, predicts well the density of hairy woodpeckers ($r^2 = 0.96$, df $= 5$, $P < 0.0001$). That is, hairy woodpeckers decline in abundance as seed mass decreases relative to cone mass. In contrast, we were unable to detect a DMIE on hairy woodpeckers. The presence and absence of pine squirrels in the seven different mountain ranges (three with pine squirrels and four without them) did not add further explanatory power (multiple regression: $P = 0.12$). In sum, the evolution of conifers in response to selection exerted by pine squirrels adversely impacts seed predators that have similar cone preferences to pine squirrels and forage mostly if not exclusively on seeds in closed rather than open cones. Those species that forage most often for seeds in open cones (i.e., after squirrels have foraged) are impacted more by exploitative competition.

Pine and other tree squirrels also have had striking effects on the ecology and evolution of the specialized finches in the red crossbill (*Loxia curvirostra*) complex. Crossbills are highly specialized for feeding on the seeds of conifer cones, and begin feeding on seeds in cones as they mature and continue to rely on seeds in cones long after pine squirrels have finished harvesting cones. Thus, crossbills potentially experience the effects of both DMIIs and TMIIs. However, which type of interaction dominates depends largely on the size of the cones (Fig. 15.3; Benkman *et al.* 2010). Like most predators, tree squirrels appear to have a preferred size of prey to forage upon (cone in this case). When cones average less than approximately 60–80 mm in length (e.g., lodgepole pine and Scots pine *P. sylvestris*), tree squirrels prefer to forage on larger cones, which favours the evolution of even smaller cones (Fig. 15.3; Benkman 1999; Summers and Proctor 1999; Benkman *et al.* 2001). We

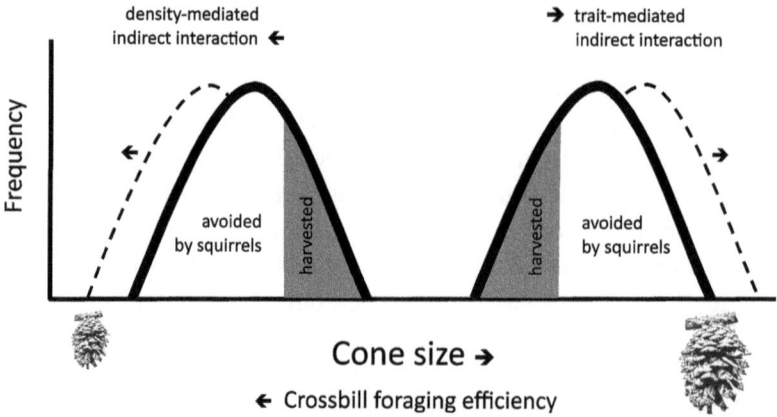

Figure 15.3 When cone size is relatively small (< 60–80 mm long), tree squirrels (*Sciurus* and *Tamiasciurus* spp.) preferentially harvest larger-sized cones, favouring the evolution of smaller cones (arrow and dashed curve to left). Because crossbill (*Loxia* spp.) foraging efficiency tends to increase with decreases in cone size, seed predation by tree squirrels causes a decrease in crossbills because of exploitative competition (a DMII). When cone size is larger, tree squirrels preferentially harvest smaller-sized cones, favouring the evolution of even larger cones. This increase in cone size reduces seed accessibility for crossbills and causes a decline in crossbills (a TMII).

include a range of cone lengths because the threshold length appears to be smaller for cones with harder scales. When cones average longer than 60–80 mm (e.g., Aleppo pine *P. halepensis* and ponderosa pine *P. p. ponderosa*), tree squirrels preferentially forage on smaller-sized cones, favouring the evolution of even larger cones (Fig. 15.3; Mezquida and Benkman 2005; Parchman and Benkman 2008). When tree squirrels preferentially consume larger than average cones and thus favour the evolution of smaller cones, crossbill foraging efficiency will likely increase (Benkman 1999). In this case, the negative indirect interaction between tree squirrels and crossbills arises because of extensive cone harvesting by tree squirrels (i.e., exploitative competition). We have found that crossbill densities are reduced by the presence of pine squirrels in lodgepole pine forests (which have relatively small cones) by up to 20-fold, which appears to be mainly if not exclusively the result of DMIIs (Benkman 1999; Siepielski and Benkman 2005). In contrast, the evolution of larger cones acts to reduce the foraging efficiency of crossbills, and cones may evolve to be larger than crossbills can efficiently handle (Mezquida and Benkman 2005; Parchman and Benkman 2008). The negative effect of tree squirrels on crossbills is then mostly if not exclusively trait-mediated (Fig. 15.3). Thus, the net effect on crossbills is negative in both cases, but it arises by different mechanisms.

Further examples and consequences of indirect interactions

The examples above show that selection exerted by pine squirrels on conifers can impact the evolution and abundance of other seed consumers. Here we consider two additional trait-mediated effects of these interactions. The first concerns the evolution of alternative seed dispersal mechanisms in pines. Most pines have wings on their seeds that cause the seeds to auto-rotate, slowing their descent and facilitating dispersal by wind. However, pines that produce seeds larger than about 90 mg are not effectively dispersed by wind because of high disc loading (Benkman 1995). Disc loading is the mass of seed divided by the square of the overall length of the seed and wing. Pines with seeds smaller than 90 mg have low values of disc loading. On the other hand, disc loading increases rapidly as seed mass approaches and exceeds 90 mg, presumably because of structural constraints in producing cones with sufficiently long scales to house the frail seed wings (Benkman 1995). Consequently, large-seeded pines rely on animals to disperse their seeds. However, which animals disperse most of the seeds appears to depend on the occurrence of tree squirrels. Large-seeded pines that occur where tree squirrels are uncommon or absent (e.g., limber pine and whitebark pine *P. albicaulis*) have relatively small cones, facilitating the harvest and dispersal of their seeds by nutcrackers and jays (Corvidae) (Benkman 1995; Siepielski and Benkman 2007b). In contrast, large-seeded pines that regularly co-occur with tree squirrels (e.g., sugar pine *P. lambertiana*) have much larger cones that effectively defend their seeds presumably because of selection exerted by tree squirrels. Such selection not only reduces the ability of birds such as nutcrackers to harvest and disperse the seeds but also enhances secondary seed dispersal by scatter-hoarding, ground-foraging rodents (Fig. 15.1; Siepielski and Benkman 2008b). Thus, the shift from mostly bird to mostly rodent dispersal in pines is the result of TMIIs.

The second indirect interaction concerns the trait-mediated effects of pine squirrels selecting against serotiny (Benkman and Siepielski 2004). Serotiny is the retention of seeds in hard woody fruits or conifer cones until high temperatures cause the fruits or cones to open and shed their seeds (Lamont *et al.* 1991). Serotiny is favoured when there are stand-replacing disturbances such as fire that occur within the average lifespan of the species (Enright *et al.* 1998). Although increasing frequencies of fire favour increasing frequencies of serotiny in plant populations, seed predators that remove a large percentage (> 50%) of the canopy seed bank select against serotiny (Fig. 15.4; Lamont *et al.* 1991; Enright *et al.* 1998). Pine squirrels are one such seed predator (e.g., Elliott 1988). The apparent result of such selection on Rocky Mountain lodgepole pine is that over 90% of the trees are serotinous in ranges without pine squirrels, but on average only about 30% of the trees are serotinous in areas with pine squirrels (Benkman and Siepielski 2004). This is particularly

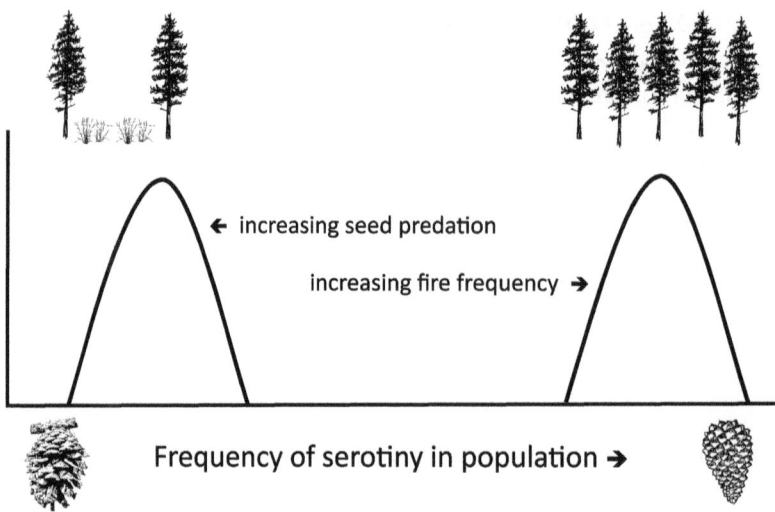

Figure 15.4 Increases in the frequency of stand-replacing fires favours an increase in the frequency of serotiny in populations of Rocky Mountain lodgepole pine (*Pinus contorta latifolia*), whereas increases in seed predation by pine squirrels favours a decrease in serotiny. The frequency of serotiny influences the size of the canopy seed bank, which in turn influences the density of pine seedlings after a stand-replacing fire with considerable consequences for community and ecosystem properties early in succession.

important because the frequency of serotiny influences the size of the canopy seed bank and thus the density of seedlings following a fire (Fig. 15.4; Tinker *et al.* 1994). For example, after the 1988 fires in Yellowstone National Park (where pine squirrels are present) the density of seedlings was only 2300 seedlings per hectare where the frequency of serotiny was 10%, but was up to 211 000 seedlings per hectare where the frequency of serotiny was 65%. In the Cypress Hills, where pine squirrels were absent and the frequency of serotiny is 92%, there were 2.5 million seedlings per hectare after a fire (Benkman *et al.* 2008). Thus, the frequency of serotiny influences tree seedling density after a fire, which has tremendous consequences for the early stages of succession. For example, the frequency of serotiny will determine the extent to which the community early in succession will be tree-dominated (Fig. 15.4), with considerable potential effects on the composition of the plant and animal communities and ecosystem processes (Tinker *et al.* 1994; Turner *et al.* 1997). Although we suspect that a sizeable amount of variation in serotiny within areas with pine squirrels owes its origin to variation in the strength of selection exerted by pine squirrels, research is needed to quantify the strength of selection exerted by both pine squirrels and fire on serotiny and how such variation affects the level of serotiny across the landscape.

Future research that examines the genetic basis of serotiny will also lead to a better understanding of the evolution of serotiny in lodgepole pine and other pines. Because serotinous conifers dominate extensive areas of North America and the Mediterranean, selection by seed predators has the potential to have widespread community and ecosystem consequences.

Although additional examples are still few, two recent sets of studies indicate that evolved trait-mediated indirect interactions may be widespread. One study shows that beavers (*Castor canadensis*) depress the abundance of cottonwood genotypes having low concentrations of condensed tannins with the potential to exert selection favouring an increase in condensed tannins in cottonwood populations (Bailey *et al.* 2004). The TMIEs arising from differences in condensed tannin concentration among cottonwoods are considerable and diverse, influencing invertebrate and endophytic communities that are associated directly and indirectly with cottonwood, and various ecosystem processes such as rates of nitrogen mineralization and decomposition (Whitham *et al.* 2006). The question remains, however, whether variation in selection by beaver can explain variation in condensed tannin concentrations among cottonwood populations and across the landscape.

A second set of studies, on Trinidadian guppies (*Poecilia reticulata*), demonstrates that variation in the distribution of their predators has both DMIEs and evolved TMIEs. First, predators depress the abundance of guppies (Reznick *et al.* 2001), which results in DMIEs on community structure and ecosystem processes (Bassar *et al.* 2010). Second, the consistent and well-known evolutionary differences in life histories between guppy populations with and without predators also have indirect effects on aquatic ecosystem structure (algal, invertebrate and detrital standing stocks) and function (gross primary productivity, leaf decomposition rates and nutrient flux; Bassar *et al.* 2010). Consequently, the ecosystem differences between stretches of stream with and without guppy predators are the result in part of predators depressing the abundance and altering the evolution of guppies (Bassar *et al.* 2010). We suspect that evolutionary responses to the selection exerted by strongly interacting species such as mammals and fish, which are well known for their disproportionately strong impacts on communities and ecosystem processes (i.e., keystone species: Carpenter *et al.* 1985; Estes and Duggins 1995), commonly have profound TMIEs and deserve further attention.

Conclusions

TMIEs are generally described as indirect effects that occur because of phenotypically plastic responses to other species (Werner and Pecor 2003). Phenotypic plasticity is widespread, and how it affects both direct and indirect species interactions has been the focus of much research (this volume).

However, traits also may vary because of evolution in response to selection exerted by other species, and such trait evolution can influence the form and strength of indirect effects. Although these evolutionary changes are less rapid than changes arising from phenotypic plasticity, the rapid rate at which evolution can occur and the divergence in selection a species may commonly experience between communities (Thompson 2005) suggest that these 'evolved' TMIIs may have a considerable impact on various processes. Our results support this inference. We have found that selection exerted by pine squirrels can depress the abundance and alter the accessibility of conifer seeds, and thereby greatly alter the abundance, the selection experienced by and the evolution of other species that rely on conifer seeds. Moreover, these indirect (and direct) interactions of pine squirrels even alter the abundance of conifers with community-wide and ecosystem consequences. Because pine squirrels are absent from some mountain ranges, this leads to strong divergent selection on the conifers between areas with and without pine squirrels, and to variation in the extent to which some species (e.g., crossbills) coevolve with conifers. These effects of pine squirrels and other tree squirrels in turn impact the adaptive radiation of crossbills (Benkman *et al.* 2010). We encourage others to consider 'evolved' TMIEs, because they are likely important in many other systems including especially those with strongly interacting species.

Acknowledgements

We thank the editors for the invitation to write this chapter, and for comments by T. Ohgushi and an anonymous reviewer. Grants from the National Science Foundation supported our research (DEB-0344503, DEB-0455705 and DEB-0515735).

References

Armbruster, W. S., Howard, J. J., Clausen, T. P. et al. (1997) Do biochemical exaptations link evolution of plant defense and pollination systems? Historical hypotheses and experimental tests with *Dalechampia* vines. *American Naturalist*, **149**, 461–484.

Armbruster, W. S., Lee, J. and Baldwin, B. G. (2009) Macroevolutionary patterns of defense and pollination in *Dalechampia* vines: adaptation, exaptation, and evolutionary novelty. *Proceedings of the National Academy of Sciences of the United States of America*, **106**, 18085–18090.

Bascompte, J. and Jordano, P. (2007) The structure of plant–animal mutualistic networks: the architecture of biodiversity. *Annual Review of Ecology, Evolution, and Systematics*, **38**, 567–593.

Bailey, J. K., Schweitzer, J. A., Rehill, B. J. et al. (2004) Beavers as molecular geneticists: a genetic basis to the foraging of an ecosystem engineer. *Ecology*, **85**, 603–608.

Bailey, J. K., Schweitzer, J. A., Koricheva, J. et al. (2009) From genes to ecosystems: synthesizing the effects of plant genetic factors across systems. *Philosophical*

Transactions of the Royal Society of London, Series B, **364**, 1607–1616.

Barbour, M. G., Burk, J. H. and Pitts, W. D. (1987) Terrestrial Plant Ecology. Menlo Park, CA: Benjamin-Cummings Publishing Company.

Bassar, R. D., Marshall, M. C., López-Sepulcre, A. et al. (2010) Local adaptation in Trinidadian guppies alters ecosystem processes. Proceedings of the National Academy of Sciences of the United States of America, **107**, 3616–3621.

Baumeister, D. and Callaway, R. M. (2006) Facilitative effects of Pinus flexilis during succession: a hierarchy of mechanisms benefits other plant species. Ecology, **87**, 1816–1830.

Benkman, C. W. (1995) Wind dispersal capacity of pine seeds and the evolution of different seed dispersal modes in pines. Oikos, **73**, 221–224.

Benkman, C. W. (1999) The selection mosaic and diversifying coevolution between crossbills and lodgepole pine. American Naturalist, **154**, S75–S91.

Benkman, C. W. and Siepielski, A. M. (2004) A keystone selective agent? Pine squirrels and the frequency of serotiny in lodgepole pine. Ecology, **85**, 2082–2087.

Benkman, C. W., Holimon, W. C. and Smith, J. W. (2001) The influence of a competitor on the geographic mosaic of coevolution between crossbills and lodgepole pine. Evolution, **55**, 282–294.

Benkman C. W., Parchman, T. L. and Mezquida, E. T. (2010) Patterns of coevolution in the adaptive radiation of crossbills. Annals of the New York Academy of Sciences, **1206**, 1–16.

Benkman, C. W., Siepielski, A. M. and Parchman, T. L. (2008) The local introduction of strongly interacting species and the loss of geographic variation in species and species interactions. Molecular Ecology, **17**, 395–404.

Benkman, C. W., Smith, J. W., Maier, M. et al. Consistency and variation in phenotypic selection exerted by a community of seed predators. Evolution, in press.

Carpenter, S. R., Kitchell, J. F. and Hodgson, J. R. (1985) Cascading trophic interactions and lake productivity. BioScience, **35**, 634–639.

Clark, C. J., Poulsen, J. R., Levey, D. J. and Osenberg, C. W. (2007) Are plant populations seed limited? A critique and meta-analysis of seed addition experiments. American Naturalist, **170**, 128–142.

Elliott, P. F. (1988) Foraging behavior of a central-place forager: field tests of theoretical predictions. American Naturalist, **131**, 159–174.

Enright, N. J., Marsula, R., Lamont, B. B. and Wissel, C (1998) The ecological significance of canopy seed storage in fire-prone environments: a model for non-sprouting shrubs. Journal of Ecology, **86**, 946–959.

Estes, J. A. and Duggins, D. O. (1995) Sea otters and kelp forests in Alaska: generality and variation in a community ecological paradigm. Ecological Monographs, **65**, 75–100.

Harmon, L. J., Matthews, B., DesRoches, S. et al. (2009) Evolutionary diversification in stickleback affects ecosystem functioning. Nature, **458**, 1167–1170.

Haloin, J. R. and Strauss, S. Y. (2008) Interplay between ecological communities and evolution: review of feedbacks from microevolutionary to macroevolutionary scales. Annals of the New York Academy of Sciences, **1133**, 87–125.

Herrera, C. M. (2002) Seed dispersal by vertebrates. In C. M. Herrera and O. Pellmyr, eds., Plant–Animal Interactions: An Evolutionary Approach. New York: Blackwell Scientific Publications, pp. 185–208.

Holt, R. D. (1994) Linking species and ecosystems: where's Darwin? In C. Jones and J. Lawton, eds., Linking Species and Ecosystems. London: Chapman and Hall, pp. 273–279.

Hulme, P. and Benkman, C. W. (2002) Granivory. In C. M. Herrera and O. Pellmyr, eds., Plant–Animal Interactions: An Evolutionary Approach. New York: Blackwell Scientific Publications, pp. 132–154.

Johnson, M. T. J. and Stinchcombe, J. R. (2007) An emerging synthesis between community ecology and evolutionary biology. *Trends in Ecology and Evolution*, **22**, 250–257.

Khalil, M. A. K. (1984) Genetics of cone morphology of black spruce (*Picea mariana* Mill, B. S. P.) in Newfoundland, Canada. *Silvae Genetica*, **33**, 101–109.

Lamont, B. B., Le Maitre, D. C., Cowling, R. M. and Enright, N. J. (1991) Canopy seed storage in woody plants. *The Botanical Review*, **57**, 277–317.

Manzaneda, A. J., Rey, P. J. and Alcántara, J. M. (2009) Conflicting selection on diaspore traits limits the evolutionary potential of seed dispersal by ants. *Journal of Evolutionary Biology*, **22**, 1407–1417.

Matziris, D. (1998) Genetic variation in cone and seed characteristics in a clonal seed orchard of Aleppo pine grown in Greece. *Silvae Genetica*, **47**, 37–41.

Mezquida, E. T. and Benkman, C. W. (2005) The geographic selection mosaic for squirrels, crossbills and Aleppo pine. *Journal of Evolutionary Biology*, **18**, 348–357.

Miller, G. E. (1986) Insects and conifer seed production in the Inland Mountain West: a review. In R. C. Shearer, compiler, *Proceedings: Conifer Tree Seed in the Inland Mountain West Symposium*, General Technical Report INT 203. Ogden, UT: US Department of Agriculture, Forest Service, pp. 225–237.

Miller, T. E. and Travis, J. (1996) The evolutionary role of indirect effects in communities. *Ecology*, **77**, 1329–1335.

Palkovacs, E. P., Marshall, M. C., Lamphere, B. A. et al. (2009) Experimental evaluation of evolution and coevolution as agents of ecosystem change in Trinidadian streams. *Philosophical Transactions of the Royal Society of London, Series B*, **364**, 1617–1628.

Parchman, T. L. and Benkman, C. W. (2008) The geographic selection mosaic for ponderosa pine and crossbills: a tale of two squirrels. *Evolution*, **62**, 348–360.

Pellmyr, O. (2002) Pollination by animals. In C. M. Herrera and O. Pellmyr, eds., *Plant–Animal Interactions: An Evolutionary Approach*.

New York: Blackwell Scientific Publications, pp. 157–184.

Reich, P. B., Hobbie, S. E., Lee, T. et al. (2006) Nitrogen limitation constrains sustainability of ecosystem response to CO_2. *Nature*, **440**, 922–925.

Reznick, D., Butler IV, M. J. and Rodd, H. (2001) Life history evolution in guppies. VII. The comparative ecology of high- and low-predation environments. *American Naturalist*, **157**, 126–140.

Siepielski, A. M. and Benkman, C. W. (2004) Interactions among moths, crossbills, squirrels, and lodgepole pine in a geographic selection mosaic. *Evolution*, **58**, 95–101.

Siepielski, A. M. and Benkman, C. W. (2005) A role for habitat area in the geographic mosaic of coevolution between red crossbills and lodgepole pine. *Journal of Evolutionary Biology*, **18**, 1042–1049.

Siepielski, A. M. and Benkman, C. W. (2007a) Convergent patterns in the selection mosaic for two North American bird-dispersed pines. *Ecological Monographs*, **77**, 203–220.

Siepielski, A. M. and Benkman, C. W. (2007b) Selection by a pre-dispersal seed predator constrains the evolution of avian seed dispersal in pines. *Functional Ecology*, **21**, 611–618.

Siepielski, A. M. and Benkman, C. W. (2008a) Seed predation and selection exerted by a seed predator influence subalpine tree densities. *Ecology*, **89**, 2960–2966.

Siepielski, A. M. and Benkman, C. W. (2008b) A seed predator drives the evolution of a seed dispersal mutualism. *Proceedings of the Royal Society of London, Series B*, **275**, 1917–1925.

Smith, C. C. (1970) The coevolution of pine squirrels (*Tamiasciurus*) and conifers. *Ecological Monographs*, **40**, 349–371.

Smith, C. C. and Balda, R. P. (1979) Competition among insects, birds and mammals for conifer seeds. *American Zoologist*, **19**, 1065–1083.

Strauss, S. Y. and Armbruster, W. S. (1997) Linking herbivory and pollination: new perspectives on plant and animal ecology and evolution. *Ecology*, **78**, 1617–1618.

Strauss, S. Y. and Irwin, R. E. (2004) Ecological and evolutionary consequences of multispecies plant–animal interactions. *Annual Review of Ecology, Evolution, and Systematics*, **35**, 435–466.

Summers, R. W. and Proctor, R. (1999) Tree and cone selection by crossbills *Loxia* sp. and red squirrels *Sciurus vulgaris* at Abernethy forest, Strathspey. *Forest Ecology and Management*, **118**, 173–182.

Thompson, J. N. (1999) Specific hypotheses on the geographic mosaic of coevolution. *American Naturalist*, **153**, S1–S14.

Thompson, J. N. (2005) *The Geographic Mosaic of Coevolution*. Chicago, IL: University of Chicago Press.

Tinker, D. B., Romme, W. H., Hargrove, W. W., Gardner, R. H. and Turner, M. G. (1994) Landscale-scape heterogeneity in lodgepole pine serotiny. *Canadian Journal of Forest Research*, **24**, 897–903.

Tomback, D. F. and Kendall, K. C. (2001) Biodiversity losses: the downward spiral. In D. F. Tomback, S. F. Arno and R. E. Keane, eds., *Whitebark Pine Communities: Ecology and Restoration*. Washington DC: Island Press, pp. 243–262.

Tomback, D. F. and Linhart, Y. B. (1990) The evolution of bird-dispersed pines. *Evolutionary Ecology*, **4**, 185–219.

Tomback, D. F., Arno, S. F. and Keane, R. E. (2001) The compelling case for management intervention. In D. F. Tomback, S. F. Arno and R. E. Keane, eds., *Whitebark Pine Communities: Ecology and Restoration*. Washington DC: Island Press, pp. 3–25.

Turner, M. G., Romme, W. H., Gardner, R. H. and Hargrove, W. W. (1997) Effects of fire size and pattern on early succession in Yellowstone National Park. *Ecological Monographs*, **67**, 411–433.

Urban, M. C., Leibold, M. A., Amarasekare, P. *et al.* (2008) The evolutionary ecology of metacommunities. *Trends in Ecology and Evolution*, **23**, 311–317.

Vander Wall, S. B. (2008) On the relative contributions of wind vs. animals to seed dispersal of four Sierra Nevada pines. *Ecology*, **89**, 1837–1849.

Waring, G. L. and Cobb, N. S. (1992) The impact of plant stress on herbivore population dynamics. In E. A. Bernays, ed., *Insect-Plant Interactions*, Vol. 4. Boca Raton, FL: CRC Press, pp. 167–226.

Werner, E. E. and Peacor, S. D. (2003) A review of trait-mediated indirect interactions in ecological communities. *Ecology*, **84**, 1083–1100.

Whitham, T. G., Bailey, J. K., Schweitzer, J. A. *et al.* (2006) A framework for community and ecosystem genetics: from genes to ecosystems. *Nature Reviews Genetics*, **7**, 510–523.

PART III

Ecosystem

Perspective: interspecific indirect genetic effects (IIGEs). Linking genetics and genomics to community ecology and ecosystem processes

GERARD J. ALLAN and STEPHEN M. SHUSTER

Department of Biological Sciences, Northern Arizona University

SCOTT WOOLBRIGHT

The Institute for Genomic Biology, University of Illinois

FAITH WALKER, NASHELLY MENESES and ARTHUR KEITH

Department of Biological Sciences, Northern Arizona University

JOSEPH K. BAILEY

Department of Ecology and Evolutionary Biology, University of Tennessee

and

THOMAS G. WHITHAM

Department of Biological Sciences, Northern Arizona University

Introduction

Trait-mediated indirect interactions (TMIIs) are important mediators of *community* diversity and structure and associated *ecosystem* processes. Elucidating the genetic basis of ecologically important phenotypic traits is the first step toward understanding the complex interactions that occur among community members. Molecular markers routinely used in *quantitative trait loci* (QTL) analyses (e.g., amplified fragment length polymorphisms (AFLPs), simple sequence repeats (SSRs)) have provided researchers with a toolbox for investigating the genetic basis of heritable traits. A goal of this research is to link genetically based traits to community interactions and ecosystem function. Ultimately, this insight can open a window onto the evolutionary dynamics that shape community structure and associated ecosystem processes (e.g., nutrient cycling). Such an approach is important as it bears on the continued development of the field of *community genetics*, which seeks to understand the genetic interactions that occur between species and their abiotic environment in complex communities (e.g., Whitham *et al.* 2003, 2006; Johnson and Agrawal 2005; LeRoy *et al.* 2006; Bangert *et al.* 2006a, b; Schweitzer *et al.* 2008; Crutsinger *et al.* 2009; Bailey *et al.* 2009).

Trait-Mediated Indirect Interactions: Ecological and Evolutionary Perspectives, eds. Takayuki Ohgushi, Oswald J. Schmitz and Robert D. Holt. Published by Cambridge University Press. © Cambridge University Press 2012.

Although approaches to community genetics differ with respect to the species studied, to date most have focused on *foundation species*, i.e., species that structure a community by creating locally stable conditions for other species, and by modulating and stabilizing fundamental ecosystem processes (Dayton 1972; Ellison *et al.* 2005). The reason for this emphasis is because most ecologists agree that not all species are equal in importance and that some are so important that they effectively 'drive' whole communities and ecosystems. Other similar terms such as core, dominant, keystone and ecosystem engineers are synonymous, but are preceded by the foundation species concept of Dayton (Ellison *et al.* 2005). Although the impacts of foundation species on the community and ecosystem are thought to be most important, species that are not thought to be foundation species can also have broader impacts (e.g., Johnson and Agrawal 2005).

Diverse examples reviewed by Bailey *et al.* (2009) and others include poplar (Whitham *et al.* 2003), seagrass (Hughes and Stachowicz 2004), goldenrod (Crutsinger *et al.* 2006), pine (Iverson *et al.* 2005), oak (Tovar-Sanchez and Oyama 2006) and eucalypt (Barbour *et al.* 2009). For example, using the common prairie herb, *Solidago*, Crutsinger *et al.* (2008, 2010) found that the most common genotypes impacted arthropod community diversity (both positively and negatively) and reduced the likelihood of habitat invasion by exotic species. In another case, involving the shrub *Baccharis*, Crutsinger *et al.* (2010) also found that its dominance on the landscape determined plant colonization success and understory diversity in a coastal dune community. Studies by Johnson and Agrawal (2005), using the perennial herb *Oenothera biennis*, have also shown that arthropod community diversity varies not only by plant genotype, but also across different environmental regimes, an important result demonstrating the combined interaction of genotype × environment. Johnson and Agrawal (2007) have also shown that genotypic diversity has a strong and pervasive impact on arthropod community diversity that extends over the lifetime of the plant, emphasizing the need to explore community-level effects both spatially and temporally. Our studies with cottonwood trees show that they create stable conditions that regulate and modify *community diversity* (Bangert *et al.* 2006a, b), *composition* (Bangert and Whitham 2007), *stability* (Keith *et al.* 2010), and ecosystem function (Schweitzer *et al.* 2008). Importantly, these community and ecosystem effects of individual genotypes are so predictable that they can be quantified as heritable plant traits that are highly repeatable over multiple years of study (e.g., Keith *et al.* 2010). Taken together, the community genetic studies described above provide a broad perspective on how genetic variation in ecologically important traits influences a wide variety of community characteristics (composition and diversity) and ecosystem properties (stability and function). All of these studies and numerous others share common ground, namely to understand the ecological and evolutionary implications that genetic variation in one species has for community structure and ecosystem processes (see reviews by Whitham *et al.* 2006; Johnson and Stinchcombe 2007; Hughes *et al.* 2008; Bailey *et al.* 2009).

Genetic and genomic information is playing a more prominent role in deciphering the genetic basis of traits believed to be involved in community and ecosystem dynamics. Figure 16.1 shows a schematic for developing and organizing genetic- and genomic-based information, which can be applied to model and non-model species. Historically, these studies have focused on the development of QTL, which have been used to understand the links between phenotypic-based trait variation and community and ecosystem processes (Whitham *et al.* 2003, 2006). More recently, however, whole genome sequencing has provided a more direct route for identifying and isolating *candidate genes* believed to be involved in trait-mediated interactions. Once a genome sequence has been determined researchers can search for single nucleotide polymorphisms (SNPs) within genes linked to ecologically important traits. Genetic maps based on AFLPs or SSRs can also be developed for defining QTL. Together, these approaches provide parallel avenues for investigating properties inherent to communities and ecosystems (e.g., community composition, diversity, stability and nutrient cycling) that are the combined response to the interactions that occur among community members and their environment, many of which involve TMIIs.

In this chapter, we focus on a combined approach to understand how genetic-based trait variation in foundation species of *Populus* and their naturally occurring hybrids indirectly influences community composition, *biodiversity* and ecological function in riparian ecosystems. First, we begin with a discussion of TMIIs from the perspective of interspecific indirect genetic effects (IIGEs; Shuster *et al.* 2006). Next, we provide examples of how IIGEs help define the complex interactions that occur between foundation species and associated community members within the context of *community selection*. Lastly, we discuss why we believe the study of IIGEs is important for both ecologists and molecular geneticists alike and how this nascent field can assist researchers interested in the molecular dissection of trait-mediated interactions.

Trait-mediated interactions and IIGEs

Trait-mediated interactions are widely regarded as being an important aspect of species interactions, operating within the context of ecological communities (Werner and Peacor 2003; Rosi-Marshall *et al.* 2007). These interactions are usually discussed in an ecological context with the terms 'direct' and 'indirect' used to describe whether the interactions involve two, or more than two species. Direct ecological interactions occur when one species influences another species in a particular way, such as when a change in predator behaviour stimulates a direct response in prey behaviour or phenotype. Whether phenotypes, fitnesses or both are affected can be variable in most definitions. In contrast, indirect ecological interactions occur when one species influences another

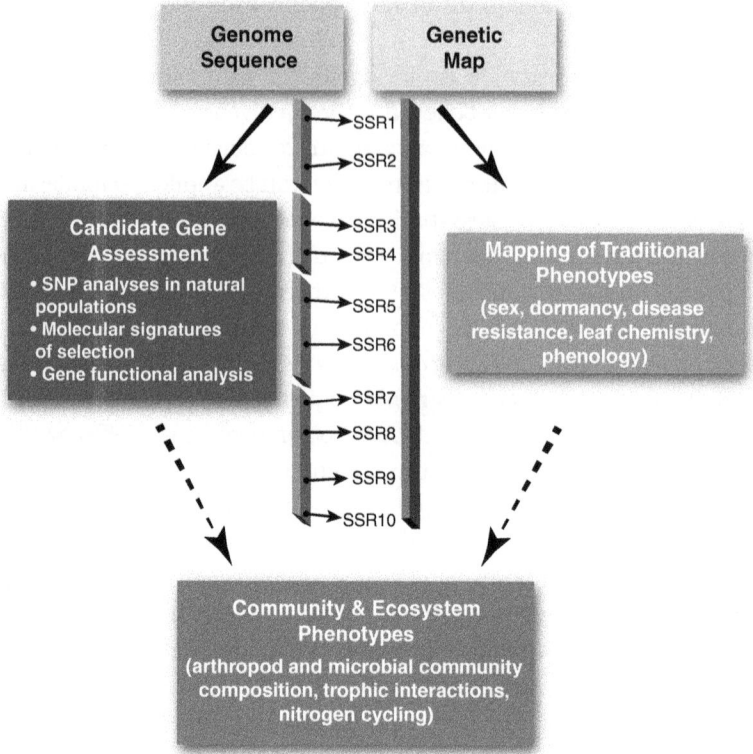

Figure 16.1 Genetic and genomic resources for *Populus* that can be applied to studies of IIGEs. Traditional phenotypes (i.e., those expressed in the individual or population) can be used as a bridge for uncovering the molecular determinants of community and ecosystem phenotypes. First, genomic regions controlling traditional phenotypes are identified with QTL analysis and/or genetic association studies. Candidate genes from these intervals are identified from the genome sequence and anchored to genetic maps with sequence-tagged markers such as simple sequence repeats (SSRs). Second, these candidate genes are assessed for their involvement in traditional tree phenotypes (e.g., tree architecture, phytochemical resistance to herbivores) with high-resolution association mapping in natural populations, searching for signatures of selection such as selective sweeps, enhanced local linkage disequilibrium, and/or rates of synonymous and nonsynonymous polymorphisms. Third, genes with evidence for associations and recent selection are then subjected to functional analysis, including transgenic overexpression and knockouts and transcriptome analysis. Fourth, genes responsible for traditional tree phenotypes can be linked to community and ecosystem phenotypes (dashed arrows) through studies in common gardens and natural populations. SNP, single-nucleotide polymorphism. Reprinted with permission from Whitham *et al.* 2008. See colour plate section.

species, but the nature of that influence can vary as a result of the influence of a third species (Wootton 1994). Although TMIIs are typically expressed as a function of the variation in phenotypic plasticity among interacting organisms, little is known about the underlying genetic basis of these traits. 'Indirect' effects in the ecological sense, for example, have traditionally been classified as either trait-based (Abrams *et al.* 1996) or density-based (Abrams 1995). We suggest that uncertainty has led to confusion among ecologists. Moreover, confusion is compounded by the fact that the terms 'direct' and 'indirect' are also used in somewhat different ways by evolutionary geneticists (Wade 2009; see below). For this reason, we argue for the recognition of a third class of interaction, namely genetic-based indirect interactions. Given that these interactions typically occur among ecologically associated species, and because genetic variation in one species appears to cause fitness variation in other associated species, we refer to this class specifically as *interspecific indirect genetic effects* (IIGEs; Shuster *et al.* 2006). We assert that identification of the genetic basis of interspecific interactions provides a way to unify both ecological and evolutionary considerations of interactions that occur within and among species.

The concept of IIGEs is drawn from a related theoretical framework introduced to understand the consequences for trait evolution via *indirect genetic effects* (IGEs) acting through interactions among individuals of the same species (Moore *et al.* 1997; Wolf *et al.* 1998; Agrawal 2001). The common conceptual link between IGEs influencing trait evolution within species and IIGEs influencing trait evolution among species (i.e., within a community context) is that the traits involved in both types of interactions have a genetic basis. In an evolutionary sense, the term 'indirect' refers to the selection that is imposed on a focal trait, via selection on traits that are genetically correlated with the focal trait in other organisms. With IGEs this refers to selection that is imposed by the existence of such genetic correlations among individuals of the same species; with IIGEs it refers to selection that is imposed by the existence of such genetic correlations among individuals in different species. This view can be contrasted with more ecological-based views that consider indirect effects without necessarily invoking a genetic basis, although it is likely inferred (e.g., Wootton 1994; Martinsen *et al.* 1998). Our goal in considering IIGEs is to expand the scope of IGEs from a social context involving individuals within a single species, to a community context consisting of populations of individuals in multiple species. The theoretical approach introduced by Shuster *et al.* 2006 and by Lonsdorf *et al.* (in review) provides a framework for measuring selection intensity within a community genetic context thus providing a quantitative method for evaluating the relative role community genetics has within a wide range of communities.

Because such interactions involve traits we suggest that trait-based indirect effects and IIGEs constitute two aspects of the same phenomenon. As with trait-based interactions, IIGEs focus on the fitness outcomes that are measurable as a result of interspecies interactions. However, an emphasis on the genetic basis for trait expression provides a means for predicting the possible evolutionary effects of interspecific interactions. For example, Shuster *et al.* (2006) confirmed that arthropod communities become distinct among cottonwood genotypes due to genetically based interactions between arthropods and their host trees. The appearance within common gardens of distinct communities among tree cross types, and among genotypes within cross types, is consistent with the hypothesis that variation in tree genotype affects variation in fitness within and among arthropod species, i.e., selection occurs within a community context, causing particular tree genotypes and arthropod genotypes to covary. The fact that similar results were produced in simulations in which the genetic basis of plant phytochemical traits and plant utilization traits of 30 arthropod species were explicitly known (see below), substantiates our common garden results and justifies empirical estimates of broad-sense *community heritability* (H^2_C) (Shuster *et al.* 2006). Because the genetic basis for tree and arthropod traits can be explicitly known, this framework provides a means for predicting the possible effects of selective herbivory by a third species, *Castor candadensis* (beaver), on cottonwood communities at all levels of genotypic and phenotypic organization.

Although the idea of IGEs is not new (e.g., Wolf *et al.* 1998), the concept of IIGEs expands on this idea by examining the among species components of genetic-based interactions. Thus, IIGEs investigate the interface at which genetic-based traits and species interactions co-occur. Moreover, because genetically based interactions among species are likely to change the population frequencies of the alleles underlying interacting traits, two outcomes of IIGEs are possible: (1) genetic differences among ecologically similar communities of interacting species are expected; and (2) the fitness consequences of these interspecific interactions are likely to cause community phenotypes to evolve. To this end, we develop a genetic-based framework that allows tracing of traditional trait-based effects back to the genetic makeup and structure of interacting community members.

Our framework has been developed within the context of *community and ecosystem genetics* (Neuhauser *et al.* 2003; Whitham *et al.* 2003) and is useful for several reasons. First, given that ecological communities comprise a web of interacting organisms it is important to understand who the 'major players' are (i.e., those primarily responsible for driving indirect interactions) and how their genetic makeup influences other community members. Foundation species (Ellison *et al.* 2005) often fill this role and typically harbour substantial genetic variation in ecologically important traits that is widespread across landscapes. For example, in their review of condensed tannins, an important phytochemical with both community and ecosystem effects, Schweitzer *et al.* (2008) reported up to 145-fold differences in concentrations among individual genotypes of the same species. Second, IIGEs are part of the evolutionary histories of interacting organisms; as such they are subject to selective pressures typical for all morphological-based traits, as well as to selection that arises as a result of interactions with other species. We argue that a genetic-based framework is the template on which indirect interactions evolve and are integral to maintaining the stability of ecological communities (e.g., Evans *et al.* 2008; Keith *et al.* 2010). Third, given that IIGEs are part of a dynamic system, they can change over space and time, a phenomenon which likely reflects genetic-based trait variation among interacting species. Fourth, biodiversity often arises as a result of genetic- and genomic-based interactions that drive population-level differences. These differences are often mirrored by genetic divergence among closely related species occupying shared niche space, or by distantly related species that are dependent on one or a few foundation species within ecological communities (Ellison *et al.* 2000).

Foundation species as mediators of IIGEs
Examples from multiple systems
Due to their dominant presence across landscapes foundation species establish the conditions necessary for developing and maintaining ecosystem diversity, function and stability (Dayton 1972; Ellison *et al.* 2005). As primary producers, forest trees are typically recognized as 'foundational' in that they define forest structure, modulate decomposition rates, regulate nutrient flux and drive carbon sequestration and energy flow within forest ecosystems. A few of the well-known examples of forest trees that perform these functions include eastern hemlock, mangrove, eucalyptus and poplar species. Eastern hemlock, for example, forms critical niche space for unique assemblages of vertebrates and invertebrates (Snyder *et al.* 2002), and is an important regulator of hydrologic and soil ecosystem processes (Ellison *et al.* 2005). Mangrove forests are among the highest net primary producers regulating intertidal food webs (Ellison and Farnsworth 2001). Eucalypts support numerous microbial, fungal, animal and plant communities (Whitham *et al.* 1994; Dungey *et al.* 2000; Barbour *et al.* 2009) and regulate carbon cycling in Australian forest ecosystems (Binkley and Ryan 1998; Pregitzer and Euskirchen 2004). Members of the genus *Populus* (commonly called 'cottonwoods') are dominant colonizers of riparian environments where they exhibit strong influences on dependent community members and associated ecosystem processes in aquatic and terrestrial habitats (Whitham *et al.* 2006). Although several foundation trees have known effects on communities and ecosystems, only one, *Populus*, has been the focus of extensive development of genetic and genomic resources (i.e., molecular-based tools), which have been used, in part, to study IIGEs.

Populus provides a genetic- and genomic-based framework for identifying IIGEs
Owing to its utility as an economically important forest tree, *Populus* has been the subject of intensive breeding programmes designed to enhance the

production of pulp and paper products and more recently, biofuel (Aylott *et al.* 2008). In addition to traditional breeding practices, marker-assisted breeding has also been employed including the development of markers for genetic fingerprinting (e.g., restriction fragment length polymorphisms (RFLPs) and SSRs). Recently, *Populus* was chosen as a model forest tree for whole genome sequencing (Tuskan *et al.* 2006), resulting in the development of numerous genetic and genomic resources (Tuskan *et al.* 2003; Brunner *et al.* 2004) including *expressed sequence tag* (EST) libraries (Sterky *et al.* 2004) for use in gene expression studies, single nucleotide polymorphism (SNP) and simple sequence repeat (SSR) databases (Tuskan *et al.* 2004) for developing QTL linkage maps (e.g., Woolbright *et al.* 2008) and marker–trait association studies (Ingvarsson 2005, 2010). Together, these resources provide a platform for the study of *community and ecosystem genomics*. Even more important has been the identification of candidate genes for examining the molecular basis of ecologically adaptive traits such as cold tolerance (Olson, pers. comm.), bud set and bud flush (Evans *et al.* unpub., Woolbright *et al.* unpub.), tree architecture (Woolbright *et al.* unpub.) and phytochemistry (Zinkgraf *et al.* unpub.), many of which have important consequences for community interactions and ecosystem processes.

IIGEs, QTLs and candidate genes

The first step toward identifying IIGEs for community and ecosystem studies is to identify specific genes or genomic regions (e.g., QTL) linked to relevant traits of interest. In foundation species such as *Populus*, these may include: defensive phytochemicals that affect food choice (Bailey *et al.* 2004; Clausen *et al.* 2005) or fitness traits of community members (Hwang and Lindroth 1995; Hemming and Lindroth 1995) and ecosystem processes such as decomposition and nutrient cycling (reviewed in Schweitzer *et al.* 2008); variation in spring bud break/leaf flush affecting food quality (Floate *et al.* 1993) or timing of gall initiation (Abbott and Withgott 2004); architectural traits such as internode length and competition between plant sinks (e.g., developing buds) and phloem-feeding arthropods (Larson and Whitham 1991), branching patterns and nest site choice for birds (Martinsen and Whitham 1994); and overall habitat complexity stemming from variation in whole tree fractal dimension (Bailey *et al.* 2004). All of these traits are highly heritable and have proven community or ecosystem effects making them ideal for mapping to one or more of the available *Populus* genetic maps.

A number of genetic maps have been created for *Populus* (e.g., Bradshaw *et al.* 1994; Wu *et al.* 2000; Cervera *et al.* 2001; Yin *et al.* 2004; Woolbright *et al.* 2008) that are useful not only for QTL mapping of ecological and commercial traits but also for comparative studies aimed at revealing the evolutionary forces behind trait divergence among species and sections (see Harding *et al.* 2005; Morreel *et al.* 2006; Tsai *et al.* 2006a, b). Such divergence has almost certainly played a role in the evolution of dependent organisms, with the effects of foundation species traits cascading through trophic levels leading to direct and indirect genetic interactions among community members (Whitham *et al.* 2003, 2006). Because recent maps are now aligned with the *Populus* genome sequence (Cervera *et al.* 2001; Yin *et al.* 2004; Woolbright *et al.* 2008), candidate gene surveys stemming from QTL for key traits can be used to link the highest levels of organization (communities and ecosystems) to variation at the simplest (DNA sequence), thus achieving a major goal in the emerging fields of community and ecosystem genetics (Mitton 2003). For example, the extended community and ecosystem effects of the defensive phytochemicals, condensed tannins (proanthocyanidins), are well established (review in Schweitzer *et al.* 2008), making them an ideal candidate for genetic mapping of an ecologically important trait. It has been found, for example, that condensed tannins affect food choice in North American beaver (*Castor canadensis*), with beaver preferring parental species and hybrids with low condensed tannin expression, avoiding higher tannin genotypes (Bailey *et al.* 2004).

The distribution and abundance of diverse species at a variety of trophic levels vary predictably among cottonwood species and hybrids (Wimp *et al.* 2005), and selective removal by beaver or other herbivores can have consequences that cascade through entire communities and ecosystems. Using a cottonwood linkage map (Woolbright *et al.* 2008), we have identified five QTL for condensed tannins. Alignment with the *Populus* genome sequence using shared microsatellite markers has revealed 10 co-locating candidate genes from the flavonoid biosynthesis pathways leading to the formation of condensed tannins. Similar analyses have been performed for other chemical, morphological, and phenological traits (Woolbright in prep and unpub.). Employing a QTL and candidate gene approach represents a first step toward uncovering the genetic basis of traits that have important ecological and evolutionary consequences for communities and ecosystems.

The conceptual framework identified in Fig. 16.1 is also relevant for exotic foundation species in which key genes have been identified that have community and ecosystem consequences. For example, in the early 1950s a female form of the dioecious aquatic plant hydrilla (*Hydrilla verticillata*) was introduced into Tampa Bay, Florida, USA, where it spread rapidly (Schmitz *et al.* 1991). This species is now the most abundant, non-indigenous aquatic plant in Florida and is one of the most serious weed problems in the southern and western USA. Due to its invasive and clonally reproducing nature *Hydrilla* is capable of rapidly replacing native plant communities via competitive exclusion, thereby upsetting the delicate balance of a previously pristine aquatic ecosystem (Colle and Shireman 1980; Schmitz and Osborne 1984; van Dijk 1985; Schmitz *et al.* 1993; Bates and Smith 1994). Efforts to control the spread of *Hydrilla* have employed fluridone (Sonar®) at a concentration of 4–12 µg/l, maintained over a period of several weeks (Netherland and Getsinger 1995; Fox *et al.* 1996). Fluridone has proven to be an effective control as it inhibits phytoene desaturase (PDS), a rate-limiting enzyme involved in the biosynthesis of carotenoids, which are critical for photosynthetic activity (Chamovitz *et al.* 1993).

Using cloning of the *pds* gene and site-directed mutagenesis, Albrecht *et al.* (2004) identified three separate and independent point mutations that were associated with resistance and susceptibility to fluridone. Furthermore, within each water body tested, genetically resistant individuals were identified and shown to be the dominant biotype. Hence, from an IIGE per-spective, a fluridone-resistant genotype could effectively eliminate IIGEs that arise through multispecies interactions within the aquatic community. Furthermore, studies have shown that when native habitat is displaced by monocultures of single species, entire ecosystems become less diverse and can have important consequences for biodiversity (Johnson *et al.* 2006).

Because IIGEs evolve in response to interactions among diverse species and their environment, we argue that understanding these interactions within a community context is essential. This approach becomes all the more impor-tant when considering the impact that foundation and even non-foundation species have on communities and ecosystems, especially from the perspective of maintaining the integrity of community diversity and ecosystem function (Whitham *et al.* 2010).

IIGEs, community interactions and ecosystem processes

Foundation species can have quantifiable impacts on the relative fitness of associated or dependent species, leading to the formation of IIGEs (Shuster *et al.* 2006; Whitham *et al.* 2008). Many of these interactions arise indirectly from the genetic variation inherent in *Populus* species and can affect other foundation species. For example, the local and geographic distribution of the gall-forming aphid *Pemphigus betae* is strongly influenced by the patterns of naturally occurring hybridization between two cottonwood species, *P. fremontii* and *P. angustifolia*, and the variation within *P. angustifolia*. This aphid is concen-trated in the naturally occurring hybrid zones where its densities average 119 times greater than in adjacent pure zones (Whitham 1989). This same pattern occurs throughout many riparian zones of western North America (Floate *et al.* 1997). At the stand scale, individual trees vary greatly in their resistance traits to these aphids such that survival ranges from 0–75%. When individual trees in the wild were asexually propagated and grown in a common garden and aphids were transferred onto the derivative clones, we found survival on the derivative clones was highly correlated with survival on the parental tree ($r=0.90$, $P<0.001$), indicating that resistance to aphid attack is genetically based.

Using a relatively novel technique called association genetic mapping (Zhu *et al.* 2008), recent studies have identified three significant QTL as well as six candidate genes that are associated with aphid resistance and susceptibility traits (Zinkgraf *et al.*, unpub). With pest and pathogens of economic importance, such studies would not be unusual except that in this system associated community genetics studies of the IIGEs have linked these QTL to community and ecosystem traits. For example, Dickson and Whitham (1996) explored how the interaction between aphid resistant and susceptible cottonwood trees affected a much larger community. They found that susceptible trees supported 31% greater species richness and 26% greater relative abundances than resistant trees (Dickson and Whitham 1996). Because this analysis included 42 arthropod taxa from 35 families and 14 orders (herbivores, predators and parasites), it represents a clear case of greater biodiversity being associated with susceptibility. Dickson and Whitham (1996) also experimentally removed the aphids from susceptible trees to see if the genetic-based interaction of the aphids with the tree was responsible for the community differences; or if independent of the aphids, the genetic differences between tree phenotypes acted directly on the community. They found that with the removal of the aphids, the communities of susceptible trees were very similar to resistant trees, arguing that IIGEs between two foundation species (i.e., tree and aphid) defined a much larger community, resulting in a *community phenotype*. When major taxonomic groups were analysed using the same data, they found that the removal of aphids from susceptible trees had a pronounced effect on arthropods, inquilines (mostly spiders that move into galls), birds and fungi (Fig. 16.2).

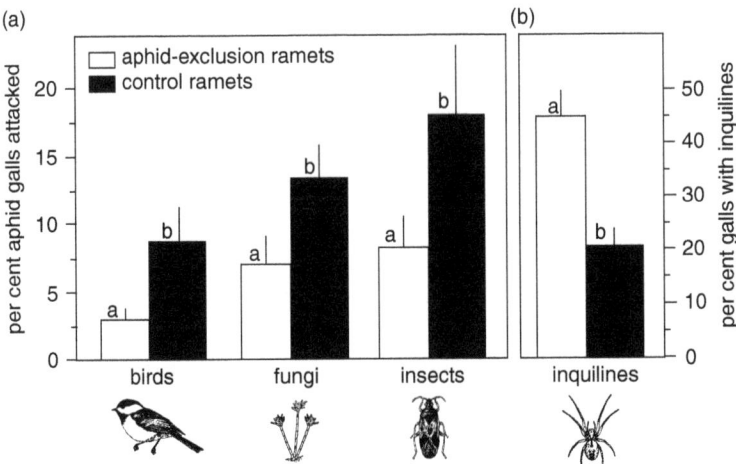

Figure 16.2 Percentage of galls attacked by birds, fungi, and insects on paired exclusion and control ramets (n = 12 pairs). Reprinted with permission from Dickson and Whitham 1996.

Such experiments involving different tree genotypes, in this case differ-ent resistant phenotypes, are crucial to the study of IIGEs because they document how plant genetic-based effects alter another species such as an aphid herbivore, which in turn can affect a much larger community of organisms. In this case, the IIGE between the plant and the community acted through a herbivore. Similar experimental findings have been found with gall-makers on goldenrod (Crawford *et al.* 2007). In another case involv-ing insect resistant and susceptible pinyon pine, which were found to sup-port different mycorrhizal communities, the removal of the herbivore had no effect on the mycorrhizal community, arguing that the effect of plant resistance on the community was direct and not mediated by a herbivore (Sthultz *et al.* 2009).

Because several species of birds prey upon the many aphids contained within a gall, susceptible trees that attract many aphids also affect higher trophic levels. Using common garden studies with clonal replicates of indi-vidual tree genotypes, Bailey *et al.* (2006) showed that aphid densities on individual tree genotypes of *P. angustifolia* were under strong genetic control ($H^2 = 0.60$). Because avian predators are very sensitive to aphid prey densities, they also found that the foraging of birds was affected by tree genotype ($H^2 = 0.53$). As aphids are directly affected by tree genotype and avian preda-tors are indirectly affected by tree genotype, this study documents a genetic basis to *trophic interactions* and energy flow through an ecosystem. Consequently, this is one of the first studies to document a heritable basis to an important ecosystem process.

Tree genotype also affects an important ecosystem process of decomposition (Schweitzer *et al.* 2005). Because *P. betae* induces the tree to produce higher levels of condensed tannins in galled leaves relative to ungalled leaves, higher tannin production results in lower decomposition rates when the leaves are abscised, drop to the ground and decomposition by microbes begins. Thus, susceptibility to aphids has an *ecosystem phenotype* that alters a fundamental ecosystem process of decomposition beneath susceptible trees relative to resistant trees that have few or no galling aphids.

Recent studies by Keith *et al.* (2010) show trees with higher aphid densities also have more stable arthropod communities over a 3-year period of study. In this study, community stability was measured as the Bray–Curtis Similarity from one year to the next. They found that as aphid density increased on a tree, community stability also increased significantly (Fig. 16.3; $r^2 = 0.37$; $P < 0.0001$). Although the mechanism(s) that results in these gall-forming aphids having a positive effect on the arthropod community is unknown, the removal experiments of Dickson and Whitham (1996) showed that the presence of aphids had a positive effect on species richness and abundance,

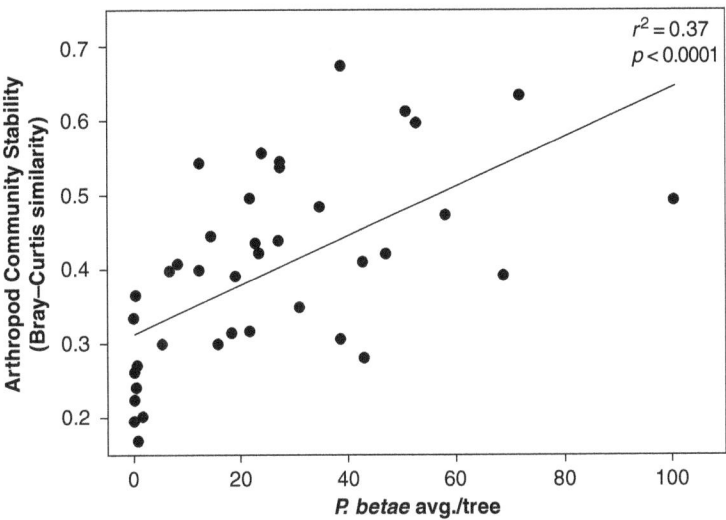

Figure 16.3 The interacting foundation species hypothesis tests how aphid abundance is correlated with arthropod community stability using Bray–Curtis similarity across years for 44 individual trees. Reprinted with permission from Keith *et al.* 2010.

which is consistent with the hypothesis that increasing diversity is associated with community stability.

The interactions discussed above often result in tree genotype-specific trophic interactions and energy flow within ecosystems (Schweitzer *et al.* 2005; Bailey *et al.* 2006). In addition, the establishment of arthropod communities on specific tree genotypes has been shown to represent heritable plant traits (Bailey *et al.* 2006; Shuster *et al.* 2006; Bangert *et al.* 2006a, b; Keith *et al.* 2010). Furthermore, in *Populus* hybrid zones, which regularly form in areas of sympatry, approximately 30–78% of the genetic variation in these trees has been found to drive community diversity (Wimp *et al.* 2004; Bangert 2004). Other traits that have a genetic basis, and therefore qualify as IIGEs that indirectly affect communities and ecosystems in *Populus* hybrid zones, include belowground carbon storage and fine root production (Fisher *et al.* 2006, 2007), water cycles (Fischer *et al.* 2004) and plant growth rate (Lojewski *et al.* 2009), all of which can be traced back to variation in tree genotype or direct estimates of genetic diversity based on molecular markers (Wimp *et al.* 2004; Bangert *et al.* 2004). These studies provide a framework for probing the links between genetically based plant phenotypes and the web of interactions that comprise riparian ecosystems. In the next section we discuss how IIGEs involving foundation cottonwood species are being used to investigate riparian community member diversity and selection within a community context.

IIGEs: cottonwoods, beavers and arthropod communities

Beavers are well known to play a major role in defining riparian communities (e.g., Johnston and Naiman 1990; Naimen *et al.* 1993; Aznar and Desrochers 2008), and are considered classic examples of foundation species or ecosystem engineers (Jones *et al.* 1994, 1997; Wright *et al.* 2002, 2003; Rosell *et al.* 2005; Hood and Bayley 2008). In the Southwest USA, this foundation mammal prefers to fell a foundation tree – cottonwood – for food, thereby eliciting a suite of far-reaching IIGEs. Here we discuss a plant trait important to beavers that is heritable and can be traced to the genomic level, and illustrates how selective foraging by beavers scales up to generate IIGEs in arthropod, and likely other, communities.

In cottonwoods, Woolbright *et al.* (2008) have identified QTL that account for a significant portion of phenotypic variation in the production of condensed tannins. Variation in concentration of these defensive phytochemicals is highly heritable (e.g., Bailey *et al.* 2006), both within pure species and their naturally occurring hybrids (*P. angustifolia* × *P. fremontii*) (Driebe and Whitham 2000; Rehill *et al.* 2005, 2007). Beavers selectively fell trees with lower condensed tannins, such as *P. fremontii* (Bailey *et al.* 2004). The impact of selective herbivory by beaver is significant due to the combined effects of beaver avoidance of high tannin genotypes and differential sprouting of different cross types (i.e., *P. angustifolia*, *P. fremontii*, hybrids; Whitham *et al.* 2006). In combination, these studies demonstrate links between a mapped trait of a foundation tree and a mammalian foundation species whose selective foraging feeds back to affect the survival of different tree cross types and genotypes within a cross type.

Beaver-driven alterations of cross type composition and tree architecture have major implications for the dependent arthropod community through IIGEs. Tree compositional changes, as a consequence of beaver herbivory, are suggested to impact arthropods, because different cottonwood cross types support significantly different arthropod communities in both field and common garden studies (Wimp *et al.* 2005; Shuster *et al.* 2006). Tree architectural changes, i.e., resprouting as a consequence of beaver herbivory, positively impact arthropod diversity. In a cottonwood hybrid zone, arthropod community richness and complexity were significantly greater on resprout growth of beaver-felled cottonwoods than on controls (Bailey and Whitham 2007). The resprout growth also attracted gall-forming sawflies (*Phyllocolpa* spp.), which solicited another set of arthropods, illustrating an even more distal link of beaver-caused indirect effects on the arthropod community (Fig. 16.4).

Another study of a pure cottonwood species also found that arthropod species richness was significantly greater on resprout growth of beaver-felled

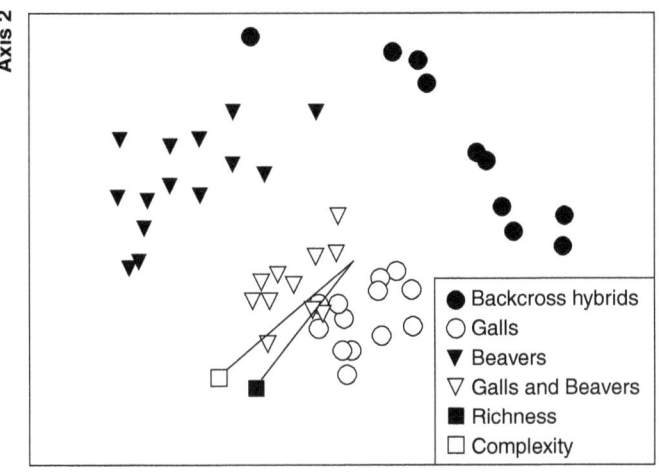

Figure 16.4 The arthropod communities of resprout growth of beaver-felled trees (solid triangles) differed from those of control backcross hybrid trees (solid circles), as shown by non-metric multidimensional scaling. The resprout growth of beaver-felled trees also attracted gall-forming sawflies (*Phyllocolpa* spp.), which solicited another set of arthropods (open circles). Open triangles show the combined effects of beavers and sawflies on the arthropod community. All groups were significantly different from one another and from the control group (ANOSIM analysis). Vector analysis (DECODA) revealed that these treatments significantly increased both arthropod richness and complexity (open and solid squares, respectively) relative to the control group. Reprinted with permission from Bailey and Whitham 2007.

trees than on juvenile growth of control trees, and importantly, was correlated with key phytochemicals (phenolic glycosides, lignin, nitrogen) (Durben *et al.* unpubl. data). As with condensed tannins, these additional phytochemicals are also likely traceable to the genome.

Cottonwood genetics, phytochemistry and beaver herbivory likely interact to influence overall biodiversity. Because arthropods are the basis of a food chain for many insectivorous birds and mammals, beaver–cottonwood–arthropod interactions provide a window on the functioning of an entire ecosystem. The IIGEs between cottonwoods, beavers and arthropods are merely the first links in complex IIGE networks resulting from the interactions between two foundation species (cottonwoods and beavers). Indeed, common garden experiments with cloned trees suggest that beavers and cottonwoods drive a multifaceted set of interactions involving multiple trophic levels and diverse taxa (mycorrhiza, insects, birds and mammals). Next, we address how selection at multiple levels can drive IIGEs within a community context.

IIGEs and multilevel selection
Selection within a community context

It is widely thought that a community genetics approach becomes less useful as species diversity increases because interspecific interactions become too diffuse to measure (see Hubbell 2001). Stated differently, if all species within a community are interacting all the time, how can coherent evolutionary change be possible? Lonsdorf *et al.* (in review) provide a mechanism by which coherent evolutionary change is possible within communities due to the disproportionate effects of foundation species. In particular, even in a species-rich community, strong and/or frequent interactions between species can greatly reduce the effective diversity within the community and cause particular species to exert disproportionate evolutionary effects. When genetic variation in one species influences fitness variation within another species, 'selection within a community context' is possible. This hypothesis is consistent with the now well-documented prediction of IIGE theory (Shuster *et al.* 2006). If IIGEs are important, and if cottonwood trees indeed represent foundation species, each cottonwood tree is expected to support a community of associated species that is distinct from that supported by other genetically distinct cottonwood trees. This prediction is abundantly borne out in studies documenting genetically based interactions between foundation cottonwood species and their associated organisms that generate particular *community phenotypes* (Whitham *et al.* 2003, 2006, 2008).

To link IIGE theory with existing data, Shuster *et al.* (2006) investigated the mechanism by which phenotypically distinct communities can be produced by genetic interactions among plants and their associated species. They modelled the evolution of community phenotypes by considering how a plant phytochemical trait, θ_j, can interact with an arthropod trait, z_i, in ways that change the fitness, allele frequencies and numerical abundances of the i-th arthropod species on the j-th cottonwood tree. These joint effects arise from the interaction of the arthropod trait, z_{ij}, with the phytochemical trait θ_j. This interaction, scaled by the magnitude of its evolutionary consequence, γ, and multiplied by the ratio of genetic variance to phenotype variance, generates a change in the average arthropod trait, or,

$$\frac{d\bar{z}_{ij}}{dt} = \frac{\sigma^2_{G_{ij}}}{\sigma^2_{z_{ij}}} \gamma \left(\theta_i - \bar{z}_{ij}\right). \tag{16.1}$$

Shuster *et al.* (2006) showed how a change in the average phenotype of the arthropod trait could be translated into changes in the numerical abundances of species, and therefore into changes in arthropod community composition, generating what they called selection 'within a community context'. In simulations, this evolutionary process produced significant differences among arthropod communities that were associated with different genotypes and hybrid cross types of cottonwood trees. Moreover, simulated communities closely resembled the actual arthropod communities on these same tree genotypes when grown in common gardens. This framework justifies empirical estimates of broad-sense community heritability (H^2_C), and provides a means for modelling the approximate effects that interspecific interactions, overall selection intensity within the community and environmental factors may have in producing community-level phenotypes.

Community-level selection

Whitham *et al.* (2003, 2006, 2008) argued that, at the community level, the interactions of all species were not equally important, and that the interactions of relatively few species, especially foundation species such as cottonwoods and beavers, could define a much larger community. Thus, if beavers tend to fell particular cottonwood genotypes and avoid others, they are expected to remove certain cottonwood-associated communities and allow other associated communities to persist. Wade (1977, 1978) showed experimentally that selection acting at the group level tends to preserve genetic interactions among traits. He argued that such epistasis could generate large among-group differences in phenotype, which permit further rapid changes in phenotype when group selection is imposed. Similar results are known from the agricultural literature when interdemic and interfamily selection is imposed (Wright 1931, 1978; Curry-Woods *et al.* 1999).

These findings inform theoretical and empirical efforts to understand the outcome of circumstances in which groups of organisms – in this case *entire communities* – are the units of selection; i.e., when *community-level selection* occurs. A central prediction of this hypothesis is: if community phenotypes arise when multiple species interact, i.e., selection within a community context occurs (Shuster *et al.* 2006), and if these interactions are preserved when some communities persist and others are removed (see Wade 1977, 1978; Goodnight and Craig 1996), then stands of cottonwoods experiencing beaver herbivory are likely to diverge from one another genetically, and show greater variance in community phenotype compared to cottonwood stands not experiencing such selection. This prediction suggests that beaver herbivory may constitute a form of community-level selection that could have a measurable effect on genetic variation among trees and among arthropods living on trees. Moreover, because arthropod communities can be analysed as quantitative traits (Shuster *et al.* 2006), community-level selection imposed by beaver herbivory could also have a measurable effect on arthropod community phenotype.

Community-level selection: a simulation approach

We explored how beaver herbivory affects arthropod community phenotype using the theoretical approach of Shuster *et al.* (2006), in which synthetic arthropod communities were created on synthetic cottonwoods. In these simulations, the genetic basis for tree traits and arthropod traits was explicitly known (see below). Also explicitly known were the number, intensity and fitness consequences of interactions among species that led to each community phenotype. This framework allowed us specifically to explore the possible genetic consequences of beaver herbivory on trees and arthropods, as well as the phenotypic consequences of beaver herbivory on arthropod communities.

Shuster *et al.* (2006; see also Ronce and Kirkpatrick 2001) created synthetic trees in which a single trait, θ, controlled by two alleles, influenced plant phytochemistry and varied among tree genotypes and cross types (e.g., pure parental and hybrid types). For these simulations they used four alleles to create 10 genotypes (four homozygotes and six possible heterozygotes). Each genotype was then replicated five times to yield 50 individual trees to approximate the scale of cottonwood common garden experiments (Whitham *et al.* 2006). Shuster *et al.* (2006) next created synthetic arthropods, in which a single trait, z, also controlled by two alleles, influenced each arthropod species' ability to use different tree genotypes as habitat. Simulated arthropods (25 species) were allowed to colonize simulated trees, resulting in particular abundances of arthropod species becoming associated with each tree due to the interaction between tree and arthropod genotypes. Non-metric multidimensional scaling (NMDS) was used to collapse arthropod multispecies abundances for each tree into a single community phenotype for genetic analysis, which demonstrated that community phenotypes can be analysed as standard quantitative traits. The analysis of Shuster *et al.* (2006) also showed that empirical estimates of broad-sense community heritability (H^2_C) arise from heritable variation in a host tree trait and the fitness consequences of IIGEs that extend from tree trait to arthropods. When arthropod traits are heritable, interspecific IGEs cause species interactions to change, and evolution within a community context occurs.

When beavers impose selection on cottonwoods by preferring particular tree genotypes (e.g., those with low tannins; Bailey *et al.* 2004), they also impose selection on the associated arthropod communities. Beavers tend to harvest trees patchily, according to their preferences for low tannin genotypes, and beaver resistant trees resprout or disproportionately contribute to the next seedling generation in heavily browsed areas at the expense of beaver susceptible trees (Bailey *et al.* 2004; Whitham *et al.* 2006). To model this process, we revised the framework of Shuster *et al.* (2006) to include simulated cottonwood stands in which simulated beavers remove particular tree genotypes. We established three treatments: (1) selective herbivory (SH) in which beavers prefer particular tree genotypes, (2) random herbivory (RH) in which beavers remove trees at random with respect to genotype, and (3) control (C) in which no trees are removed. Arthropod community phenotypes appear to arise due to the fitness consequences of genetic interactions between trees and arthropods (Shuster *et al.* 2006). Because these interactions are preserved when beaver herbivory causes community-level selection, changes in community phenotypes arising from community-level selection could be *disproportionate* to changes in allele and in trait frequency arising from individual selection on trees and arthropods alone. This effect is likely to be identifiable within the SH cottonwood stands but not within the RH and C stands because community-level selection is likely to increase the among-stand component of phenotypic variation (see Wade 1977).

We measured the variance of arthropod community phenotype as determined from NMDS scores of arthropod abundances. If community-level selection as described above occurred, we expected the among-stand component of the total variance in community phenotype to be large. To illustrate this prediction, we simulated beaver herbivory on five stands of 25 trees ($n = 125$). Each stand contained five replicates of five tree phenotypes whose resistance to beavers differed among stands. We allowed beavers to cause 80% mortality of trees in each forest and allowed recolonization by the resistant tree genotypes in each forest. In this simulation, beaver herbivory increased the among-stand component of the total variance in community phenotype by 76 fold (Fig. 16.5a) and increased the among-stand fraction of the total variance by 83% (Fig. 16.5b).

This study indicates that selection acting in a community context (see Shuster *et al.* 2006) combined with the differential survival and proliferation of particular communities, i.e., community-level selection (see Goodnight and Craig 1996) can cause measurable increases in the among-stand component of

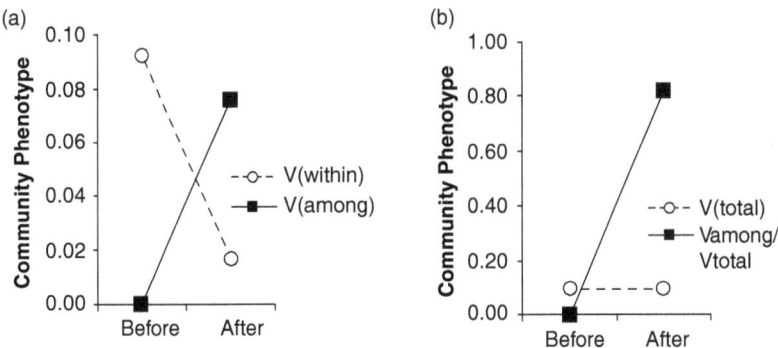

Figure 16.5 Simulated beaver herbivory in a 125-tree forest consisting of five stands of genetically heterogeneous trees (A) increased the among-stand component of the total variance in community phenotype 76-fold, and (B) increased the fraction of the total variance in community phenotype that existed among stands from 0 to 83%; simulation methods in Shuster *et al.* 2006.

variance in community phenotype. This outcome shows not only that community-level selection is possible but also provides an empirical framework (see Wade 1977) for documenting the existence of community-level selection in natural populations.

The relevance of IIGEs for future research

Our discussion of IIGEs and their relevance to community and ecosystem processes begets two critical questions: (1) Why should ecologists care about identifying IIGEs? and 2) Why should molecular geneticists be concerned about the community and ecosystem consequences of IIGEs? Given that IIGEs have both ecological and evolutionary implications we hope the answer to the first question is obvious: ecologists should care about IIGEs because species interactions do not operate within a vacuum. Rather, the numerous connections among species are essentially knit together via the fitness consequences of genetic-based interactions among diverse taxa occupying multiple trophic levels. Furthermore, these consequences can apply to both foundation and non-foundation species' interactions (Hughes *et al.* 2008) as in the case of *Hydrilla* (Michel *et al.* 2004) and studies of other non-foundation species (e.g., Johnson *et al.* 2006). We argue that molecular geneticists should also be concerned about IIGEs because they can have important consequences when manipulated, as in the case of genetically transformed organisms (i.e., transgenics). For example, agricultural crops that have been genetically altered, such as maize, have been shown to have community and ecosystem phenotypes arising from indirect interactions. Rosi-Marshall *et al.* (2007) showed that Bt corn residues in a terrestrial agricultural community negatively impacted a non-target species in an adjacent aquatic community. Such studies emphasize that transgenics and their associated IIGEs need to be further evaluated within the context of target and non-target communities. Doing so will allow us to understand better how changes in the genetic makeup of organisms, especially those considered to be 'community drivers', influence the function, stability and diversity of ecosystems.

Whether the study of IIGEs is undertaken by ecologists or molecular geneticists, it is also important to recognize that studies of IIGEs provide opportunities to examine the relative contributions of genes versus the environment. This is particularly relevant for understanding the genetic basis of TMIIs because the degree to which a trait is influenced by a gene (or genes) often depends on both the environment and the genes examined. Environmental inputs, for example, often influence the expression of particular genes, which makes it all the more critical to separate genetic from environmental effects. Calculating the heritability of interacting traits is clearly one way of isolating genetic from environmental effects, and is an approach that is inherent in the study of IIGEs (Shuster *et al.* 2006). Furthermore, for highly heritable traits, it may even be possible to search for individual genes that contribute to variation in particular traits, especially for organisms whose genomes have been sequenced and characterized. For example, are there genes, or suites of genes that affect key traits in a foundation species and subsequently promote greater diversity and community stability? Such traits have typically been viewed as emergent properties of systems not deducible from lower-order processes, but recent studies by Keith *et al.* (2010) show that they can be under strong genetic control and highly heritable.

Given that the study of IIGEs is in the early developmental stages it is also important to recognize where the strengths and potential pitfalls are when it comes to launching a programme designed around IIGEs. First, we emphasize that while a genetics-based approach is both useful and informative, it does not represent the 'be all and end all' of studying trait-mediated indirect effects (TMIEs) in communities and ecosystems. Hence, we are not advocating that researchers abandon traditional approaches to studying TMIIs (e.g., via characterization and analysis of behavioural and morphological traits), but only to consider where the connections between genes and the environment lie. Researchers need to evaluate and distinguish carefully between genetic-mediated and environment-mediated variation. Such an approach, for example, is facilitated by rearing individuals in common environments in the laboratory, or through the use of common gardens in the field. Second, it is clear that the genetics and genomics 'revolution' is advancing at a rapid pace and that the emerging field of *functional genomics* (e.g., Goga-Vukmirovic and Tilghman 2000) combined with its sister discipline, *comparative genomics*, will likely play a bigger and more important role in the integration of a genes to ecosystem programme. This is particularly important because it will take us from a 'gene-by-gene' approach to one that synthesizes information from whole genomes. Such information can ultimately contribute to insight into the interactions that arise from gene transcription, translation and protein–protein interactions, all of which result in the expression of traits that are the focus of TMIIs.

We suggest that researchers carefully consider the focal species and associated genetic and genomic resources available for studying IIGEs. Our studies suggest that maximum information can be obtained from focusing on those species that tend to drive community and ecosystem processes, and for which some genetic or genomic resources are available. Fortunately, a number of genetic resources are widely available for a large majority of organisms (e.g., consider AFLPs for genetic mapping) and genomic resources (SNPs, SSRs) are obtainable via partial DNA sequencing of individual genomic regions, or by new technologies such as *next generation sequencing*, which is quickly becoming accessible to members of the scientific community (Mardis 2008). Despite these new methods and technologies, however, we recognize that researchers new to community genetics may not have extensive genomic resources available for their system of interest. In this case, more traditional molecular methods such as QTL and genome-wide association studies using conventional markers (e.g., AFLPs or SSRs) can be initiated within a reasonable time frame and have the benefit of a well-established framework for molecular quantitative analyses (e.g., Beavis 1997; Kearsey 1998; Stinchcombe and Hoekstra 2008). Finally, as demonstrated by our simulation studies, we also call attention to the need to integrate modelling of community and ecosystem interactions, an approach also advocated by Goudard and Loreau (this volume). Combined with genetic or genomic information, the synthesis of an IIGE and modelling framework can provide a powerful way to investigate experimentally theoretical predictions, which can be empirically tested in the field. Hence, by combining molecular-based methodologies with theoretical models and simulation studies, researchers can begin to understand, at a deeper level, the complex web of interactions that drive community diversity and the associated ecosystem processes that arise from multispecies interactions.

Acknowledgements

Our most current research has been supported by an NSF Frontiers of Integrated Biological Research (NSF-FIBR) grant (DEB-0425908). We thank the US Bureau of Reclamation for additional funding and support for our common garden studies as well as the Ogden Nature Center and the Utah Department of Natural Resources.

Glossary

Candidate gene   A gene located on a chromosome that is suspected of being involved in the expression of a particular trait (e.g., phytochemical defence).

Community   An association of interacting species that live in a particular area.

Community and ecosystem genetics   The study of the genetic interactions that occur among species and their abiotic environment in complex communities.

Community and ecosystem genetics and genomics   The study of the composition and function of whole ecosystems using genetic and genomic data.

Community and ecosystem heritability   The tendency for related individuals to support similar communities of organisms and ecosystem processes.

Community and ecosystem phenotypes   The effects of genes at levels higher than the population.

Community composition   The distribution of individual species within a community. Different communities can have the same diversity or number of species and abundance of individual species, but still differ in their composition.

Community diversity   The number of species in a local area (alpha diversity) or region (gamma diversity). Also, a measure of the number of species that includes the relative abundances of each species.

Community-level selection   Selection that occurs when relative fitness depends on the characteristics of communities to which individuals belong.

Community stability   The similarity in community composition across years for individual tree genotypes, which can be quantified using average Bray–Curtis similarity estimates.

Comparative genomics   The study of the structure and function of whole genomes derived from genome sequencing of different species.

Ecosystem   A biotic community and its abiotic environment.

Expressed sequence tags (ESTs)   Short sections of DNA sequence (usually 200 to 500 nucleotides long) that are generated by sequencing either one or both ends of an expressed gene. Sequences of expressed genes are then used as 'tags' to fish out genes of interest by comparing matching sequences obtained from DNA from different organisms and identify the location(s) of a gene within a genome.

Foundation species   Species that structure a community by creating locally stable conditions for other species, and by modulating and stabilizing fundamental ecosystem processes.

Functional genomics   A field of molecular biology that incorporates genetic and genomic information derived from genome sequencing projects in order to characterize gene function and gene–gene interactions.

Indirect genetic effects (IGEs)   An environmental influence on the phenotype of one individual that is due to the expression of genes by another individual of the same species.

Interspecific indirect genetic effects (IIGEs)   An environmental influence on the phenotype of an individual in one species that is due to the expression of genes by another individual of a different species.

Next generation sequencing   A relatively new method of DNA sequencing capable of rapidly producing millions of DNA sequence reads for a wide variety of organisms.

Quantitative trait locus (QTL)   A genetic locus that is identified through statistical associations between mapped genetic markers and complex traits (such as growth rate or body form).

Simple sequence repeats   Hypervariable repeating sequences of 1–6 base pairs of DNA commonly found throughout the genome.

Single nucleotide polymorphisms   DNA sequence variation that occurs when a single nucleotide – A, T, C, or G – in the genome differs between members of a species or paired chromosomes in an individual.

Trait-mediated indirect interactions   An interaction that arises when changes in a particular trait (e.g., behavioural or morphological) in one species cause a change in traits in another species, which then indirectly affects one or more additional species.

Trophic interactions   A community of interacting species each of which occupies a particular level in a food chain. Interaction among trophic levels represents the transfer of energy from primary producers to predators of herbivores.

References

Abbott, P. and Withgott, J. H. (2004) Phylogenetic and molecular evidence for allochronic speciation in gall-forming aphids (*Pemphigus*). *Evolution*, **58**, 539–553.

Abrams, P. A. (1995) Implications of dynamically variable traits for identifying, classifying and measuring direct and indirect effects in ecological communities. *American Naturalist*, **146**, 112–134.

Abrams, P. A., Menge, B. A., Mittelbach, G. G., Spiller, D. and Yodzis, P. (1996) The role of indirect effects in food webs. In G. A. Polis and K. O. Winemiller, eds., *Food Webs: Integration of Patterns and Dynamics*. New York: Chapman and Hall, pp. 371–395.

Agrawal, A. A. (2001) Phenotypic plasticity in the interactions and evolution of species. *Science*, **294**, 321–326.

Albrecht, P., Bode J., Buiting, K., Prashanth, A. K. and Lohmann, D. R. (2004) Recurrent deletion of a region containing exon 24 of the RB1 gene caused by non-homologous recombination between a LINE-1HS and MER21B element. *Journal of Medical Genetics*, **41**, e122.

Aylott, M. J., Casella, E., Tubby, I. *et al.* (2008) Yield and spatial supply of bioenergy poplar and willow short-rotation coppice in the UK. *New Phytologist*, **178**, 358–370.

Aznar, J.-C. and Desrochers, A. (2008) Building for the future: abandoned beaver ponds promote bird diversity. *Ecoscience*, **15**, 250–257.

Bailey, J. K. and Whitham, T. G. (2007) Biodiversity is related to indirect interactions among species of large effect. In T. Ohgushi, T. P. Craig and P. W. Price, eds., *Ecological Communities: Plant Mediation in Indirect Interaction Webs*. Cambridge: Cambridge University Press, pp. 306–328.

Bailey, J. K., Bangert, R. K., Schweitzer, J. A. *et al.* (2004). Fractal geometry is heritable in trees. *Evolution*, **58**, 2100–2102.

Bailey, J. K., Hendry, A., Kennison, M. *et al.* (2009) From genes to ecosystems: an emerging synthesis of eco-evolutionary dynamics. *New Phytologist*, **184**, 746–749.

Bailey, J. K., Schweitzer, J. A., Rehill, B. J. *et al.* (2004). Beavers as molecular geneticists: a genetic basis to the foraging of an ecosystem engineer. *Ecology*, **85**, 603–608.

Bailey, J. K., Wooley, S., Lindroth, R. L. and Whitham, T. G. (2006) Importance of species interactions to community heritability: a genetic basis to trophic-level interactions. *Ecology Letters*, **9**, 78–85.

Bangert, R. K. (2004) Macroecology of a genetic assembly rule: cottonwood genes structure the leaf-modifying arthropod community. PhD thesis, Northern Arizona University.

Bangert, R. K. and Whitham, T. G. (2007). Genetic assembly rules and community phenotypes. *Evolutionary Ecology*, **21**, 549–560.

Bangert, R. K., Allan, G. J., Turek, R. J. *et al.* (2006a) From genes to geography: a genetic similarity rule for arthropod community structure at multiple geographic scales. *Molecular Ecology*, **15**, 4215–4228.

Bangert, R. K., Turek, R. J., Rehill, B. *et al.* (2006b). A genetic similarity rule determines arthropod community structure. *Molecular Ecology*, **15**, 1379–1392.

Barbour, R. C., O'Reilly-Wapstra, J. M., De Little, D. W. *et al.* (2009) A geographic mosaic of genetic variation within a foundation tree species and its community-level consequences. *Ecology*, **90**, 1762–1772.

Bates, A. L. and Smith, C. S. (1994) Submersed plant invasions and declines in the southeastern United States. *Lake Reservoir Management*, **10**, 53–55.

Beavis, W. D. (1997) QTL analyses: power, precision, and accuracy. In A. H. Paterson, ed., *Molecular Dissection of Complex Traits*. Boca Raton, FL: CRC Press, pp. 145–162.

Binkley, D. and Ryan, M. (1998) Net primary production and nutrient cycling in replicated stands of *Eucalyptus saligna* and *Albizia falcataria*. *Forest Ecology and Management*, **112**, 79–85.

Bradshaw, H. D. Jr, Villar, M., Watson. B. D. *et al.* (1994) Molecular genetics of growth and development in *Populus*. III. A genetic linkage map of a hybrid poplar composed of RFLP, STS, and RAPD markers. *Theoretical Applied Genetics*, **89**, 167–178.

Brunner, A. M., Busov, V. B. and Strauss, S. H. (2004) Poplar genome sequence: functional genomics in an ecologically dominant plant species. *Trends in Plant Science*, **9**, 49–56.

Cervera, M., Storme, V., Ivens, B., *et al.* (2001) Dense genetic linkage maps of three *Populus* species (*Populus deltoides, P. nigra,* and *P. trichocarpa*) based on AFLP and microsatellite markers. *Genetics*, **158**, 787–809.

Chamovitz, D., Sandmann, G. and Hirschberg, J. (1993) Molecular and biochemical characterization of herbicide-resistant mutants of cyanobacteria reveals that phytoene desaturation is a rate limiting step in carotenoid biosynthesis. *Journal of Biological Chemistry*, **268**, 17348–17353.

Clausen, T. P., Reichardt, P. B., Bryant, J. P. and Sinclaire, A. R. E. (2005) Chemical defense of *Populus balsamifera*: a clarification. *Journal of Chemical Ecology*, **18**, 1505–1510.

Colle, D. E. and Shireman, J. V. (1980) Coefficents of condition for largemouth bass, bluegill, and red ear sunfish in *Hydrilla*-infested lakes. *Transactions of the American Fisheries Society*, **109**, 521–531.

Crawford, K. M., Crutsinger, G. M. and Sanders, N. J. (2007) Host- plant genotypic diversity mediates the distribution of an eco-system engineer. *Ecology*, **88**, 2114–2120.

Crutsinger, G. M., Cadotte, M. W. and Sanders N. J. (2009) Plant genetics shapes inquiline community structure across spatial scales. *Ecology Letters*, **12**, 285–292.

Crutsinger, G. M., Collins, M. D., Fordyce, J. A. *et al.* (2006) Plant genotypic diversity predicts community structure and governs an ecosystem process. *Science*, **313**, 966–968.

Crutsinger, G. M., Collins, M. D., Fordyce, J. A. and Sanders, N. J. (2008) Temporal dynamics in non-additive responses of arthropods to host-plant genotypic diversity. *Oikos*, **117**, 255–264.

Crutsinger, G. M., Strauss S. Y., and Rudgers, J. A. (2010) Genetic variation within a dominant

shrub species determines plant invasion resistance in a coastal dune system. *Ecology*, **91**, 1237–1243.

Curry-Woods, III L., Halleman, E. M., Douglass, L. and Harrell, R. M. (1999) Variation in growth rate within and among stocks and families of striped bass. *North American Journal of Aquaculture*, **61**, 8–12.

Dayton, P. K. (1972) Toward an understanding of community resilience and the potential effects of enrichments to the benthos at McMurdo Sound, Antarctica. In B. C. Parker, ed., *Proceedings of the Colloquium on Conservation Problems*. Lawrence, KS: Allen Press, pp. 81–96.

Dickson, L. L. and Whitham, T. G. (1996). Genetically-based plant resistance traits affect arthropods, fungi, and birds. *Oecologia*, **106**, 400–406.

Driebe, E. and Whitham, T. G. (2000) Cottonwood hybridization affects tannin and nitrogen content of leaf litter and alters decomposition. *Oecologia*, **123**, 99–107.

Dungey, H. S., Potts, B. M., Whitham, T. G. and Li, H. F. 2000. Plant genetics affects arthropod community richness and composition: evidence from a synthetic eucalypt hybrid population. *Evolution*, **54**, 1939–1946.

Ellison, A. M. and Farnsworth, E. J. (2001) Mangrove communities. In M. D. Bertness, S. D. Gaines and M. E. Hay, eds., *Marine Community Ecology*. Sunderland, MA: Sinauer Associates, pp. 423–442.

Ellison, A. M., Bank, M. S., Clinton, B. D., *et al.* (2005) Loss of foundation species: consequences for the structure and dynamics of forested ecosystems. *Frontiers in Ecology and the Environment*, **9**, 479–486.

Ellison, A. M., Mukherjee, B. B. and Karim, A. (2000) Testing patterns of zonation in mangroves: scale dependence and environmental correlates in the Sundarbans of Bangladesh. *Journal of Ecology*, **88**, 813–824.

Evans, L. M., Allan, G. J., Shuster, S. M., Woolbright, S. A. and Whitham, T. G. (2008) Tree hybridization and genotypic variation drive cryptic speciation of a specialist mite herbivore. *Evolution*, **62**, 3027–3040.

Fischer, D. G., Hart, S. C., LeRoy, C. J. and Whitham, T. G. (2007) Variation in belowground carbon fluxes along a *Populus* hybridization gradient. *New Phytologist*, **176**, 415–425.

Fischer, D. G., Hart, S. C., Rehill, B. J. *et al.* (2006) Do high tannin leaves require more roots? *Oecologia*, **149**, 668–675.

Fischer, D. G., Hart, S. C., Whitham, T. G., Martinsen, G. D. and Keim, P. (2004) Ecosystem implications of genetic variation in water-use of a dominant riparian tree. *Oecologia*, **139**, 288–297.

Floate, K. D., Kearsley, M. J. C. and Whitham, T. G. (1993) Elevated herbivory in plant hybrid zones: *Chrysomela confluens, Populus*, and phenological sinks. *Ecology*, **74**, 2056–2065.

Floate, K. D., Martinsen, G. D. and Whitham, T. G. (1997) Cottonwood hybrid zones as centres of abundance for gall aphids in western North America: importance of relative habitat size. *Journal of Animal Ecology*, **66**, 179–188.

Fox, A. M., Haller, W. T. and Shilling, D. G. (1996) *Hydrilla* control with split treatments of fluridone in Lake Harris, Florida. *Hydrobiologia*, **340**, 235–239.

Goodnight, C. J. and Craig, D. M. (1996) The effect of coexistence on competitive outcome in *Tribolium castaneum* and *T. confusum*. *Evolution*, **50**, 1241–1250.

Goga-Vukmirovic, O. and Tilghman, S. M. (2000) Exploring genome space. *Nature*, **405**, 820–822.

Harding, S. A., Jiang, H., Jeong, M. L. *et al.* (2005) Functional genomics analysis of foliar condensed tannins and phenolic glycoside regulation in natural cottonwood hybrids. *Tree Physiology*, **25**, 1475–1486.

Hemming, J. and Lindroth, R. L. (1995) Intraspecific variation in aspen phytochemistry: effects on performance of gypsy moths and forest tent caterpillars. *Oecologia*, **103**, 79–88.

Hood, A. H. and Bayley, S. E. (2008) Beaver (*Castor canadensis*) mitigate the effects of climate on the area of open water in boreal wetlands in western Canada. *Biological Conservation*, **141**, 556–567.

Hubbell, S. P. (2001) *The Unified Neutral Theory of Biodiversity and Biogeography*. Princeton, NJ: Princeton University Press.

Hughes, A. R. and Stachowicz J. J. (2004) Genetic diversity enhances the resistance of a seagrass ecosystem to disturbance. *Proceedings of the National Academy of Sciences of the United States of America*, **101**, 8998–9002.

Hughes, A. R., Inouye, B. D., Johnson, M. T. J. *et al.* (2008) Ecological consequences of genetic diversity. *Ecology Letters*, **11**, 609–623.

Hwang, S. Y. and Lindroth, R. L. (1995) Clonal variation in foliar chemistry of aspen: effects on gypsy moths and forest tent caterpillars. *Oecologia*, **111**, 99–108.

Ingvarsson, P. K. (2010) Nucleotide polymorphism, linkage disequilibrium and complex trait dissection in *Populus*. In S. Jansson, R. Bhalerao and A. T. Groover, eds., *Genetics and Genomics of* Populus. New York: Springer, pp. 91–112.

Iverson, L. R., Prasad, A. M. and Schwartz, M. W. (2005). Predicting potential changes in suitable habitat and distribution by 2100 for tree species of the eastern United States. *Journal of Agricultural Meteorology*, **61**, 29–37.

Johnson, M. T. J. and Agrawal, A. A. (2005) Plant genotype and the environment interact to shape a diverse arthropod community on evening primrose. *Ecology*, **86**, 874–885.

Johnson, M. T. J. and Agrawal, A. A. (2007) Covariation and composition of arthropod species across plant genotypes of evening primrose (*Oenothera biennis*). *Oikos*, **116**, 941–956.

Johnson, M. T. J. and Stinchcombe, J. R. (2007) An emerging synthesis between community ecology and evolutionary biology. *Trends in Ecology and Evolution*, **22**, 250–257.

Johnson, M. T. J., Lajeunesse, M. J. and Agrawal, A. A. (2006) Additive and interactive effects of plant genotypic diversity on arthropod communities and plant fitness. *Ecology Letters*, **9**, 24–34.

Johnston, C. A. and Naiman, R. J. (1990) Browse selection by beaver: effects on riparian forest composition. *Canadian Journal of Forest Research*, **20**, 1036–1043.

Jones, C. G., Lawton, J. H. and Shachak, M. (1994) Organisms as ecosystem engineers. *Oikos*, **69**, 373–386.

Jones, C. G., Lawton, J. H. and Shachak, M. (1997) Positive and negative effects of organisms as physical ecosystem engineers. *Ecology*, **78**, 1946–1957.

Kearsey, M. J. (1998) The principles of QTL analysis (a minimal mathematics approach). *Journal of Experimental Botany*, **49**, 1619–1623.

Keith, A. R., Bailey, J. K. and Whitham, T. G. (2010) A genetic basis to community repeatability and stability. *Ecology*, **91**, 3398–3406.

Larson, K. C. and Whitham, T. G. (1991) Manipulation of food resources by a gall-forming aphid: the physiology of sink-source interactions. *Oecologia*, **88**, 15–21.

LeRoy, C. J., Whitham, T. G., Keim, P. and Marks, C. J. (2006) Plant genes link forests and streams. *Ecology*, **87**, 255–261.

Lojewski, N., Fischer, D. G., Bailey, J. K. et al. (2009) Genetic basis of aboveground productivity in two native *Populus* species and their hybrids. *Tree Physiology*, **29**, 1133–1142.

Mardis, E. R. (2008) The impact of next-generation sequencing technology on genetics. *Trends in Genetics*, **24**, 133–41.

Martinsen, G. D. and Whitham, T. G. (1994) More birds nest in hybrid cottonwood trees. *Wilson Bulletin*, **106**, 474–481.

Martinsen, G. D., Driebe, E. M. and Whitham, T. G. (1998) Indirect interactions mediated by changing plant chemistry: beaver browsing benefits beetles. *Ecology*, **79**, 192–200.

Michel, A., Arias, R. S., Scheffler, B. E. et al. (2004) Somatic mutation-mediated evolution of herbicide resistance in the nonindigenous invasive plant hydrilla (*Hydrilla verticillata*). *Molecular Ecology*, **13**, 3229–3237.

Mitton, J. B. (2003) The union of ecology and evolution: extended phenotypes and community genetics. *BioScience*, **53**, 208–209.

Moore, A. J., Brodie, III E. D. and Wolf, J. B. (1997) Interacting phenotypes and the evolutionary process. I. Direct and indirect effects of social interactions. *Evolution*, **51**, 1352–1362.

Morreel, K., Goeminne, G., Storme, V. et al. (2006) Genetical metabolomics of flavonoid biosynthesis in *Populus*: a case study. *The Plant Journal*, **47**, 224–237.

Naiman, R. J., Decamps, H. and Pollock, M. (1993) The role of riparian corridors in maintaining regional biodiversity. *Ecological Applications*, **3**, 209–212.

Netherland, M. D. and Getsinger, K. D. (1995) Laboratory evaluation of threshold fluridone concentrations under static conditions for controlling hydrilla and Eurasian watermilfoil. *Journal of Aquatic Plant Management*, **33**, 33–36.

Neuhauser, C., Andow, D. A., Heimpel, G. et al. (2003) Community genetics: expanding the synthesis of ecology and genetics. *Ecology*, **84**, 545–558.

Pregitzer, K. S. and Euskirchen, E. S. (2004) Carbon cycling and storage in world forests: biome patterns related to stand age. *Global Change Biology*, **10**, 2052–2077.

Rehill, B., Clauss, A., Wieczorek, L., Whitham, T. G. and Lindroth, R. L. (2005) Foliar phenolic glycosides from *Populus fremontii*, *Populus angustifolia*, and their hybrids. *Biochemical Systematics and Ecology*, **33**, 125–131.

Ronce, O. and Kirkpatrick, M. (2001) When sources become sinks: migrational meltdown in heterogeneous habitats. *Evolution*, **55**, 1520–1531.

Rosell, F., Bozser, O., Collen, P. and Parker, H. (2005) Ecological impacts of beavers *Castor fiber* and *Castor canadensis* and their ability to modify ecosystems. *Mammal Review*, **35**, 248–276.

Rosi-Marshall, E. J., Tank, J. L., Royer, T. V. et al. (2007) Toxins in transgenic crop byproducts may affect headwater stream ecosystems. *Proceedings of the National Academy of Sciences of the United States of America*, **104**, 16204–16208.

Schmitz, D. C., and Osborne, J. A.s (1984) Zooplankton densities in a *Hydrilla* infested lake. *Hydrobiologia*, **111**, 127–132.

Schmitz, D. C., Nelson, B. V., Nall, L. E. and Schardt, J. D. (1991) Exotic aquatic plants in Florida: a historical perspective and review of the present aquatic plant regulation program. In T. D. Center *et al.*, eds., *Proceedings of the Symposium on Exotic Pest Plants: November 2–4, 1988, University of Miami, Rosenstiel School of Marine and Atmospheric Science, Miami, FL*. Washington DC: United States Department of the Interior, National Park Service Document, pp. 303–323.

Schmitz, D. C., Schardt, J. D., Leslie, A. J., *et al.* (1993) The ecological impact and management history of three invasive alien aquatic plant species in Florida. In B. N. McKnight, ed., *Biology Pollution: The Control and Impact of Invasive Exotic Species*. Indianapolis, IN: Indiana Academy of Science, pp. 173–194.

Shuster, S. M., Lonsdorf, E. V., Wimp, G. M., Bailey, J. K. and Whitham, T. G. (2006) Community heritability measures the evolutionary consequences of indirect genetic effects on community structure. *Evolution*, **60**, 991–1003.

Schweitzer, J. A., Bailey, J. K., Hart, S. C., *et al.* (2005) The interaction of plant genotype and herbivory decelerates leaf litter decomposition and alter nutrient dynamics. *Oikos*, **110**, 133–145.

Schweitzer, J. A., Madritch, M. D., Bailey, J. K. *et al.* (2008) From genes to ecosystems: the genetic basis of condensed tannins and their role in nutrient regulation in a *Populus* model system. *Ecosystems*, **11**, 1005–1020.

Snyder, C. D., Young, J. A., Lemarie, D. P. and Smith, D. R. (2002) Influence of eastern hemlock (*Tsuga canadensis*) forests on aquatic invertebrate assemblages in headwater streams. *Canadian Journal of Fisheries and Aquatic Science*, **59**, 262–275.

Sterky, F., Bhalerao, R. R., Unneberg, P. *et al.* (2004) A *Populus* EST resource for plant functional genomics. *Proceedings of the National Academy of Sciences of the United States of America*, **101**, 13951–13956.

Sthultz, C. M., Whitham, T. G., Kennedy, K., Deckert, R. and Gehring, C. A. (2009) Genetically based susceptibility to herbivory influences the ectomycorrhizal fungal communities of a foundation tree species. *New Phytologist*, **184**, 657–667.

Stinchcombe, J. R. and Hoekstra, H. E. (2008) Combining population genomics and quantitative genetics: finding the genes underlying ecologically important traits. *Heredity*, **100**, 158–170.

Tovar-Sanchez, E. and Oyama, K. (2006) Effect of hybridization of the *Quercus crassifolia* × *Quercus crassipes* complex on the community structure of endophagous insects. *Oecologia*, **147**, 702–713.

Tsai, C-J., Harding, S. A., Tschaplinski, T. J., Lindroth, R. L. and Yuan, Y. (2006a) Genome-wide analysis of the structural genes regulating defense phenylpropanoid metabolism in *Populus*. *New Phytologist*, **172**, 47–62.

Tsai, C.-J., Kayal, W. E. and Harding, S. A. (2006b) *Populus*, the new model system for investigating phenylpropanoid complexity. *International Journal of Applied Science and Engineering*, **4**, 221–223.

Tuskan, G. A., DiFazio, S., Jansson, S. *et al.* (2006) The genome of black cottonwood, *Populus trichocarpa* (Torr. and Gray). *Science*, **313**, 1596–1604.

Tuskan, G. A., DiFazio, S. P. and Teichmann, T. (2003) Poplar genomics is getting popular: the impact of the poplar genome project on tree research. *Plant Biology*, **5**, 1–3

Tuskan, G. A., Gunter, L. E., Yang, Z. K. *et al.* 2004. Characterization of microsatellites revealed by genomic sequencing of *Populus trichocarpa*. *Canadian Journal of Forest Research*, **34**, 85–93.

van Dijk, G. (1985) *Vallisneria* and its interactions with other species. *Aquatics*, **7**, 6–10.

Wade, M. J. (1977). An experimental study of group selection. *Evolution*, **31**, 134–153.

Wade, M. J. (1978) A critical review of models of group selection. *Quarterly Review of Biology*, **53**, 101–114.

Wade, M. J., Priest, N. K. and Cruickshank, T. E. (2009) A theoretical overview of genetic maternal effects: evolutionary predictions and empirical tests with mammalian data. In D. Maestripieri and J. M. Mateo, eds., *Maternal Effects in Mammals*. Chicago, IL: University of Chicago Press, pp. 38–63

Werner, E. E. and Peacor, S. D. (2003) A review of trait-mediated indirect interactions in ecological communities. *Ecology*, **84**, 1083–1100.

Whitham, T. G. (1989) Plant hybrid zones as sinks for pests. *Science*, **244**, 1490–1493

Whitham, T. G., Bailey, J. K., Schweitzer, J. A. *et al.* (2006) A framework for community and ecosystem genetics: from genes to ecosystems. *Nature Reviews Genetics*, **7**, 510–523.

Whitham, T. G., Gehring, C. A., Evans, L. M., *et al.* (2010) A community and ecosystem genetics approach to conservation biology and management. In A. DeWoody, J. Bickham, C. Michler, K. Nichols, G. Rhodes and K. Woeste, eds., *Molecular Approaches in Natural Resource Conservation*. Cambridge: Cambridge University Press, pp. 50–73.

Whitham, T. G., Morrow, P. A. and Potts, B. M. (1994) Plant hybrid zones as centers of biodiversity: the herbivore community of two endemic Tasmanian eucalypts. *Oecologia*, **97**, 481–490.

Whitham, T. G., Young, W. P., Martinsen, G. D. *et al.* (2003) Community and ecosystem genetics: a consequence of the extended phenotype. *Ecology*, **84**, 559–573.

Whitham, T. G., DiFazio, S. P., Schweitzer, J. A. *et al.* (2008) Extending genomics to natural communities and ecosystems. *Science*, **320**, 492–495.

Wimp, G. M., Martinsen, G. D., Floate, K. D., Keim, P. S. and Whitham, T. G. (2005) Plant genetic determinants of arthropod community structure and diversity. *Evolution*, **59**, 61–69.

Wimp, G. M., Young, W. P., Woolbright, S. A. *et al.* (2004). Conserving plant genetic diversity for dependent animal communities. *Ecology Letters*, **7**, 776–780.

Woolbright, S. A., DiFazio, S. P., Yin, T. *et al.* (2008) A dense linkage map of a hybrid (*Populus fremontii* x *P. angustifolia*) BC$_1$ family contributes to long-term ecological research and comparison mapping in a model forest tree. *Heredity*, **100**, 59–70.

Wootton, J. T. (1994) The nature and consequences of indirect effects in ecological communities. *Annual Review of Ecology and Systematics*, **25**, 443–466.

Wright, J. P., Flecker, A. S. and Jones, C. G. (2003) Local vs. landscape controls on plant species richness in beaver meadows. *Ecology*, **84**, 3162–3173.

Wright, J. P., Jones, C. G. and Flecker, A. S. (2002) An ecosystem engineer, the beaver, increases species richness at the landscape scale. *Oecologia*, **132**, 96–101.

Wolf, J. B., Brodie, E. D. III., Cheverud, J. M., Moore, A. J. and Wade, M. J. (1998) Evolutionary consequences of indirect genetic effects. *Trends in Ecology and Evolution*, **13**, 64–69.

Wu, R. L., Han, H. F., Fang, J. J. *et al.* (2000) An integrated genetic map of *Populus deltoides* based on amplified fragment length polymorphisms. *Theoretical Applied Genetics*, **100**, 1249–1256.

Yin, T. M., DiFazio, S. P., Gunter, L. E., Riemenschneider, D. and Tuskan, G. A. (2004) Large-scale heterospecific segregation distortion in *Populus* revealed by a dense genetic map. *Theoretical Applications in Genetics*, **109**, 451–463.

Zhu, C., Gore, M., Buckler, E. S. and Yu, J. (2008) Status and prospects of association mapping in plants. *The Plant Genome*, **1**, 5–20.

CHAPTER SEVENTEEN

Species functional traits, trophic control and the ecosystem consequences of adaptive foraging in the middle of food chains

GEOFFREY C. TRUSSELL

Marine Science Center and Department of Biology, Northeastern University

and

OSWALD J. SCHMITZ

School of Forestry and Environmental Studies, Yale University

Introduction

An ecosystem is often defined simply as a community of organisms interacting with each other and their biophysical environment. This definition arose from early conceptions of how the natural world is organized and is elegant in its simplicity because it captures the basic elements of a functioning system (Tansley 1935; Leopold 1939; Lindeman 1942). But those trying to develop a synthetic, empirical understanding of how ecosystems function and how they will respond to environmental change are abundantly aware that there is much inherent complexity implied by this seemingly simple definition. To cope with this complexity, ecologists have traditionally abstracted one part of the definition and elaborated the other. For example, ecosystem ecologists have long assumed that interacting organisms can be simply assigned to different compartments (e.g., producer, primary and secondary consumer, decomposer) and focused on environmental and biophysical aspects that dictate the transformation and flow of materials and energy among various compartments (Lindeman 1942; Odum 1969; Likens *et al.* 1970). In contrast, community ecologists have downplayed the biophysical aspects of materials and energy transfer and focused on organismal populations (Shelford 1913; Elton 1927; Hutchinson 1957; Paine 1966; MacArthur 1972), their diversity and the myriad interactions (e.g., predation, competition, facilitation) that determine their distribution and abundance (Reiners 1986; DeAngelis 1992).

Modern efforts to integrate organismal and abiotic factors into the study of ecosystems arguably were inspired by Hairston, Slobodkin and Smith's (HSS)

Trait-Mediated Indirect Interactions: Ecological and Evolutionary Perspectives, eds. Takayuki Ohgushi, Oswald J. Schmitz and Robert D. Holt. Published by Cambridge University Press. © Cambridge University Press 2012.

classic paper (Hairston *et al.* 1960), which sought to merge Lindeman's trophic dynamic perspective (Lindeman 1942) and MacArthur's population ecology perspective (MacArthur 1958) to explain why, in the face of putatively abundant herbivores, the world is still largely green rather than denuded by herbivory. HSS made the simple argument that the world is green because predators limit the impact of herbivores on plants. This paper highlighted the ecological significance of indirect effects by viewing the biological component of ecosystems as being comprised of linear food chains where interacting species (who eats whom) determine the flow of materials and energy through the ecosystem (Paine 1988; Cohen *et al.* 1990).

HSS provided a decidedly 'top-down' controlled view of trophic interactions and ecosystem dynamics. Initial reception of this paper was not terribly enthusiastic because it challenged a deeply held 'bottom-up' control of the world that nutrient delivery to plants (Elton 1927; Lindeman 1942; and subsequently White 1978; Mattson 1980) was the primary factor limiting the transfer efficiency of materials and energy up the food chain and that plant structural and chemical defences deterred herbivores from acquiring the nutrients (Murdoch 1966; Ehrlich and Birch 1967). The idea of bottom-up control also extends to conceptions of detritus-based food chains because most primary production ends up as detritus, and detritus decomposition provides the resources that ultimately dictate plant production. Moreover, detritivore populations often subsidize predator populations that ultimately influence plant-based chains (Bengtsson *et al.* 1996; Wise *et al.* 1999; Scheu 2001; Wardle 2002; Moore *et al.* 2004; Hättenschweiler *et al.* 2005). This debate over the relative importance of top-down versus bottom-up control to ecosystem function continues.

We propose here that envisioning ecosystem control in terms of the top-down versus bottom-up duality will, however, fail to yield a robust understanding of the processes driving the structuring and functioning of ecosystems. We argue instead that trophic control of ecosystem structure and dynamics ultimately arises from the middle of the food chain. Our argument is principally based on the universal fact that intermediate species within food chains must balance the trade-off between eating and being eaten (Lawton and McNeill 1979; Abrams 1984; Bernays 1998; Schmitz *et al.* 2004; Singer *et al.* 2004), regardless of whether they belong to plant-based or detritus-based chains. This trade-off is shaped by adaptive trait plasticity in middle consumers (herbivores, arthropod detritivores) that maximizes fitness in different environmental contexts. This view highlights the need to consider the evolutionary ecology of trophic interactions, particularly in linking plasticity of individual behavioural and physiological traits to community and ecosystem level processes, including food chain length, stoichiometry and nutrient cycling. Because trait plasticity makes solutions to trade-offs context dependent, paying attention to complexity arising from phenotypic responses in the middle of food chains should enhance the ability to predict ecosystem functioning in different environmental contexts.

The middle of food webs

Food web diversity is dominated by middle trophic levels (60% of the total species) whereas top and basal trophic levels harbour approximately 15% of the total species (Williams and Martinez 2000). Species in the middle trophic level of plant-based food webs are herbivores and there is diversity in their feeding modes (a functional trait). The common feeding modes are grazing and sap-feeding (Meyer and Root 1993; Schmitz 2010) but other modes such as root boring, stem boring, leaf mining and galling exist (Hawkins 1988; Schoonhoven et al. 2005). Within the grazing and sap-feeding modes, there are differences in the degree of specialization and generalization in utilization of plant resources (Bernays 1998). These feeding modes differ in the degree to which they mediate top-down control to plants. Most sap feeders and specialist leaf chewers tend to have very weak per capita effects on plants (Schmitz 2010) and so it is less likely that predation or predation risk effects will propagate from carnivores through to plants. Indeed, comparison between experiments where sap feeding species were absent to those where the full herbivore community was present reveals that the generalist grazer appears to have the dominant – even overriding – effect on ecosystem structure and function (Schmitz 2010). Whether or not this is generally true remains to be determined by experimentation that combines herbivores by their feeding modes and measures the collective net interaction strengths.

Ecosystems can also be envisioned as sets of parallel food chains rather than completely reticulate food webs because individual consumer species tend to interact with a small fraction of available resource species within a system (Schoener 1989; Martinez 1995; Beckerman et al. 2006). In addition, identifying those chains and ecological contexts that are driven by strong direct and indirect interactions may be a useful way to begin abstracting trophic diversity in the middle of food webs to identify interaction modules for experimentation. This is not to suggest that weak interactions are unimportant, but it is clear that a few strongly interacting species and many weakly interacting species comprise ecosystems (Paine 1992; Hall and Raffaelli 1993; Wootton and Emerson 2005).

The evolutionary ecology and consequences of foraging decisions in the middle

One consequence of being in the middle of the food chain is that one consumes resources as well as serves as a resource for other species. This point was central to HSS, which led to a widespread appreciation of indirect effects transmitted via trophic cascades. Nevertheless, this classical view characterized species in the middle as passive players in the consumptive activities of predators. This unrealistic perspective has since motivated consideration of the evolutionary ecology of how species in the middle balance the benefits of feeding against the risk of being food for someone else (Brown *et al.* 1999; Werner and Peacor 2003; Brown and Kotler 2004; Schmitz *et al.* 2004; Trussell *et al.* 2003). Understanding interactions from the perspective of fear of being eaten (rather than being eaten) provides a logical way to integrate evolutionary, community and ecosystem ecology.

The fear of being eaten has driven the evolution of numerous behavioural and morphological adaptations in prey that reduce their vulnerability to predators (Lima 1998). For example, prey subjected to predation risk (i.e., cues that indicate predator presence rather than direct predation) often reduce their foraging activity or retreat to safer habitats to reduce their vulnerability to predation. This non-consumptive effect of predators on prey behaviour can cascade to influence prey–resource interactions in a way that is analogous to that caused by predators that are consuming prey. Indeed, the strength of such trait-mediated cascades often rivals or exceeds that of more widely appreciated consumptive cascades (Schmitz *et al.* 2004; Trussell *et al.* 2006a) and many textbook examples of the consumptive effects of predators have been re-evaluated in this light (Peckarsky *et al.* 2008).

The strength and ubiquity of trait-mediated cascades is not surprising simply because predators can scare many prey individuals at once and the behavioural responses of these individuals are often immediate. Hence the effects of fear can emerge quickly (Peacor and Werner 2001) and over large landscapes (Brown *et al.* 1999; Ripple and Beschta 2004; Creel *et al.* 2005; Fortin *et al.* 2005) whereas for many predators considerable time is required to capture and consume one individual prey. Fear effects are now widely known to influence community structure and dynamics. They are also important to ecosystem ecology because the fear of being eaten can influence ecosystem processes such as the transfer of materials and energy between biotic compartments and nutrient cycling (Trussell *et al.* 2006b, 2008; Schmitz 2008, 2010). Yet, there has been reluctance to consider behavioural or physiological trait plasticity at the ecosystem level. This is because population level processes that occur over years, let alone behavioural effects, which putatively occur within hours or days, are presumed to attenuate on the decadal or longer timescale of ecosystem functioning. But, recent evidence suggests that (1) chronic responses to predator-induced fear and stress in prey persist for long time periods; and (2) ecosystem level effects can manifest rapidly, over months to years (Hawlena and Schmitz 2010b). Thus risk effects and ecosystem functioning can and do operate on contemporary timescales (Schmitz 2008). Incorporating individual traits into community and ecosystem ecology will likely lead to more robust understanding of ecosystem dynamics and function. Doing so may also shed light on some longstanding ecological puzzles such as Elton's classic observation (Elton 1927) that many food chains are short, despite the availability of nutrients and energy to support longer food chains.

Ecosystem consequences of the middle: energy flow in food chains and food chain length

Trophic interactions involve the transfer of materials and energy from basal (plant) trophic levels through middle levels up to the top (predator) trophic level. This interaction occurs whether the effects of predators on prey are driven largely by consumptive or largely by fear effects. However, the fear of being eaten can affect the nature and amount of energy flow in ecosystems in two ways. First, predators will be less successful at hunting prey that have evolved effective avoidance strategies and thus will expend more energy to capture them than when hunting prey with less effective avoidance behaviours. Thus, anti-predator behaviours diminish predator foraging success that in turn reduces energy transfer from the middle to the top of the food chain. Predator avoidance strategies also often require prey to reduce their foraging activity. Consequently, predation risk can also reduce energy transfer between the basal and middle trophic levels of food chains and thus constrain the flow of materials and energy up the entire food chain (Trussell *et al.* 2006b, 2008; Schmitz *et al.* 2008).

The exact way that the flow of materials and energy becomes altered by risk depends on prey traits as well as predator traits, such as their hunting mode. For example, in a meadow ecosystem, generalist grasshopper herbivores that face persistent risk cues from sit-and-wait ambush spider predators shift their habitat and diet to include a higher proportion of herbs and a lower proportion of grasses (Schmitz 2008). Grasses tend to have higher protein (N) and lower soluble carbohydrate (useable C) than herbs. Consequently, the presence of these predators can alter the amount of C and N transmitted to herbivores and thus the elemental stoichiometry of both herbivore body tissue and the remaining plant tissue that enters the detrital pool to become decomposed (Hawlena and Schmitz 2010a). Alternatively, the same grasshoppers facing widely roaming, active hunting spiders do not undergo diet or habitat shifts owing to weak and ephemeral risk cues from these predators. Thus, trophic transfer will remain unaltered in the presence of such predators (Schmitz 2008).

Predation risk can also create significant physiological stress in prey and thus strongly shape their energy budgets (McPeek *et al.* 2001; Stoks *et al.* 2005; Trussell *et al.* 2006b, 2008) by influencing, among other things, metabolic rates (Rovero *et al.* 1999), hormone production levels (Boonstra *et al.* 1998; Creel *et al.* 2007, 2009) and the expression of heat-shock proteins (Kagawa and Mugiya 2002; Pauwels *et al.* 2005) and antioxidant enzymes (Slos and Stoks

2008). These physiological responses can be energetically costly and thus reduce the ability of prey to convert ingested resource biomass into secondary production (i.e., alter growth efficiency) that is available to other trophic levels. Risk effects essentially produce 'trophic heat': energy loss from the system that otherwise would be retained and consumed by other species. The transfer of energy between the basal and the middle trophic levels may remain unchanged if prey consume the same amount of food in the presence and absence of predation risk (McPeek 2004; Stoks et al. 2005), but trophic heat created by risk can strongly attenuate energy transfer from the middle to the top trophic level.

This underscores that it may not be reasonable to assume that trophic transfer efficiency is constant in ecosystems. Indeed, prey foraging plasticity may cause transfer efficiency to become a dynamic variable over the long-term. A prey's decision to avoid predation risk may becomes less sustainable over time if the energy reserves that prey rely on to delay the onset of starvation eventually become depleted due to risk effects (see Dahlgren et al. 2009 for a case where starvation may be important in predator-free situations). The use of stored energy reserves may temporarily reduce energy transfer from the bottom to the middle trophic levels. But, once starvation thresholds set in, prey will need to adopt more predation risk prone behaviour because the risk of mortality due to starvation can exceed the risk of mortality from predation. This in turn increases the rate of energy and materials transfer between the bottom and middle trophic levels. The time required for such starvation-induced shifts to occur will likely depend on prey metabolic traits and may emerge more quickly in endotherms than ectotherms. Thus, predation risk induced trophic heat may strongly determine the energetic efficiency of trait versus density-mediated trophic cascades with important consequences for ecosystems. This connection between predation risk and prey metabolism is intriguing and may enhance ongoing efforts that seek to understand the predictive power of metabolism (sensu Brown et al. 2004) as it relates to many fundamental ecological questions (see also Hawlena and Schmitz 2010b).

A longstanding question in ecology has been 'what determines the length of food chains?' This question emerged from Elton's (1927) classic observation that most food chains are short. Understanding variation in food chain length is fundamentally important because it can explain variation in ecosystem productivity (Carpenter and Kitchell 1993) and nutrient cycling and stability (DeAngelis et al. 1989). One classical hypothesis (the energy-flow hypothesis) suggests that energy transfer between trophic levels is generally inefficient (20–50% for invertebrate ectotherms, ~10% for vertebrate ectotherms and < 2% for endotherms; May 1983) leading to short food chains because inefficiency dramatically attenuates the energy available to support more, successive trophic levels (Hutchinson 1959; Slobodkin 1960). This reasoning implies that longer food chains should be more prevalent in systems that are highly productive (Jenkins et al. 1992; Kaunzinger and Morin 1998) or that contain species (i.e., invertebrate ectotherms) with high energetic conversion efficiencies (May 1983; Yodzis 1984).

This hypothesis has been challenged (Pimm and Lawton 1977; Pimm 1982; Hairston and Hairston 1993; Post 2002) because it fails to explain, for example, the existence of long food chains in oligotrophic ocean systems and short food chains in productive upwelling systems (Ryther 1969; Azam et al. 1983; Fenchel 1988). Instead, others suggest that the combined effects of ecosystem size and productivity or stability are most important in determining food chain length (Hutchinson 1959; Pimm and Lawton 1977; Pimm 1982; Briand and Cohen 1987; Schoener 1989; Kaunzinger and Morin 1998; Post 2002). But, consideration of trophic heat suggests that it may be premature to dismiss the energy flow hypothesis (Trussell et al. 2006b, 2008; Schmitz et al. 2008). The low energy transfer efficiency between trophic levels and resulting effects on food chain length may be partly explained by decreased energy conversion into production that is driven by the physiological consequences of predation risk. For example, the mere presence of cues signalling risk from an invasive predatory crab (Carcinus maenas) reduced the energy acquisition (by 31%), biomass production (by 60–83%) and growth efficiency (by 44–76%) of the snail Nucella lapillus foraging on basal resources (barnacles, Semibalanus balanoides) (Trussell et al. 2006b). Here again, predation risk effects on metabolism may, in addition to the more widely appreciated effects of temperature and body size (Brown et al. 2004), explain changes in thermodynamic constraints on energy transfer with important ecosystem consequences.

The bottom of food chains: traits and their connection with the middle

Plant traits (e.g., metabolic pathways for fixing carbon, the ability to fix nitrogen, cell structure, etc.) determine their functional role in ecosystems (Hooper et al. 2005; McGill et al. 2006; Violle et al. 2007). But, the expression of plant traits and their ensuing function in an ecosystem may also depend on the impacts of consumers. For example, in a meadow ecosystem, goldenrod Solidago rugosa outcompetes other plants for light and resources and its dominant status would lead one to believe that it is a prominent driver of ecosystem function (production and elemental cycling) in this system. However, top-down trophic interactions ultimately determine whether Solidago dominates in the first place. Nonconsumptive indirect effects (Beckerman et al. 1997) strongly drive this system because the architectural attributes that make it a good competitor (erect and leafy structure) also provide ideal habitat for grasshopper herbivores seeking refuge from spider predation risk. Experimentation also revealed that responses of the snail *Nucella lapillus* consumers to predation risk by green crabs *Carcinus maenas* varied with basal resources: barnacles (*Semibalanus balanoides*) or mussels (*Mytilus edulis*). Barnacle patches provided habitat with little structure whereas mussel patches provided abundant structure including refuges (crevices) between individual mussels where snails are able to hide and feed (Trussell *et al.* 2008). In the absence of predation risk, snail growth efficiency and thus energy transfer was similar on both resources. But predator addition dramatically reduced growth efficiency for snails foraging on barnacles but not on mussels. These resource identity effects may have emerged because the structure provided by mussels greatly diminished the risk perceived by snails, and attendant physiological stress, when crabs were present. Thus, basal species can be functionally important because they create complex architecture that provides prey species refuge from predation (Hawkins 1988; Denno *et al.* 2005). Consideration of plant traits in this context may help to explain contingency in the nature of trophic control (Schmitz 2010). Indeed, there is evidence that plant architectural traits can be a source of bottom-up forcing that can feed back to influence the nature and strength of top-down control in ecosystems (Aquilino *et al.* 2005; Pearson 2009).

The top of food chains: traits and their connection with the middle

Paying attention to predator and herbivore traits can further illuminate the connection between how middle species respond to environmental context and ecosystem function (Schmitz *et al.* 2004). Predator identities and attendant traits are important because they can determine the nature and strength of indirect predator effects (Peckarsky and McIntosh 1998; Bernot and Turner 2001; Schmitz and Suttle 2001; Finke and Denno 2005; Straub and Snyder 2006). Schmitz *et al.* (2004) considered the implications of the interaction between predator hunting mode (sit and wait, sit and pursue, active) and predator and prey habitat domain, which is the proportion of overall habitat that is used by the predator relative to its prey. In a nutshell, the concept sets up contingent outcomes for trophic interactions because the hunting mode and habitat domain of predators in relation to the mobility and resource choice of herbivores (specialist sap-feeding being tied to host plants versus generalist sap-feeding or grazing herbivores have greater flexibility to shift feeding locations) influences both the type (time budget shifts versus habitat shifts) and intensity of prey avoidance behaviours. Consequently, the type of cascade (trait- versus density-mediated) caused by a given predator may be largely shaped by how its traits influence the risk and stress levels perceived by herbivore prey (see Schmitz *et al.* 2004).

Nutrient and energy demands and C:N:P cycling

The theory of ecological stoichiometry suggests that energy and material transfer (i.e., resource uptake, production and excretion) at the individual level is constrained by the imbalance between the content and molar ratio of essential chemical elements (e.g., nitrogen [N], phosphorus [P] and carbon [C]) available in resources and needed by consumers for maintenance and production (Sterner and Elser 2002). Consumers are limited by various elements, depending on their metabolic and somatic demands (Sterner and Elser 2002).

For example, stoichiometric models assume that C is used primarily to fuel respiration with some allocation to production, whereas N or P is used exclusively for production. These models further assume that N or P for maintenance is recycled within the body (Sterner 1997). Consumers regulate their homoeostatic balance of C:N:P by respiring excess C or excreting excess inorganic N and P (Sterner 1997). Theory suggests that context dependency in the extent to which elements are allocated to production arises from differences in both resource quality (C:N:P content) and quantity among environments. These factors may also determine which of the elements become most limiting to production. For example, food quality may be immaterial in environments with very low food quantity because most, if not all, resource intake will be devoted to supplying C for respiration with negligible amounts left over for production (Sterner 1997). Consequently, there will be minimal, if any, demand for N or P. Higher resource quantity allows excess intake to be allocated to production, at which time balancing elemental ratios influences or shapes resource intake. The mechanism determining this plasticity is encapsulated by the concept of the threshold elemental ratio (TER) – the dietary mixture where growth limitation switches from one element to another (Sterner 1997).

Assuming that resources are in high quantity, we can then explore how TER is expected to change using the following formula based on species' physiology (Frost et al. 2006). TER for C and N reflects both animal body C:N ratios and the proportion of ingested C used for growth (i.e., gross growth efficiency of C, GGE_C). GGE_C can be expressed as the percentage of ingested C that was assimilated into new growth (Frost et al. 2006):

$$GGE_C = \frac{(I_C \times A_C) - R_C}{I_C} \tag{17.1}$$

where A_C is the assimilation efficiency of C, I_C is the mass-specific ingestion rate above a saturating food level, and R_C is the mass-specific respiration rate. The $TER_{C:N}$ can be expressed as the product of physiological nutrient efficiencies and body elemental composition:

$$TER_{C:N} = \frac{A_N}{GGE_C} \times \frac{Q_C}{Q_N} \qquad (17.2)$$

where A_N is the assimilation efficiency of N, and Q_C and Q_N are the proportion of animal dry mass in C and N.

The stress induced by predation risk (Lima 1998; Creel and Christiansen 2008) can elevate respiration (Rovero *et al.* 1999; Hawlena and Schmitz 2010a). Accordingly GGE_C will become lower and $TER_{C:N}$ will become higher (Hawlena and Schmitz 2010b). This implies that there should be a shift in nutrient demand from N-rich proteins that support growth and reproduction toward carbohydrates (C-rich, N-poor) that fuel the heightened respiratory demands of antipredator behaviour, which it seems to do (Stoks *et al.* 2005; Trussell *et al.* 2006b, 2008; Hawlena and Schmitz 2010b). Accordingly, C:N:P ratios within prey and in excreta or egesta may also become altered by predator physiological effects (Hawlena and Schmitz 2010a). Such plasticity in C:N:P uptake and body composition in response to altered metabolic rate suggests that our understanding of physiological plasticity of consumers to environmental changes must be improved if we are to develop more robust predictions of nutrient dynamics in natural systems (see also Hawlena and Schmitz 2010b).

Plant material and detritus can be highly nutrient limited (i.e., high C and low N and P concentrations: Wardle (2002); Hättenschweiler *et al.* (2005); Martinson *et al.* (2008)) thereby placing important constraints on resource uptake, stoichiometric balance, and hence fitness of individual herbivores as well as detritivores (Elser *et al.* 2000; Fagan *et al.* 2002; Martinson *et al.* 2008).

Summary
The factors determining the flow of materials and energy through food webs are fundamentally important to the structuring and functioning of natural ecosystems. We have argued that trait plasticity in intermediate species can strongly influence the energetics and stoichiometry of food chains because such plasticity is key to resolving the trade-off between eating and being eaten. This connection emerges because the decision to feed or hide is partly driven by the energetic status of prey when confronted with risk. Moreover, the stress imposed by predation risk can further modify prey energy budgets by influencing aspects of their physiology, such as metabolism. To this end, predation risk effects contribute toward understanding the general role of energy metabolism in ecosystems. The metabolic theory of ecology (Brown *et al.* 2004) suggests that the effects of body size and temperature on individual metabolism ultimately drive the structure and dynamics of natural ecosystems. For example, there are strong associations between individual

metabolism and population growth (Savage *et al.* 2004), patterns of biodiversity (Allen *et al.* 2002), biomass accumulation (Anderson *et al.* 2006) and the global carbon cycle (Allen *et al.* 2005). We suggest that predation risk also has substantial effects on prey metabolism. In turn, risk-induced changes in prey stochiometry and the production of trophic heat can strongly dictate how energy and materials passes through food chains.

Acknowledgements

This work was supported by the generous support of the National Science Foundation to G. C. Trussell (OCE-0648525, OCE-0727628) and O. J. Schmitz (DEB-0816504). This is contribution number 278 from the Marine Science Center.

References

Abrams, P. A. (1984) Foraging time optimization and interactions in food webs. *American Naturalist*, **124**, 80–96.

Allen, A. P., Brown, J. H. and Gillooly, J. F. (2002) Global biodiversity, biochemical kinetics, and the energetic-equivalence rule. *Science*, **297**, 1545–1548.

Allen, A. P., Gillooly, J. F. and Brown, J. H. (2005) Linking the global carbon cycle to individual metabolism. *Functional Ecology*, **19**, 202–213.

Anderson, K. J., Allen, A. P., Gillooly, J. F. and Brown, J. H. (2006) Temperature-dependence of biomass accumulation rates during secondary succession. *Ecology Letters*, **9**, 673–682.

Aquilino, K. M., Cardinale, B. J. and Ives, A. R. (2005) Reciprocal effects of host plant and natural enemy diversity on herbivore suppression: an empirical study of a model tritrophic system. *Oikos*, **108**, 275–282.

Azam, F., Fenchel, T., Field, J. G., Gray, J. S., Meyer-Reil, L. A. and Thingstad, F. (1983) The ecological role of water-column microbes in the sea. *Marine Ecology Progress Series*, **10**, 257–263.

Beckerman, A. P., Petchey, O. L. and Warren, P. H. (2006) Foraging biology predicts food web complexity. *Proceedings of the National Academy of Sciences of the United States of America*, **103**, 13745–13749.

Beckerman, A. P., Uriarte, M. and Schmitz, O. J. (1997) Experimental evidence for a behavior-mediated trophic cascade in a terrestrial food chain. *Proceedings of the National Academy of Sciences of the United States of America*, **94**, 10735–10738.

Bengtsson, J., Setälä, H. and Zheng, W. D. (1996) Food webs and nutrient cycling in soils: interactions and positive feedbacks. In G. Polis and K. Winemiller, eds., *Food Webs: Integration of Patterns and Dynamics*. London: Chapman and Hall, pp. 30–38.

Bernays, E. A. (1998) Evolution of feeding behavior in insect herbivores. *BioScience*, **48**, 35–44.

Bernot, R. J. and Turner, A. M. (2001) Predator identity and trait-mediated indirect effects in a littoral food web. *Oecologia*, **129**, 139–146.

Boonstra, R., Krebs, C. J. and Chr. Stenseth, N. (1998) Population cycles in small mammals: the problem of explaining the low phase. *Ecology*, **79**, 1479–1488.

Briand, F. and Cohen, J. E. (1987) Environmental correlates of food chain length. *Science*, **238**, 956–960.

Brown, J. H., Gillooly, J. F., Allen, A. P., Savage, V. M. and West, G. B. (2004) Toward a metabolic theory of ecology. *Ecology*, **85**, 1771–1789.

Brown, J. S. and Kotler, B. P. (2004) Hazardous duty pay and the foraging cost of predation. *Ecology Letters*, **7**, 999–1014.

Brown, J. S., Laundre, J. W. and Gurung, M. (1999) The ecology of fear: optimal foraging, game theory and trophic interactions. *Journal of Mammalogy*, **80**, 385–399.

Carpenter, S. R. and Kitchell, J. F. (1993) *The Trophic Cascade in Lakes*. Cambridge: Cambridge University Press.

Cohen, J. E., Briand, F. and Newman, C. M. (1990) *Community Food Webs: Data and Theory*. New York: Springer Verlag.

Creel, S. and Christianson, D. (2008) Relationship between direct predation and risk effects. *Trends in Ecology and Evolution*, **23**, 194–201.

Creel, S., Christianson, D., Liley, S. and Winnie Jr., J. A. (2007) Predation risk affects reproductive physiology and demography of elk. *Science*, **315**, 960.

Creel, S., Winnie Jr., J. A. and Christianson, D. (2009) Glucocorticoid stress hormones and the effect of predation risk on elk production. *Proceedings of the National Academy of Sciences of the United States of America*, **106**, 12388–12393.

Creel, S., Winnie, J., Maxwell, B., Hamlin, K. and Creel, M. (2005) Elk alter habitat selection as an antipredator response to wolves. *Ecology*, **86**, 3387–3397.

Dahlgren, J., Oksanen, L., Oksanen, T. et al. (2009) Plant defenses to no avail? Responses of plants with varying edibility to food web manipulations in a low arctic scrubland. *Evolutionary Ecology Research*, **11**, 1189–1203.

DeAngelis, D. L. (1992) *Dynamics of Nutrient Cycling and Food Webs*. New York: Chapman and Hall.

DeAngelis, D. L., Bartell, S. M. and Brenkert, A. L. (1989) Effects of nutrient recycling and food chain length on resilience. *American Naturalist*, **134**, 778–805.

Denno, R. F., Finke, D. L. and Langellotto, G. A. (2005) Direct and indirect effects of vegetation structure and habitat complexity on predator–prey and predator-predator interactions. In P. Barbosa and I. Castellanos,

eds., *Ecology of Predator–Prey Interactions*. Oxford: Oxford University Press, pp. 211–239.

Ehrlich, P. R. and Birch, L. C. (1967) Balance of nature and population control. *American Naturalist*, **101**, 97–107.

Elser, J. J., Sterner, R. W., Gorokhova, E., et al. (2000) Biological stoichiometry from genes to ecosystems. *Ecology Letters*, **3**, 540–550.

Elton, C. S. (1927) *Animal Ecology*. New York: Macmillan Publishing.

Fagan, W. F., Siemann, E., Mitter, C. et al. (2002) Nitrogen in insects: implications for trophic complexity and species diversification. *American Naturalist*, **160**, 784–802.

Finke, D. L. and Denno, R. F. (2005) Predator diversity and the functioning of ecosystems: the role of intraguild predation in dampening trophic cascades. *Ecology Letters*, **8**, 1299–1306.

Fortin, D., Beyer, H. L., Boyce, M. S. et al. (2005) Wolves influence elk movements: behavior shapes a trophic cascade in Yellowstone National Park. *Ecology*, **86**, 1320–1330.

Fenchel, T. (1988) Marine plankton food chains. *Annual Review of Ecology and Systematics*, **19**, 19–38.

Frost, P. C., Benstead, J. P., Cross, W. F. et al. (2006) Threshold elemental ratios of carbon and phosphorus in aquatic consumers. *Ecology Letters*, **9**, 774–779.

Hairston Jr., N. G. and Hairston Sr., N. G. (1993) Cause-effect relationships in energy flow, trophic structure, and interspecific interactions. *American Naturalist*, **142**, 379–411.

Hairston, N. G., Smith, F. E. and Slobodkin, L. B. (1960) Community structure, population control and competition. *American Naturalist*, **94**, 421–425.

Hall, S. J. and Raffaelli, D. G. (1993) Food webs: theory and reality. *Advances in Ecological Research*, **24**, 187–239.

Hättenschweiler, S., Tiunov, A. V. and Scheu, S. (2005) Biodiversity and litter decomposition in terrestrial ecosystems. *Annual Review of*

Ecology, Evolution, and Systematics, **36**, 191–218.

Hawkins, B. A. (1988) Species diversity in the third and fourth trophic levels: patterns and mechanisms. *Journal of Animal Ecology*, **57**, 137–162.

Hawlena, D. and Schmitz, O. J. (2010a) Herbivore physiological response to fear of predation and implications for ecosystem nutrient dynamics. *Proceedings of the National Academy of Sciences of the United States of America*, **107**, 15503–15507.

Hawlena, D. and Schmitz, O. J. (2010b) Physiological stress as a fundamental mechanism linking predation to ecosystem processes. *American Naturalist*, **176**, 537–556.

Hooper, D. U., Chapin, F. S., Ewel, J. J. et al. (2005) Effects of biodiversity on ecosystem functioning: a consensus of current knowledge. *Ecological Monographs*, **75**, 3–35.

Hutchinson, G. E. (1957) Concluding remarks. Population studies: animal ecology and demography. *Cold Spring Harbor Symposium on Quantitative Biology*, **22**, 415–427.

Hutchinson, G. E. (1959) Homage to Santa Rosalia or why are there so many kinds of animals? *American Naturalist*, **93**, 145–159.

Jenkins, B., Kitching, R. L. and Pimm, S. L. (1992) Productivity, disturbance, and food web structure at a local spatial scale in experimental container habitats. *Oikos*, **65**, 249–255.

Kagawa, N. and Mugiya, Y. (2002) Brain Hsp70 mRNA expression is linked with plasma cortisol levels in goldfish (*Carassius auratus*) exposed to a potential predator. *Zoological Science*, **19**, 735–740.

Kaunzinger, C. M. K. and Morin, P. J. (1998) Productivity controls food-chain properties in microbial communities. *Nature*, **395**, 495–497.

Laska, M. S. and Wootton, J. T. (1998) Theoretical concepts and empirical approaches to measuring interaction strength. *Ecology*, **79**, 461–476.

Lawton, J. H. and McNeill, S. (1979) Between the devil and the deep blue sea: on the problem of being a herbivore. In R. M. Anderson, B. D. Turner and L. R. Taylor, eds., *Population Dynamics*. Oxford: Blackwell Publishing, pp. 223–244.

Leopold, A. (1939) A biotic view of the land. *Journal of Forestry*, **37**, 727–730.

Likens, G. E., Borman, F. H., Johnson, N. M., Fisher, D. W. and Pierce, R. S. (1970) Effects of forest cutting and herbicide treatment on nutrient budgets in the Hubbard Brook watershed-ecosystem. *Ecological Monographs*, **40**, 23–47.

Lima, S. L. (1998) Nonlethal effects in the ecology of predator–prey interactions: what are the ecological effects of anti-predator decision-making? *BioScience*, **48**, 25–34.

Lindeman, R. L. (1942) The trophic-dynamic aspect of ecology. *Ecology*, **22**, 399–418.

MacArthur, R. H. (1958) Population ecology of some warblers of northeastern coniferous forests. *Ecology*, **39**, 599–619.

MacArthur, R. H. (1972) *Geographical Ecology*. Princeton, NJ: Princeton University Press.

Martinez, N. D. (1995) Unifying ecological subdisciplines with ecosystem food webs. In C. Jones and J. Lawton, eds., *Linking Species and Ecosystem*. New York: Chapman Hall, pp. 166–176.

Martinson, H. M., Schneider, K., Gilbert, J. et al. (2008) Detritivory: stoichiometry of a neglected trophic level. *Ecological Research*, **23**, 487–491.

Mattson, W. J. (1980) Herbivory in relation to plant nitrogen content. *Annual Review of Ecology and Systematics*, **11**, 119–61.

May, R. M. (1983) The structure of food webs. *Nature*, **301**, 566–568.

McGill, B. J., Enquist, B. J., Weiher, E. and Westoby, M. (2006) Rebuilding community ecology from functional traits. *Trends in Ecology and Evolution*, **21**, 178–185.

McPeek, M. A. (2004) The growth/predation risk trade-off: so what is the mechanism? *American Naturalist*, **163**, E88–E111.

McPeek, M. A., Grace, M. and Richardson, J. M. L. (2001) Physiological and behavioral responses to predators shape the growth/predation risk trade-off in damselflies. *Ecology*, **82**, 1535–1545.

Meyer, G. B. and Root, R. B. (1993) Effects of herbivorous insects and soil fertility on reproduction of goldenrod. *Ecology*, **74**, 1117–1124.

Moore, J. C., Berlow, E. L., Coleman, D. C. *et al.* (2004) Detritus, trophic dynamics and biodiversity. *Ecology Letters*, **7**, 584–600.

Murdoch, W. W. (1966) Community structure, population control and competition: a critique. *American Naturalist*, **100**, 219–226.

Odum, E. P. (1969) The strategy of ecosystem development. *Science*, **164**, 262–270.

Paine, R. T. (1966) Food web complexity and species diversity. *American Naturalist*, **100**, 65–73.

Paine, R. T. (1988) Food webs: road maps of interactions or grist for theoretical development? *Ecology*, **69**, 1648–1654.

Paine, R. T. (1992) Food-web analysis through field measurements of per capita interaction strength. *Nature*, **355**, 73–75.

Pauwels, K., Stoks, R. and DeMeester, L. (2005) Coping with predator-induced stress: interclonal differences in induction of heat-shock proteins in the water flea *Daphnia magna*. *Journal of Evolutionary Biology*, **18**, 867–872.

Peacor, S. D. and Werner, E. E. (2001) The contribution of trait-mediated indirect effects to the net effects of a predator. *Proceedings of the National Academy of Sciences of the United States of America*, **98**, 3904–3908.

Pearson, D. E. (2009) Invasive plant architecture alters trophic interactions by changing predator abundance and behavior. *Oecologia*, **159**, 549–558.

Peckarsky, B. L. and McIntosh, A. R. (1998) Fitness and community consequences of avoiding multiple predators. *Oecologia*, **113**, 565–576.

Peckarsky, B. L., Abrams, P. A., Bolnick, D. I. *et al.* (2008) Revisiting the classics: considering nonconsumptive effects in textbook examples of predator–prey interactions. *Ecology*, **89**, 2416–2425.

Pimm, S. L. (1982) *Food Webs*. London: Chapman and Hall.

Pimm, S. L. and Lawton, J. H. (1977) Number of trophic levels in ecological communities. *Nature*, **268**, 329–331.

Post, D. M. (2002) The long and short of food-chain length. *Trends in Ecology and Evolution*, **17**, 269–277.

Reiners, W. A. (1986) Complementary models for ecosystems. *American Naturalist*, **127**, 59–73.

Ripple, W. J. and Beschta, R. L. (2004) Wolves and the ecology of fear: can predation risk structure ecosystems? *BioScience*, **54**, 755–766.

Rovero, F., Hughes, R. N. and Chelazzi, G. (1999) Cardiac and behavioural responses of mussels to risk of predation by dogwhelks. *Animal Behaviour*, **58**, 707–714.

Ryther, J. H. (1969) Photosynthesis and fish production in the sea. *Science*, **166**, 72–76.

Savage, V. M., Gillooly, J. F., Brown, J. H., West, G. B. and Charnov, E. L. (2004) Effects of body size and temperature on population growth. *American Naturalist*, **163**, 429–441.

Scheu, S. (2001) Plants and generalist predators as links between the below-ground and above-ground system. *Basic and Applied Ecology*, **2**, 3–13.

Schmitz, O. J. (2008) Herbivory from individuals to ecosystems. *Annual Review of Ecology, Evolution, and Systematics*, **39**, 133–152.

Schmitz, O. J. (2010) *Resolving Ecosystem Complexity*. Princeton, NJ: Princeton University Press.

Schmitz, O. J. and Suttle, K. B. (2001) Effects of top predator species on direct and indirect interactions in a food web. *Ecology*, **82**, 2072–2081.

Schmitz, O. J., Grabowski, J. H., Peckarsky, B. L. *et al.* (2008) From individuals to ecosystem function: toward an integration of evolutionary and ecosystem ecology. *Ecology* **89**, 2436–2445.

Schmitz, O. J., Ovadia, O. and Krivan, V. (2004) Trophic cascades: the primacy of trait-mediated indirect interactions. *Ecology Letters*, **7**, 153–163.

Schoener, T. W. (1989) Food webs from the small to the large. *Ecology*, **70**, 1559–1589.

Schoonhoven, L. M., van Loon, J. J. A. and Dicke, M. (2005) *Insect–Plant Biology*. Oxford: Oxford University Press.

Shelford, V. E. (1913) *Animal Communities in Temperate America*. Chicago, IL: University of Chicago Press.

Singer, M. S., Carrière, Y., Theuring, C. and Hartmann, T. (2004) Disentangling food quality from resistance against parasitoids: diet choice by a generalist caterpillar. *American Naturalist*, **164**, 423–429.

Slobodkin, L. B. (1960) Ecological energy relationships at the population level. *American Naturalist*, **94**, 213–236.

Slos, S. and Stoks, R. (2008) Predation risk induces stress proteins and reduces antioxidant defense. *Functional Ecology*, **22**, 637–642.

Sterner, R. W. (1997) Modelling interactions of food quality and quantity in homeostatic consumers. *Freshwater Biology*, **38**, 473–481.

Sterner, R. W. and Elser, J. J. (2002) *Ecological Stoichiometry: The Biology of Elements from Molecules to the Biosphere*. Princeton, NJ: Princeton University Press.

Stoks, R., De Block, M. and McPeek, M. A. (2005) Alternative growth and energy storage responses to mortality threats in damselflies. *Ecology Letters*, **8**, 1307–1316.

Straub, C. S. and Snyder, W. E. (2006) Species identity dominates the relationship between predator biodiversity and herbivore suppression. *Ecology*, **87**, 277–282.

Tansley, A. G. (1935) The use and abuse of vegetational concepts and terms. *Ecology*, **16**, 284–307.

Trussell, G. C., Ewanchuk, P. J. and Bertness, M. D. (2003) Trait-mediated interactions in rocky intertidal food chains: predator risk cues alter prey feeding rates. *Ecology*, **84**, 629–640.

Trussell, G. C., Ewanchuk, P. J. and Matassa, C. M. (2006a) Habitat effects on the relative importance of trait- and density-mediated indirect interactions. *Ecology Letters*, **9**, 1245–1252.

Trussell, G. C., Ewanchuk, P. J. and Matassa, C. M. (2006b) The fear of being eaten reduces energy transfer in a simple food chain. *Ecology*, **87**, 2979–2984.

Trussell, G. C., Ewanchuk, P. J. and Matassa, C. M. (2008) Resource identity modifies the influence of predation risk on ecosystem function. *Ecology*, **89**, 2798–2807.

Violle, C., Navas, M-L., Vile, D. *et al.* (2007) Let the concept of trait be functional! *Oikos*, **116**, 882–892.

Wardle, D. A. (2002) *Communities and Ecosystems: Linking the Aboveground and Belowground Components*. Princeton, NJ: Princeton University Press.

Werner, E. E. and Peacor, S. D. (2003) A review of trait-mediated indirect interactions in ecological communities. *Ecology*, **84**, 1083–1100.

White, T. C. R. (1978) The importance of a relative shortage of food in animal ecology. *Oecologia*, **33**, 71–86.

Williams, R. J. and Martinez, N. D. (2000) Simple rules yield complex food webs. *Nature*, **404**, 180–183.

Wise, D. H., Snyder, W. E., Tuntibunpakul, P. and Halaj, J. (1999) Spiders in decomposition food webs of agroecosystems: theory and evidence. *Journal of Arachnology*, **27**, 363–370.

Wootton, J. T. and Emmerson, M. (2005) Measurement of interaction strength in nature. *Annual Review of Ecology, Evolution, and Systematics*, **36**, 419–444.

Yodzis, P. (1984) Energy flow and the vertical structure of real ecosystems. *Oecologia*, **65**, 86–88.

CHAPTER EIGHTEEN

Effects of herbivores on terrestrial ecosystem processes: the role of trait-mediated indirect effects

MARK D. HUNTER

Department of Ecology and Evolutionary Biology, University of Michigan

BARBARA C. REYNOLDS

Department of Environmental Studies, University of North Carolina-Asheville

MYRA C. HALL

Georgia Perimeter College

and

CHRISTOPHER J. FROST

Warnell School of Forest Resources, University of Georgia

Introduction

Primary production in terrestrial environments generates about 100 gigatons of biomass annually (Gessner *et al.* 2010). While on average 90% of terrestrial plant biomass escapes herbivory (Cebrian 2004), herbivores nonetheless exert a pervasive influence on the quality of all plant tissues in ecological and evolutionary time (Hunter 2001; Dethier 1954; Ehrlich and Raven 1964; Thompson 1994; Karban and Baldwin 1997). By inducing chemical changes in plant tissues in ecological time, or by acting as agents of natural selection favouring defended tissues in evolutionary time, herbivores have a significant impact on plant traits. Accordingly, terrestrial herbivores may engage in density-mediated indirect effects (DMIEs) with other organisms by their consumption of 10% of plant biomass, and in trait-mediated indirect effects (TMIEs) by their ecological and evolutionary effects on the 90% of the biomass that is not consumed. Finally, assuming average assimilation efficiencies of around 20% (Speight *et al.* 2008), herbivores may convert around 2% of terrestrial plant biomass into animal biomass. If herbivores are not consumed by their own predators, their cadavers are subsequently available for decomposition by the soil microbial community. Cadaver inputs and burrowing or trampling (Hunter 1992) are the only direct effects (DEs) of herbivores on soil processes of which we are aware. Based on these numbers alone, we might expect TMIEs of herbivores on other organisms to be relatively more important than DMIEs or DEs. Simply put, the effects of terrestrial herbivores

Trait-Mediated Indirect Interactions: Ecological and Evolutionary Perspectives, eds. Takayuki Ohgushi, Oswald J. Schmitz and Robert D. Holt. Published by Cambridge University Press. © Cambridge University Press 2012.

on plant quality may often be more important ecologically than their effects on plant biomass.

TMIEs induced by herbivores may be particularly important in soil food webs. Communities of decomposers and detritivores in the soil rely largely upon energy derived from dead plant material and are fundamental players in the global carbon cycle (Cornwell *et al.* 2008). If 90% of land plant biomass enters the decomposer food web without being consumed (Cebrian 2004), a focus on DMIEs or DEs alone might lead us to assume that herbivores have relatively minor effects on the population and community ecology of decomposers and detritivores. However, the activities of herbivores can lead to dramatic changes in the nutritional, chemical and structural characteristics of plant tissues (Speight *et al.* 2008) that subsequently mediate their rates of decomposition (Choudhury 1988; Pastor and Naiman 1992; Findlay *et al.* 1996; Hunter 2001; Chapman *et al.* 2006). Plant traits matter greatly to rates of nutrient cycling and decomposition and, on a global scale, variation in litter quality explains more variation in rates of decomposition than does climate (Cornwell *et al.* 2008). Accordingly, TMIEs between herbivores and decomposers may play a fundamental role in the transfer of energy and cycling of matter in soil food webs, the global carbon cycle and subsequent feedbacks to global climate.

In detrital systems, we consider microbial species (generally fungi and bacteria) that can catabolize recalcitrant organic carbon (C) completely to CO_2 as decomposers. Other organisms that consume dead organic material (and the microbes on that material) without complete catabolism of that material are referred to as detritivores (Coleman *et al.* 2004). What mechanisms might drive TMIEs between herbivores and these members of the detrital food web? Any changes in litter traits (but not amount of litter) caused by herbivore species that then influence litter-feeding species (bacteria, fungi, soil invertebrates, etc.) should be viewed as TMIEs. Accordingly, herbivore-induced changes in litter toughness, chemistry or stoichiometry qualify as potential mechanisms underlying TMIEs. Likewise, changes in the phenology of litter production (e.g., wound-induced premature leaf abscission, green-fall) could also mediate TMIEs, if those changes in phenology cause subsequent changes in the detrital food web. In studies linking herbivores to detrital food webs, changes in rates of decomposition and nutrient cycling are often used as proxies for effects on decomposers and detritivores; it is generally easier to measure the results of decomposer and detritivore activity than it is to measure their abundances directly. Strictly speaking, TMIEs should occur among species, not between species (e.g., herbivores) and processes (e.g., decomposition), although measuring microbial identity, abundance and diversity remains challenging. Here, we will generally consider TMIEs of herbivores on soil processes (decomposition, nutrient cycling) as strong evidence for TMIEs among species.

In this chapter, we explore potential mechanisms underlying TMIEs between herbivores and members of the soil food web. We consider the relative importance of such TMIEs compared to other ecological factors in determining variation in decomposition processes. We describe a case study from our own work in oak-dominated ecosystems and conclude with some thoughts on areas for future work. One of us reviewed this area of study about a decade ago (Hunter 2001). Here, we try to focus primarily on insights in the literature from 2001 onwards, although we refer to older literature where necessary to provide historical context or to give examples that remain the best in their area. We also focus on aboveground herbivores in terrestrial environments; effects of root herbivores on ecosystem processes have been reviewed elsewhere (Bardgett and Wardle 2003; Hunter 2008) and including literature on TMIEs among herbivores and decomposers/detritivores in aquatic systems would require more space than is available.

Background

The influence of herbivores on nutrient cycling and primary production has been studied since the 1960s and 1970s (Pitelka 1964; Chew 1974; Mattson and Addy 1975) initially from the perspective that herbivores could influence the rates of matter and energy transfer among compartments in ecosystems (Schowalter 2000). We might consider these seminal contributions as the first estimates of DMIEs of herbivores on soil communities. However, herbivores are not simply compartments through which matter and energy are transferred – rather, they can shape the quality of materials passed among compartments, as well as the rate and quantity of materials transferred (McNaughton *et al.* 1988; Ritchie *et al.* 1998; Hunter 2001; Bardgett and Wardle 2003; Chapman *et al.*, 2006). There are seven broad mechanisms by which the activity of herbivores can cause changes in decomposition and nutrient availability in soils (Hunter 2001). These are (1) deposition of faeces, (2) inputs of cadavers, (3) defoliator-mediated changes in the chemistry of precipitation (through-fall), (4) changes in the quality and quantity of litter inputs, (5) changes in nutrient uptake by the plant community, (6) effects upon root exudation and root/mutualist interactions and (7) effects upon the physical structure of plant canopies and subsequent changes in soil microclimate (Table 18.1). The relative importance of these pathways will determine the overall impacts of herbivores on ecosystem processes (Bardgett and Wardle 2003; Wardle and Bardgett 2004) and the relative importance of DMIEs and TMIEs.

The mechanisms by which herbivores influence soil processes (Hunter 2001) operate at different rates. As a result, a given species of herbivore may increase rates of decomposition and nutrient cycling in the short term while decreasing those same rates in the long term (Uriarte 2000). 'Fast-cycle' effects (McNaughton *et al.* 1988) are generally those that respond to labile inputs to the soil derived from herbivore activity. Because fast-cycle inputs are labile, they tend to increase rates of decomposition and nutrient cycling, consistent with the 'acceleration hypothesis' (Ritchie *et al.* 1998), which posits that herbivores

Table 18.1 *Pathways by which herbivores can influence terrestrial ecosystem processes. Pathways are channels for direct effects (DEs), density-mediated indirect effects (DMIEs) or trait-mediated indirect effects (TMIEs) that will influence the fast or slow decomposition cycle. Expectations are for acceleration (A) or deceleration (D) of biogeochemical cycling. The overall effects of herbivores in any particular ecosystem will depend on the relative importance of each of these pathways. As a starting point for future discussion, we provide our own view of the need for future research on each pathway and its likely importance to soil processes (ranked from 1 to 5, where 1 is high priority or high importance). The pathways shown here are expanded from the seven major mechanisms described by Hunter (2001)*

Pathway	Interaction type	Cycle	Expectation	Research priority	Process importance
Cadavers	DE	Fast	A	5	5
Faeces and urine	TMIE	Fast	A	3	1
Herbivore-modified through-fall	TMIE	Fast	A	4	3
Green-fall and premature abscission	TMIE	Fast	A	2	2
Induced root exudation	TMIE/DMIE	Fast	A	1	2
Reductions in plant litter inputs	DMIE	Slow	D	3	3
Selective foraging	TMIE/DMIE	Slow	D or A	2	1
Induced litter recalcitrance	TMIE	Slow	D	2	2
Increased litter quality	TMIE	Slow	A	2	2
Changes in microclimate	TMIE	N/A	D or A	3	4
Changes in resource uptake	TMIE	N/A	D or A	1	2

increase rates of biogeochemical cycling. Inputs of faeces, green-fall, through-fall, cadavers and root exudates tend to cause rapid responses in microbial activity and accelerate rates of decomposition and nutrient cycling (Hamilton and Frank 2001; Reynolds and Hunter 2001; Yang 2004; Madritch *et al.* 2007). In contrast, 'slow-cycle' effects (McNaughton *et al.* 1988) are those that influence the long-term decomposition rate of senesced plant material. It turns out that herbivores can either increase or decrease the rate at which plant litter decomposes, depending on the plant species involved (Chapman *et al.* 2006). It follows that slow cycle TMIEs of herbivores on decomposition can support both 'acceleration' and 'deceleration' hypotheses (Ritchie *et al.* 1998); slow cycle effects will accelerate biogeochemical cycles when herbivores make the dominant form of senesced plant litter less recalcitrant (Belovsky and Slade 2000; Classen *et al.*

2006; Sariyildiz *et al.* 2008) and decelerate biogeochemical cycles when herbivores make the dominant form of senesced plant litter more recalcitrant (Pastor and Naiman 1992; Findlay *et al.* 1996; Schweitzer *et al.* 2005b).

A key point here is that differences among ecosystems in their responses to herbivore activity may represent differential expression of fast-cycle and slow-cycle effects. Moreover, a single system measured at different points in time may exhibit both increased and decreased rates of organic matter turnover caused by herbivores (Uriarte 2000). The overall result will depend upon a dynamic balance among the alternative pathways by which herbivores act on ecosystem processes (Pineiro *et al.* 2010; Bardgett and Wardle 2003; Hunter 2001). For example, reductions in litter quality (and its decomposition rate) associated with herbivore browsing in African savanna may be compensated for by increased deposition of urine and faeces in highly browsed sites (Fornara and Du Toit 2008).

In this section, we describe the major pathways by which TMIEs of herbivores can influence members of the soil detrital food web (Table 18.1), and consider their relative importance. The distinction between DMIEs, TMIEs and DEs is not always clear-cut. For example, when herbivores stimulate increases in C exudation from plant roots (Hamilton and Frank 2001), they change both the quantity and average quality of C entering the rhizosphere. Similarly, should inputs of herbivore faeces be considered a TMIE or DE? Certainly, the 'traits' of plants have been altered by their passage through individual herbivores to become faeces. Here, we use the term TMIE quite broadly to encompass any major qualitative changes in plant tissues caused by herbivores that may influence soil decomposers and detritivores.

Fast-cycle pathways and the acceleration of biogeochemical cycles: faeces, cadavers, green-fall, through-fall and root exudates

We should expect TMIEs of herbivores on soil communities to accelerate biogeochemical cycles when they are mediated by the deposition of high-quality substrates that decompose rapidly, such as herbivore faeces. Faeces and urine represent high-quality litter materials deposited onto soil, often at a time of year when warm temperatures favour microbial activity (Hunter 2001; Domisch *et al.* 2009). Faeces contain relatively high concentrations of nutrients that plants would otherwise have resorbed prior to leaf senescence (e.g., N, P) and faeces C may be much more labile than that in litter from the same plant species (Hunter *et al.* 2003b; van der Wal *et al.* 2004; Madritch *et al.* 2007). Soil microbes are often C limited (Schimel and Weintraub 2003), and the C:N ratio of faeces can be higher than that of fresh foliage (le Mellec *et al.* 2009), though generally less than that of senesced litter. Nutrient inputs to soils in herbivore faeces can exceed nutrient return in leaf litter (Fogal and Slansky 1985; Grace 1986), and can double N return rates from plants to soil

(Hollinger 1986). In a recent study, endemic herbivore densities contributed up to 33% and 58% of soil N and P inputs, respectively, of those in annual foliar senescence (Domisch *et al.* 2009). Moreover, because invertebrate herbivores generally feed during warmer months (Speight *et al.* 2008), at least in temperate regions, microbial decomposers may be much more active during periods of insect frass fall than during periods of litter fall. Although the N in faeces can be rapidly immobilized by soil microbes (Lovett and Ruesink 1995; Christenson *et al.* 2002), it can also be remobilized by the activities of fungivorous, bacterivorous and coprophagous soil fauna (Zimmer and Topp 2002; Reynolds *et al.* 2003) and subsequently available for export or assimilation by plants (Reynolds *et al.* 2000; Frost and Hunter 2004; Frost and Hunter 2007). As a result of internal cycling in the soil, waste-based TMIEs of herbivores on soil communities can be maintained over significant periods of time. For example, leaf-cutting ants consume up to 10% of canopy leaves in the foraging area of their colonies and represent important herbivores in many tropical forests. Waste plant and faecal material discarded by leaf-cutter ants contains large concentrations of nitrogen and phosphorus in both total and soluble forms (Hudson *et al.* 2009). Although this waste material decomposes within one year after nests are abandoned, soil nitrate and ammonium concentrations remain high for multiple years (Hudson *et al.* 2009).

Honeydew, the sugar-rich excrement of phloem-feeding insects such as aphids, generally has a higher C:N ratio than does the faeces of most herbivores and can represent a substantial input of labile C to soils (Choudhury 1985; Stadler *et al.* 1998). In recent work, honeydew addition has been shown to promote fungal-based over bacterial-based decomposition channels in soils (Wardle *et al.* 2010), which may explain why honeydew can reduce the rate of litter decomposition in soils (Wardhaugh and Didham 2005). In at least some systems, inputs of labile C in honeydew appear to reduce the availability of N in soil solution (Stadler and Michalzik 1998).

There are, of course, limits on faeces-based TMIEs between herbivores and soil organisms. Although foliage may be modified in important ways as it passes through the herbivore gut, the chemistry of faeces may nonetheless reflect the chemistry of the foliage that was consumed to produce it. For example, genetic variation among aspen clones in foliar chemistry is maintained in frass produced by caterpillars feeding on those genotypes (Madritch *et al.* 2007). Moreover, frass-related changes in soil respiration and decomposer enzyme activity reflect the plant-genotypic origin of the frass. In other words, plant traits are not entirely eradicated by TMIEs of herbivores; effects of herbivore faeces on soil ecosystem structure and function therefore represent a combination of TMIEs and DMIEs (Madritch *et al.* 2007).

The cadavers of herbivores can also influence ecosystem processes in soils. Although herbivore biomass represents the processing and repackaging of plant material, cadaver inputs are categorized here as DEs (Table 18.1). Herbivore tissues have lower C:N and C:P ratios than do plant tissues (Elser *et al.* 2000; Woods *et al.* 2004) and lack lignin and cellulose, so the C in herbivore tissues is generally much more labile than that in plants (Hunter 2001). As a result, herbivore cadavers can have significant impacts on soil microbes and detritivores. For example, the carcasses of periodical cicadas represent enormous resource pulses into the detrital food web of North American forest soils every 13 or 17 years. Cicada cadavers cause increases in soil microbial biomass and nutrient availability that are measurable in subsequent plant growth (Yang 2004) and subsequent patterns of herbivory (Yang 2008). Likewise, large ungulate carcasses generate biogeochemical 'hotspots' in forest soils that influence ecological interactions in space and time (Bump *et al.* 2009).

Green-fall ('orts') and premature leaf abscission represent important herbivore-derived inputs into many soil systems and important pathways of TMIEs. As leaf-chewing herbivores eat plant tissue, they may drop significant quantities of leaf material onto the soil (Risley 1986); it is not uncommon to observe a 'green carpet' of green-fall on the forest floor during outbreaks of canopy-feeding insects. Likewise, many plants respond to herbivore attack by abscising their leaves prematurely (Risley 1986; Chapman *et al.* 2006) before the resorption of sugars and nutrients is complete. Green leaf tissue generally decomposes faster than does senesced litter, presumably because green-fall has higher concentrations of N and P, and lower lignin:N ratios than does senesced litter (Fonte and Schowalter 2004). As a result, green-fall and premature leaf abscission tend to increase rates of soil respiration, carbon turnover and nutrient cycling in soils (Risley and Crossley 1988; Risley and Crossley 1992; Reynolds and Hunter 2001). Effects of premature leaf abscission may be particularly important, though still underappreciated, in coniferous forests; damaged conifer needles often fall earlier and with higher nutrient concentrations than do undamaged needles (Chapman *et al.* 2003; Stadler *et al.* 2006; Sariyildiz *et al.* 2008).

Through-fall is precipitation that passes through the canopy of plants before reaching the soil. Herbivore-derived changes in through-fall result from leaching of nutrients from damaged leaves and dissolution of herbivore faeces from plant surfaces (Tukey and Morgan 1963). The through-fall associated with insect outbreaks can significantly increase rates of nutrient transfer from the canopy to the forest floor. For example, concentrations of dissolved organic C and dissolved organic N increase in through-fall associated with outbreaks of the hemlock woolly adelgid (Stadler *et al.* 2006). In this system, TMIEs of adelgid herbivores on through-fall and needle litter combine to increase rates of N cycling in forest soils (Orwig *et al.* 2008). In oak hardwood forests, insect defoliation also increases substantially the fluxes of P from the canopy to the soil (Reynolds and Hunter, 2001).

Finally, fast-cycle TMIEs of herbivores on soil communities may result from increases in root exudation following defoliation, which can have dramatic impacts on soil processes. For example, defoliation-induced root exudation of labile C in grasslands can increase the biomass of rhizosphere microbes, increase daily rates of N mineralization by five-fold and facilitate re-uptake of N lost to herbivore activity (Hamilton and Frank 2001; Hamilton *et al*. 2008). This particular pathway is perhaps the most difficult to study experimentally, although stable isotope tracers provide great potential (Frost and Hunter 2008a). At present, it is the least understood fast-cycle pathway and merits a significant increase in research effort.

Members of the microbial community that respond to recent and labile carbon inputs will likely mediate acceleration of the fast-cycle effects described above. However, much of the organic matter in soil is much older and comprised in large part of recalcitrant material. Interestingly, labile carbon inputs typical of the TMIEs described above may either increase ('prime') or decrease subsequent decomposition of older, recalcitrant material (Kuzyakov *et al*. 2000; Fontaine *et al*. 2004; Bradford *et al*. 2008a). Decreases are predicted based on preferential substrate utilization (Fontaine and Barot 2005) in which labile inputs favour the soil microbes that specialize on recent and labile C. These labile C specialists then outcompete the microorganisms that utilize recalcitrant C, and decrease overall rates of decomposition in soils (Bradford *et al*., 2008a). To our knowledge, there have been no direct studies of the effects of herbivore-derived inputs on competition among soil microbes that specialize on different soil C pools.

Slow-cycle pathways and acceleration or deceleration of biogeochemical cycles: selective foraging, defence induction and incomplete nutrient resorption

We should expect TMIEs of herbivores on soil communities to modify biogeochemical cycles by slow-cycle pathways (McNaughton *et al*. 1988) when they influence the quality of the dominant forms of senesced plant litter that enter the detrital pathway. Variation in the quality of litter, both within and among plant species, is a fundamental driver of decomposition dynamics (Madritch and Hunter 2002; Cornelissen *et al*. 2004; Madritch and Hunter 2004; Schweitzer *et al*. 2005a) and the most important determinant of variation in decomposition rate on a global scale (Cornwell *et al*. 2008). Litter traits such as condensed tannin concentration consistently slow the rate of litter decomposition in terrestrial and aquatic systems (Schweitzer *et al*. 2008; Coq *et al*. 2010). Likewise, litter toughness, phenolic:N and lignin:N ratios are generally correlated negatively with decomposition rate (Cornelissen *et al*. 2004; Kurokawa and Nakashizuka 2008).

Because herbivores forage selectively, they can change the relative abundance of plant species in communities and therefore influence the relative contributions of litter species available to decomposers (Pastor and Naiman 1992; Wardle *et al.* 2002; Persson *et al.* 2005). Over time, herbivores will typically reduce the availability of more palatable species and increase the availability of less palatable species. In general, plant traits that are correlated with palatability to herbivores are also correlated with rates of litter decomposition (Cornelissen *et al.* 2004; Yule and Gomez 2009; Semmartin *et al.* 2010). These traits include C:N ratio, C:P ratio, lignin:N ratio and tannin concentrations (Cornelissen *et al.* 2004; Schweitzer *et al.* 2008; Kurokawa *et al.* 2010). As a result of this slow-cycle effect, selective foraging will generally act to decelerate rates of decomposition. In a classic example, moose in North America tend to avoid browsing on conifer trees in favour of deciduous trees, resulting in the gradual replacement of deciduous species with evergreens in heavily browsed areas. Conifer needles decompose more slowly than do the leaves of deciduous trees, with the result that selective foraging by moose tends to reduce rates of nutrient cycling and organic matter turnover (Pastor *et al.* 1993; Pastor and Cohen 1997). Other studies support the hypothesis that selective foraging by herbivores reduces rates of nutrient cycling. The removal of exotic herbivores (sheep) from Santa Cruz Island, California, is resulting in recovery of the habitat from exotic grass species (less palatable, more recalcitrant to decomposition) to native shrubs (palatable, faster decomposers). Re-colonization by native shrubs therefore increases rates of nutrient cycling and pools of nitrogen and phosphorus in the soil (Yelenik and Levine 2010). However, the pattern is stronger in some systems than others (Kurokawa and Nakashizuka 2008). For example, experimental removal of insect herbivores from old fields may not increase the average palatability or decomposition rate of plant species (Schadler *et al.* 2003). Moreover, herbivore-induced changes in plant community composition may act to increase or decrease subsequent rates of decomposition, depending upon the community's location on a gradient in precipitation (Semmartin *et al.* 2004). This study serves to emphasize that environmental context will ultimately determine the relative importance of slow-cycle TMIEs on soil carbon dynamics and nutrient cycling (Bardgett and Wardle 2003; Wardle and Bardgett 2004).

Why does grazing not always shift plant community structure towards unpalatable species that resist decomposition? In some cases, although herbivores influence plant community composition, there are no concomitant changes in litter decomposition rates (Golodets *et al.* 2010). At the opposite end of the spectrum, herbivores may sometimes favour plants with litter that decomposes rapidly. For example, grazing tends to promote plant communities with fast decomposing litter in Argentinean grasslands. Moreover, soils within highly-grazed sites seemed to be primed for faster decomposition of litter, whatever the quality of that litter might be (Garibaldi *et al.* 2007). Similarly, grazing by reindeer in northern Europe appears to favour plant species that decompose more rapidly (Olofsson and Oksanen 2002). One possible reason for these counterintuitive results is that, in some environments, herbivores may select for 'tolerant' plants instead of 'resistant' plants (Hunter 2001). Tolerance is a strategy by which plants invest in rapid re-growth (= compensation) following damage at the expense of investing in resistance-based defence mechanisms (Van Der Meijden *et al.* 1988). Because tolerant plants have more palatable tissues than do resistant plants, selection for tolerance may also select for tissues that decompose rapidly (Cornelissen *et al.* 2004). If tolerance to herbivory is a strategy that is favoured in high resource environments, it may explain why herbivores tend to accelerate nutrient cycling in productive environments and decelerate nutrient cycling in unproductive environments (Bardgett and Wardle 2003). Establishing links among plant tolerance, resistance, productivity and TMIEs of herbivores on soil communities remains an understudied area of research and there are some systems that do not fall neatly into current paradigms. For example, in Montana prairie, grasshoppers preferentially consume plants of low N content that decompose slowly – their selective grazing causes an increase in rates of nutrient cycling by favouring plants with high N content that decompose rapidly (Belovsky and Slade 2000). Presumably, the selective foraging of Montana grasshoppers reflects resource requirements different from those typically considered important for herbivore species. Recent work suggests that some herbivores may select plants with high digestible C content (and therefore high C:N ratios) when under predation stress, with concomitant effects on TMIEs and ecosystem processes (Hawlena and Schmitz 2010a, b).

Selective foraging by herbivores is not just expressed among plant species, but also among individual plants within species. Herbivores that forage selectively for palatable individuals or genotypes within plant species may cause replacement of palatable individuals with less palatable individuals over time (Uriarte 2000). Because the traits that determine palatability are often genetically based (Speight *et al.* 2008), herbivores should generate selection pressure for decreasing palatability over evolutionary time. Again, because unpalatable plant tissues are often recalcitrant to decomposition (Cornelissen *et al.* 2004; Kurokawa *et al.* 2010) natural selection imposed by herbivores on levels of plant defence should reduce rates of nutrient cycling and organic matter turnover. We know of only one study that has demonstrated within-species variation in litter recalcitrance related to selection by herbivores. Using a chronosequence of goldenrod, Uriarte (2000) was able to show that goldenrod clones become increasingly resistant to defoliation by the beetle *Trirhabda virgata* over successional time. Associated with this increasing resistance, plants become increasingly recalcitrant to decomposition, suggesting that selection among goldenrod clones by beetles reduces rates of decomposition over time. Unfortunately, the intriguing patterns reported by Uriarte (2000) have not been studied sufficiently in other systems. We are left to assume that the long-term coevolutionary processes between plants and their herbivores have generated plant genotypes and species that vary in their rates of decomposition precisely because of adaptation and counter-adaptation with herbivores (reciprocal bouts of genetic change that form the basis of coevolution). This is an area of research in much need of further study although there are glimpses from other systems to suggest that such processes operate. Studies by Whitham and colleagues (Chapter 16, this volume) have certainly established genetically based variation in plant defence that influences ecosystem processes. However, replacement of geno-types over time, and therefore evolution per se, is much more difficult to observe. Meta-analysis suggests that plants are particularly susceptible to novel suites of herbivores (Parker et al. 2006), suggesting that evolutionary history plays a major role in patterns of plant damage and subsequent TMIEs on soil communities.

In addition to effects caused by selective foraging, herbivores can influ-ence soil processes by inducing chemical changes in the foliage of plants that they damage. When those chemical changes persist into the leaf litter, and influence subsequent rates of decomposition, they are described as 'afterlife effects' (Choudhury 1988; Findlay et al. 1996; Chapman et al. 2006). It turns out that 'afterlife effects' of leaf litter on decomposition processes that are mediated by herbivory can either accelerate or decelerate rates of decomposition and nutrient cycling, depending on the plant species and ecosystem in which the herbivory takes place. For example, defoliation of oriental spruce trees by the beetle Ips typographus is associated with increases in the quality of spruce needle litter; damaged needles have higher N concentrations and lower C:N and lignin:N ratios. As a consequence, spruce needles from damaged stands decompose more rapidly than do needles from undamaged stands (Sariyildiz et al. 2008). Mass loss rates are significantly and positively correlated with initial nitrogen concentration and negatively with C:N and lignin:N ratios. These results are consistent with some other studies of conifer herbivores. Defoliation of pinon pine by the stem-boring moth, Dioryctria albovittella, increases needle litter N content by 16% and increases the subsequent rate of litter decomposition (Classen et al. 2007). Surprisingly, bulk soil microbial communities are 52% more abundant under trees in the absence of herbivores than in their presence, suggesting that microbial abundance is not directly coupled to high quality litter inputs from herbivores. Rather, seasonal and microclimatic variation in soil may have more important effects on soil C and microbial activity than does insect defoliation (Classen et al. 2005, 2006).

In contrast, lace bug damage to bur oak trees induces increases in the lignin content of oak litter, with subsequent declines in the rate of litter decomposition (Kay *et al.* 2008). In the same system, aphid populations on bur oak have no apparent effect on either litter chemistry or subsequent decomposition rates, illustrating that TMIEs of herbivores on decomposer systems are herbivore-specific. In the bur oak system, lace bug herbivory is much more common in burned areas than in unburned areas. Frequent fires may therefore promote oak–herbivore interactions that decelerate decomposition, amplifying other influences of fire that slow nitrogen cycling (Kay *et al.* 2008). The paper by Kay *et al.* (2008) raises the unpleasant spectre of species-specificity in the direction and magnitude of TMIEs of herbivores on decomposer systems, which would strongly limit our ability to make generalizations. Such species-specificity may be common. In Argentinean grasslands, some plant litters decompose more rapidly following long-term exposure to grazing whereas some decompose more slowly (Semmartin *et al.* 2008). Rabbit grazing does not change the quality of litter in English grasslands, although it does change the quantity of litter and rates of N mineralization (Olofsson *et al.* 2007). Unless we can recognize plant or herbivore traits that are associated with their positive or negative effects on decomposition rates, our ability to generalize will be limited.

In the most comprehensive analysis to date, Chapman *et al.* (2006) have argued that TMIEs of herbivores on decomposition should differ between deciduous and evergreen trees. In their own work, and in much of the literature that they cite, defoliation of evergreen trees results in premature leaf abscission, less nutrient resorption from those abscissing leaves, higher litter quality and increased rates of litter turnover (Chapman *et al.*, 2003). In contrast, defoliation of deciduous trees results in the induction of defensive compounds that increase the recalcitrance of litter and reduce rates of litter decomposition (Choudhury 1988; Findlay *et al.* 1996; Schweitzer *et al.* 2005b). There is certainly some support for this model. For example, hemlock needle litter contains higher N concentrations under trees infested with hemlock woolly adelgid than under uninfested trees (Stadler *et al.* 2006). Likewise, scale insect attack on pinon pine increases needle N and P by 50% (Classen *et al.* 2006). Again in support of Chapman *et al.*'s model (though for a perennial herb, not a tree), galled tissues from goldenrod plants release nitrogen back to the soil more slowly than do ungalled tissues, although overall effects on decomposition are modest (Crutsinger *et al.* 2008). Also with galling insects, litter from defoliated or gall-infested cottonwood trees decomposes more slowly than does litter from undefoliated trees, associated with herbivore-induced increases in condensed tannins (Schweitzer *et al.* 2005a, b). The strength of the TMIE of cottonwood herbivores on decomposition varies among plant genotypes, a theme that is developed further in the chapter by Allen *et al.* in this volume. Galling insects may generally induce defensive (and recalcitrant) compounds in associated plant tissues because the lack of mobility of larvae in galls sets a premium on defending the local tissues that provide them with nutrition (C. J. Frost, unpublished data). However, such TMIEs are not limited to plant damage by insect herbivores. Grazing by red deer in the Scottish Highlands reduces dissolved organic C, nitrate, ammonium and N mineralization in soils (Harrison and Bardgett 2004) in part because of browsing-induced reductions in the quality of birch litter and its rate of decomposition (Harrison and Bardgett 2003).

At the same time, there are clear counterexamples to Chapman et al.'s (2006) hypothesis; defoliation can actually increase directly the rate at which deciduous plant tissues decompose. For example, reindeer browsing of birch in Finland tends to decrease concentrations of secondary metabolites in birch litter, increase rates of decomposition, and increase rates of N cycling in soils (Stark et al. 2007). Likewise, phenolic concentrations are generally lower in damaged oak litter than in undamaged oak litter, but with negligible consequences for subsequent decomposition (Hall et al. 2005a, 2006; Frost and Hunter 2008b). There are also some idiosyncratic responses to plant damage that influence the decomposition of senesced tissues. In the Argentinean grass Paspalum dilatatum, grazing causes a morphological change that favours the production of leaf blade over leaf sheath. Leaf blades have higher N and lower lignin:N ratios than do leaf sheaths, so that long-term grazing increases rates of litter decomposition and nutrient cycling (Semmartin and Ghersa 2006).

Perhaps more importantly, the multiple fast-cycle and slow-cycle mechanisms underlying TMIEs of herbivores on soil food webs (Hunter 2001) may combine to reverse Chapman et al.'s (2006) predictions. Some defoliators of deciduous forests, such as the gypsy moth, are such messy eaters that they drop as high-quality green-fall about one-third of all the leaf material that they remove (Risley 1986); green-fall decomposes rapidly in comparison to leaf litter (Fonte and Schowalter 2004) and, in combination with frass inputs, defoliation of deciduous trees can be positively correlated with rates of litter decomposition and soil nutrient flux (Fonte and Schowalter 2005). Moreover, increases in soil nitrate and the export of N from defoliated watersheds in deciduous forests argue for higher rates of nutrient cycling following herbivore outbreaks in at least some systems (Reynolds et al. 2000; Reynolds and Hunter 2001; Hunter et al. 2003b). It may be that defoliation-induced increases in the recalcitrance of deciduous litter are counteracted by associated inputs of frass and green-fall (Frost and Hunter 2004, 2007), or of only short-term importance during decomposition (Hall et al. 2005b, 2006; Frost and Hunter 2008b).

For completeness, there are two additional mechanisms by which herbivores can influence soil processes that do not fall neatly into either fast-cycle or slow-cycle effects. Defoliation can causes changes in soil microclimate that influence subsequent rates of decomposition and nutrient cycling, whether those rates be driven by fast- or slow-cycle effects (Classen *et al.* 2005). For example, when plant canopies are damaged by herbivores, soil moisture levels can decline as a result of increased sunlight reaching (and drying out) the ground, so that rates of litter decomposition (be it leaves, frass, green-fall, etc.) decline (Cobb *et al.* 2006). Additionally, herbivore attack can change the rate at which plant communities utilize soil resources. For example, damaged plants may reduce their per capita N uptake (Frost and Hunter 2007), leaving N resources for microbial use, long-term storage, or export from the ecosystem. Selective foraging by herbivores (above) may also result in long-term changes in plant community structure that influence the use of soil resources (Pastor and Naiman 1992).

Effects on decomposer and detritivore communities

Strictly speaking, for the interactions described above to qualify as TMIEs among species, herbivores should influence population-level processes among members of the soil detrital food web. In addition to effects on the processes that they mediate (decomposition, nutrient cycling) what is the evidence that soil microbes and soil fauna respond to TMIEs of herbivores? Effects are summarized for decomposers and detritivores below, but it is clear that much work remains to be done to characterize TMIEs of herbivores on microbial and detritivore population processes and community structure.

Decomposers (microbes)

At large spatial scales of measurement (field studies as opposed to laboratory microcosms), effects of TMIEs of herbivores on soil microbial processes are not always strong. For example, intensive summer browsing by reindeer has apparently negligible effects on microbial activity, enzyme activity and decomposition in Finnish subarctic mountain birch forests (Stark *et al.* 2008). The authors hypothesized that in areas with intensive summer browsing by reindeer, the C flow to belowground would be reduced, but that the C inputs would be more labile. However, despite large differences in vegetation structure, the reindeer ranges did not differ in microbial biomass or the abundance of nematodes. Furthermore, there was no difference in microbial phospholipid fatty acid composition (an index of community structure) or in the trophic groups of soil nematodes. Likewise, there were few effects on the organic fractions (non-polar extractives, water-soluble extractives, acid-soluble fraction and acid-insoluble residue) of the soil organic matter. The authors suggest that the high stocks of organic matter in subarctic systems may weaken the link between herbivore activity and microbial processes (Stark *et al.* 2008). Unfortunately, there are too few comparable studies to assess the generality of this idea. In other systems, potential TMIEs of herbivores on soil microbes may be swamped by other factors. For example, attack of pinon pine trees by scale insects and moths generally increases the quality and quantity of needle litter entering the soil (Classen *et al.* 2005, 2006). However, soil microbial populations are lower under insect-susceptible trees, and appear to be driven largely by other ecological factors, including microclimate and season (Classen *et al.* 2005, 2006).

It is clear, however, that herbivores can influence soil microbial populations in some ecosystems. Microbial biomass is lower in the presence than in the absence of large grazers in Kenyan grasslands. The herbivore-induced reduction in microbial biomass appears to be driven by reductions in the quantity of plant litter inputs (a DMIE), rather than changes in litter quality (a TMIE). Overall, levels of plant production appear to have a greater influence on soil microbes than does grazing in this system (Sankaran and Augustine 2004). In stark contrast, cattle grazing increases soil microbial biomass in the sub-tropical grasslands of Florida (Wang *et al.* 2006). Given that overall C inputs to the soil must be lower, the increase in microbial biomass likely occurs because of herbivore-mediated changes in the quality or timing of litter inputs. The deposition of reindeer faeces in tundra ecosystems increases soil microbial C and N, enhances the standing crop of grasses (van der Wal *et al.* 2004) and may contribute to overall levels of higher N cycling under reindeer grazing (Olofsson *et al.* 2004). However, effects may be localized to patches of faeces rather than across broad geographic regions (Stark *et al.* 2008, above). In systems with insect herbivores, defoliation may also influence soil microbial populations. For example, bacterial populations increase on hemlock litter subjected to defoliation by hemlock woolly adelgid, although fungi show no consistent response (Stadler *et al.* 2006). Fungal populations can increase rapidly on insect frass, with consequences for nutrient dynamics (Lovett and Ruesink 1995; Christenson *et al.* 2002). The relative expression of extracellular enzymes released by soil microbes during the decomposition process is also influenced by frass inputs (Madritch *et al.* 2007).

Detritivores

It is possible that herbivore-mediated impacts on soil microbes are sometimes hard to measure because of rapid responses by the detritivore community that graze in part on soil microbes (Mikola *et al.* 2001; Coleman *et al.* 2004). For example, defoliation of legumes increases the density of bacterial-feeding protozoa even though there are no observable changes in microbial biomass (Saj *et al.* 2008); increased microbial productivity may be balanced by increased consumption of microbes by higher trophic levels. Populations of soil microarthropods and nematodes can also increase markedly following defoliation to plants. For example, defoliation by Lepidoptera in the canopies of Douglas fir trees influences the relative abundance of soil arthropods

beneath defoliated trees (Schowalter and Sabin 1991). Likewise, mite populations are higher in the litter under insect-susceptible pinon pines trees than in
the litter under resistant trees (Classen *et al.* 2006) and inputs of herbivore-
derived frass, green-fall and through-fall in temperate forests can cause substantial increases in populations of Collembola and nematodes (Reynolds *et al.*
2003). Soil nematode community structure also responds to cattle grazing in
sub-tropical pasture (Wang *et al.* 2006), although the relative influence of
TMIEs and DMIEs is unknown. The relationship between herbivore faeces
and some detritivores can be sufficiently predictable that the fossil remains
of soil detritivores (Oribatid mites) can be used to estimate herbivore densities
and land use patterns over millennial time periods (Chepstow-Lusty *et al.*
2007). While fast-cycle effects of herbivore TMIEs on detritivores have been
reported several times, slow-cycle TMIEs of herbivores on detritivores remain
relatively rare in the literature. In one study, green-fall from oaks that had
been defoliated by Lepidoptera in the previous growing season exhibited
legacy effects that reduced the biomass and N uptake of soil Collembola in
the following year (Bradford *et al.* 2008b). The relative importance of fast-cycle
and slow-cycle TMIEs of herbivores on soil detritivores remains unclear at this
time. We should note that, as with microbes, detritivores do not always
exhibit TMIEs from herbivores. For example, despite significant changes in
herbivore community structure and levels of defoliation (Stiling *et al.* 2003;
Hall *et al.* 2005a), there has been no measurable change in the detritivore
community associated with Florida scrub oaks after 11 years of continuous
CO_2 exposure (Stiling *et al.* 2010).

Some perspective on relative importance

Our growing interest in TMIEs should not encourage us to overstate their
importance in all systems. Rather, we should seek to understand how and why
their relative importance appears to differ among systems. In some studies,
experimental removal of herbivores has negligible effects on decomposition
processes (Verchot *et al.* 2002; Schadler *et al.* 2004). At the very low levels of
productivity typical of arid environments, rates of herbivory may be
extremely low such that TMIEs of herbivores on decomposition may be trivial.
Rather, organic matter dynamics may be driven by trophic cascades from
predators to detritivores within the brown food web (Ayal 2007). In the boreal
forest, long-term rates of carbon storage and decomposition may be driven
primarily by patterns of hydrology and soil temperature (Turetsky *et al.* 2005).
TMIEs caused by herbivores may be rather minor in comparison, although
they may also contribute to the proximate mechanisms by which hydrology
and temperature exert their influences on organic matter dynamics. In other
words, temperature and hydrology may provide the template upon which
other abiotic and biotic forces (vegetation type, fire, herbivory) may operate

(Turetsky *et al.* 2005). In the Trans-Himalayas, grazing by exotic grazers reduces C sequestration in soils. In this case, selective grazing by exotic grazers favours plants of lower overall productivity that simply deposit less litter into the soil system – C storage declines primarily because of DMIEs, not TMIEs (Bagchi and Ritchie 2010). This is an important point – it is not enough to measure changes in litter quality and to measure how fast that litter decomposes. Herbivores will almost always have impacts on the quantity of materials reaching soil systems, not just the quality. Long-term reductions in C inputs that result from chronic herbivory may often have effects greater than those caused by changes in resource quality (Persson *et al.* 2005; Sorensen *et al.* 2008). For example, grazing by geese in the Arctic tundra reduces carbon pools in the soil primarily by reducing rates of photosynthesis and therefore C sequestration from the atmosphere (Sjogersten *et al.* 2008). At least in some cases, DMIEs (reductions in the quantity of plant material) appear to overwhelm any TMIEs (changes in the quality of plant material) on soil C dynamics.

A case study with *Quercus*

Assessing the overall importance of TMIEs of herbivores on soil communities and processes requires long-term studies that investigate a wide variety of potential TMIE and DMIE pathways. Since the mid 1990s, we have been studying the effects of herbivores on soil processes and ecosystem function in oak-dominated landscapes (Fig. 18.1). Oaks (genus *Quercus*) represent suitable systems to study herbivore–ecosystem interactions because they are dominant members of forest communities throughout the northern hemisphere. With about 600 species recognized worldwide, oaks are distributed from northern to tropical latitudes and from the Americas throughout Europe to Asia. They support abundant and diverse herbivore faunas and are foundation species in many forest ecosystems throughout the world.

In 1998, we observed a serendipitous outbreak of sawfly larvae, *Periclista*, on red and white oaks at the Coweeta Hydrologic Laboratory in western North Carolina, USA. We subsequently compared concentrations of through-fall nutrients, soil nutrients, and nutrient export between defoliated and undefoliated watersheds at Coweeta (Reynolds *et al.* 2000). During the defoliation event (May 1998), frass inputs were over three-fold higher in defoliated than in undefoliated areas and by June, 1998, nitrate-N concentrations in through-fall were five times greater in defoliated than in undefoliated areas. By June and July of 1998, soil nitrate-N concentrations were seven-fold higher under defoliated oak canopies than under undefoliated canopies. Between June and August of the same year, streams draining defoliated areas contained twice the concentration of nitrate-N than did streams draining undefoliated areas. These results suggested to us that oak herbivore outbreaks could have

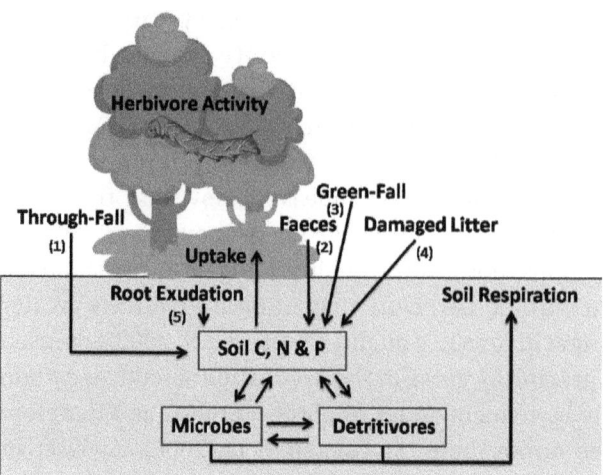

Figure 18.1 Pathways by which oak herbivores influence ecosystem processes.
(1) Through-fall increases soil P by up to 16-fold, Collembola biomass up to 1.5-fold, and
nematode biomass up to 7.5-fold; (2) faeces increase soil N pools over 2-fold,
Collembola biomass up to 3-fold, and nematode biomass up to 14-fold; (3) green-fall
increases soil N by 1.2-fold and soil respiration by 1.27-fold; (4) herbivore-induced
changes in litter quality have negligible effects on soil processes; (5) C exudates from
roots are maintained at constant levels despite herbivore-induced declines in C
allocation to roots. During outbreaks, fast-cycle mechanisms combine to increase soil
N by 7-fold and N export in streams by 2-fold in oak-dominated watersheds, while trees
recover about 16% of the N initially lost in herbivore faeces. See colour plate section.

dramatic impacts on nutrient dynamics at the watershed scale, including
nutrient export in streams (Reynolds *et al.* 2000).

As a result of these initial observations, we began to study the mechanisms
by which oak herbivores influence ecosystem processes. Working at Coweeta,
we manipulated herbivore-derived inputs of frass, green-fall and simulated
through-fall in replicated field plots (Reynolds and Hunter 2001). Soil respiration was 27% higher in treatments with green-fall than in treatments excluding green-fall, suggesting that labile compounds in green litter stimulate the
activity of soil organisms; herbivore inputs were second only to soil temperature in driving rates of soil respiration. Soil nitrate-N was 20% higher in
treatments with green-fall than in treatments excluding green-fall, suggesting
that green-fall inputs from herbivores represent an important input of N from
canopy to soil. Moreover, concentrations of phosphate-P in the soil were up to
16-fold higher in the presence of herbivore-simulated through-fall than in its
absence. We had not expected herbivores to have such a significant impact on
P flux, yet they were the most important predictor of soil phosphate-P concentrations in our study. Given the increases in nitrate-N availability that we
observed following the sawfly outbreak (above), we were initially surprised to

observe declines in soil nitrate-N following experimental frass and through-fall additions (Reynolds and Hunter 2001). However, other work has reported rapid immobilization of insect frass N by microbial decomposers (Lovett and Ruesink 1995; Christenson *et al.* 2002) and that may have occurred in our study as well.

We had doubts that the increases in soil and stream nitrate-N observed following the *Periclista* outbreak could have been driven by increases in green-fall deposition alone. In particular, we thought that frass fall was strongly implicated in N flux from canopy to soil and subsequent stream export. In continued sampling at Coweeta, we had observed strong positive correlations between frass fall and soil nitrate-N and ammonium-N concentrations, even at endemic densities of insect herbivores (Hunter *et al.* 2003b). For example, at one site, variation in frass deposition explained a remarkable 62% of the variation in soil nitrate-N availability. At the same time, we observed pro-found spatial and temporal variability in the relationship between frass inputs and soil mineral N availability. Given that frass N can be immobilized rapidly by decomposers (Lovett and Ruesink 1995; Christenson *et al.* 2002), we began to explore factors that might re-mobilize frass N in some sites and make it available in soil solution. Again, using experimental additions of frass and through-fall in field plots at Coweeta, we analysed changes in soil fauna that resulted from our treatments. In response to frass and through-fall additions, we observed blooms of Collembola, fungal-feeding nematodes and bacterial-feeding nematodes (Reynolds *et al.* 2003), all of which can re-mobilize nutrients that are immobilized by microbial populations (Coleman *et al.* 2004). Importantly, we also observed substantial spatial and temporal varia-tion in population densities of soil fauna (Reynolds *et al.* 2003), which may explain in part why correlations between insect frass and soil mineral N are so variable in space and time. In studying larger members of the soil fauna, we also found that exclusion of predators (particularly ants and spiders) could influence the physical and chemical trajectories of litter decomposition (Hunter *et al.* 2003a). We therefore concur with others who argue that spatial and temporal variation in trophic interactions within soil food webs may determine in part where and when herbivores influence soil mineral N avail-ability (Mikola *et al.* 2001; Schmitz 2008).

By this point, we had established that frass, green-fall and through-fall inputs could increase soil mineral N and P availability and soil respiration under field conditions in oak-dominated ecosystems. We then began to explore the impacts of herbivore-derived inputs on soil processes under more controlled conditions to study the mechanisms in more detail. We developed single-tree mesocosm experiments in which we could manip-ulate defoliation levels and herbivore inputs independently with a high degree of replication (Frost and Hunter 2004). As before, we found that the experimental addition of insect frass to soil caused increases in soil mineral N pools and export of nitrate-N from the system. An important question then arises; can trees regain any of the N lost to herbivory by uptake from the soil? We used ^{15}N-enriched insect frass to address this question (Frost and Hunter 2007). After depositing ^{15}N-enriched frass on soil during spring, we observed mineralization of frass N and some N export within 1 week of frass deposition. However, within 1 month, oak trees had recovered some of the ^{15}N from insect frass in their foliage. Moreover, we recovered ^{15}N from the tissues of insects feeding on the same trees in late summer. In other words, the ^{15}N had cycled from insect frass to the soil, back into the trees and into a second generation of insect herbivores within a single growing season. This represents the fastest reported cycling of N from one herbivore back to another herbivore of which we are aware (Frost and Hunter 2007) and provides strong support for the acceleration hypothesis (Ritchie *et al.* 1998). Because our experimental mesocosms allowed us to manipulate frass addition and defoliation independently, we were able to measure whether damage by herbivores influenced the ability of trees to reacquire N from frass. Overall, trees recovered about 16% of the N that they had lost in insect frass. Moreover, in the second year of the study, defoliated trees recovered less ^{15}N from frass than did undefoliated trees, illustrating a significant cost of herbivory to N recovery from soil. Insects fed experimentally from these trees during the second year also accumulated less ^{15}N if the trees had been damaged in the previous year, and natural insect colonization was lower on damaged trees than on undamaged trees. We have therefore illustrated the potential for negative feedback in this system – high levels of defoliation result in high levels of N loss in frass and lower levels of N recovery from nutrient uptake. Lower N uptake results in poorer-quality foliage for herbivores, lower herbivore densities and less defoliation in the subsequent year (Frost and Hunter 2007). Such negative feedback processes between generations of insect herbivores have the potential to generate cyclic dynamics in herbivore populations (Baltensweiler *et al.* 1977; Haukioja *et al.* 1985; Hunter 1994; Hunter *et al.* 1997; Hunter 2002) via TMIEs on soil processes. Herbivores also contributed significantly to soil N storage, with a substantial proportion of ^{15}N from frass remaining bound in the top 5 cm of soil (Frost and Hunter 2007).

In our mesocosm experiments, we also established that soil respiration increased following leaf damage by caterpillars, even in the absence of the typical inputs (frass, through-fall, green-fall) that would generally accompany defoliation (Frost and Hunter 2004). Based on work in other systems (Hamilton and Frank 2001; Hamilton *et al.* 2008), we predicted that defoliated oaks would exude C from their roots and stimulate microbial activity in the rhizosphere. In a microcosm experiment, we labelled oak trees with $^{13}CO_2$ while exposing them to insect herbivory (Frost and Hunter 2008a). Allocation of ^{13}C to root tissue was substantially reduced following damage to leaves by insects. However, the total amount of ^{13}C exuded to the rhizosphere was equivalent between damaged and undamaged trees. This means that relatively more of the C allocated below ground was allocated to root exudation in damaged than in undamaged trees, in partial support of our prediction. Future work with larger oaks might establish a better picture of how root exudation responds to defoliation in the canopy.

Finally, we have explored potential 'afterlife' effects of insect damage on oak leaves by comparing the chemical composition and decomposition rates of damaged and undamaged oak litter. The green leaves of oaks generally respond to defoliation with higher concentrations of tannins and lower concentrations of N (Rossiter et al. 1988; Hunter and Schultz 1995), but subsequent effects on litter quality and decomposition had not been established previously. In experiments with red oak, Q. rubra, trees, insect damage induced increases in tannins and decreases in N in green foliage as expected. However, these changes in green leaves did not translate into higher tannin litter after leaf senescence – damaged oak litter was actually lower in tannin than was undamaged litter (Frost and Hunter 2008b). Moreover, rates of litter decomposition were unaffected by the presence or absence of insect damage arguing against TMIEs of litter damage on the decomposer community. In a second series of experiments with scrub oak trees in Florida, we found similar results. These experiments were conducted to study interactive effects of insect herbivores and elevated atmospheric concentrations of CO_2 on oak leaf litter quality and subsequent decomposition (Hall et al. 2005a, 2006). We focused on three dominant oak species in the Florida scrub oak community; Q. geminata, Q. myrtifolia and Q. chapmanii. As with red oak trees, phenolic concentrations were actually lower in herbivore-damaged litter than in undamaged litter, whereas lignin concentrations and lignin:N ratios were higher. Importantly, changes in litter chemistry from year to year, unrelated to our experimental treatments, were far larger than any effects of either CO_2 or insect damage on litter chemistry (Hall et al. 2005a). As a result, damage by insects had no effect on subsequent rates of litter decomposition. Rather, the location in which the decomposing litter was placed had the greatest impact on decomposition rates (Hall et al. 2006). Our data show that, unlike some other systems (Findlay et al. 1996; Chapman et al. 2006), the leaf litter of four different oak species does not transmit 'afterlife' effects of insect damage to decomposer communities via TMIEs.

In summary (Fig. 18.1), more than a decade of research on interactions between oak herbivores and soil processes has established that: (1) inputs of insect frass, through-fall and green-fall increase N inputs into soil; (2) while these inputs may be rapidly immobilized by microbial decomposers, they can be remobilized by soil faunal interactions in the brown (= detrital) food web;

(3) effects are significant at endemic densities of insect herbivores as well as during outbreaks; (4) subsequent increases in the soil mineral N pool can result in N export from the ecosystem, storage in soils, or re-uptake by plants; (5) rapid recycling of N can occur from plants through herbivore frass, into soil solution, back into plants, and into a second generation of herbivores within a single growing season; (6) insect defoliation of oaks reduces uptake of soil N in subsequent years, reducing foliage quality for herbivores, and resulting in negative feedback that reduces herbivore densities on trees; (7) oak trees appear to maintain C exudation into the rhizosphere, even when defoliation reduces C allocation to roots, which may explain herbivore-induced increases in soil respiration over and above those caused by frass, through-fall or green-fall; and (8) afterlife effects of damaged litter appear to have negligible effects on soil processes in oak-dominated ecosystems. Overall, the substantial TMIEs between oak herbivores and decomposer communities in the soil appear to be driven in large part by 'fast-cycle' inputs of frass and green-fall, and support the hypothesis that oak herbivores accelerate rates of biogeochemical cycling in oak-dominated ecosystems.

Conclusions and future directions

There seems little doubt that herbivores can have a profound influence on terrestrial ecosystem processes through TMIEs on soil communities. At the same time, the relative impacts of TMIEs and DMIEs have yet to be established with any quantitative rigour in most systems. In the oak dominated land-scapes described above, we have reported significant effects of TMIEs on soil processes that likely overwhelm DMIEs in all but outbreak years of forest defoliators. That being said, it is much harder to assess the relative importance of different TMIEs. For example, estimating the consequences of herbivore-induced root exudation from mature forest trees remains logistically challenging. Until such methodological challenges are overcome, it may be that structural equation modelling to compare alternative causal pathways (Cronin *et al.* 2010) may represent a useful approach. Moreover, unequivocal links between herbivores and the population processes of decomposers and detritivores remain much weaker than they should. With an increasing variety of molecular tools at our disposal (Zak *et al.* 2006), there should be a concerted effort to measure the changes in population dynamics and community structure of soil microorganisms that result from TMIEs of herbivores.

There are other obvious gaps in our knowledge. While we have learned much about the effects of herbivores on the N cycle, much less is known about potential TMIEs of herbivores on P cycling. Herbivores can change the N:P stoichiometry of soils and plants (Carline *et al.* 2005; Anderson *et al.* 2007) and we have observed up to 16-fold increases in soil P levels after experimental manipulation of herbivore-derived through-fall (Reynolds and Hunter 2001). Potential TMIEs of herbivores on P availability in soil communities should be a fruitful area for future research, especially given recent results implicating C:P ratios as fundamental drivers of decomposition across plant groups (Kurokawa *et al.* 2010).

It is also important to 'complete the loop' and to study feedback processes between decomposers and herbivores aboveground. Because decomposers recycle nutrients of key importance to plant growth and defence, it should be no surprise that decomposer abundance may feedback to influence herbivore performance and subsequent plant damage (Wurst *et al.* 2004; Lohmann *et al.* 2009; Eisenhauer *et al.* 2010; Megias and Muller 2010). In other words, there will be important TMIEs between decomposers and herbivores, just as there are between herbivores and decomposers. How much such feedback loops matter to the major biogeochemical cycles is completely unknown.

Perhaps most importantly, we still lack an adequate theoretical framework with which to predict the magnitude and direction of TMIEs of herbivores on ecosystem processes. Despite important attempts at synthesis (Lovett *et al.* 2002; Bardgett and Wardle 2003; Chapman *et al.* 2006) our ability to generalize among systems remains relatively poor and key questions remain. Do vertebrate and invertebrate herbivores have similar effects upon ecosystem processes? Are there biome-specific patterns that we might explore? When do top-down effects of consumers in brown food webs matter more than bottom-up effects of plant quality, including those 'bottom-up' effects that are actually TMIEs of herbivores on soil processes (Dunham 2008; Schmitz 2008; Srivastava *et al.* 2009)? Under what circumstances do abiotic factors and ecosystem productivity mediate the effects of herbivores on soil communities? Although herbivores sometimes accelerate biogeochemical cycling in more productive ecosystems and decelerate them in less productive systems (Bardgett and Wardle 2003), there are enough counterexamples (Persson *et al.* 2005; Garibaldi *et al.* 2007; Stark *et al.* 2007) to suggest that we still lack the appropriate theoretical framework to predict how and when herbivores will influence ecosystem processes.

Prediction is becoming ever more urgent. Recent research illustrates that multiple facets of global environmental change have the potential to alter in major ways the TMIEs of herbivores on ecosystem processes. Expanding herbivore distributions that result from global warming can feed back to further increase warming trends (Kurz *et al.* 2008) or influence nutrient cycling (Christenson *et al.* 2010). Likewise, anthropogenic increases in atmospheric N deposition will likely increase plant foliar N content, rates of herbivory and subsequent effects on ecosystem processes (Madritch and Hunter 2003; Throop *et al.* 2004; Throop and Lerdau 2004; Zehnder and Hunter 2008). We know that elevated concentrations of atmospheric CO_2 can sometimes (Allard *et al.* 2004) but not always (Hall *et al.* 2006) interact with herbivore activity to influence rates of litter decomposition. Likewise, invasive herbivore species can have profound effects on carbon and nitrogen dynamics in some systems (Gandhi and Herms 2010). Perhaps the single greatest challenge will be to understand the complex interactions among drivers of global environmental change (Tylianakis *et al.* 2008) and how they combine to determine the TMIEs of herbivores on ecosystem processes.

Acknowledgements

Dac Crossley and Dave Coleman provided invaluable guidance as we started our studies of herbivore–ecosystem interactions. Generous colleagues at the Coweeta LTER site provided intellectual guidance and logistical support. We gratefully acknowledge the National Science Foundation (DEB-9527522, DEB-9632854, DEB-9815133, DEB-0342750, DEB-0404876, DEB-0814340) and the US Department of Energy (DE-FC03–90ER61010) who paid the bills.

References

Allard, V., Newton, P. C. D., Lieffering, M. *et al.* (2004) Elevated CO_2 effects on decomposition processes in a grazed grassland. *Global Change Biology*, **10**, 1553–1564.

Anderson, T. M., Ritchie, M. E. and McNaughton, S. J. (2007) Rainfall and soils modify plant community response to grazing in Serengeti National Park. *Ecology*, **88**, 1191–1201.

Ayal, Y. (2007) Trophic structure and the role of predation in shaping hot desert communities. *Journal of Arid Environments*, **68**, 171–187.

Bagchi, S. and Ritchie, M. E. (2010) Introduced grazers can restrict potential soil carbon sequestration through impacts on plant community composition. *Ecology Letters*, **13**, 959–968.

Baltensweiler, W., Benz, G., Bovey, P. and Delucchi, P. (1977) Dynamics of larch bud moth populations. *Annual Review of Entomology*, **22**, 79–100.

Bardgett, R. D. and Wardle, D. A. (2003) Herbivore-mediated linkages between aboveground and belowground communities. *Ecology*, **84**, 2258–2268.

Belovsky, G. E. and Slade, J. B. (2000) Insect herbivory accelerates nutrient cycling and increases plant production. *Proceedings of the National Academy of Sciences of the United States of America*, **97**, 14412–14417.

Bradford, M. A., Fierer, N. and Reynolds, J. F. (2008a) Soil carbon stocks in experimental mesocosms are dependent on the rate of labile carbon, nitrogen and phosphorus inputs to soils. *Functional Ecology*, **23**, 627–636.

Bradford, M. A., Gancos, T. and Frost, C. J. (2008b) Slow-cycle effects of foliar herbivory alter the nitrogen acquisition and population size of Collembola. *Soil Biology and Biochemistry*, **40**, 1253–1258.

Bump, J. K., Webster, C. R., Vucetich, J. A. *et al.* (2009) Ungulate carcasses perforate ecological filters and create biogeochemical hotspots in forest herbaceous layers allowing trees a competitive advantage. *Ecosystems*, **12**, 996–1007.

Carline, K. A., Jones, H. E. and Bardgett, R. D. (2005) Large herbivores affect the stoichiometry of nutrients in a regenerating woodland ecosystem. *Oikos*, **110**, 453–460.

Cebrian, J. (2004) Role of first-order consumers in ecosystem carbon flow. *Ecology Letters*, **7**, 232–240.

Chapman, S. K., Hart, S. C., Cobb, N. S., Whitham, T. G. and Koch, G. W. (2003) Insect herbivory increases litter quality and decomposition: an extension of the acceleration hypothesis. *Ecology*, **84**, 2867–2876.

Chapman, S. K., Schweitzer, J. A. and Whitham, T. G. (2006) Herbivory differentially alters plant litter dynamics of evergreen and deciduous trees. *Oikos*, **114**, 566–574.

Chepstow-Lusty, A. J., Frogley, M. R., Bauer, B. S. *et al.* (2007) Evaluating socio-economic change in the Andes using oribatid mite abundances as indicators of domestic animal densities. *Journal of Archaeological Science*, **34**, 1178–1186.

Chew, R. M. (1974) Consumers as regulators of ecosystems: an alternative to energetics. *Ohio Journal of Science*, **74**, 359–370.

Choudhury, D. (1985) Aphid honeydew: a re-appraisal of Owen and Wiegert's hypothesis. *Oikos*, **45**, 287–290.

Choudhury, D. (1988) Herbivore induced changes in leaf litter resource quality: a neglected aspect of herbivory in ecosystem dynamics. *Oikos*, **51**, 389–393.

Christenson, L. C., Lovett, G. M., Mitchell, M. J. and Groffman, P. G. (2002) The fate of nitrogen in gypsy moth frass deposited to an oak forest floor. *Oecologia*, **131**, 444–452.

Christenson, L. M., Mitchell, M. J., Groffman, P. M. and Lovett, G. M. (2010) Winter climate change implications for decomposition in northeastern forests: comparisons of sugar maple litter with herbivore fecal inputs. *Global Change Biology*, **16**, 2589–2601.

Classen, A. T., DeMarco, J., Hart, S. C. *et al.* (2006) Impacts of herbivorous insects on decomposer communities during the early stages of primary succession in a semi-arid woodland. *Soil Biology and Biochemistry*, **38**, 972–982.

Classen, A. T., Hart, S. C., Whitman, T. G., Cobb, N. S. and Koch, G. W. (2005) Insect infestations linked to shifts in microclimate: important climate change implications. *Soil Science Society of America Journal*, **69**, 2049–2057.

Classen, A. T., Overby, S. T., Hart, S. C., Koch, G. W. and Whitham, T. G. (2007) Season mediates herbivore effects on litter and soil microbial abundance and activity in a semi-arid woodland. *Plant and Soil*, **295**, 217–227.

Cobb, R. C., Orwig, D. A. and Currie, S. (2006) Decomposition of green foliage in eastern hemlock forests of southern New England impacted by hemlock woolly adelgid infestations. *Canadian Journal of Forest Research*, **36**, 1331–1341.

Coleman, D. C., Crossley, D. A. and Hendrix, P. F. (2004) *Fundamentals of Soil Ecology*. Amsterdam, The Netherlands: Elsevier Academic Press.

Coq, S., Souquet, J.-M., Meudec, E., Cheynier, V. and Hättenschwiler, S. (2010) Interspecific variation in leaf litter tannins drives decomposition in a tropical rain forest of French Guiana. *Ecology*, **91**, 2080–2091.

Cornelissen, J. H. C., Quested, H. M., Gwynn-Jones, D. *et al.* (2004) Leaf digestibility and litter decomposability are related in a wide range of subarctic plant species and types. *Functional Ecology*, **18**, 779–786.

Cornwell, W. K., Cornelissen, J. H. C., Amatangelo, K. *et al.* (2008) Plant species traits are the predominant control on litter decomposition rates within biomes worldwide. *Ecology Letters*, **11**, 1065–1071.

Cronin, J. P., Tonsor, S. J. and Carson, W. P. (2010) A simultaneuous test of trophic interaction models: which vegetation characteristic explains herbivore control over plant community mass? *Ecology Letters*, **13**, 202–212.

Crutsinger, G. M., Habenicht, M. N., Classen, A. T., Schweitzer, J. A. and Sanders, N. J. (2008) Galling by *Rhopalomyia solidaginis* alters *Solidago altissima* architecture and litter nutrient dynamics in an old-field ecosystem. *Plant and Soil*, **303**, 95–103.

Dethier, V. G. (1954) Evolution of feeding preferences in phytophagous insects. *Evolution*, **8**, 33–54.

Domisch, T., Finer, L., Neuvonen, S. *et al.* (2009) Foraging activity and dietary spectrum of wood ants (*Formica rufa* group) and their role in nutrient fluxes in boreal forests. *Ecological Entomology*, **34**, 369–377.

Dunham, A. E. (2008) Above and below ground impacts of terrestrial mammals and birds in a tropical forest. *Oikos*, **117**, 571–579.

Ehrlich, P. R. and Raven, P. H. (1964) Butterflies and plants: a study in coevolution. *Evolution*, **18**, 586–608.

Eisenhauer, N., Horsch, V., Moeser, J. and Scheu, S. (2010) Synergistic effects of microbial and animal decomposers on plant and herbivore performance. *Basic and Applied Ecology*, **11**, 23–34.

Elser, J. J., Sterner, R. W., Gorokhova, E. *et al.* (2000) Biological stoichiometry from genes to ecosystems. *Ecology Letters*, **3**, 540–550.

Findlay, S., Carreiro, M., Krischik, V. and Jones, C. G. (1996) Effects of damage to living plants on leaf litter quality. *Ecological Applications*, **6**, 269–275.

Fogal, W. H. and Slansky, F., Jr. (1985) Contribution of feeding by European pine sawfly larvae to litter production and element flux in Scots pine plantations. *Canadian Journal of Forest Research*, **15**, 484–487.

Fontaine, S. and Barot, S. (2005) Size and functional diversity of microbe populations control plant persistence and long-term soil carbon accumulation. *Ecology Letters*, **8**, 1075–1087.

Fontaine, S., Bardoux, G., Abbadie, L. and Mariotti, A. (2004) Carbon input to soil may decrease soil carbon content. *Ecology Letters*, **7**, 314–320.

Fonte, S. J. and Schowalter, T. D. (2004) Decomposition of greenfall vs. senescent foliage in a tropical forest ecosystem in Puerto Rico. *Biotropica*, **36**, 474–482.

Fonte, S. J. and Schowalter, T. D. (2005) The influence of a neotropical herbivore (*Lamponius portoricensis*) on nutrient cycling and soil processes. *Oecologia*, **146**, 423–431.

Fornara, D. A. and Du Toit, J. T. (2008) Browsing-induced effects on leaf litter quality and decomposition in a southern African savanna. *Ecosystems*, **11**, 238–249.

Frost, C. J. and Hunter, M. D. (2004) Insect canopy herbivory and frass deposition affect soil nutrient dynamics and export in oak mesocosms. *Ecology*, **85**, 3335–3347.

Frost, C. J. and Hunter, M. D. (2007) Recycling of nitrogen in herbivore feces: plant recovery, herbivore assimilation, soil retention, and leaching losses. *Oecologia*, **151**, 42–53.

Frost, C. J. and Hunter, M. D. (2008a) Herbivore-induced shifts in carbon and nitrogen allocation in red oak seedlings. *New Phytologist*, **178**, 835–845.

Frost, C. J. and Hunter, M. D. (2008b) Insect herbivores and their frass affect *Quercus rubra*, leaf quality and initial stages of subsequent litter decomposition. *Oikos*, **117**, 13–22.

Gandhi, K. J. K. and Herms, D. A. (2010) Direct and indirect effects of alien insect herbivores on ecological processes and interactions in forests of eastern North America. *Biological Invasions*, **12**, 389–405.

Garibaldi, L. A., Semmartin, M. and Chaneton, E. J. (2007) Grazing-induced changes in plant composition affect litter quality and nutrient cycling in flooding Pampa grasslands. *Oecologia*, **151**, 650–662.

Gessner, M. O., Swan, C. M., Dang, C. K. *et al.* (2010) Diversity meets decomposition. *Trends in Ecology and Evolution*, **25**, 372–380.

Golodets, C., Kigel, J. and Sternberg, M. (2010) Recovery of plant species composition and ecosystem function after cessation of grazing in a Mediterranean grassland. *Plant and Soil*, **329**, 365–378.

Grace, J. R. (1986) The influence of gypsy moth on the composition and nutrient content of litter fall in a Pennsylvania oak forest. *Forest Science*, **32**, 855–870.

Hall, M. C., Stiling, P., Hungate, B. A., Drake, B. G. and Hunter, M. D. (2005a) Effects of elevated CO_2 and herbivore damage on litter quality in a scrub oak ecosystem. *Journal of Chemical Ecology*, **31**, 2343–2356.

Hall, M. C., Stiling, P., Moon, D. C., Drake, B. G. and Hunter, M. D. (2005b) Effects of elevated CO_2 on foliar quality and herbivore damage in a scrub oak ecosystem. *Journal of Chemical Ecology*, **31**, 267–286.

Hall, M. C., Stiling, P., Moon, D. C., Drake, B. G. and Hunter, M. D. (2006) Elevated CO_2 increases the long-term decomposition rate of *Quercus myrtifolia* leaf litter. *Global Change Biology*, **12**, 568–577.

Hamilton, E. W. and Frank, D. A. (2001) Can plants stimulate soil microbes and their own nutrient supply? Evidence from a grazing tolerant grass. *Ecology*, **82**, 2397–2402.

Hamilton, E. W., Frank, D. A., Hinchey, P. M. and Murray, T. R. (2008) Defoliation induces root exudation and triggers positive rhizospheric feedbacks in a temperate grassland. *Soil Biology and Biochemistry*, **40**, 2865–2873.

Harrison, K. A. and Bardgett, R. D. (2003) How browsing by red deer impacts on litter decomposition in a native regenerating woodland in the Highlands of Scotland. *Biology and Fertility of Soils*, **38**, 393–399.

Harrison, K. A. and Bardgett, R. D. (2004) Browsing by red deer negatively impacts on soil nitrogen availability in regenerating native forest. *Soil Biology and Biochemistry*, **36**, 115–126.

Haukioja, E., Niemela, P. and Siren, S. (1985) Foliage phenols and nitrogen in relation to growth, insect damage, and ability to recover after defoliation, in the mountain birch *Betula pubescens* ssp. *tortuosa*. *Oecologia*, **65**, 214–222.

Hawlena, D. and Schmitz, O. J. (2010a) Herbivore physiological response to predation risk and implications for ecosystem nutrient dynamics. *Proceedings of the National Academy of Sciences of the United States of America*, **107**, 15503–15507.

Hawlena, D. and Schmitz, O. J. (2010b) Physiological stress as a fundamental mechanism linking predation to ecosystem functioning. *American Naturalist*, **176**, 537–556.

Hollinger, D. Y. (1986) Herbivory and the cycling of nitrogen and phosphorus in isolated California oak trees. *Oecologia*, **70**, 291–297.

Hudson, T. M., Turner, B. L., Herz, H. and Robinson, J. S. (2009) Temporal patterns of nutrient availability around nests of leaf-cutting ants (*Atta colombica*) in secondary moist tropical forest. *Soil Biology and Biochemistry*, **41**, 1088–1093.

Hunter, M. D. (1992) Interactions within herbivore communities mediated by the host plant: the keystone herbivore concept. In M. D. Hunter, T. Ohgushi and P. W. Price, eds., *Effects of Resource Distribution on Animal-Plant Interactions*. San Diego, CA: Academic Press, pp. 287–325.

Hunter, M. D. (1994) The search for pattern in pest outbreaks. In S. R. Leather, A. D. Watt, N. A. C. Kidd and N. J. Mills, eds., *Individuals, Populations and Patterns in Ecology*. Andover, UK: Intercept, pp. 443–448.

Hunter, M. D. (2001) Insect population dynamics meets ecosystem ecology: effects of herbivory on soil nutrient dynamics. *Agricultural and Forest Entomology*, **3**, 77–84.

Hunter, M. D. (2002) Ecological causes of pest outbreaks. In D. Pimentel, ed., *Encyclopedia of Pest Management*. New York: Marcel Dekker Inc., pp. 214–220.

Hunter, M. D. (2008) Root herbivory in forest ecosystems. In S. N. Johnson and P. J. Murray, eds., *Root Feeders, an Ecosystem Perspective*. Ascot, UK: CAB Biosciences, pp. 68–95.

Hunter, M. D. and Schultz, J. C. (1995) Fertilization mitigates chemical induction and herbivore responses within damaged oak trees. *Ecology*, **76**, 1226–1232.

Hunter, M. D., Adl, S., Pringle, C. M. and Coleman, D. C. (2003a) Relative effects of macro invertebrates and habitat on the chemistry of litter during decomposition. *Pedobiologia*, **47**, 101–115.

Hunter, M. D., Linnen, C. R. and Reynolds, B. C. (2003b) Effects of endemic densities of canopy herbivores on nutrient dynamics along a gradient in elevation in the southern Appalachians. *Pedobiologia*, **47**, 231–244.

Hunter, M. D., Varley, G. C. and Gradwell, G. R. (1997) Estimating the relative roles of top-down and bottom-up forces on insect herbivore populations: a classic study revisited. *Proceedings of the National Academy of Sciences of the United States of America*, **94**, 9176–9181.

Karban, R. and Baldwin, I. T. (1997) *Induced Responses to Herbivory*. Chicago, IL: University of Chicago Press.

Kay, A. D., Mankowski, J. and Hobbie, S. E. (2008) Long-term burning interacts with herbivory to slow decomposition. *Ecology*, **89**, 1188–1194.

Kurokawa, H. and Nakashizuka, T. (2008) Leaf herbivory and decomposability in a Malaysian tropical rain forest. *Ecology*, **89**, 2645–2656.

Kurokawa, H., Peltzer, D. A. and Wardle, D. A. (2010) Plant traits, leaf palatability and litter decomposability for co-occurring woody species differing in invasion status and nitrogen fixation ability. *Functional Ecology*, **24**, 513–523.

Kurz, W. A., Dymond, C. C., Stinson, G. *et al.* (2008) Mountain pine beetle and forest carbon feedback to climate change. *Nature*, **452**, 987–990.

Kuzyakov, Y., Friedel, J. K. and Stahr, K. (2000). Review of mechanisms and quantification of priming effects. *Soil Biology and Biochemistry*, **32**, 1485–1498.

le Mellec, A., Habermann, M. and Michalzik, B. (2009) Canopy herbivory altering C to N ratios and soil input patterns of different organic matter fractions in a Scots pine forest. *Plant and Soil*, **325**, 255–262.

Lohmann, M., Scheu, S. and Muller, C. (2009) Decomposers and root feeders interactively affect plant defence in *Sinapis alba*. *Oecologia*, **160**, 289–298.

Lovett, G. M. and Ruesink, A. E. (1995) Carbon and nitrogen mineralization from decomposing gypsy moth frass. *Oecologia*, **104**, 133–138.

Lovett, G. M., Christenson, L. M., Groffman, P. M. *et al.* (2002) Insect defoliation and nitrogen cycling in forests. *BioScience*, **52**, 335–341.

Madritch, M. D. and Hunter, M. D. (2002) Phenotypic diversity influences ecosystem functioning in an oak sandhills community. *Ecology*, **83**, 2084–2090.

Madritch, M. D. and Hunter, M. D. (2003) Intraspecific litter diversity and nitrogen deposition affect nutrient dynamics and soil respiration. *Oecologia*, **136**, 124–128.

Madritch, M. D. and Hunter, M. D. (2004) Phenotypic diversity and litter chemistry affect nutrient dynamics during litter decomposition in a two species mix. *Oikos*, **105**, 125–131.

Madritch, M. D., Donaldson, J. R. and Lindroth, R. L. (2007) Canopy herbivory can mediate the influence of plant genotype on soil processes through frass deposition. *Soil Biology and Biochemistry*, **39**, 1192–1201.

Mattson, W. J. and Addy, N. D. (1975) Phytophagous insects as regulators of forest production. *Science*, **190**, 515–521.

McNaughton, S. J., Ruess, R. W. and Seagle, S. W. (1988) Large mammals and process dynamics in African ecosystems: herbivorous mammals affect primary productivity and regulate recycling balances. *BioScience*, **38**, 794–800.

Megias, A. G. and Muller, C. (2010) Root herbivores and detritivores shape above-ground multitrophic assemblage through plant-mediated effects. *Journal of Animal Ecology*, **79**, 923–931.

Mikola, J., Yeates, G. W., Barker, G. M., Wardle, D. A. and Bonner, K. I. (2001) Effects of defoliation intensity on soil food-web properties in an experimental grassland community. *Oikos*, **92**, 333–343.

Olofsson, J. and Oksanen, L. (2002) Role of litter decomposition for the increased primary production in areas heavily grazed by reindeer: a litterbag experiment. *Oikos*, **96**, 507–515.

Olofsson, J., de Mazancourt, C. and Crawley, M. J. (2007) Contrasting effects of rabbit exclusion on nutrient availability and primary production in grasslands at different timescales. *Oecologia*, **150**, 582–589.

Olofsson, J., Stark, S. and Oksanen, L. (2004) Reindeer influence on ecosystem processes in the tundra. *Oikos*, **105**, 386–396.

Orwig, D. A., Cobb, R. C., D'Amato, A. W., Kizlinski, M. L. and Foster, D. R. (2008) Multi-year ecosystem response to hemlock woolly adelgid infestation in southern New England forests. *Canadian Journal of Forest Research*, **38**, 834–843.

Parker, J. D., Burkepile, D. E. and Hay, M. E. (2006) Opposing effects of native and exotic herbivores on plant invasions. *Science*, **311**, 1459–1461.

Pastor, J. and Cohen, Y. (1997) Herbivores, the functional diversity of plant species, and the cycling of nutrients in ecosystems. *Theoretical Population Biology*, **51**, 165–179.

Pastor, J. and Naiman, R. J. (1992) Selective foraging and ecosystem processes in boreal forests. *American Naturalist*, **139**, 690–705.

Pastor, J., Dewey, B., Naiman, R. J., McInnes, P. F. and Cohen, Y. (1993) Moose browsing and soil fertility in the boreal forests of Isle-Royale National Park. *Ecology*, **74**, 467–480.

Persson, I. L., Pastor, J., Danell, K. and Bergstrom, R. (2005) Impact of moose population density on the production and composition of litter in boreal forests. *Oikos*, **108**, 297–306.

Pineiro, G., Paruelo, J. M., Oesterheld, M. and Jobbagy, E. G. (2010) Pathways of grazing effects on soil organic carbon and nitrogen. *Rangeland Ecology and Management*, **63**, 109–119.

Pitelka, F. A. (1964) The nutrient-recovery hypothesis for arctic microtine cycles. I. Introduction. In D. J. Crisp, ed., *Grazing in*

Terrestrial and Marine Environments. Oxford: Blackwell Science, pp. 55–56.

Reynolds, B. C. and Hunter, M. D. (2001) Responses of soil respiration, soil nutrients, and litter decomposition to inputs from canopy herbivores. *Soil Biology and Biochemistry*, **33**, 1641–1652.

Reynolds, B. C., Crossley, D. A. and Hunter, M. D. (2003) Response of soil invertebrates to forest canopy inputs along a productivity gradient. *Pedobiologia*, **47**, 127–139.

Reynolds, B. C., Hunter, M. D. and Crossley, D. A., Jr. (2000) Effects of canopy herbivory on nutrient cycling in a northern hardwood forest in western North Carolina. *Selbyana*, **21**, 74–78.

Risley, L. S. (1986) The influence of herbivores on seasonal leaf-fall: premature leaf abscission and petiole clipping. *Journal of Agricultural Entomology*, **3**, 152–162.

Risley, L. S. and Crossley, D. A. (1988) Herbivore-caused greenfall in the southern Appalachians. *Ecology*, **69**, 1118–1127.

Risley, L. S. and Crossley, D. A. (1992) Contribution of herbivore-caused greenfall to litterfall: N flux in several southern Appalachian forested watersheds. *American Midland Naturalist*, **129**, 67–74.

Ritchie, M. E., Tilman, D. and Knops, J. M. H. (1998) Herbivore effects on plant and nitrogen dynamics in oak savanna. *Ecology*, **79**, 165–177.

Rossiter, M. C., Schultz, J. C. and Baldwin, I. T. (1988) Relationships among defoliation, red oak phenolics, and gypsy moth growth and reproduction. *Ecology*, **69**, 267–277.

Saj, S., Mikola, J. and Ekelund, F. (2008) Legume defoliation affects rhizosphere decomposers, but not the uptake of organic matter N by a neighbouring grass. *Plant and Soil*, **311**, 141–149.

Sankaran, M. and Augustine, D. J. (2004) Large herbivores suppress decomposer abundance in a semiarid grazing ecosystem. *Ecology*, **85**, 1052–1061.

Sariyildiz, T., Akkuzu, E., Kucuk, M., Duman, A. and Aksu, Y. (2008) Effects of *Ips typographus*

(L.) damage on litter quality and decomposition rates of Oriental Spruce *Picea orientalis* (L.) Link. in Hatila Valley National Park, Turkey. *European Journal of Forest Research*, **127**, 429–440.

Schadler, M., Alphei, J., Scheu, S., Brandl, R. and Auge, H. (2004) Resource dynamics in an early-successional plant community are influenced by insect exclusion. *Soil Biology and Biochemistry*, **36**, 1817–1826.

Schadler, M., Jung, G., Auge, H. and Brandl, R. (2003) Palatability, decomposition and insect herbivory: patterns in a successional old-field plant community. *Oikos*, **103**, 121–132.

Schimel, J. P. and Weintraub, M. N. (2003) The implications of exoenzyme activity on microbial carbon and nitrogen limitation in soil: a theoretical model. *Soil Biology and Biochemistry*, **35**, 549–563.

Schmitz, O. J. (2008) Herbivory from individuals to ecosystems. *Annual Review of Ecology, Evolution, and Systematics*, **39**, 133–152.

Schowalter, T. D. (2000) *Insect Ecology: An Ecosystem Approach*. San Diego, CA: Academic Press.

Schowalter, T. D. and Sabin, T. E. (1991) Litter microarthropod responses to canopy herbivory, season and decomposition in litterbags in a regenerating conifer ecosystem in western Oregon. *Biology and Fertility of Soils*, **11**, 93–96.

Schweitzer, J. A., Bailey, J. K., Hart, S. C. and Whitham, T. G. (2005a) Nonadditive effects of mixing cottonwood genotypes on litter decomposition and nutrient dynamics. *Ecology*, **86**, 2834–2840.

Schweitzer, J. A., Bailey, J. K., Hart, S. C. *et al.* (2005b) The interaction of plant genotype and herbivory decelerate leaf litter decomposition and alter nutrient dynamics. *Oikos*, **110**, 133–145.

Schweitzer, J. A., Madritch, M. D., Bailey, J. K. *et al.* (2008) From genes to ecosystems: the genetic basis of condensed tannins and their role in nutrient regulation in a *Populus* model system. *Ecosystems*, **11**, 1005–1020.

Semmartin, M. and Ghersa, C. M. (2006) Intraspecific changes in plant morphology,

associated with grazing, and effects on litter quality, carbon and nutrient dynamics during decomposition. *Austral Ecology*, **31**, 99–105.

Semmartin, M., Aguiar, M. R., Distel, R. A., Moretto, A. S. and Ghersa, C. M. (2004) Litter quality and nutrient cycling affected by grazing-induced species replacements along a precipitation gradient. *Oikos*, **107**, 148–160.

Semmartin, M., Di Bella, C. and de Salamone, I. G. (2010) Grazing-induced changes in plant species composition affect plant and soil properties of grassland mesocosms. *Plant and Soil*, **328**, 471–481.

Semmartin, M., Garibaldi, L. A. and Chaneton, E. J. (2008) Grazing history effects on above- and belowground litter decomposition and nutrient cycling in two co-occurring grasses. *Plant and Soil*, **303**, 177–189.

Sjogersten, S., van der Wal, R. and Woodin, S. J. (2008) Habitat type determines herbivory controls over CO_2 fluxes in a warmer arctic. *Ecology*, **89**, 2103–2116.

Sorensen, L. I., Kytoviita, M. M., Olofsson, J. and Mikola, J. (2008) Soil feedback on plant growth in a sub-arctic grassland as a result of repeated defoliation. *Soil Biology and Biochemistry*, **40**, 2891–2897.

Speight, M. R., Hunter, M. D. and Watt, A. D. (2008) *The Ecology of Insects: Concepts and Applications*. Oxford: Blackwell Scientific.

Srivastava, D. S., Cardinale, B. J., Downing, A. L. *et al.* (2009) Diversity has stronger top-down than bottom-up effects on decomposition. *Ecology*, **90**, 1073–1083.

Stadler, B. and Michalzik, B. (1998) Linking aphid honeydew, throughfall and forest floor solution chemistry of Norway spruce. *Ecology Letters*, **1**, 13–16.

Stadler, B., Michalzik, B. and Muller, T. (1998) Linking aphid ecology with nutrient fluxes in a coniferous forest. *Ecology*, **79**, 1514–1525

Stadler, B., Muller, T. and Orwig, D. (2006) The ecology of energy and nutrient fluxes in hemlock forests invaded by hemlock woolly adelgid. *Ecology*, **87**, 1792–1804.

Stark, S., Julkunen-Tiitto, R. and Kumpula, J. (2007) Ecological role of reindeer summer browsing in the mountain birch (*Betula pubescens* ssp. *czerepanovii*) forests: effects on plant defense, litter decomposition, and soil nutrient cycling. *Oecologia*, **151**, 486–498.

Stark, S., Kytoviita, M. M., Mannisto, M. K. and Neumann, A. B. (2008) Soil microbial and microfaunal communities and organic matter quality in reindeer winter and summer ranges in Finnish subarctic mountain birch forests. *Applied Soil Ecology*, **40**, 456–464.

Stiling, P., Forkner, R. and Drake, B. (2010) Long-term exposure to elevated CO_2 in a Florida scrub oak forest increases herbivore densities but has no effect on other arthropod guilds. *Insect Conservation and Diversity*, **3**, 152–156.

Stiling, P., Moon, D. C., Hunter, M. D. *et al.* (2003) Elevated CO_2 lowers relative and absolute herbivore density across all species of a scrub-oak forest. *Oecologia*, **134**, 82–87.

Thompson, J. N. (1994) *The Coevolutionary Process.* Chicago, IL: University of Chicago Press.

Throop, H. L. and Lerdau, M. T. (2004) Effects of nitrogen deposition on insect herbivory: implications for community and ecosystem processes. *Ecosystems*, **7**, 109–133.

Throop, H. L., Holland, E. A., Parton, W. J., Ojima, D. S. and Keough, C. A. (2004) Effects of nitrogen deposition and insect herbivory on patterns of ecosystem-level carbon and nitrogen dynamics: results from the CENTURY model. *Global Change Biology*, **10**, 1092–1105.

Tukey, H. B. and Morgan, J. V. (1963) Injury to foliage and its effects upon the leaching of nutrients from above-ground plant parts. *Physiologia Planta*, **16**, 557–564.

Turetsky, M. R., Mack, M. C., Harden, J. W. and Manies, K. L. (2005) Spatial patterning of soil carbon storage across boreal landscapes. In G. M. Lovett, C. G. Jones, M. G. Turne and K. C. Weathers, eds., *Ecosystem Function in Heterogeneous Landscapes*. New York: Springer, pp. 229–256.

Tylianakis, J. M., Didham, R. K., Bascompte, J. and Wardle, D. A. (2008) Global change and species interactions in terrestrial ecosystems. *Ecology Letters*, **11**, 1351–1363.

Uriarte, M. (2000) Interactions between goldenrod (*Solidago altissima* L.) and its insect herbivore (*Trirhabda virgata*) over the course of succession. *Oecologia*, **122**, 521–528.

Van Der Meijden, E., Wijn, M. and Verkaar, H. J. (1988) Defense and regrowth: alternative plant strategies in the struggle against herbivores. *Oikos*, **51**, 355–363.

van der Wal, R., Bardgett, R. D., Harrison, K. A. and Stien, A. (2004) Vertebrate herbivores and ecosystem control: cascading effects of faeces on tundra ecosystems. *Ecography*, **27**, 242–252.

Verchot, L. V., Groffman, P. M. and Frank, D. A. (2002) Landscape versus ungulate control of gross mineralization and gross nitrification in semi-arid grasslands of Yellowstone National Park. *Soil Biology and Biochemistry*, **34**, 1691–1699.

Wang, K. H., McSorley, R., Bohlen, P. and Gathumbi, S. M. (2006) Cattle grazing increases microbial biomass and alters soil nematode communities in subtropical pastures. *Soil Biology and Biochemistry*, **38**, 1956–1965.

Wardhaugh, C. W. and Didham, R. K. (2005) Preliminary evidence suggests that beech scale insect honeydew has a negative effect on terrestrial litter decomposition rates in *Nothofagus* forests of New Zealand. *Journal of Ecology*, **30**, 279–284.

Wardle, D. A. and Bardgett, R. D. (2004) Human-induced changes in large herbivorous mammal density: the consequences for decomposers. *Frontiers in Ecology and the Environment*, **2**, 145–153.

Wardle, D. A., Bonner, K. I. and Barker, G. M. (2002) Linkages between plant litter decomposition, litter quality, and vegetation responses to herbivores. *Functional Ecology*, **16**, 585–595.

Wardle, D. A., Karl, B. J., Beggs, J. R. *et al.* (2010) Determining the impact of scale insect honeydew, and invasive wasps and rodents on the decomposer subsystem in a New Zealand beech forest. *Biological Invasions*, **12**, 2619–2638.

Woods, H. A., Fagan, W. F., Elser, J. J. and Harrison, J. F. (2004) Allometric and phylogenetic variation in insect phosphorus content. *Functional Ecology*, **18**, 103–109.

Wurst, S., Dugassa-Gobena, D. and Scheu, S. (2004) Earthworms and litter distribution affect plant-defensive chemistry. *Journal of Chemical Ecology*, **30**, 691–701.

Yang, L. H. (2004) Periodical cicadas as resource pulses in North American forests. *Science*, **306**, 1565–1567.

Yang, L. H. (2008) Pulses of dead periodical cicadas increase herbivory of American bellflowers. *Ecology*, **89**, 1497–1502.

Yelenik, S. G. and Levine, J. M. (2010) Native shrub reestablishment in exotic annual grasslands: do ecosystem processes recover? *Ecological Applications*, **20**, 716–727.

Yule, C. M. and Gomez, L. N. (2009) Leaf litter decomposition in a tropical peat swamp forest in Peninsular Malaysia. *Wetlands Ecology and Management*, **17**, 231–241.

Zak, D. R., Blackwood, C. B. and Waldrop, M. P. (2006) A molecular dawn for biogeochemistry. *Trends in Ecology and Evolution*, **21**, 288–295.

Zehnder, C. B. and Hunter, M. D. (2008) Effects of nitrogen deposition on the interaction between an aphid and its host plant. *Ecological Entomology*, **33**, 24–30.

Zimmer, M. and Topp, W. (2002) The role of coprophagy in nutrient release from feces of phytophagous insects. *Soil Biology and Biochemistry*, **34**, 1093–1099.

CHAPTER NINETEEN

Functional and heritable consequences of plant genotype on community composition and ecosystem processes

JENNIFER A. SCHWEITZER and JOSEPH K. BAILEY

Department of Ecology and Evolutionary Biology, University of Tennessee

DYLAN G. FISCHER and CARRI J. LEROY

Environmental Studies Program, The Evergreen State College

THOMAS G. WHITHAM

Department of Biological Sciences, Northern Arizona University

and

STEPHEN C. HART

School of Natural Sciences and Sierra Nevada Research Institute, University of California – Merced

Introduction

Foundation species represent excellent model systems for understanding the broad consequences of variation on community and ecosystem processes as they provide a focal resource upon which associated interacting species depend. As foundation species (Dayton 1972; Ellison *et al.* 2005), trees and other dominant plants often create stable conditions via plant traits that allow dependent communities to assemble regularly and influence ecosystem processes such as net primary productivity (NPP) and soil fertility (i.e., nutrient cycling, via accumulations of leaf or root organic matter or root exudates; Zinke 1962; Zak *et al.* 1986; Binkley and Giardina 1998; Bartelt-Ryser *et al.* 2005; Wardle 2006). Recent studies in both terrestrial and aquatic habitats have shown that intraspecific genetic variation (defined at multiple genetic scales, including introgression [movement of genes from one species to another], genotypic diversity [studies manipulating the number of genotypes in a population] and genotypic variation [variation among genotypes]) in foundation plants can have community-wide consequences. Intraspecific variation affects associated vertebrate, arthropod and microbial community composition or activity and ecosystem level processes (recently reviewed in Johnson and Stinchcombe 2007; Hughes *et al.* 2008; Whitham *et al.* 2008; Bailey *et al.* 2009). For example, genetic variation resulting from the introgression of genes from one species to another through

Trait-Mediated Indirect Interactions: Ecological and Evolutionary Perspectives, eds. Takayuki Ohgushi, Oswald J. Schmitz and Robert D. Holt. Published by Cambridge University Press. © Cambridge University Press 2012.

Plant Hybridization Gradient

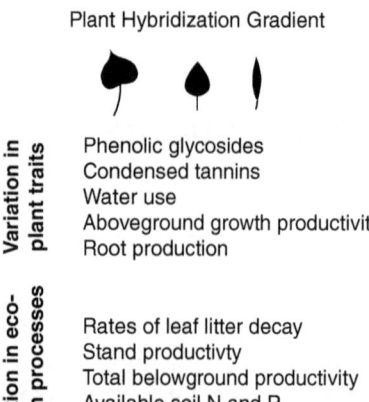

Variation in plant traits

Phenolic glycosides
Condensed tannins
Water use
Aboveground growth productivity
Root production

Variation in eco-system processes

Rates of leaf litter decay
Stand productivty
Total belowground productivity
Available soil N and P
Rates of net N mineralization/nitrification

Feedback to fitness

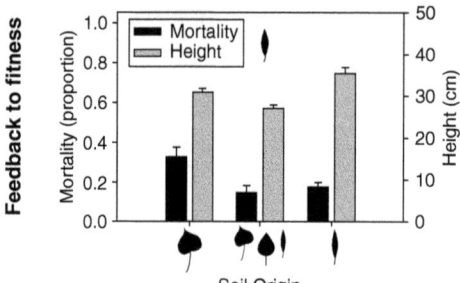

Figure 19.1 Schematic illustrating the effects of genetic introgression of *Populus* (*P. fremontii* × *P. angustifolia*) on plant functional traits, ecosystem processes and feedbacks to plant fitness (data from Whitham *et al.*, 2006 and references therein; Pregitzer *et al.*, 2010).

the process of hybridization has been shown to have important consequences for associated species, communities and ecosystem processes in multiple hybridizing plant species, including *Salix* spp., *Eucalyptus* spp., *Quercus* spp. and *Populus* spp. (Fritz *et al.* 1994; Dungey *et al.* 2000; Hochwender and Fritz 2004; Ito and Ozaki 2005; Wimp *et al.* 2005; Tovar-Sanchez and Oyama 2006; Bangert *et al.* 2008). In the *Populus* system specifically, recent field and common garden studies have shown that genetic variation across a hybridizing system (*P. fremontii, P. angustifolia* and their natural F_1 and backcross hybrids) results in shifts in plant traits, including secondary chemistry, plant water use and above- and below-ground productivity (Fischer *et al.* 2004; Rehill *et al.* 2006; Schweitzer *et al.* 2008a; Lojewski *et al.* 2009). Whether due directly or indirectly to these plant traits, rates of leaf litter decomposition, total belowground carbon (C) allocation and pools of soil nitrogen (N) and rates of net N mineralization also shift along this genetic gradient (Schweitzer *et al.* 2004, 2008a, b; LeRoy *et al.* 2006; Whitham *et al.* 2006; Lojewski *et al.* 2009; Fischer *et al.* 2007, 2010). Moreover, variation in the microbial community and the nutrient mineralization processes they mediate creates a positive feedback whereby *Populus angustifolia* seedlings grown in the greenhouse with soil from *P. angustifolia* field sites survive nearly 50% more and grow 24% taller than when grown in *P. fremontii* soils or soils from hybrids (Fig. 19.1; Pregitzer *et al.* 2010). While negative feedback is thought to maintain species diversity due to the buildup of pathogens (Kulmatiski *et al.* 2008; Mangon *et al.* 2010), positive feedback such as this is also common, especially with clonal plants and may contribute to maintaining genetic variation within populations (Bever *et al.* 1997; Van Breeman and Finzi 1998; Smith *et al.* 2012).

Patterns in community composition and ecosystem processes due to genotypic variation in plants are often directly or indirectly linked to specific traits (or suites of traits) of plants. Genetic individuals (i.e., genotypes) within a species vary in a multitude of traits including, but not limited to, growth rates, secondary chemistry, physiological processes (i.e., photosynthesis and water use/loss) and development times that have been shown to impact other species (Hughes *et al.* 2009; Bossdorf *et al.* 2009; Clark 2010). This should not be surprising as variation among individuals is the cornerstone of population ecology and is the fundamental raw material for evolution by natural selection. However, this variation in traits can directly or indirectly influence associated species creating unique communities on individuals such as gut microbes in humans (Zoetendal *et al.* 2001; Eckburg *et al.* 2005), herbivores and their predators on plants (McIntyre and Whitham 2003; Shuster *et al.* 2006; Bailey *et al.* 2006; Simchuk 2008; Vorburger *et al.* 2010) and fungi and bacteria associated with a plant genotype or the soils beneath them (Kasurinen *et al.* 2005; Schweitzer *et al.* 2008b; Holeski *et al.* 2009; Karst *et al.* 2009). These community interactions can extend to affect ecosystem processes associated with individual genotypes (Schweitzer *et al.* 2005; Madritch *et al.* 2006, 2009). Long-term studies with aspen genotypes (i.e., clones) have shown that variation in foliar (and bark) chemistry often interacts with the environment and can influence patterns in resistance or susceptibility to specific herbivores and pathogens (Hwang and Lindroth 1997; Osier and Lindroth 2004; Müller *et al.* 2006; Wooley *et al.* 2008; Diner *et al.* 2009). Differences among individual genotypes in chemistry can also alter associated understory communities below the canopy of trees (Iason *et al.* 2005); and differences in growth rates among genotypes can result in differential soil N uptake rates, that may determine plant community co-occurrences (Hughes *et al.* 2009). Variation in growth rates or productivity among genotypes may be large and have profound consequences for the performance of associated species as well as affecting whole community dynamics and resistance to invasion by exotic species (Crutsinger *et al.* 2008b; Vellend *et al.* 2010).

The regularity of similarities in plant traits and their linkages to other species have led researchers to consider a 'genetic similarity rule', whereby closely related plant species or different races within a single species may have similar traits (e.g., chemical profiles) that regularly structure similar associated communities of organisms (Bangert *et al.* 2006; Barbour *et al.* 2009). Overall, the direct and indirect effects of plant traits on community and ecosystem processes are clear

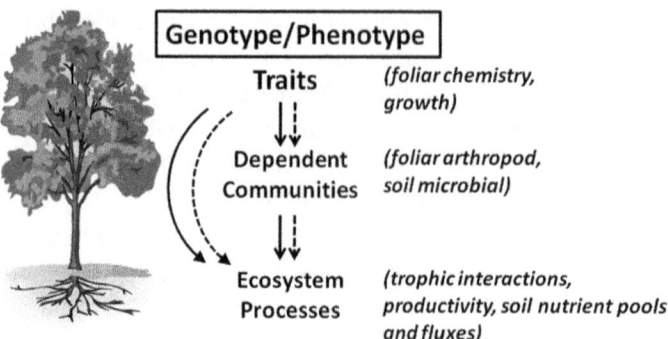

Genotype/Phenotype

Traits   *(foliar chemistry,*
                growth)

Dependent   *(foliar arthropod,*
Communities   *soil microbial)*

Ecosystem   *(trophic interactions,*
Processes   *productivity, soil nutrient pools*
                and fluxes)

Figure 19.2 Conceptual diagram showing the direct and indirect interactions that may occur between plant genotypes, associated communities and ecosystem processes. It indicates how heritable variation among plant genotypes (or phenotypes) results in the differential expression of traits (or suites of traits) such as above- or belowground productivity or foliar chemistry that can either directly (black lines) or indirectly (dashed lines) influence associated communities, such as foliar arthropods or soil microbial communities (e.g., by changing the quantity or quality of leaf or litter resources). Variation in heritable plant traits can also influence ecosystem processes directly by changing biotic or abiotic conditions that influence ecosystem processes, such as trophic interactions or pools (or turnover) of soil nutrients, or indirectly affect these processes via changes to the composition or activity of the microbial community. See colour plate section.

and indicate the putative linkage of functional traits of individuals to community and ecosystem phenotypes (i.e., the extended phenotypes of plant traits on community composition and ecosystem processes, sensu Whitham *et al.* 2006).

Understanding the cascading effects of genotypes on plant trait variation for plant–arthropod or plant–soil interactions is the link that places the study of communities and ecosystems in the realm of evolutionary biology. This linkage demonstrates that selective events that alter plant traits may also alter associated dependent communities and the processes they mediate (Whitham *et al.* 2006, 2008). In this chapter we elucidate the role of individuals in systems which suggests that variation among genotypes within populations with increasing global change may affect the services that ecosystems provide (e.g., biodiversity, productivity or soil fertility). Figure 19.2 illustrates how variation among genotypes (or phenotypes) may result in the differential expression of suites of traits that can either directly influence associated communities and ecosystem processes (black lines) by changing biotic or abiotic conditions that species experience or indirectly impact associated communities and ecosystem processes (dashed lines) by changing biotic or abiotic conditions that affect one species and cascade to impact others. With a suite of previously published and original data, focusing on *Populus* spp. as model species, we examine the extended effects of individual plant genotypes on associated communities and ecosystem processes. Specifically, in this chapter we will: (1) illustrate the impacts of individual plant genotypes on associated communities; (2) explore the role of individual genotypes on ecosystem processes; (3) provide examples of genotype-mediated linkages between species and above- and belowground processes; and (4) explore the role of non-additive responses in genotype mixtures.

Individual genotypes and communities

Individual genotypes vary in many traits that have functional consequences for directly or indirectly influencing associated communities. Common garden studies with replicate clones of plants often show strong, heritable variation in plant traits that vary widely in their plasticity (Geber and Griffen 2003). Furthermore, the genomic architecture of many of these traits is rapidly being identified (Tsai et al. 2006; Chen et al. 2009). Within the genus *Populus*, significant variation exists among genotypes both within and among species in above- and belowground productivity, mineral content (i.e., nitrogen, carbon, phosphorus), lignin, phenolic glycosides and polyphenols of leaf, bark and root tissues (and the ability to induce these compounds) as well as susceptibility and resistance to dominant herbivores (Havill and Raffa 1999; Ralph et al. 2006; Lojewski et al. 2009). For example, genotypes of *P. angustifolia* demonstrate a 10-fold difference in aboveground productivity, a 4.6-fold difference in total below-ground C allocation, and a 1.8-fold difference in foliar N content and a 13.8-fold difference in condensed tannins concentrations (Rehill et al. 2006; Fischer et al. 2007; Lojewski et al. 2009). These alterations may directly or indirectly influence their susceptibility to dominant herbivores such as *Pemphigus betae* (a gall-forming herbivore; Bailey et al. 2006) or whole communities of arthropods and microbes (Shuster et al. 2006; Schweitzer et al. 2008b; Keith et al. 2010). This variation among genotypes within a species in plant traits can then have strong influences on associated community members that depend on plant productivity or forage quality for their fitness.

A rich literature has shown that variation in plant resistance, susceptibility or tolerance to specific herbivores, pathogens or disease may be direct or indirect in nature (e.g., due to the interactions with other species such as fungal endo-phytes or mycorrhizae; Gehring and Whitham 1994; Jani et al. 2010). These studies have shown that individual variation in traits influence their associa-tions with other species and demonstrate the importance of trait-based approaches to understanding the effects of species interactions within popula-tions. That individual genotypes or phenotypes may predictably structure the composition or activity of interacting communities is much rarer and only recently has been shown to occur in replicated experimental settings (Johnson et al. 2006; Shuster et al. 2006; Crutsinger et al. 2006, 2008a, b, c, 2009).

Repeatability in community structure (or ecosystem processes) can be quantified by calculating the broad-sense heritability of community structure (defined in Shuster *et al.* 2006). Broad-sense heritability is simply a measure of the amount of phenotypic variation in community structure that is explained by genetic variation among clones. Broad-sense heritability, or clonal repeatability, is calculated as the among-genotype component of variation in community composition [σ^2_s] divided by the total variance in the community composition for all replicate trees [σ^2_{total}], where the total variance in the phenotype is the sum of the among genotype and the error component [σ^2_e]; Falconer and McKay 1996; Conner and Hartl 2004; Shuster *et al.* 2006). For example, Keith *et al.* (2010) have documented that individual replicate genotypes (i.e., clones) of *P. angustifolia* in a common garden have unique and stable arthropod communities associated with each genotype. With an arthropod community of 103 species, they found, over a 3-year time period, that replicate clones of the same nine genotypes tended to support the same arthropod community and that different genotypes supported different communities. The calculated broad-sense arthropod community heritability of individual tree genotypes was high and similar in each of 3 years of study ($H^2_C = 0.68$, 0.68 and 0.63, respectively; Keith *et al.* 2010) indicating high repeatability in arthropod communities that are stable through time.

Studies, to date, that have examined the effects of individual genotypes on communities in the context of manipulations of genotypic diversity (i.e., comparing genotype monocultures to varying genotype mixtures) have found large effects of individual genotypes. For example, differences between individual replicated genotypes of multiple plant species results in a 1.6-fold (Johnson *et al.* 2006) and 2.6-fold (Crutsinger *et al.* 2006) difference in total arthropod richness and up to a 4-fold difference in arthropod abundances (Johnson *et al.* 2006). With pollinator communities specifically, the richness and abundance of flower visitors among replicated genotypes can differ 8- and 12-fold, respectively (Genung *et al.* 2010). These data indicate that individual genotypes within a species have unique traits that can influence associated species.

Many ecosystem processes are mediated by the activities of microbial communities, therefore elucidating the effects of intraspecific variation in plant species on microbes or their activities is important to understanding the role of biotic interactions on ecosystem function. Microbial communities or their activities are regulated by many factors including abiotic variables such as pH, moisture and temperature but also by biotic factors such as the composition of the microbial community itself as well as the organic matter quantity or quality of the substrates they decompose (Waldrop *et al.* 2000; Fierer and Jackson 2006; Morris and Blackwood 2007; Fierer *et al.* 2009). Variation in local environmental factors and the contribution of organic matter that varies in quantity and quality (i.e., concentration of C, N or secondary compounds such as lignin and polyphenols) by individual plant genotypes has the potential to influence microbial communities or their activities (Hättenschwiler and Vitousek 2000; Kraus *et al.* 2004). Recent studies finding a 'home-field' advantage for leaf litter decomposed under its specific 'home' or native species conditions versus 'away' indicate that microbial communities can be selected for under a range of selective environments including plant species (Gholz *et al.* 2000; Ayres *et al.* 2009; Strickland *et al.* 2009) and in the following examples, plant genotype (Madritch *et al.* 2006, 2009; Madritch and Lindroth 2011). For example, a study that examined the effects of both above- and belowground inputs (i.e., leaf litter and roots) across replicates of two clones found significant clonal effects on total microbial communities (as measured with phospholipid fatty acid biomarkers [PLFA]) and on ectomycorrhizal colonization (averaged across the experimental treatments; Kasurinen *et al.* 2005). These data support the hypothesis that clonal variation in plant traits may impact soil microbial communities and may even mediate responses to global change (see review in Bradley and Pregitzer 2007). Similarly, after 13 years of plants 'conditioning' the soils beneath them, Schweitzer *et al.* (2008b) found that intraspecific plant genetic variation affected soil microbial C and N pools and microbial community composition. Overall, they found that plant genotype influenced microbial N pools beneath replicated genotypes of *Populus* spp. (*P. angustifolia*, hybrids, but not *P. fremontii* genotypes) and demonstrated significant broad-sense heritability of the microbial community. Plant genotype explained 61%, 23% and 41% of the variation in microbial N for *P. angustifolia*, F_1 and backcross hybrids, respectively. Total PLFA, as a measure of microbial C, demonstrated significant genotype effects in soils associated with *P. angustifolia* and backcross hybrid genotypes. Plant genotype was a significant predictor of soil microbial community composition within genotypes of *P. angustifolia*, with plant genotype explaining 70% of the variation in community composition, as assessed by PLFA. Nine, clonally replicated, *P. angustifolia* individual genotypes demonstrated a 3.2-fold difference in microbial C pools, 3.0-fold differences in microbial N pools and a 1.5-fold difference in microbial PLFA richness (Fig. 19.3). These data indicate that plant genotypes can be an important ecological factor affecting soil microbial dynamics, as the range of variation between genotypes can be as high as that between plant species (sensu Harrison and Bardgett 2010). Data such as these suggest that the specific traits that vary by plant genotype can act as agents of selection for microbial communities to create locally adapted soil communities that can lead to a 'home-field advantage' at fine genetic scales (Pregitzer *et al.* 2010; Smith *et al.* 2012).

Individual genotypes and ecosystem processes

Individual trees have long been known to influence soils and soil processes within the shadow of their canopies by altering soil microclimate and organic matter inputs creating 'single-tree influence circles' (Zinke 1962;

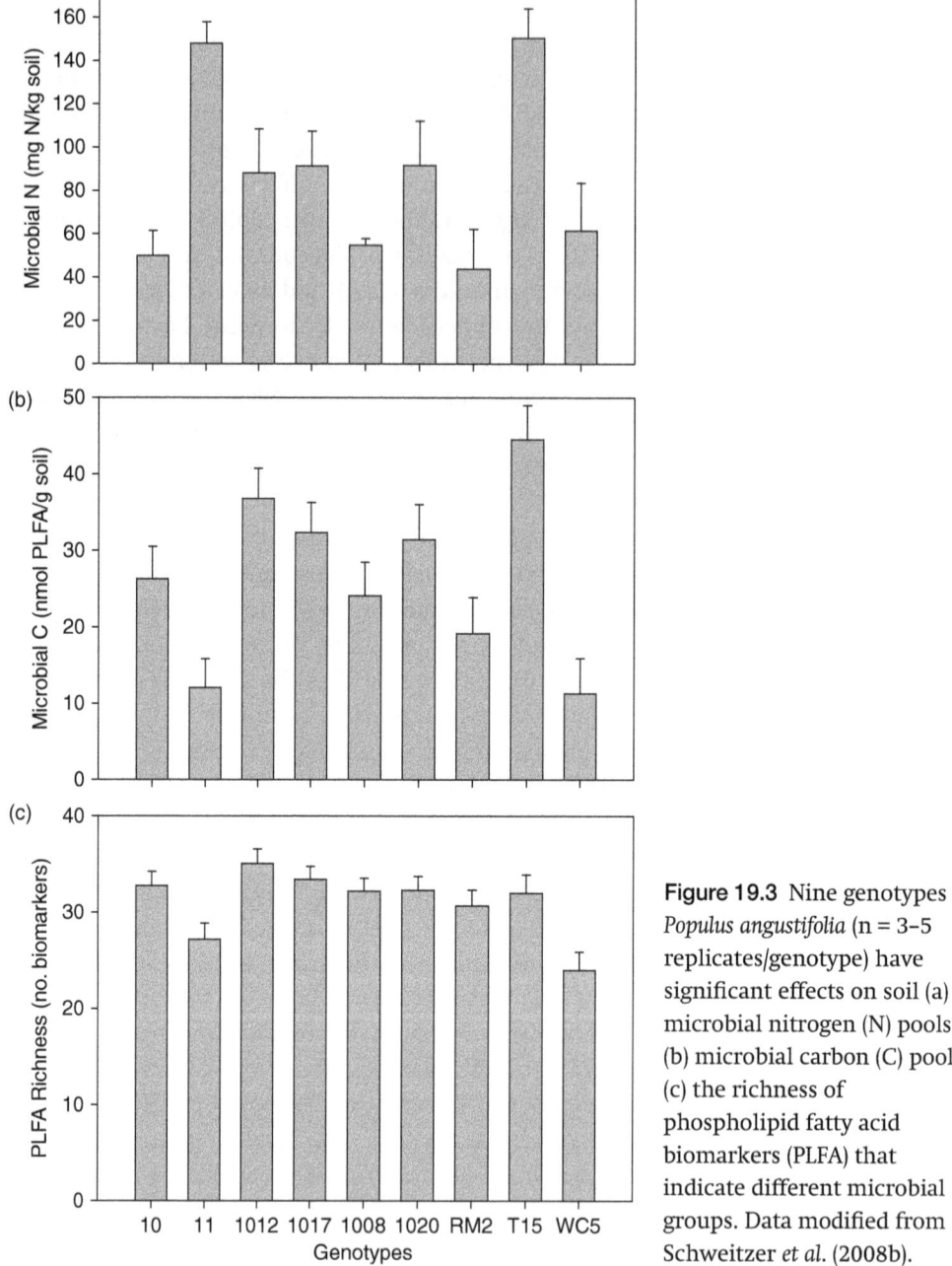

Figure 19.3 Nine genotypes of *Populus angustifolia* (n = 3–5 replicates/genotype) have significant effects on soil (a) microbial nitrogen (N) pools; (b) microbial carbon (C) pools; (c) the richness of phospholipid fatty acid biomarkers (PLFA) that indicate different microbial groups. Data modified from Schweitzer *et al.* (2008b).

Boettcher and Kalisz 1990; Rhoades 1997; Amioti *et al.* 2000). For example, in a study examining nine phenotypes of oak (*Quercus leavis*), Madritch and Hunter (2002) found that tree phenotypes differ in litter quality traits that had significant effects on litter mass loss and C and N dynamics. However, few studies have examined, with replicate clones, the strength of this individual (i.e., genotype) effect on ecosystem processes, such as leaf litter decomposition or total aboveground productivity.

A common garden study with 21 genotypes of *Solidago altissma* (grown either in monocultures as well as in various mixtures) demonstrated that replicated genotypes grown in monoculture varied 1.75-fold in total aboveground productivity (Crutsinger *et al.* 2006). Studies on seven replicated genotypes of *Populus angustifolia* grown in three different common garden environments (across a 350 m elevation gradient) demonstrated strong environment and genotype effects on annual rates of aboveground NPP but no interaction between environment and genotype (Lojewski *et al.* 2009). Similarly, Madritch *et al.* (2006) grew replicates of five *P. tremuloides* genotypes in either a high or low nutrient environment and decomposed the leaf litter in a common environment. They found strong effects of genotype, nutrient environment and their interaction on initial litter quality and rates of mass loss over time. Leaf litter from genotypes grown under high nutrient environments varied 2.8-fold in decomposition after 12 months in the field while genotypes grown under low nutrient environments varied 1.6-fold, indicating the strong interactive effects with environmental variation. These studies demonstrate that replicate genotypes can substantially influence ecosystem processes that are detectable even in combination with environmental variation, even though genetic effects may be weaker than other effects.

Genotypic variation in microbial pools and community composition suggests that variation among plant genotypes in traits can directly or indirectly (through the processes that microbes mediate) impact ecosystem processes such as rates of mineralization. On the same trees as described in Schweitzer *et al.* (2008b and above), we examined the role of genotypic variation on ecosystem processes utilizing 19 replicated (n = 3–5) genotypes of *Populus* spp. randomly planted and then grown in a common environment for 15 years to quantify the clonal repeatability (Falconer and McKay 1996; Conner and Hartl 2004) of soil net N transformations associated with individual genotypes (see Schweitzer *et al.* 2008b for details on experimental design and site description). From October 2005–October 2006, we utilized in-field incubations across the four seasons (three randomly placed soil cores were placed within 0.25 m of the trunks) to assess annual rates of soil net N mineralization and net nitrification from beneath the canopy of individual trees (following the specific field methods outlined in Schweitzer *et al.* 2004, except using 2.5 cm × 15 cm soil cores).

When tree genotype was nested within *Populus* species or crosstype (i.e., *P. fremontii*, *P. angustifolia* and their natural F_1 hybrids), we found a significant effect of plant genotype on annual rates of net N mineralization (Fig. 19.4; ANOVA, F = 2.08, P = 0.026); however, there were no differences between the crosstypes (F = 0.44, P = 0.646). The lack of differences between species may be

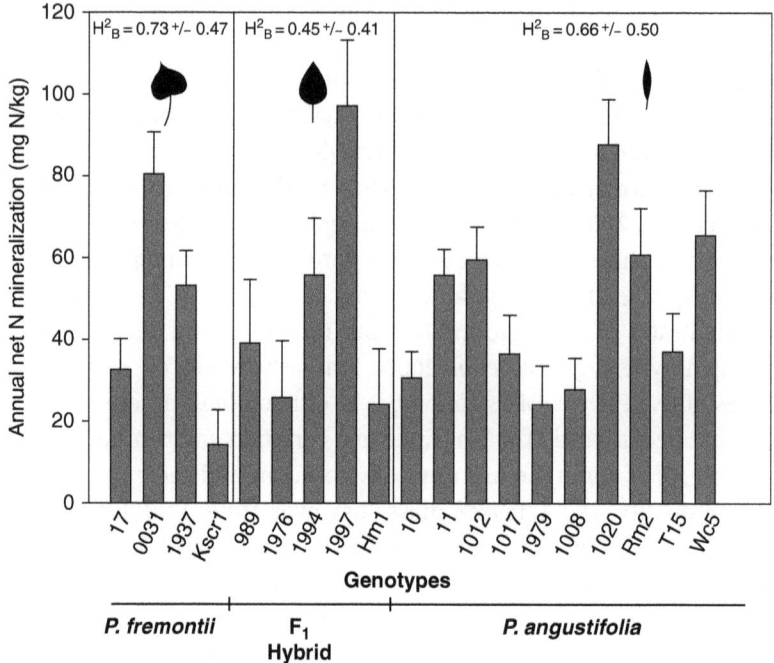

Figure 19.4 Genotypic variation in annual rates of net nitrogen (N) mineralization (n = 3–5 replicates/genotype) beneath genotypes of *Populus fremontii*, *P. angustifolia* and their F1 hybrids in a 15-year-old common environment. Significant repeatable variation exists among genotypes in all tree types. Broad-sense heritability scores are listed with 95% confidence intervals. See colour plate section.

due to high genotypic variation and the specific genotypic identity within each species (or F_1 hybrid) and the limited number of replicated genotypes available to study in the 15-year-old common garden. When we examined the differences in annual rates of soil net N mineralization among genotypes within each group with univariate ANOVAs we found significant genotypic differences within all tree types. From the variance component estimates from the univariate ANOVA analyses, we calculated the broad-sense heritability of annual net N mineralization. We found that plant genotype explained 73%, 45% and 66% of the variation in annual net N mineralization for *P. fremontii* ($H^2 = 0.73 \pm 0.47$), F_1 hybrids ($H^2 = 0.45 \pm 0.41$) and *P. angustifolia* ($H^2 = 0.66 \pm 0.50$), respectively (Fig. 19.4).

In contrast, when we examined rates of annual net nitrification (i.e., net conversion of ammonium to nitrate) with the same analysis as above, we found no significant overall effect of plant genotype or tree crosstype ($F = 1.41$, $P = 0.178$; $F = 0.44$, $P = 0.646$, respectively). We found no relationship between plant tree type or crosstype with soil organic C or total N in either a nested model ($F = 0.66$, $P = 0.68$; $F = 0.87$, $P = 0.43$, respectively) or individual ANOVAs (Table 19.1). The lack of a genotypic effect on rates of net nitrification suggests

Table 19.1 *Individual analysis of variance (ANOVA) results for the effects of plant genotype on soil nitrogen (N) transformation rates for* Populus fremontii, P. angustifolia *and their natural F_1 hybrids. F statistics are reported with P values in parentheses; values in bold are statistically significant at $\alpha = 0.05$.*

	P. fremontii	F_1 hybrid	P. angustifolia
Annual net N mineralization	8.89 **(0.01)**	3.89 **(0.03)**	5.97 **(0.02)**
Annual net nitrification	0.36 (0.79)	0.83 (0.53)	1.78 (0.22)
Soil total nitrogen	0.53 (0.67)	0.34 (0.84)	3.11 (0.08)
Soil organic carbon	0.29 (.83)	0.49 (0.74)	2.98 (0.09)

that the comparatively small community of soil microorganisms capable of nitrification may be less sensitive to underlying plant genotypic variation than the broad community of ammonifying soil microorganisms. Nevertheless, the net N mineralization data indicate not only that individual genotypes can create 'single-tree influence circles' but that the plant–soil linkages that create this influence are repeatable among clones of the same genotype.

The 3.4–4.9-fold difference in annual rates of net N mineralization that we recorded among genotypes of two *Populus* species and their F_1 hybrids indicates that genotypic variation may indeed have large effects on plant traits that extend to soil properties. Overall these data suggest that the effects of plant genotype on biogeochemical cycles demonstrate tight linkages from plants to soils at the level of individual tree, the effects of which may be as large as between plant species (sensu Wedin and Tilman 1990; Harrison and Bardgett 2010). Such repeatable, ecosystem phenotypes from plant genotypes to their associated soils indicate that there is genetic variation for some mechanistic plant phenotypic trait (or a combination of traits). While the specific plant trait(s) associated with rates of net N mineralization are unknown in the study described here, these data support previous research demonstrating the importance of both plant population genetic variation and genotypic variation for ecosystem processes (Madritch and Hunter 2002, 2005; Schweitzer *et al.* 2004, 2005; Madritch *et al.* 2006, 2007; Bailey *et al.* 2009). These data indicate that heritable variation in ecosystem phenotypes can occur. However, much more research is needed to understand the extended and long-term effects of plant genotype on the soil environment, especially in the context of plant genetic interactions with the environment ($G \times E$) to determine under what conditions genotypic variation may have a substantial influence.

Genotype-mediated linkages
Genotypes may mediate linkages between species and between above- and belowground components of an ecosystem. Variation in traits may have specific

effects on associated communities that may have predictable outcomes for trophic interactions and ecosystem processes, such as leaf litter decomposition. For example, genotypes of *Populus angustifolia* (and their closely related back-cross hybrids) demonstrate heritable variation in the susceptibility of genotypes to a gall-forming herbivore, *Pemphigus betae*, and the incidence of gall attack by birds (Bailey *et al.* 2006). These results (and others) suggest that plant genotypes can mediate tritrophic interactions and energy flow through an ecosystem, via plant chemistry (Müller *et al.* 2006). Similarly, there are extended consequences of susceptibility to *P. betae* on rates of leaf litter decomposition (Schweitzer *et al.* 2005). Observationally and with experimental removals, leaf litter from genotypes with *P. betae* galls decomposes 34–40% slower than non-galled leaf litter. Moreover, genotypes with and without herbivory have different nutrient dynamics such that genotypes release and immobilize nutrients at different rates, depending on the presence or absence of the gall. Plant genotype may also mediate the effects of herbivory through frass deposition, influencing the soil microbial community and their activities. For example, Madritch *et al.* (2007) found that two herbivores (gypsy moth and forest tent caterpillars) that were fed leaf tissue from multiple aspen genotypes produced frass that reflected the specific nutrient content of the genotype's foliage. When the leaf litter and frass were incubated in microcosms, they found that plant genotype also influences the frass-induced patterns of soil respiration in the same way that leaf litter of genotypes influenced soil respiration. Taken together, these examples indicate the importance of plant genotype on mediating herbivore dynamics that has extended consequences for energy flow and ecosystem processes.

Non-additivity and populations with varying genotypic composition

The above sections demonstrate the effects that individual plant genotypes have on structuring associated species, arthropod communities, soil microbial communities and ecosystem processes such as leaf litter decomposition, rates of net N mineralization and aboveground NPP. Evidence also clearly indicates that environmental variation is an important factor regulating the expression of these extended phenotypes. This is particularly relevant when environmental variation is biotic in nature. Genotypes rarely occur in isolation in nature and unpredictable outcomes may occur when multiple genotypes co-occur either within the same species or across species, relative to single genotype effects on community and ecosystem processes. These unpredictable outcomes of genetically based species interactions are referred to as non-additive responses (Conner and Hartl 2004). Non-additivity occurs when the total response of a variable is greater or less than the sum of the partitioned responses generated by the individual components. Non-additive outcomes may have important implications for our understanding of population-level genetic variance and the evolutionary trajectory of populations with perturbation (sensu Bradley and Pregitzer 2007; Hughes *et al.* 2008). While much is still to be determined about the relative strength of genotype effects across systems and with environmental variation, the extended effects of genotypic variation (i.e., mixtures of individual genotypes) on communities and ecosystems may result in effects that are dependent upon the identity of genotypes present in the population, much like species effects are often dependent upon the species identity (Hooper *et al.* 2005).

Based on current knowledge, genotypic diversity (i.e., genotype mixtures) appears to have stronger effects at the community level than at the ecosystem level, based on relative effect sizes (Bailey *et al.* 2009). Mixtures of plant genotypes have been shown to impact positively the richness of arthropods with mixtures of *Solidago altissima* genotypes (Crutsinger *et al.* 2006, 2008a) and positively affect the abundance of omnivorous arthropods (but perhaps not specialist guilds; Johnson *et al.* 2006) in *Oenothera biennis*. Genotypic diversity also appears to have weak effects on ground-dwelling arthropod communities (Madritch and Hunter 2005 [phenotypic diversity]; Crutsinger *et al.* 2008c; Kanaga *et al.* 2009). However, genotypic mixtures can affect the resilience of plant communities with disturbance or stress in marine intertidal zones (Hughes and Stachowicz 2004; Reusch *et al.* 2005) and impact the success of invasive species across plant systems (Crutsinger *et al.* 2008b; Vellend *et al.* 2010). Genotypic identity, not just the number of genotypes in a population, may also have important consequences for mediating the outcomes of competition in a community and affect overall species diversity (Booth and Grime 2003; Fridley *et al.* 2007; Lankau and Strauss 2007).

At the ecosystem level, non-additivity in leaf litter decomposition with species mixtures commonly occurs when different species of leaf litter are mixed (Gartner and Cardon 2004; Ball *et al.* 2008), however, non-additive responses with genotype mixtures occur with some response variables but not others. For example, increasing genotypic diversity in *Solidago altissima*, from 1 to 24 randomly chosen genotypes (Crutsinger *et al.* 2006, 2008a) and *Zostera marina* impacted overall plant productivity over time (Reusch *et al.* 2005). Similarly, planted mixtures of *Arabadopsis thaliana* gentoypes resulted in higher overall productivity regardless of fertilization treatment or density (Kotowska *et al.* 2010). Rates of leaf litter decomposition, however, show varying effects in genotype mixtures. For example, Schweitzer *et al.* (2005) found that mixtures of five *Populus* spp. genotypes (with and without herbivory) demonstrated non-additive effects on rates of leaf litter decay such that mixtures of genotypes decayed faster than would be predicted based on the same genotypes decomposed in isolation. In contrast, Madritch *et al.* (2006) and Crutsinger *et al.* (2009) found that manipulating numbers of genotypes in litterbags of *Populus tremuloides* (three and five genotype mixtures) and *Solidago altissma* (three, six or nine genotype mixtures), respectively, had little effect on overall rates of decay, but that genotype identity (or nutrient environmental treatment) had larger effects on rates of decomposition. These data, across both community and ecosystem response variables, suggest that the identity and specific traits of individual genotypes may matter more than genotypic diversity, much like has been found at the species level with increasing species richness.

Implications and conclusions

The results from studies of genotypes to date clearly show that there is genetic variation for many plant phenotypic traits such as physiology, growth, morphology and phytochemistry, although these traits differ in their plasticity (Geber and Griffen 2003). The effects of variation in plant traits may have extended effects that directly or indirectly: (1) influence associated arthropod and microbial communities; (2) influence ecosystem processes such as energy flow, productivity and nutrient cycling; (3) mediate linkages among species and above- and belowground processes; however, (4) mixtures of genotypes may result in unpredictable outcomes that may be dependent upon the composition and identity of the genotypes. Together these results indicate that variation among individual genotypes within a species can have large and sometimes predictable effects on associated communities and ecosystem processes. While much is still to be determined about the strength and scale of genotype-mediated effects with environmental variation (both biotic and abiotic), these data indicate that intraspecific variation may be as important as interspecific variation in understanding and predicting ecological patterns in nature.

    Understanding the extended effects of individual variation in plant traits on plant–arthropod or plant–soil interactions places the study of communities and ecosystems in the realm of evolutionary biology as it demonstrates that genetic variation in plant traits can influence communities and the processes they mediate (Van Breeman and Finzi 1998; Whitham *et al.* 2006, 2008). While community composition or an ecosystem process cannot be passed from one plant generation to another generation, the genetic factors responsible for structuring the community or ecosystem process in a specific way can be passed across generations. It is the trait(s) of the plant that may be heritable resulting in differential repeatability of associated species composition or processes among genotypes. One mechanistic hypothesis for these processes involves interspecific indirect genetic effects (i.e., IIGEs; outlined in Shuster *et al.* 2006; Johnson *et al.* 2009; Allen *et al.*, this volume, Chapter 16), whereby the genotype of one individual affects the phenotype and fitness of associated interacting species, resulting in variation in the flow of energy or the cycling of nutrients. Therefore, selective regimes that alter the variation of genotypes, within and across populations, have the potential to strongly influence the composition of communities, ecosystem processes and the evolution of associated species.

Acknowledgements

We thank the Ogden Nature Center and the Utah Department of Natural Resources for supporting our restoration and common garden studies and the Dept. of Forestry at Northern Arizona University for support. Thanks to Dan Guido, Morgan Luce, Clara Pregitzer and Michelle Stritar for assistance in the field and laboratory. Support was provided from the National Science Foundation, Frontiers in Integrated Biological Research program (FIBR; DEB-0425908) and (DEB-0743437) to JAS and JKB.

References

Amioti, N. M., Zalba, P., Sanchez, L. F. and Peinemann, N. (2000) The influence of single trees on properties associated with loess-derived grassland soils in Argentina. *Ecology*, **81**, 3283–3290.

Ayres, E., Steltzer, H., Simmons, B. L. *et al.* (2009) Homefield advantage accelerates leaf litter decomposition in forests. *Soil Biology and Biochemistry*, **41**, 606–610.

Bailey, J. K., Schweitzer, J. A., Koricheva, J. *et al.* (2009) Community and ecosystem consequences of gene flow and genotypic diversity across systems and environments: a meta-analysis. *Philosophical Transactions of the Royal Society of London, Series B*, **364**, 1607–1616.

Bailey, J. K., Wooley, S. C., Lindroth, R. L. and Whitham, T. G. (2006) Importance of species interactions to community heritability: a genetic basis to trophic-level interactions. *Ecology Letters*, **9**, 78–85.

Ball, B. A., Hunter, M. A., Kominoski, J. S., Swan, C. M. and Bradford, M. A. (2008) Consequences of non-random species loss for decomposition dynamics: experimental evidence for additive and non-additive effects. *Journal of Ecology*, **96**, 306–313.

Bangert, R. K., Lonsdorf, E. V., Wimp, G. M. *et al.* (2008) Genetic structure of a foundation species: scaling community phenotypes from the individual to the region. *Heredity*, **100**, 121–131.

Bangert, R. K., Turek, R., Rehill, J. B. *et al.* (2006) A genetic similarity rule determines arthropod community structure. *Molecular Ecology*, **15**, 1379–1391.

Barbour, R. C., O'Reilly-Wapstra, J. M., De Little, D. W. *et al.* (2009) A geographic mosaic of genetic variation within a foundation tree species and its community-level consequences. *Ecology*, **90**, 1762–1772.

Bartelt-Ryser, J., Joshi, J., Schmid, B., Brandl, H. and Balser, T. (2005) Soil feedbacks of plant diversity on soil microbial communities and subsequent plant growth. *Perspectives in Plant Ecology, Evolution and Systematics*, **7**, 27–49.

Bever, J. D., Westover, K. M. and Antonovics, J. (1997) Incorporating the soil community into plant population dynamics: the utility of the feedback approach. *Journal of Ecology*, **85**, 561–573.

Binkley, D. and Giardina, C. (1998) Why do tree species affect soils? The warp and woof of tree–soil interactions. *Soil Biology and Biochemistry*, **42**, 89–106.

Boettcher, S. E. and Kalisz, P. J. (1990) Single tree influence on soil properties in the mountains of Kentucky. *Ecology*, **71**, 1365–1372.

Booth, R. E. and Grime, J. P. (2003) Effects of genetic impoverishment on plant community diversity. *Journal of Ecology*, **91**, 721–730.

Bossdorf, O., Shuja, Z. and Banta, J. A. (2009) Genotype and maternal environment affect belowground interactions between *Arabadopsis thaliana* and its competitors. *Oikos*, **118**, 1541–1551.

Bradley, K. L. and Pregitzer, K. S. (2007) Ecosystem assembly and terrestrial carbon balance under elevated CO_2. *Trends in Ecology and Evolution*, **22**, 538–547.

Chen, F., Liu, C.-J., Tschlapinski, T. and Zhao, N. (2009) Genomics of secondary metabolism in *Populus*: interactions with biotic and abiotic environments. *Critical Reviews in Plant Science*, **28**, 375–392.

Clark, J. S. (2010) Individuals and the variation needed for high species diversity in forest trees. *Science*, **327**, 1129–1132.

Conner, J. K. and Hartl, D. L. (2004) *A Primer of Ecological Genetics*. Sunderland, MA: Sinauer.

Crutsinger, G. M., Collins, M. D., Fordyce, J. A. *et al.* (2006) Plant genotypic diversity predicts community structure and governs an ecosystem process. *Science*, **313**, 966–968.

Crutsinger, G. M., Collins, M. D., Fordyce, J. A. and Sanders, N. J. (2008a) Temporal dynamics in non-additive responses of arthropods to host-plant genotypic diversity. *Oikos*, **117**, 255–264.

Crutsinger, G. M., Reynolds, N., Sanders, N. J. and Classen, A. T. (2008c) Disparate effects of host plant genotypic diversity on above- and belowground communities. *Oecologia*, **158**, 65–75.

Crutsinger, G. M., Sanders, N. J. and Classen, A. T. (2009) Contrasting intra- and interspecific variation on litter dynamics. *Basic and Applied Ecology*, **10**, 535–543.

Crutsinger, G. M., Souza, L. and Sanders, N. J. (2008b) Intraspecific diversity and dominant genotypes resist invasion. *Ecology Letters*, **11**, 16–23.

Dayton, P. K. (1972) Toward an understanding of community resilience and the potential effects of enrichments to the benthos at McMurdo Sound, Antarctica. In B. C. Parker, ed., *Proceedings of the Colloquium on Conservation problems in Antarctica*. Lawrence, KS: Allen Press, 81–96.

Diner, B., Berteaux, D., Fyles, J. and Lindroth, R. L. (2009) Behavioral archives link the chemistry and clonal structure of trembling aspen to the food choice of North American porcupine. *Oecologia*, **160**, 687–695.

Dungey, H. S., Potts, B. M., Whitham, T. G. and Li, H. F. (2000) Plant genetics affects arthropod community richness and composition: evidence from a synthetic eucalypt hybrid population. *Evolution*, **54**, 1938–1946.

Ellison, A. M., Bank, M. S., Clinton, B. D. *et al.* (2005) Loss of foundation species: consequences for the structure and dynamics of forested ecosystems. *Frontiers in Ecology and the Environment*, **9**, 479–486.

Eckburg, P. B., Bik, E. M., Bernstein, C. N. *et al.* (2005) Diversity of the human intestine microbial flora. *Science*, **308**, 1635–1638.

Falconer, D. S., McKay, T. F. C. (1996) *Introduction to Quantitative Genetics*, 4th edn. New York: Pearson, Prentice Hall, Longman.

Fierer, N. and Jackson, R. (2006) The diversity and biogeography of soil bacterial communities. *Proceedings of the National Academy of Sciences of the United States of America*, **103**, 626–631.

Fierer, N., Grandy, N. S., Six, J. and Paul, E. A. (2009) Searching for unifying principles in soil biology. *Soil Biology and Biochemistry*, **41**, 2249–2256.

Fischer, D. G., Hart, S. C., LeRoy, C. J. and Whitham, T. G. (2007) Variation in belowground carbon fluxes along a *Populus* hybridization gradient. *New Phytolologist*, **176**, 415–425.

Fischer, D. G., Hart, S. C., Schweitzer, J. A., Selmants, P. C. and Whitham, T. G. (2010) Soil nitrogen availability varies with plant genetics across diverse river drainages. *Plant and Soil*, **331**, 391–400.

Fischer, D. G., Hart, S. C., Whitham, T. G., Martinsen, G. D. and Keim P. (2004) Ecosystem implications of genetic variation in water use of a dominant riparian tree. *Oecologia*, **139**, 188–197.

Fridley, J. D., Grime, J. P. and Bilton, M. (2007) Genotypic identity of interspecific neighbours mediates plant responses to competition and environmental variation in a species rich grassland. *Journal of Ecology*, **95**, 908–915.

Fritz, R. S., Nichols-Orians, C. M. and Brunsfeld, S. J. (1994) Interspecific hybridization of plants and resistance to herbivores: hypotheses, genetics, and variable responses in a diverse herbivore community. *Oecologia*, **97**, 106–117.

Gartner, T. B. and Cardon, Z. G. (2004) Decomposition dynamics in mixed-species leaf litter. *Oikos*, **104**, 230–246.

Geber, M. A. and Griffen, L. R. (2003) Inheritance and natural selection on functional traits. *International Journal of Plant Science*, **164** (Suppl.), S21–S42.

Gehring, C. A. and Whitham T. G. (1994) Interactions between aboveground herbivores and the mycorrhizal mutualists of plants. *Trends in Ecology and Evolution*, **9**, 251–255.

Genung, M. A., Lessard, J. P., Brown C. B. et al. (2010) Non-additive effects of genotypic diversity increase floral abundance and abundance of floral visitors. *Public Library of Sciences One* (doi:10.1371/journal. pone.0008711).

Gholz, H. L., Wedin, D. A., Smitherman, S. M. et al. (2000) Longterm dynamics of pine and hardwood litter in contrasting environments: toward a global model of decomposition. *Ecosystems*, **6**, 751–765.

Harrison, K. A. and Bardgett, R. D. (2010) Influence of plant species and soil conditions on plant-soil feedback in mixed grassland communities. *Journal of Ecology*, **98**, 384–395.

Hättenschwiler, S. and Vitousek, P. M. (2000) The role of polyphenols in terrestrial ecosystem nutrient cycling. *Trends in Ecology and Evolution*, **15**, 238–243.

Havill, N. P. and Raffa, K. F. (1999) Effects of eliciting treatment and genotypic variation on induced resistance in *Populus*: impacts on gypsy moth development and feeding behavior. *Oecologia*, **120**, 295–303.

Hochwender, C. G. and Fritz, R. S. (2004) Plant genetic differences influence herbivore community structure: evidence from a hybrid willow system. *Oecologia*, **138**, 547–557.

Holeski, L. M., Vogelzang, A., Stanosz, G. and Lindroth, R. L. (2009) Incidence of *Venturia* shoot blight damage in aspen (*Populus tremuloides* Michx.) varies with tree chemistry and genotype. *Biochemical Systematics and Ecology*, **37**, 139–145.

Hooper, D. U., Chapin, F. S., Ewel, J. J. et al. (2005) Effects of biodiversity on ecosystem functioning: a consensus of current knowledge. *Ecological Mongraphs*, **75**, 3–35.

Hughes, A. R. and Stachowicz, J. J. (2004) Genetic diversity enhances the resistance of a seagrass ecosystem to disturbance. *Proceedings of the National Academy of Sciences of the United States of America*, **101**, 8998–9002.

Hughes, A. R., Inouye, B., Johnson, M. T. J., Underwood, N. and Vellend, M. (2008) Ecological consequences of genetic diversity. *Ecology Letters*, **11**, 609–623.

Hughes, A. R., Stachowicz, J. J. and Williams, S. L. (2009) Morphological and physiological variation among seagrass (*Zostera marina*) genotypes. *Oecologia*, **159**, 725–733.

Hwang, S.-Y. and Lindroth, R. L. (1997) Clonal variation in foliar chemistry of aspen: effects on gypsy moths and forest tent caterpillars. *Oecologia*, **111**, 99–108.

Iason, G. R., Lennon, J. J., Pakeman, R. J. et al. (2005) Does chemical composition of individual Scots pine trees determine the biodiversity of their associated ground vegetation? *Ecology Letters*, **8**, 364–369.

Ito, M. and Ozaki, K. (2005) Response of a gall wasp community to genetic variation in the host plant *Quercus crispula*: a test using half-sib families. *Acta Oecologica*, **27**, 17–24.

Jani, A. J., Faeth, S. H. and Gardner, H. (2010) Asexual endophytes and associated alkaloids alter arthropod community structure and increase herbivore abundances on a native grass. *Ecology Letters*, **13**, 106–117.

Johnson, M.T.J. and Stinchcombe, J.R. (2007) An emerging synthesis between community ecology and evolutionary biology. *Trends in Ecology and Evolution*, **22**, 250–257.

Johnson, M.T.J., Lajeunesse, M.J. and Agrawal, A.A. (2006) Additive and interactive effects of plant genotypic diversity on arthropod communities and plant fitness. *Ecology Letters*, **9**, 24–34.

Johnson, M.T.J., Vellend, M. and Stinchcombe, J.R. (2009) Evolution in plant populations as a driver of ecological changes in arthropod communities. *Philosophical Transactions of the Royal Society of London, Series B*, **364**, 1593–1605.

Kanaga, M., Latta IV, L.C., Mock, K.E. *et al.* (2009) Plant genotypic diversity and environmental stress interact to negatively affect arthropod community diversity. *Arthropod–Plant Interactions*, **3**, 249–258.

Karst, J., Jones, M.D. and Turkington, R. (2009) Ectomycorrhiza colonization and intraspecific variation in growth responses to lodgepole pine. *Plant Ecology*, **200**, 161–165.

Kasurinen, A., Keinänen, M.M., Kaipainen, S. *et al.* (2005) Belowground response of silver birch trees exposed to elevated CO_2 and O_3 for three growing seasons. *Global Change Biology*, **11**, 1167–1179.

Keith, A.R., Bailey, J.K. and Whitham, T.G. (2010) A genetic basis to community repeatability and stability. *Ecology*, **91**, 3398–3406.

Kotowska, A.M., Cahill, Jr., A.F. and Keddie, B.A. (2010) Plant genetic diversity yields increased plant productivity and herbivore performance. *Journal of Ecology*, **98**, 237–245.

Kraus, T.E.C., Zasoski, R.J., Dahlgren, R.A., Horwath, W.R. and Preston, C.M. (2004) Carbon and nitrogen dynamics in a forest soil amended with purified tannins from different plant species. *Soil Biology and Biochemistry*, **36**, 309–321.

Kulmatiski, A., Beard, K.H., Stevens, J. and Cobbold, S.M. (2008) Plant-soil feedbacks: a meta-analytical review. *Ecology Letters*, **11**, 980–992.

Lankau, R.A. and Strauss, S.Y. (2007) Mutual feedbacks maintain both genetic and species diversity in a plant community. *Science*, **317**, 1561–1563.

LeRoy, C.J., Whitham, T.G., Keim, P. and Marks, C.J. (2006) Plant genes link forests and streams. *Ecology*, **87**, 255–261.

Lojewski, N.R., Fischer, D.G., Bailey, J.K. *et al.* (2009) Genetic basis of aboveground productivity in two native *Populus* species and their hybrids. *Tree Physiology*, **29**, 1133–1142

Madritch, M.D. and Hunter, M.D. (2002) Phenotypic diversity influences ecosystem functioning in an oak sandhills community. *Ecology*, **83**, 2084–2090.

Madritch, M.D. and Hunter, M.D. (2005) Phenotypic variation in oak litter influences short- and long-term nutrient cycling through litter chemistry. *Soil Biology and Biochemistry*, **37**, 319–327.

Madritch, M.M. and Lindroth, R.L. (2011) Soil microbial communities adapt to genetic variation in leaf litter inputs. *Oikos*, **120**,1696–1704.

Madritch, M.D., Donaldson, J.R. and Lindroth, R.L. (2006) Genetic identity of *Populus tremuloides* litter influences decomposition and nutrient release in a mixed forest stand. *Ecosystems*, **9**, 528–537.

Madritch, M.D., Donaldson, J.R. and Lindroth, R.L. (2007) Canopy herbivory mediates the influence of plant genotype on soil processes through frass deposition. *Soil Biology and Biochemistry*, **39**, 1192–1201.

Madritch, M.D., Greene, S.L. and Lindroth, R.L. (2009) Genetic mosaics of ecosystem functioning across aspen-dominated landscapes. *Oecologia*, **160**, 119–127.

Mangan, S.A., Schnitzer S.A., Herre, E.A. *et al.* (2010) Negative plant-soil feedback predicts tree species relative abundance in a tropical forest. *Nature*, **466**, 752–755.

McIntyre, P.J. and Whitham, T.G. (2003) Plant genotype affects long-term herbivore population dynamics and extinction: conservation implications. *Ecology*, **84**, 311–322.

Morris, S. J. and Blackwood, C. B. (2007) The ecology of soil organisms. In E. A. Paul, ed., *Soil Microbiology, Ecology and Biochemistry*, 3rd edn. Amsterdam, The Netherlands: Academic Press, pp. 195–226.

Müller, M. S., McWilliams, S. R., Podlesak, D. *et al.* (2006) Tri-trophic effects of plant defenses: chickadees consume caterpillars based on host leaf chemistry. *Oikos*, **114**, 507–517.

Osier, T. L. and Lindroth, R. L. (2004) Long-term effects of defoliation on quaking aspen in relation to genotype and nutrient availability: plant growth, phytochemistry and insect performance. *Oecologia*, **139**, 55–65.

Pregitzer, C. P., Bailey, J. K., Hart, S. C. and Schweitzer, J. A. (2010) Soils as agents of selection: feedbacks between plants and soils alter seedling survival and performance. *Evolutionary Ecology*, **24**, 1045–1059.

Ralph, S., Oddy, C., Cooper, D. *et al.* (2006). Genomics of hybrid poplar (*Populus trichocarpa deltoides*) interacting with forest tent caterpillars (*Malacosoma disstria*): normalized and full-length cDNA libraries, expressed sequence tags, and a cDNA microarray for the study of insect-induced defences in poplar. *Molecular Ecology*, **15**, 1275–1297.

Rehill, B. J., Whitham, T. G., Martinsen, G. D. *et al.* (2006) Developmental trajectories in cottonwood phytochemistry. *Journal of Chemical Ecology*, **32**, 2269–2285.

Reusch, T. B. H., Ehlers, A., Hämmereli, I. and Worm, B. (2005) Ecosystem recovery after climatic extremes enhanced by genotypic diversity. *Proceedings of the National Academy of Sciences of the United States of America*, **102**, 2826–2831.

Rhoades, C. C. (1997) Single-tree influences on soils properties in agroforestry: lessons from natural forestry and savannah ecosystems. *Agroforestry Systems*, **35**, 71–94.

Schweitzer, J. A., Bailey, J. K., Fischer, D. G. *et al.* (2008b) Soil microorganism–plant interactions: heritable relationship between plant genotype and associated microorgansims. *Ecology*, **89**, 773–781.

Schweitzer, J. A., Bailey, J. K., Hart, S. C. *et al.* (2005) The interaction of plant genotype and herbivory decelerate leaf litter decomposition and alter nutrient dynamics. *Oikos*, **110**, 133–145.

Schweitzer, J. A., Bailey, J. K., Rehill, B. J. *et al.* (2004) Genetically based trait in dominant tree affects ecosystem processes. *Ecology Letters*, **7**, 127–134.

Schweitzer, J. A., Madritch, M. D., Bailey, J. K. *et al.* (2008a) From genes to ecosystems: the genetic basis of condensed tannins and their role in nutrient regulation in a *Populus* model system. *Ecosystems*, **11**, 1005–1020.

Shuster, S. M., Lonsdorf, E. V., Wimp, G. M., Bailey, J. K. and Whitham, T. G. (2006) Community heritability measures the evolutionary consequences of indirect genetic effects on community structure. *Evolution*, **60**, 991–1003.

Simchuk, A. (2008) The effect of fodder plant genotype on the variation of larval fitness traits in genotype classes of green oak leafroller moth. *Russian Journal of Genetics*, **44**, 418–424.

Smith, D. S., J. A. Schweitzer, J. K. Bailey *et al.* (2012) Soil-mediated local adaptation alters seedling survival and performance. *Plant and Soil*, **352**, 243–251.

Strickland, M. S, Lauber, C., Fierer, N. and Bradford, M. A (2009) Testing the functional significance of microbial community composition. *Ecology*, **90**, 441–451.

Tovar-Sanchez, E. and Oyama, K. (2006) Effect of hybridization of the *Quercus crassifolia–Quercus crassipes* complex on the community structure of endophagous insects. *Oecologia*, **147**, 702–713.

Tsai, C-J, Harding, S. A, Tschaplinski, T. J., Lindroth, R. L. and Yuan, Y. (2006) Genome-wide analysis of the structural genes regulating defense phenylpropanoid metabolism in *Populus. New Phytologist,* **172**, 47–62.

Van Breeman, N. and Finzi, A. C (1998) Plant-soil interactions: ecological aspects and evolutionary implications. *Biogeochemistry,* **42**, 1–19.

Vellend, M., Drummond, E. B. M. and Tomimatsu, H. (2010) Effects of genotype identity and diversity on the invasiveness and invasibility of plant populations. *Oecologia,* **162**, 371–381.

Vorburger, C., Eugster, B., Villiger, J. and Wimmer, C. (2010) Host genotype affects the relative success of competing lines of aphid parasitoids under superparasitism. *Ecological Entomology,* **35**, 77–83.

Waldrop, M. P., Balser, T. C. and Firestone, M. K. (2000) Linking microbial community composition to function in a tropical soil. *Soil Biology and Biochemistry,* **32**, 1837–1846.

Wardle, D. A. (2006) The influence of biotic interactions on soil biodiversity. *Ecology Letters,* **9**, 870–886.

Wedin, D. A. and Tilman, D. (1990) Species effects on nitrogen cycling: a test with perennial grasses. *Oecologia,* **84**, 433–441.

Whitham, T. G., Bailey, J. K., Schweitzer, J. A. *et al.* (2006) A framework for community and ecosystem genetics: from genes to ecosystems. *Nature Review Genetics,* **7**, 510–523.

Whitham, T. G., DiFazio, S. P., Schweitzer, J. A. *et al.* (2008) Extending genomics to natural communities and ecosystems. *Science,* **320**, 492–495.

Wimp, G. M., Martinsen, G. D., Floate, K. D., Keim, P. S., Whitham, T. G. (2005) Plant genetic determinants of arthropod community structure and diversity. *Evolution,* **59**, 61–69.

Wooley, S. C., Walker, S., Vernon, J. and Lindroth, R. L. (2008) Aspen decline, aspen chemistry, and elk herbivory: are they linked? *Rangelands,* **30**, 17–21.

Zak, D. R, Pregitzer, K. S. and Host, G. E. (1986) Landscape variation in nitrogen mineralization and nitrification. *Canadian Journal of Forest Research,* **16**, 1258–1263.

Zinke, P. J. (1962) The pattern of individual forest trees on soil properties. *Ecology,* **43**, 130–133.

Zoetendal, E. G., Akkermans, A. D. L., Akkermans-van Vliet, W. M., de Visser, J. A. G. M. and de Vos, W. M. (2001) The host genotype affects the bacterial community in the human gastrointestinal tract. *Microbial Ecology in Health and Disease,* **13**, 129–134.

CHAPTER TWENTY

Microbial mutualists and biodiversity in ecosystems

JENNIFER A. RUDGERS

Department of Biology, University of New Mexico

and

KEITH CLAY

Department of Biology, Indiana University

Defining and detecting microbial trait-mediated indirect interactions

Microbial trait-mediated indirect interactions (TMIIs) occur when a microbe (species A) changes a trait of its host species (species B) that consequently affects another species in the community (species C, or the enemy of the host in a protection mutualism). This trait-mediated effect is distinguished from a density-mediated effect, in which the impact of the microbe would spread to other species exclusively through changes in the density of the host organism (e.g., via increased host mortality caused by a pathogen, density-mediated indirect interactions, DMII). Importantly, models reveal that TMIIs have different dynamical consequences than DMIIs (Chapter 1), thus it is important to distinguish between these effects.

Food web diagrams can bring clarity to the characterization of TMIIs. In contrast to typical predator–prey interactions, microbially-mediated protection mutualisms can involve a microbial partner at the same, or very similar, trophic level as the host organism (Fig. 20.1). Furthermore, symbioses involve the exchange of resources, for example photosynthetic carbon exchanged for soil nutrients in the symbiosis between plants and mycorrhizal fungi; this exchange is depicted explicitly in our diagrams, which include trait-mediated effects that are transmitted to a responding resource species (Fig. 20.1e) or to a consumer species (Fig. 20.1f).

In microbial TMIIs, the presence of the symbiont acts as an agent of phenotypic plasticity within the host, whereby hosts that acquire symbionts experience a shift in trait expression. This shift is classified as constitutive if the microbe changes a host trait with no additional flexibility in trait expression. Alternatively, microbial symbiosis may alter trait plasticity. We include in our definition of microbial TMIIs cases where the trait is microbial in origin as

Trait-Mediated Indirect Interactions: Ecological and Evolutionary Perspectives, eds. Takayuki Ohgushi, Oswald J. Schmitz and Robert D. Holt. Published by Cambridge University Press. © Cambridge University Press 2012.

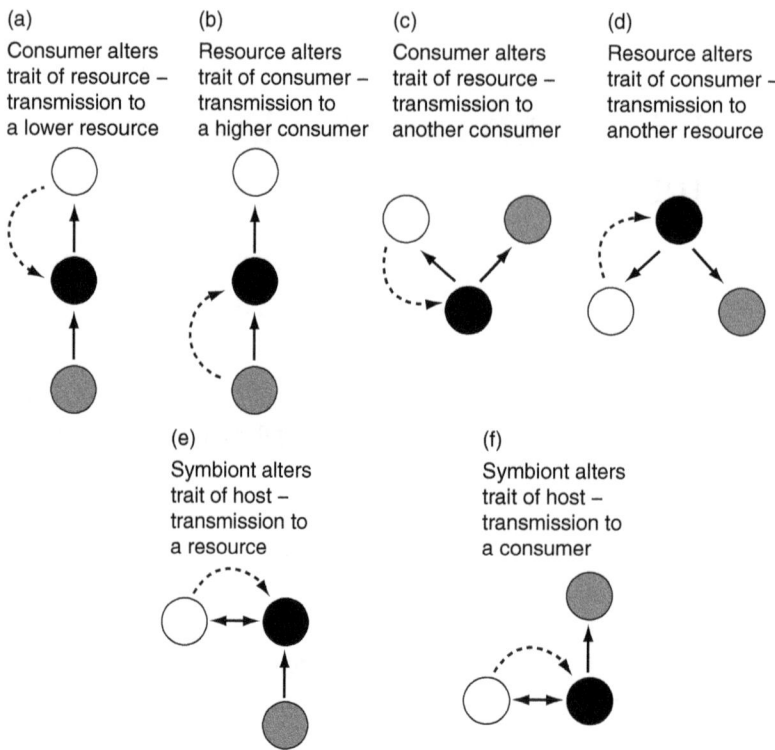

(a)
Consumer alters trait of resource – transmission to a lower resource

(b)
Resource alters trait of consumer – transmission to a higher consumer

(c)
Consumer alters trait of resource – transmission to another consumer

(d)
Resource alters trait of consumer – transmission to another resource

(e)
Symbiont alters trait of host – transmission to a resource

(f)
Symbiont alters trait of host – transmission to a consumer

Figure 20.1 Examples of microbial and non-microbial agents of TMIIs using the notation of Werner and Peacor (2003). Open circles indicate agents of TMIIs, and filled circles indicate the reacting species. Solid arrows indicate energy transfer, through consumption or resource exchange. Dotted arrows indicate trait change and point to the species undergoing trait change (shown as black fill). Note that microbial TMIIs have the microbe and host species exchanging resources (double arrows, e.g., carbon traded for physically protected growing space) often at the same trophic level.

well as cases where the microbe induces a change in a trait of the host because the trait is expressed as part of the host phenotype.

In evaluating examples of microbial TMIIs, it is useful to consider the question: How does one separate trait-mediated effects from density-mediated effects? In some cases, the symbiont will have no direct effect on host density and will only change traits that are expressed through the host. The simplest experiment would thus compare interactions for symbiotic versus symbiont-free host populations. However, because symbiotic host lineages may differ genetically from symbiont-free lineages, it is also important to manipulate symbiont presence in order to decouple microbial effects from host genotype. Ideally, these manipulations would occur in both symbiotic host genotypes (removal of the symbiont) and in symbiont-free host genotypes (addition of the symbiont). To determine that the interaction is indirect, experiments need to compare the performance of hosts in both the presence and the absence of natural enemies in order to detect any direct benefits of the microbial symbiont to the host that are independent of the indirect protection against natural enemies.

It is less clear how the appropriate manipulations can be achieved if the microbe affects both host traits and host density. TMII experiments with predators typically expose prey to predator cues (e.g., volatiles or chemically modified water) or eliminate predation ability (e.g., gluing spider mouthparts) to separate trait-mediated effects of the predator (e.g., predator-avoidance morphologies or behaviours) from density-mediated effects. A similar approach may be applied when specific traits are known to be altered by microbial interactions. For example, in symbioses between fungal endo-phytes and grasses, the fungi produce a well-characterized suite of alkaloids, which have defensive properties. Synthetic alkaloids could be added to plant leaves in quantities mimicking natural levels to manipulate the trait directly (Clay and Cheplick 1989). Another approach, particularly for relatively immo-bile macro-organisms, involves press experiments that repeatedly adjust host densities, such that only host traits differ between treatments. For example, in experiments manipulating endosymbionts in aphids, aphid densities could be adjusted by repeatedly adding new individuals to maintain equivalent den-sities in symbiotic and symbiont-free populations.

Examples of TMIIs: protection mutualisms via microbial symbiosis

Here, we briefly review an array of examples involving microbes that may be unfamiliar to many researchers working on macro-organisms (Table 20.1). We focus on studies that have experimentally manipulated symbiont presence and that describe trait changes associated with symbiosis. Our goal is to consider these examples in light of the TMII framework and to illustrate where additional TMIIs might occur in nature. Several microbial systems that have been explored in great detail with experimental tests of microbial effects clearly represent TMIIs (e.g., endophytic fungi in grasses, bacteria in bryozoans). A common theme is that host traits are often chemical and of microbial origin, reflecting the considerable biochemical capabilities of microorganisms. However, some of the microbial interactions described here have yet to be fully characterized, and in particular, key trait(s) altered by microbial symbiosis have yet to be identified (Table 20.1).

Advances in current understanding of microbial TMIIs are most likely to come from studies that utilize manipulative experiments in natural environ-ments that allow for a diversity of indirect interactions with other species. Vertically transmitted symbionts provide especially tractable opportunities for experimental investigations because contagious spread is often absent or rare. If the symbionts can be eliminated, it is unlikely that they would be

Table 20.1 *Examples and reviews of microbially mediated TMIIs in aquatic and terrestrial ecosystems. Included are cases where a community member has been documented to respond to the presence of a microbial symbiont. In some examples, traits of the microbe have been identified, in other cases traits have been suggested, but not documented, to mediate the interaction*

Host	Microbial symbiont	Trait	Responding species	Reference
Aquatic/marine				
ciliates (*Euplotidium* spp.)	epibiotic bacteria (Verrucomicrobia)	a subset of bacteria make a ribbon-like apparatus extruded in response to predators	predatory ciliate (*Amphileptus marina*)	Görtz et al. (2009)
corals (e.g., *Oculina patagonica, Acropora palmata*)	diverse bacterial community, particularly in surface mucus	*suggested:* antibiotic compounds	pathogenic bacteria (*Vibrio* spp.)	Reshef et al. (2006); Ritchie (2006); Rosenberg et al. (2007)
bryozoan larvae (*Bugula neritina*)	bacteria (*Endobugula sertula*)	cytotoxic compounds (bryostatins)	predatory fish (*Lagodon rhomboides, Monocanthus ciliatus*)	Lindquist and Hay (1996); Davidson et al. (2001); Lopanik et al. (2004)
amphipod (*Gammarus roeseli*)	microsporidian (*Dictyocoela* sp. *roeselum*)	reduced susceptibility to behavioural manipulation by the worm	parasitic worm (*Polymorphus minutus*)	Haine et al. (2005)
marine isopods (*Santia* spp.)	episymbiotic cyanobacteria (*Synechococcus*-type)	red colouration, *suggested:* chemical defence	predatory reef fish (*Chromis, Amblyglyphidodon, Dascyllus, Pomacentrus*)	Lindquist et al. (2005)
brine shrimp (*Artemia* spp.)	bacteria	not identified	pathogenic bacteria (*Vibrio proteolyticus* CW8T2)	Verschuere et al. (2000)
lobster (*Homarus americanus*)	bacteria (Gram negative)	antifungal compound (tyrosol)	pathogenic fungus (*Lagenidium callinectes*)	Gil-Turnes and Fenical (1992)

Host	Symbiont	Mechanism	Enemy	Reference
shrimp (*Palaemon macrodactylus*)	bacteria (*Alteromonas* sp.)	antifungal compound (isatin)	pathogenic fungus (*Lagenidium callinectes*)	Gil-Turnes et al. (1989)
squid (*Euprymna scolopes*)	bacteria in light organ (Vibrionales)	production of light to eliminate shadow	predatory fish and seals (*not well studied*)	Ruby (1996); Stabb and Millikan (2009)

Terrestrial – animal

Host	Symbiont	Mechanism	Enemy	Reference
entomopathogenic nematodes	bacteria	anti-microbial compounds (bacteriocins)	microbes and invertebrates	Reviewed by Koppenhofer and Gaugler (2009)
locust (*Schistocerca gregaria*)	diversity of bacteria in gut	*suggested*: toxic phenols	pathogenic fungus (*Metarhizium anisopliae*) pathogenic bacteria (*Serratia marcescens*)	Dillon and Charnley (1988); Dillon et al. (2005)
pea aphid (*Acythrosiphon pisum*)	'secondary' endosymbiont (*Regiella insecticola*)	not identified	pathogenic fungus (*Pandora neoaphidis*)	Scarborough et al. (2005)
pea aphid (*Acythrosiphon pisum*)	'secondary' endosymbiont (*Hamiltonella defensa*)	toxin-producing genes present in a bacteriophage	parasitoid wasp (*Aphidius ervi*)	Oliver et al. (2003); Moran et al. (2005)
fruit fly (*Drosophila hydei*)	bacteria (*Spiroplasma*)	not identified	parasitoid wasp (*Leptopilina heterotoma*)	Xie et al. (2010)
fruit fly (*Drosophila neotestacea*)	bacteria (*Spiroplasma*)	not identified	parasitic nematode (*Howardula aoronymphium*)	Jaenike et al. (2010)
rove beetles (*Paederus* spp.)	bacteria (*Pseudomonas* relative)	chemical (pederin)	predatory spiders (*Pardosa*, *Pirata*, *Evarcha*)	Kellner and Dettner (1996); Piel (2002)

Table 20.1 (cont.)

Host	Microbial symbiont	Trait	Responding species	Reference
fruit flies (*Drosophila melanogaster* Culex), mosquito (*Culex quinquefasciatus*)	bacteria (*Wolbachia pipientis*)	not identified	West Nile virus	Osborne et al. (2009); Glaser and Meola (2010)
common dog tick (*Dermacentor variabilis*)	endosymbiotic fungus (*Scopulariopsis brevicaulis*)	not identified	pathogenic fungus (*Metarhizum anisopliae*)	Yoder et al. (2008)
leaf-cutter (attine) ants	actinobacteria (*Pseudocardinia*)	antibiotic protection of fungal garden	parasitic fungus (*Escovopsis*)	Currie et al. (1999a); Currie et al. (1999b); Currie et al. (2003)
digger wasps (Crabronidae)	actinobacteria in antennal glands (*Streptomyces* sp.)	antibiotics	pathogenic fungi in brood cells	Kaltenpoth et al. (2005); Kaltenpoth et al. (2010)
hoopoe bird (*Upupa epops*)	bacteria in uropygial gland used for preening (*Enterococcus faecalis*)	bacteriocins	feather decomposing bacteria (*Bacillus licheniformis*)	Ruiz-Rodriguez et al. (2009)
lab mouse (*Mus musculus*)	latent herpes virus	up-regulation of host innate immune response	pathogenic bacteria (*Listeria monocytogenes, Yersinia pestis*)	Barton et al. (2007)
lab mouse (*Mus musculus*)	bacteria (*Bacteroides fragilis*)	not identified	pathogenic bacteria (*Helicobacter hepaticus*)	Mazmanian et al. (2008)

Host	Symbiont	Toxin/compound	Antagonist	References
lab mouse (*Mus musculus*)	bacteria (*Lactobacillus reuteri*)	not identified	pathogenic bacteria (*Helicobacter hepaticus*)	Pena *et al.* (2005)
Terrestrial – plant				
tall fescue grass (*Lolium arundinaceum*)	endophytic fungus (*Neotyphodium coenophialum*)	alkaloids	herbivorous insects, arthropods, plant diversity, plant succession	Clay and Holah (1999); Clay *et al.* (2005); Rudgers and Clay (2007); Rudgers *et al.* (2007); Rudgers and Clay (2008)
several grasses (Poaceae)	endophytic fungi (epichloae)	alkaloids	herbivorous insects, pathogenic fungi	Clay and Schardl (2002); Koh and Hik (2007); Crawford *et al.* (2010)
sedges (*Cyperus virens*, *Cyperus pseudovegetus*, Cyperaceae)	epiphytic fungus (*Balansia cyperi*)	*suggested*: alkaloids	herbivorous insects, pathogenic fungi	Clay *et al.* (1985); Stovall and Clay (1991)
legumes (*Astragalus*, *Oxytropis* spp., Fabaceae)	endophytic fungi (*Embellisia* spp.)	indolizine swainsonine alkaloids	herbivorous mammals	Ralphs *et al.* (2008); Cook *et al.* (2009)
morning glory (*Ipomoea asarifolia*, Convolvulaceae)	endophytic fungi (*Periglandula* sp.)	ergot alkaloids	*suggested*: herbivores	Steiner *et al.* (2006)
shrub (*Baccharis coridifolia*, Asteraceae)	epiphytic fungus (Hypocreales)	mycotoxins (trichothecenes)	*suggested*: pathogenic fungi	Bertoni *et al.* (1997); Rosso *et al.* (2000)

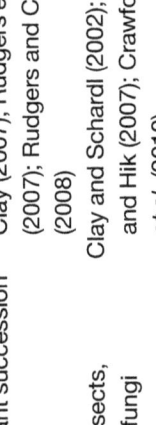

Table 20.1 (cont.)

Host	Microbial symbiont	Trait	Responding species	Reference
clover (*Trifolium repens*, Fabaceae)	rhizobia bacteria	*suggested:* cyanogenic defences	herbivorous insect (*Spodoptera littoralis*)	Kempel et al. (2009)
soybean (*Glycine max*, Fabaceae)	rhizobia bacteria (naturally occurring strains)	not identified	herbivorous insect (*Aphis glycines*)	Dean et al. (2009)
legumes	rhizobia bacteria	several traits	herbivorous insects	Reviewed by Hartley and Gange (2009)
plants	mycorrhizal fungi	several traits	herbivorous insects	Reviewed by Hartley and Gange (2009); Koricheva et al. (2009); Vannette and Hunter (2009)
plants	mycorrhizal fungi	several traits	pathogenic fungi, bacteria	Reviewed by Garrido (2009)

re-acquired horizontally. However, because most macro-organisms support diverse microbial communities, understanding how multiple symbionts interact within hosts will be important to deciphering their community consequences. For example, when pea aphids were co-infected by the secondary bacterial symbionts *Hamiltonella defensa* and *Serratia symbiotica*, the aphids showed enhanced resistance to parasitoid wasps relative to infection by *H. defensa* alone; however, titres of *S. symbiotica* increased 20-fold in co-infections, which may be detrimental to host aphids (Oliver *et al.* 2006). A central issue for microbial TMIIs is connecting microbial community composition to function for hosts with diverse microbial symbiota. In systems lacking a single, dominant microbial symbiont, such as in sponges or mammalian guts, metagenomic approaches may prove useful for elucidating both microbial diversity and the diversity of genes controlling key traits, such as secondary chemistry. Terrestrial systems are better characterized at this point, but ultimately aquatic systems, especially marine organisms, may prove to be a richer source of microbial TMIIs given the prevalence of symbiotic interactions in the sea.

Aquatic ecosystems

Symbiotic interactions are common in aquatic environments where they provide hosts with an array of services including nutrition, dispersal and protection. Marine systems are particularly diverse in symbioses involving invertebrates, emphasizing that symbiosis is evolutionary ancient and ecologically significant. Examples of TMIIs from aquatic systems illustrate symbiont-mediated protection via multiple pathways including chemical defence, morphological changes and behavioural alterations.

Symbiont-mediated protection has been documented for both macroscopic and microscopic hosts. For example, experiments have shown that episymbiotic bacteria of both shrimp and lobster embryos produced antifungal compounds that significantly reduced attack by fungal pathogens (Gil-Turnes *et al.* 1989; Gil-Turnes and Fenical 1992). Similarly, unicellular *Euplotidium* ciliates host epibiotic bacteria that differentiate into a ribbon-like apparatus that is extruded in response to predators, disrupting prey recognition (Görtz *et al.* 2009). In perhaps the best experimentally documented example, larvae of the bryozoan *Bugula neritina* often host the bacterium *Endobugula sertula*, which produces the macrocyclic polyketides, bryostatins (Davidson *et al.* 2001). Interestingly, extracts from larval *Bugula*, but not from adults, were unpalatable to fish, and bryostatin concentrations in adults (which are less vulnerable to predators than larvae) were very low (Lindquist and Hay 1996; Lopanik *et al.* 2004). The role of symbiont-produced bryostatins as a TMII was demonstrated by experimental elimination of *Endobugula* with antibiotics where the loss of bryostatins in treated larvae increased their palatability to fish (Lopanik *et al.* 2004).

In addition to the few experimental studies (Table 20.1), anti-fouling and anti-biotic compounds have been documented in several corals, tunicates, seaweeds and sponges (Piel 2004; Ritchie 2006). Corals, in particular, have been proposed to gain microbially mediated protection against pathogens from microbes present in the surface mucus layer (the coral probiotic hypothesis: Reshef *et al.* 2006; Rosenberg *et al.* 2007). A number of other marine organisms have a high probability of microbial TMIIs, ranging from toxin-producing dinoflagellates (Görtz *et al.* 2009) to squid with light-generating bacteria that may camouflage their shadows (Stabb and Millikan 2009). For example, marine sponges a host a wide variety of archaea, bacteria, fungi and algae, and bacteria alone can constitute as much as 40–60% of sponge biomass (Schmitt *et al.* 2007). In aquatic organisms with complex morphologies, gut-associated bacteria likely play protective roles and may have applications for managing disease resistance. For example, probiotic bacteria are commonly added to cultures to protect shrimp, oysters, fish and scallops (reviewed by Tinh *et al.* 2008). However, for the majority of cases, whether compounds are produced by the host, by a symbiont or by the host in response to a symbiont has yet to be determined. If symbionts commonly alter the defensive traits of foundational species, such as corals, there could be large impacts on a diverse community of associated macrobiota.

Terrestrial ecosystems

Terrestrial systems differ from aquatic habitats not only in the particular species involved but also in more general characteristics such as the localization and movement of nutrients in the soil versus the water column, the greater dominance of insects and angiosperms on land versus clonal invertebrates and algae in the water and the higher prevalence of fungi in terrestrial than aquatic habitats. Despite these fundamental differences, a diversity of TMIIs can be found in both habitat types that illustrate similar underlying microbial mechanisms.

Animals: protective symbioses in invertebrates

Invertebrates present many examples of microbial TMIIs – many of them involving chemical traits (Table 20.1). Invertebrate examples illustrate that not only the presence, but also the diversity, of microbes can play a protective function for hosts. For example, an experiment manipulating the diversity of gut bacteria in *Schistocerca* locusts significantly reduced the density of a pathogenic bacterium (Dillon *et al.* 2005). Some of these protective effects may result from competitive interactions between microbes colonizing the shared host, rather than from TMIIs. Thus, further work will be needed to determine whether these interactions are in fact trait-mediated.

The two best-documented examples in terrestrial invertebrates are arguably leaf-cutter ants with protected fungal gardens and aphids with endosymbiotic bacteria that defend against natural enemies. Currie and colleagues (1999a, 1999b, 2003) have reported that leaf-cutter ants harbour antibiotic-producing actinobacteria (*Pseudocardinia*) in specialized structures, and that bacterial compounds are inhibitory to the parasites of their fungal gardens. While experimental manipulations of leaf-cutter ant nests are difficult to conduct in nature, laboratory bioassays clearly indicate this is a TMII in which antibiotics produced by symbiotic bacteria protect fungal gardens from their pathogens (Currie *et al.* 2003). Distinct from extracellular antibiotic-producing bacteria, bacterial symbionts occur internally in aphids and include 'primary' endosymbionts, which are obligate and nutritional, as well as 'secondary' endosymbionts, which are facultative and sometimes protective. Secondary symbionts of the pea aphid, *Acythrosiphon pisum*, have received the most study (reviewed by Oliver and Moran 2009). For example, *Regiella insecticola* improved aphid survival when challenged with an entomopathogenic fungus (Scarborough *et al.* 2005), *Hamiltonella defensa* protected aphids against the parasitoid wasp *Aphidius ervi* (Oliver *et al.* 2003), and a *Rickettsiella* changed aphid body colour from red to green – which may provide protection as well (Tsuchida *et al.* 2010). Interestingly, the protection conferred by *H. defensa* appears to arise from toxin-producing genes present in a bacteriophage, adding a fourth player to the interaction (Moran *et al.* 2005). Complete genome sequencing has revealed that pea aphids lack many of the genes for microbial recognition and immune response that are present in other insect groups, suggesting that symbionts may have co-opted these defensive traits in aphids (Gerardo *et al.* 2010). Clearly, bacterial symbionts can alter aphid traits in ways that affect aphids' natural enemies – whether there are consequences for host plants and the broader food web remains to be investigated.

Putative TMIIs involving symbiont-mediated protection of arthropod vectors against pathogens of animals and humans are being reported with increasing frequency and may have significant biocontrol value. For example, a transcriptional profile of malaria-transmitting *Anopheles gambiae* mosquitoes suggested that the gut microbiota caused an up-regulation of mosquito immune responses, including anti-*Plasmodium* factors, resulting in a decreased capacity to sustain *Plasmodium* infection (Dong *et al.* 2009). Thus, the microbiota of vectors could potentially reduce disease levels in human and wildlife populations by altering the vector traits and immunities.

Animals: protective symbioses in vertebrates
As in invertebrates, gut bacteria likely play protective roles against disease in many vertebrate taxa. Laboratory studies on rodents have been numerous. For example, latent herpes virus defended laboratory mice against the bacteria *Listeria monocytogenes* (a food-borne pathogen causing listeriosis) and *Yersinia pestis* (agent of plague) by up-regulating the host's innate immune responses

(Barton *et al.* 2007), highlighting that this particular effect was due to a change in the host traits. Similar effects on innate immune responses have been documented for probiotic bacteria, such as *Bifidobacterium* (e.g., Sonnenburg *et al.* 2006). Experimental studies on model vertebrate systems combined with observational data from the Human Microbiome Project (Turnbaugh *et al.* 2007) have high potential to inform our understanding of microbial TMIIs in humans.

Protective symbioses in vertebrates are not limited to gut microbiota. For example, in birds, bacteria present in secretions used during preening are known to protect against feather degradation, thereby suppressing decomposers in laboratory experiments (Rodriguez *et al.* 2009). While it seems unlikely that the cascading effects to the broader community are strong in avian systems (feathers comprising a small fraction of animal detritus), this case raises the question, how many other symbioses may alter decomposer communities only after the host organism has died (or excreted or shed)?

Plants: aboveground symbioses

Over the past several decades, toxin production by a number of well-known poisonous plants has been shown to result from microbial infection. Clay (1988) suggested that many of these toxic associations represent a defensive mutualism in which microbes produce physiologically active compounds that protect host plants against herbivory. The best documented of these TMIIs involve endophytic fungi in the family Clavicipitaceae, which grow systemically in aboveground plant tissues, are vertically transmitted through seeds, and form symbioses with grasses, sedges and morning glories (Clay and Schardl 2002; Steiner *et al.* 2006). Protection of plants from herbivores mediated by toxin-producing fungal symbionts is not limited to fungal endophytes in the Clavicipitaceae, and includes Asteraceae and Fabaceae as well as lichens and conifers (Table 20.1). Fungal and bacterial endophyte symbioses appear to be ubiquitous in plants, and many of these endophytes may also play a defensive role (Arnold *et al.* 2003; Rodriguez *et al.* 2009).

The TMII mediated by fungal endophyte symbiosis is clearly documented by experimental elimination of the fungus, resulting in the loss of both alkaloids and herbivore resistance. For example, endophyte-infected tall fescue grass (*Lolium arundinaceum*), which has been especially well studied given its economic importance, exhibited increased resistance to herbivores compared to uninfected fescue (Rudgers and Clay 2005) due to the production of alkaloids by the endophyte (Bush *et al.* 1997). Similar increases in herbivore resistance have been reported in native grasses (Table 20.1). In the case of tall fescue, endophyte symbiosis also has cascading consequences – reducing plant community diversity (Clay and Holah 1999; Rudgers *et al.* 2010b), suppressing plant succession (Rudgers and Clay 2007) and altering food web interactions (Rudgers and Clay 2008) and ecosystem processes, including invasion resistance and decomposition (Lemons *et al.* 2005; Rudgers *et al.* 2005).

Plants: belowground symbioses
In addition to hosting aboveground endophytes, plants associate with a number of bacterial and fungal symbionts present in the soil; these organisms also have potential to interact with plant enemies in TMIIs (Table 20.1). Symbioses with mycorrhizal fungi are present in > 80% of flowering plant species (Brundrett 2002), and while mycorrhizal fungi are generally considered nutritional mutualists, they have long been known to protect host plants against pathogens and have more recently been implicated in herbivore resistance (Table 20.1). The degree to which these effects are trait mediated remains unresolved, as another mechanism of protection may simply be competition for colonization sites on the roots. However, some plant traits have been documented to be affected by mycorrhizal symbioses, including defensive chemistry, tolerance to herbivory and plant signalling pathways. Thus far, studies suggest stronger negative impacts of mycorrhizal associations on generalist, chewing insect herbivores relative to specialists and sucking insects (Koricheva *et al.* 2009). In addition, a few reports highlight effects on the third trophic level, including increased attraction of parasitoids of herbivores (Guerrieri *et al.* 2004) as well as reduced production of extrafloral nectaries (Laird and Addicott 2007), demonstrating the potential of these TMIIs to influence additional community members. While far fewer studies have investigated TMII for rhizobia than for mycorrhizal fungi (Table 20.1), in both symbioses it appears that effects on herbivores, pathogens and even the enemies of herbivores can vary considerably. Understanding variation caused by plant identity (e.g., Sikes *et al.* 2009), symbiont identity (e.g., Maherali and Klironomos 2007) and the type and degree of specialization of the plant's natural enemy (Koricheva *et al.* 2009) may improve the ability to predict the strength and direction of these TMIIs.

Other terrestrial examples
While we have reviewed the widespread distribution of TMIIs involving plants associated with microorganisms (bacteria and fungi), recent work suggests that these microbes themselves may form symbiotic associations for the production of secondary metabolites for protection or pathogenesis. For example, Partida-Martinez *et al.* (2007) demonstrated that 'mycotoxin' production by the fungus *Rhizopus microsporus* was actually the result of infection by bacteria in the genus *Burkholderia*. In general, mycotoxins are thought to be secondary compounds produced for defence or for competition with other microbes or animals. A wide range of fungi are associated with bacterial

endosymbionts (Bertaux *et al.* 2005), raising the question of which partner is producing the defensive chemistry.

Hypotheses for community and ecosystem impacts
Pairwise species interactions

The simplest microbial TMIIs consist of changes to pairwise interactions with the host organism, and these constitute the majority of examples that we reviewed above. While we have focused on microbes altering host–consumer interactions (Fig. 20.1f), TMIIs can also occur in other types of pairwise species interactions. First, the presence of one microbial symbiont could alter host traits in ways that influence host interactions with a mutualist. For example, studies on grasses have shown that plants with aboveground fungal endophytes have reduced colonization of roots by mycorrhizal fungi, relative to endophyte-free plants (Omacini *et al.* 2006). Such antagonisms between mutualistic symbionts could be trait mediated, for instance, if the endophyte altered host root exudates in ways that disrupted mycorrhizas. Conversely, pairwise mutualisms could be enhanced by microbial TMIIs. For example, mycorrhizal fungi could increase plant attractiveness to pollinators by altering plant traits, such as floral size or nectar production (Cahill *et al.* 2008). Second, the presence of a microbial symbiont could alter host competitive interactions. Indeed, the roles of soil mutualists, such as arbuscular mycorrhizal fungi, are well investigated in this context (e.g., Hoeksema *et al.* 2010). Similarly, experiments have demonstrated that the outcome of competition between two *Steinernema* nematode species depended on the presence and identity of their bacterial symbionts (Sicard *et al.* 2006). Third, some microbial symbionts can improve the host's ability to capture prey, thereby altering host–resource interactions (Fig. 20.1e). For example, the deep-sea anglerfishes host bioluminescent bacteria in lures to attract prey (Haygood and Distel 1993). Altogether, these examples serve to highlight a number of pathways through which microbes may generate TMIIs by altering pairwise species interactions.

Community structure and assembly

TMIIs are likely to have consequences that cascade to other members of the community, particularly when one (or more) of the organisms involved is a keystone species, a foundation species (e.g., corals) or a dominant member of the community (e.g., grasses in a grassland). In such cases, microbial TMIIs may alter community diversity. For example, our work with tall fescue grass and its *Neotyphodium* endophyte has shown reduced plant and arthropod diversity in the presence of the symbiosis (Clay and Holah 1999; Rudgers and Clay 2008). In order to classify a community response as a TMII, it is important to show (1) that the interaction is indirect and therefore does not occur (or is substantially weaker) in the absence of the intermediary species (here, host natural enemies) and (2) that the interaction is trait mediated, such that direct manipulation of the trait alone would produce a similar community response. For example, in the tall fescue system, we have documented that the benefit of *Neotyphodium* to the host grass is indirect: experimental exclusion(s) of natural enemies (insects and mammals) resulted in weaker increases in symbiotic tall fescue relative to plots where herbivores were present (Clay *et al.* 2005). Furthermore, although experiments to manipulate directly the traits (alkaloids) would be technically difficult on a large, community-wide scale, we have shown that a fungal genotype lacking the full complement of alkaloid toxins had weaker effects on the community than the wild-type fungal genotype (Rudgers *et al.* 2010b) and in laboratory trials, adding alkaloids to endophyte-free leaves reduced consumption by herbivores (Clay and Cheplick 1989). Depending on the system, the effects of a microbially altered host trait could attenuate or accumulate (e.g., biomagnification) at higher trophic levels, and more work will be needed to elucidate when attenuation versus accumulation may be most likely to occur. We hypothesize that effects could accumulate when particular species are resistant to microbial toxins and sequester them for their own defence, similar to insect herbivores that sequester plant secondary compounds. In contrast, attenuating effects may be more common for non-resistant, generalist natural enemies that lack sequestration abilities. Furthermore, TMIIs that cascade to a large number of community members have the potential to alter the temporal and spatial patterns of community assembly. For example, the tall fescue endophyte, by increasing dominance of the host grass, slowed community succession from grassland to forest (Rudgers *et al.* 2007). For hosts that commonly persist as mosaics of symbiotic and symbiont-free individuals or populations, the presence/absence of the symbiont could produce alternative stable states in the system, one associated with the presence of the symbiont and one with the absence of the symbiont. For example, Koh and Hik (2007) demonstrated that grassland patches close to predator refuges had high mammalian herbivory and high levels of endophyte symbiosis, whereas patches subject to low herbivory had lower levels of endophyte symbiosis. Alternatively, symbiont-free populations may be transitory and subject to rapid extinction.

Ecosystem-level consequences

Microbial TMIIs could have ecosystem-level consequences, particularly in cases where the host organism is a foundation or a dominant species in the ecosystem. These effects may include direct impacts on ecosystems due to microbially mediated changes in host traits, such as the presence of fungal alkaloids altering the rate of litter decomposition (Lemons *et al.* 2005). In addition, indirect effects could occur due to microbially-mediated shifts in community structure. For example, in tall fescue grass, the strong reductions in plant diversity in plots with the endophyte (Clay and Holah 1999) could subsequently alter pools of C, N and other nutrients in the soil. Leaf-cutter ants, whose fungal farms can be protected from pathogens by symbiotic, antibiotic-producing bacteria (Currie *et al.* 1999b), may have significant effects on forest ecosystems. For example, Folgarait (1998) showed that leaf-cutter ants can reduce 17% of the annual leaf production in a tropical forest. These ecosystem impacts could be significantly reduced if more ant nests were lost to pathogens. Nutritional mutualists, such as plant-associated rhizobia and mycorrhizal fungi, have obvious ecosystem consequences by making N or P available to host plants, which in turn may form the basis of the TMII. Herbivores, such as aphids, could also have ecosystem-level impacts by differentially feeding on plant species. Those impacts would be stronger when aphids are protected from predation and parasitism by their bacterial endosymbionts (Oliver and Moran 2009). In marine environments, foundation species such as corals may affect ecosystem properties, such as productivity and nutrient dynamics.

Predicting the direction and strength of microbial TMII

In general, we expect stronger effect sizes of microbial TMIIs that cause greater magnitudes of change in host traits. Importantly, the rate of change in traits may be much faster than the rate of change in the densities of host organisms, thus resulting in stronger immediate impacts on communities than density effects alone (see also Werner and Peacor 2003). We suggest that several additional factors may also influence the strength and direction of microbial TMIIs.

First, the *ecological dominance of the symbiosis* should increase the strength of its impacts on the community and ecosystem. For example, if traits are modified in a keystone species, foundation species or ecosystem engineer (Gribben *et al.* 2009), these changes are more likely to have strong impacts at the community and ecosystem level than changes in species with a smaller ecosystem footprint. We are not aware of any published studies comparing the impacts of microbial TMIIs for species that vary in their ecological dominance.

Second, we predict that the *specificity of the host–microbe association* will influence the direction of its impact on the surrounding community and ecosystem. If the microbial symbiont only alters traits of a single host, this should promote host dominance and thereby reduce the biodiversity of the system. While we are unaware of any direct tests of this idea, explorations of pairwise species interaction networks show very different structural patterns for networks of specialized and symbiotic mutualists compared to generalist, nonsymbiotic mutualists (Guimaraes *et al.* 2007). Furthermore, an example of the influence of specificity comes from symbioses between grasses and fungal endophytes, which are highly host specific. In tall fescue grass, the endophyte promotes host dominance to the exclusion of other species in the community, with effects not only on competing plants (at the same trophic level as the host, Clay and Holah 1999; Rudgers *et al.* 2007) but also on consumers and on their consumers (Rudgers and Clay 2008). Under this scenario, the host becomes more of a keystone species in its community and ecosystem impacts relative to other species in the community as a result of the microbial symbiont. It is also possible that tightly coupled host–microbe associations could select for an arms race where all species need a symbiont (or a better symbiont) in order to compete within the community. Such an arms race has been suggested for leaf-cutter ants where fungal gardens are attacked by pathogens and protected by symbiotic bacterial antibiotics (Poulsen *et al.* 2010). In contrast, if microbial symbionts simultaneously alter the traits of several co-occurring hosts, for example during the formation of a mycorrhizal network connecting plant roots belowground, this interaction may level the playing field, reducing the competitive dominance of any one host species, and thereby, increasing biodiversity (van der Heijden *et al.* 1998).

Third, the *evolutionary novelty of the symbiosis with respect to the surrounding community* is additionally likely to influence the strength, and possibly the direction of community and ecosystem impacts. In a community of highly coevolved species, there will have been time for co-occurring species to adapt to any dramatic, symbiont-mediated changes in host traits, whereas the introduction of a novel, non-native host–symbiont association to a community may have stronger immediate impacts on resident, native species (e.g., a novel weapon, Callaway and Ridenour 2004). Evolutionary novelty has been proposed as an explanation for the strong community impacts of the tall fescue–*Neotyphodium* symbiosis (Saikkonen *et al.* 2006), although the data required to test this hypothesis, including responses in replicated non-native and native symbiotic systems, are currently lacking.

Fourth, as has been suggested for non-microbial TMIIs (Werner and Peacor 2003), the *duration and timing of change in the trait* likely play a role in the strength of impacts. Host traits that microbes alter constitutively are expected to have longer lasting and stronger impacts than more ephemeral or plastic changes. In assessing the host response, we suggest that adopting a population dynamics approach, such as projection matrix models, is likely to provide the most comprehensive assessment of the consequences of the TMII for the host species because changes in host traits are likely to affect multiple aspects of host demographic rates (Damiani 2005; Rudgers *et al.* 2010a).

Finally, taking a *community genetics perspective* (Whitham *et al.* 2003), it is likely that even the genotype of the species involved in symbiosis could alter community composition. We have found that two fungal endophyte genotypes differing in alkaloid chemistries had differential impacts on the plant community (Rudgers *et al.* 2010b). Similarly, recent work in a barley–aphid–parasitoid system showed that > 10% of the variation in the size of parasitic wasps could be explained by complex interactions among aphid genotypes, plant genotypes and the presence of rhizobacteria (Zytynska *et al.* 2010).

Conclusions

In summary, new studies are required better to characterize the diversity and function of microbial TMIIs in a wider array of systems. The research reviewed here emphasizes the need to consider whether ecologically important plant and animal traits may actually be of direct microbial origin or are indirectly affected by microbial symbiosis. Molecular tools such as pyrosequencing and metagenomics provide new opportunities for characterizing cryptic host-associated microbial communities and their functional capabilities. Combined with experimental manipulations of microbe presence/absence or genotypic variation, the ecological impacts of microbial TMIIs can be better elucidated. While several studies have successfully characterized the ecological impacts of single microbial symbionts, a major challenge for future research will be to understand the functional significance of the diverse microbial communities within hosts. In addition, the wholesale movement of plants, animals and microbes around the world, the potential constitution of new host–microbial associations, and the widespread use of antibiotics in agriculture and medicine all suggest that microbial TMIIs will be dynamic and have unexpected consequences. While certain systems have been well characterized (e.g., endophytes of grasses, bacterial endosymbionts of aphids, leaf-cutter ants and their associated communities), a wider taxonomic and habitat diversity needs to be explored. Finally, relatively few systems have been subject to experimental manipulation and measurement of consequences, especially in real-world environments. These gaps in our knowledge provide many opportunities for enhancing our understanding of TMIIs and the complex interactions between microbial symbionts and their hosts.

References

Arnold, A. E., Mejia, L. C., Kyllo, D. *et al.* (2003) Fungal endophytes limit pathogen damage in a tropical tree. *Proceedings of the National Academy of Sciences of the United States of America*, **100**, 15649–15654.

Barton, E. S., White, D. W., Cathelyn, J. S. *et al.* (2007) Herpes virus latency confers symbiotic protection from bacterial infection. *Nature*, **447**, 326–327.

Bertaux, J., Schmid, M., Hutzler, P. *et al.* (2005) Occurrence and distribution of endobacteria in the plant-associated mycelium of the ectomycorrhizal fungus *Laccaria bicolor* S238N. *Environmental Microbiology*, **7**, 1786–1795.

Bertoni, M. D., Romero, N., Reddy, P. V. and White, J. F. (1997) A hypocrealean epibiont on meristems of *Baccharis coridifolia*. *Mycologia*, **89**, 375–382.

Brundrett, M. C. (2002) Coevolution of roots and mycorrhizas of land plants. *New Phytologist*, **154**, 275–304.

Bush, L. P., Wilkinson, H. H. and Schardl, C. L. (1997) Bioprotective alkaloids of grass–fungal endophyte symbioses. *Plant Physiology*, **114**, 1–7.

Cahill, J. F., Elle, E., Smith, G. R. and Shore, B. H. (2008) Disruption of a belowground mutualism alters interactions between plants and their floral visitors. *Ecology*, **89**, 1791–1801.

Callaway, R. M. and Ridenour, W. M. (2004) Novel weapons: invasive success and the evolution of increased competitive ability. *Frontiers in Ecology and the Environment*, **2**, 436–443.

Clay, K. (1988) Fungal endophytes of grasses: a defensive mutualism between plants and fungi. *Ecology*, **69**, 10–16.

Clay, K. and Cheplick, G. P. (1989) Effect of ergot alkaloids from fungal endophyte-infected grasses on fall armyworm (*Spodoptera frugiperda*). *Journal of Chemical Ecology*, **15**, 169–182.

Clay, K. and Holah, J. (1999) Fungal endophyte symbiosis and plant diversity in successional fields. *Science*, **285**, 1742–1744.

Clay, K. and Schardl, C. (2002) Evolutionary origins and ecological consequences of endophyte symbiosis with grasses. *American Naturalist*, **160**, S99–S127.

Clay, K., Hardy, T. N. and Hammond, A. M., Jr. (1985) Fungal endophytes of *Cyperus* and their effect on an insect herbivore. *American Journal of Botany*, **72**, 1284–1289.

Clay, K., Holah, J. and Rudgers, J. A. (2005) Herbivores cause a rapid increase in hereditary symbiosis and alter plant community composition. *Proceedings of the National Academy of Sciences of the United States of America*, **102**, 12465–12470.

Cook, D., Gardner, D. R., Ralphs, M. H. *et al.* (2009) Swainsoninine concentrations and endophyte amounts of *Undifilum oxytropis* in different plant parts of *Oxytropis sericea*. *Journal of Chemical Ecology*, **35**, 1272–1278.

Crawford, K. M., Land, J. M. and Rudgers, J. A. (2010) Fungal endophytes of native

grasses decrease insect herbivore preference and performance. *Oecologia*, **164**, 431–444.

Currie, C. R., Bot, A. N. M. and Boomsma, J. J. (2003) Experimental evidence of a tripartite mutualism: bacteria protect ant fungus gardens from specialized parasites. *Oikos*, **101**, 91–102.

Currie, C. R., Mueller, U. G. and Malloch, D. (1999a) The agricultural pathology of ant fungus gardens. *Proceedings of the National Academy of Sciences of the United States of America*, **96**, 7998–8002.

Currie, C. R., Scott, J. A., Summerbell, R. C. and Malloch, D. (1999b) Fungus-growing ants use antibiotic-producing bacteria to control garden parasites. *Nature*, **398**, 701–704.

Damiani, C. C. (2005) Integrating direct effects and trait-mediated indirect effects using a projection matrix model. *Ecology*, **86**, 2068–2074.

Davidson, S. K., Allen, S. W., Lim, G. E., Anderson, C. M. and Haygood, M. G. (2001) Evidence for the biosynthesis of bryostatins by the bacterial symbiont 'Candidatus Endobugula sertula' of the bryozoan *Bugula neritina*. *Applied and Environmental Microbiology*, **67**, 4531–4537.

Dean, J. M., Mescher, M. C. and De Moraes, C. M. (2009) Plant–rhizobia mutualism influences aphid abundance on soybean. *Plant and Soil*, **323**, 187–196.

Dillon, R. J. and Charnley, A. K. (1988) Inhibition of *Metarhizium anisopliae* by the gut bacterial flora of the desert locust: characterization of antifungal toxins. *Canadian Journal of Microbiology*, **34**, 1075–1082.

Dillon, R. J., Vennard, C. T., Buckling, A. and Charnley, A. K. (2005) Diversity of locust gut bacteria protects against pathogen invasion. *Ecology Letters*, **8**, 1291–1298.

Dong, Y. M., Manfredini, F. and Dimopoulos, G. (2009) Implication of the mosquito midgut microbiota in the defense against malaria parasites. *PLoS Pathogens*, **5**.

Folgarait, P. J. (1998) Ant biodiversity and its relationship to ecosystem functioning: a review. *Biodiversity and Conservation*, **7**, 1221–1244.

Garrido, J. M. G. (2009) Arbuscular mycorrhizae as defense against pathogens. In J. F. White, Jr. and M. S. Torres, eds., *Defensive Mutualism in Microbial Symbiosis*. Boca Raton, FL: CRC Press, pp. 183–198.

Gerardo, N. M., Altincicek, B., Anselme, C. *et al.* (2010) Immunity and other defenses in pea aphids, *Acyrthosiphon pisum. Genome Biology*, **11**, R21.

Gil-Turnes, M. S. and Fenical, W. (1992) Embryos of *Homarus americanus* are protected by epibiotic bacteria. *Biological Bulletin*, **182**, 105–108.

Gil-Turnes, M. S., Hay, M. E. and Fenical, W. (1989) Symbiotic marine bacteria chemically defend crustacean embryos from a pathogenic fungus. *Science*, **246**, 116–118.

Glaser, R. L. and Meola, M. A. (2010) The native *Wolbachia* endosymbionts of *Drosophila melanogaster* and *Culex quinquefasciatus* increase host resistance to West Nile virus infection. *PLoS ONE*, **5**, e11977.

Görtz, H., Rosati, G., Schweikert, M., Schrallhammer, M., Omura, G. and Suzaki, T. (2009) Microbial symbionts for defense and competition among ciliate hosts. In J. F. White, Jr. and M. S. Torres, eds., *Defensive Mutualism in Microbial Symbiosis*. Boca Raton, FL: CRC Press, pp. 45–64.

Gribben, P. E., Byers, J. E., Clements, M., McKenzie, L. A., Steinberg, P. D. and Wright, J. T. (2009) Behavioural interactions between ecosystem engineers control community species richness. *Ecology Letters*, **12**, 1127–1136.

Guerrieri, E., Lingua, G., Digilio, M. C., Massa, N. and Berta, G. (2004) Do interactions between plant roots and the rhizosphere affect parasitoid behaviour? *Ecological Entomology*, **29**, 753–756.

Guimaraes, P. R., Rico-Gray, V., Oliveira, P. S. *et al.* (2007) Interaction intimacy affects structure and coevolutionary dynamics in mutualistic networks. *Current Biology*, **17**, 1797–1803.

Haine, E. R., Boucansaud, K. and Rigaud, T. (2005) Conflict between parasites with different transmission strategies infecting an amphipod host. *Proceedings of the Royal Society of London, Series B*, **272**, 2505–2510.

Hartley, S. E. and Gange, A. C. (2009) Impacts of plant symbiotic fungi on insect herbivores: mutualism in a multitrophic context. *Annual Review of Entomology*, **54**, 323–342.

Haygood, M. G. and Distel, D. L. (1993) Bioluminescent symbionts of flashlight fishes and deep sea anglerfishes form unique lineages related to the genus *Vibrio. Nature*, **363**, 154–156.

Hoeksema, J. D., Chaudhary, V. B., Gehring, C. A. *et al.* (2010) A meta-analysis of context-dependency in plant response to inoculation with mycorrhizal fungi. *Ecology Letters*, **13**, 394–407.

Jaenike, J., Unckless, R., Cockburn, S. N., Boelio, L. M. and Perlman, S. J. (2010) Adaptation via symbiosis: recent spread of a *Drosophila* defensive symbiont. *Science*, **329**, 212–215.

Kaltenpoth, M., Gottler, W., Herzner, G. and Strohm, E. (2005) Symbiotic bacteria protect wasp larvae from fungal infestation. *Current Biology*, **15**, 475–479.

Kaltenpoth, M., Schmitt, T., Polidori, C., Koedam, D. and Strohm, E. (2010) Symbiotic streptomycetes in antennal glands of the South American digger wasp genus *Trachypus* (Hymenoptera, Crabronidae). *Physiological Entomology*, **35**, 196–200.

Kellner, R. L. L. and Dettner, K. (1996) Differential efficacy of toxic pederin in deterring potential arthropod predators of *Paederus* (Coleoptera: Staphylinidae) offspring. *Oecologia*, **107**, 293–300.

Kempel, A., Brandl, R. and Schadler, M. (2009) Symbiotic soil microorganisms as players in aboveground plant–herbivore interactions: the role of rhizobia. *Oikos*, **118**, 634–640.

Koh, S. and Hik, D. S. (2007) Herbivory mediates grass–endophyte relationships. *Ecology*, **88**, 2752–2757.

Koppenhofer, H. S. and Gaugler, R. (2009) Entomopathogenic nematode and bacterial mutualism. In J. F. White, Jr. and M. S. Torres, eds., *Defensive Mutualism in Microbial Symbiosis*. Boca Raton, FL: CRC Press, pp. 99–116.

Koricheva, J., Gange, A. C. and Jones, T. (2009) Effects of mycorrhizal fungi on insect herbivores: a meta-analysis. *Ecology*, **90**, 2088–2097.

Laird, R. A. and Addicott, J. F. (2007) Arbuscular mycorrhizal fungi reduce the construction of extrafloral nectaries in *Vicia faba*. *Oecologia*, **152**, 541–551.

Lemons, A., Clay, K. and Rudgers, J. A. (2005) Connecting plant-microbial interactions above and belowground: a fungal endophyte affects decomposition. *Oecologia*, **145**, 595–604.

Lindquist, N. and Hay, M. E. (1996) Palatability and chemical defense of marine invertebrate larvae. *Ecological Monographs*, **66**, 431–450.

Lindquist, N., Barber, P. H. and Weisz, J. B. (2005) Episymbiotic microbes as food and defence for marine isopods: unique symbioses in a hostile environment. *Proceedings of the Royal Society of London, Series B*, **272**, 1209–1216.

Lopanik, N., Lindquist, N. and Targett, N. (2004) Potent cytotoxins produced by a microbial symbiont protect host larvae from predation. *Oecologia*, **139**, 131–139.

Maherali, H. and Klironomos, J. N. (2007) Influence of phylogeny on fungal community assembly and ecosystem functioning. *Science*, **316**, 1746–1748.

Mazmanian, S. K., Round, J. L. and Kasper, D. L. (2008) A microbial symbiosis factor prevents intestinal inflammatory disease. *Nature*, **453**, 620–625.

Moran, N. A., Degnan, P. H., Santos, S. R., Dunbar, H. E. and Ochman, H. (2005) The players in a mutualistic symbiosis: insects, bacteria, viruses, and virulence genes. *Proceedings of the National Academy of Sciences of the United States of America*, **102**, 16919–16926.

Oliver, K. M. and Moran, N. A. (2009) Defensive symbionts in aphids and other insects. In J. F. White, Jr. and M. S. Torres, eds., *Defensive Mutualism in Microbial Symbiosis*. Boca Raton, FL: CRC Press, pp. 129–148.

Oliver, K. M., Moran, N. A. and Hunter, M. S. (2006) Costs and benefits of a superinfection of facultative symbionts in aphids. *Proceedings of the Royal Society of London, Series B*, **273**, 1273–1280.

Oliver, K. M., Russell, J. A., Moran, N. A. and Hunter, M. S. (2003) Facultative bacterial symbionts in aphids confer resistance to parasitic wasps. *Proceedings of the National Academy of Sciences of the United States of America*, **100**, 1803–1807.

Omacini, M., Eggers, T., Bonkowski, M., Gange, A. C. and Jones, T. H. (2006) Leaf endophytes affect mycorrhizal status and growth of co-infected and neighbouring plants. *Functional Ecology*, **20**, 226–232.

Osborne, S. E., Leong, Y. S., O'Neill, S. L. and Johnson, K. N. (2009) Variation in antiviral protection mediated by different *Wolbachia* strains in *Drosophila simulans*. *PLoS Pathogens*, **5**.

Partida-Martinez, L. P., de Looss, C. F., Ishida, K. et al. (2007) Rhizonin, the first mycotoxin isolated from the Zygomycota, is not a fungal metabolite but is produced by bacterial endosymbionts. *Applied and Environmental Microbiology*, **73**, 793–797.

Pena, J. A., Rogers, A. B., Ge, Z. M. et al. (2005) Probiotic *Lactobacillus* spp. diminish *Helicobacter hepaticus*-induced inflammatory bowel disease in interleukin-10-deficient mice. *Infection and Immunity*, **73**, 912–920.

Piel, J. (2002) A polyketide synthase-peptide synthetase gene cluster from an uncultured bacterial symbiont of *Paederus* beetles. *Proceedings of the National Academy of Sciences of the United States of America*, **99**, 14002–14007.

Piel, J. (2004) Metabolites from symbiotic bacteria. *Natural Product Reports*, **21**, 519–538.

Poulsen, M., Cafaro, M. J., Erhardt, D. P. et al. (2010) Variation in *Pseudonocardia* antibiotic defence helps govern parasite-induced morbidity in *Acromyrmex* leaf-cutting ants. *Environmental Microbiology Reports*, **2**, 534–540.

Ralphs, M. H., Creamer, R., Baucom, D. *et al.* (2008) Relationship between the endophyte *Embellisia* spp. and the toxic alkaloid swainsonine in major locoweed species (*Astragalus* and *Oxytropis*). *Journal of Chemical Ecology*, **34**, 32–38.

Reshef, L., Koren, O., Loya, Y., Zilber-Rosenberg, I. and Rosenberg, E. (2006) The coral probiotic hypothesis. *Environmental Microbiology*, **8**, 2068–2073.

Ritchie, K. B. (2006) Regulation of microbial populations by coral surface mucus and mucus-associated bacteria. *Marine Ecology-Progress Series*, **322**, 1–14.

Rodriguez, R. J., White, J. F., Arnold, A. E. and Redman, R. S. (2009) Fungal endophytes: diversity and functional roles. *New Phytologist*, **182**, 314–330.

Rosenberg, E., Koren, O., Reshef, L., Efrony, R. and Zilber-Rosenberg, I. (2007) The role of microorganisms in coral health, disease and evolution. *Nature Reviews Microbiology*, **5**, 355–362.

Rosso, M. L., Maier, M. S. and Bertoni, M. D. (2000) Macrocyclic trichothecene production by the fungus epibiont of *Baccharis coridifolia*. *Molecules*, **5**, 345–347.

Ruby, E. G. (1996) Lessons from a cooperative, bacterial–animal association: the *Vibrio fischeri–Euprymna scolopes* light organ symbiosis. *Annual Review of Microbiology*, **50**, 591–624.

Rudgers, J. A. and Clay, K. (2005) Fungal endophytes in terrestrial communities and ecosystems. In E. J. Dighton, P. Oudemans and J. F. J. White, eds., *The Fungal Community*. New York: M. Dekker, pp. 423–442.

Rudgers, J. A. and Clay, K. (2007) Endophyte symbiosis with tall fescue: how strong are the impacts on communities and ecosystems? *Fungal Biology Reviews*, **21**, 107–124.

Rudgers, J. A. and Clay, K. (2008) An invasive plant–fungal mutualism reduces arthropod diversity. *Ecology Letters*, **11**, 831–840.

Rudgers, J. A., Davitt, A. J., Clay, K., Gundel, P. and Omacini, M. (2010a) Searching for evidence against the mutualistic nature of hereditary symbiosis: a comment on Faeth (2009). *American Naturalist*, **76**, 99–103.

Rudgers, J. A., Fischer, S. and Clay, K. (2010b) Managing plant symbiosis: fungal endophyte genotype alters plant community composition. *Journal of Applied Ecology*, **47**, 468–477.

Rudgers, J. A., Holah, J., Orr, S. P. and Clay, K. (2007) Forest succession suppressed by an introduced plant-fungal symbiosis. *Ecology*, **88**, 18–25.

Rudgers, J. A., Mattingly, W. B. and Koslow, J. M. (2005) Mutualistic fungus promotes plant invasion into diverse communities. *Oecologia*, **144**, 463–471.

Saikkonen, K., Lehtonen, P., Helander, M., Koricheva, J. and Faeth, S. H. (2006) Model systems in ecology: dissecting the endophyte-grass literature. *Trends in Plant Science*, **11**, 428–433.

Scarborough, C. L., Ferrari, J. and Godfray, H. C. J. (2005) Aphid protected from pathogen by endosymbiont. *Science*, **310**, 1781–1781.

Schmitt, S., Wehrl, M., Bayer, K., Siegl, A. and Hentschel, U. (2007) Marine sponges as models for commensal microbe-host interactions. *Symbiosis*, **44**, 43–50.

Sicard, M., Hinsinger, J., Le Brun, N. *et al.* (2006) Interspecific competition between entomopathogenic nematodes (*Steinernema*) is modified by their bacterial symbionts (*Xenorhabdus*). *BMC Evolutionary Biology*, **6**, 68.

Sikes, B. A., Cottenie, K. and Klironomos, J. N. (2009) Plant and fungal identity determines pathogen protection of plant roots by arbuscular mycorrhizas. *Journal of Ecology*, **97**, 1274–1280.

Sonnenburg, J. L., Chen, C. T. L. and Gordon, J. I. (2006) Genomic and metabolic studies of the impact of probiotics on a model gut symbiont and host. *PLoS Biology*, **4**, 2213–2226.

Stabb, E. V. and Millikan, D. S. (2009) Is the *Vibrio fischeri–Euprymna scolopes* symbiosis a defensive mutualism? In J. F. White, Jr. and M. S. Torres, eds., *Defensive Mutualism in Microbial Symbiosis*. Boca Raton, FL: CRC Press, pp. 85–98.

Steiner, U., Ahimsa-Muller, M. A., Markert, A. *et al.* (2006) Molecular characterization of a seed transmitted clavicipitaceous fungus occurring on dicotyledonous plants (Convolvulaceae). *Planta*, **224**, 533–544.

Stovall, M. and Clay, K. (1991) Fungitoxic effects of *Balansia cyperi* (Clavicipitaceae). *Mycologia*, **83**, 288–295.

Tinh, N. T. N., Dierckens, K., Sorgeloos, P. and Bossier, P. (2008) A review of the functionality of probiotics in the larviculture food chain. *Marine Biotechnology*, **10**, 1–12.

Tsuchida, T., Koga, R., Horikawa, M. *et al.* (2010) Symbiotic bacterium modifies aphid body color. *Science*, **330** 1102–1104.

Turnbaugh, P. J., Ley, R. E., Hamady, M. *et al.* (2007) The Human Microbiome Project. *Nature*, **449**, 804–810.

van der Heijden, M. G. A., Klironomos, J. N., Ursic, M. *et al.* (1998) Mycorrhizal fungal diversity determines plant biodiversity, ecosystem variability and productivity. *Nature*, **396**, 69–72.

Vannette, R. L. and Hunter, M. D. (2009) Mycorrhizal fungi as mediators of defence against insect pests in agricultural systems. *Agricultural and Forest Entomology*, **11**, 351–358.

Verschuere, L., Heang, H., Criel, G., Sorgeloos, P. and Verstraete, W. (2000) Selected bacterial strains protect *Artemia* spp. from the pathogenic effects of *Vibrio proteolyticus* CW8T2. *Applied and Environmental Microbiology*, **66**, 1139–1146.

Werner, E. E. and Peacor, S. D. (2003) A review of trait-mediated indirect interactions in ecological communities. *Ecology*, **84**, 1083–1100.

Xie, J. L., Vilchez, I. and Mateos, M. (2010) *Spiroplasma* bacteria enhance survival of *Drosophila hydei* attacked by the parasitic wasp *Leptopilina heterotoma*. *PLoS ONE*, **5**, e12149.

Yoder, J. A., Benoit, J. B., Denlinger, D. L., Tank, J. L. and Zettler, L. W. (2008) An endosymbiotic conidial fungus, *Scopulariopsis brevicaulis*, protects the American dog tick, *Dermacentor variabilis*, from desiccation imposed by an entomopathogenic fungus. *Journal of Invertebrate Pathology*, **97**, 119–127.

Zytynska, S. E., Fleming, S., Tetard-Jones, C., Kertesz, M. A. and Preziosi, R. F. (2010) Community genetic interactions mediate indirect ecological effects between a parasitoid wasp and rhizobacteria. *Ecology*, **91**, 1563–1568.

CHAPTER TWENTY-ONE

Integrating trait-mediated effects and non-trophic interactions in the study of biodiversity and ecosystem functioning

ALEXANDRA GOUDARD

Lycée Champollion, Grenoble

and

MICHEL LOREAU

Station d'Ecologie Expérimentale du CNRS

The need to consider trait-mediated effects and non-trophic interactions in the study of biodiversity and ecosystem functioning

The relationship between biodiversity and ecosystem functioning has emerged as a central issue in ecology during the past two decades because of increasing concerns about the potential ecological and societal consequences of current biodiversity loss (Loreau *et al.* 2001; Naeem *et al.* 2009; Loreau 2010). Biological invasions are one of the main drivers of this biodiversity loss as human activities contribute to accelerate introductions of exotic species in many of the world's ecosystems, with considerable economic impacts (Williamson 1996; Mooney and Hobbs 2000). Although the effects of biodiversity loss on ecosystem functioning have now been studied in a wide range of organisms and ecosystems (Hooper *et al.* 2005; Balvanera *et al.* 2006; Cardinale *et al.* 2006; Naeem *et al.* 2009), most experimental and theoretical studies have considered either competitive communities at a single trophic level or relatively simple food webs (Duffy *et al.* 2007). The only direct species interactions considered in these studies are trophic interactions, and the only indirect species interactions considered are density-mediated interactions (for instance, exploitation competition). Similarly, the impacts of biological invasions on native ecosystems and the relationship between biodiversity and invasion resistance have been mainly studied in experimental plant communities (e.g., Kennedy *et al.* 2002) and models that consider a single type of species interactions, either competition (e.g., Case 1990; Byers and Noonburg 2003) or trophic interactions (e.g., Law and Morton 1996).

Although there is growing recognition of the significance of non-trophic and trait-mediated interactions in communities and ecosystems (Bolker *et al.* 2003; Bruno *et al.* 2003), these interactions are still poorly studied theoretically, and we still know little about general patterns and mechanisms. Ecosystems are much more than mere food webs; they are complex systems that involve multiple

Trait-Mediated Indirect Interactions: Ecological and Evolutionary Perspectives, eds. Takayuki Ohgushi, Oswald J. Schmitz and Robert D. Holt. Published by Cambridge University Press. © Cambridge University Press 2012.

forms of interactions and multiple ecological networks (Ings *et al.* 2009; Olff *et al.* 2009). Recent experiments have showed that non-trophic interactions, such as facilitation and habitat modification (Mulder *et al.* 2001; Cardinale *et al.* 2002; Rixen and Mulder 2005), play an important role in ecosystem functioning, and that different kinds of species interactions typically co-occur in natural ecosystems (Callaway and Walker 1997). Very few theoretical studies have explored the role of non-trophic interactions in ecosystems, and most of these studies consider only one kind of species interaction, in particular mutualism. But simple models of mutualism do not respect the physical principle of mass conservation, and hence often lead to unrealistic explosive systems (Ringel *et al.* 1996). Therefore, an important current challenge is to develop theories and models to provide generalizations on the role of non-trophic and trait-mediated interactions in the maintenance of biodiversity, in ecosystem functioning, and in the relationship between biodiversity and ecosystem functioning.

Here we present an interaction web model that includes all types of direct species interactions, both trophic and non-trophic (interference competition, mutualism, exploitation, commensalism, amensalism), as well as all types of indirect effects, be they density-, trait- or habitat-mediated. Our model also satisfies mass balance constraints, which allows study of aggregate ecosystem properties in a consistent manner. Non-trophic interactions are added to a food web through trophic interaction modifications, a form of trait-mediated effects. We analyse this model numerically, and mimic a community assembly process of successive species introductions, following Loreau *et al.*'s (2001) recommendation to study the relationship between biodiversity and ecosystem functioning with a dynamical approach. We then study the relationships that emerge from community assembly between species diversity, the strength and prevalence of non-trophic effects, and a range of ecosystem properties, including the biomass and production of the various trophic levels, invasion resistance, and robustness to resident species extinctions due to invasions. We compare the results obtained for full interaction webs with those obtained for simple food webs without non-trophic interactions, and we further compare our theoretical predictions with some empirical data from competitive plant communities and food webs. The description of the model and the results on the effects of non-trophic interactions on ecosystem functioning and its relationship with biodiversity are summarized from Goudard and Loreau (2008); all the other sections present entirely new material.

Different kinds of indirect effects in ecosystems
Density-mediated indirect effects versus trait-mediated indirect effects

Indirect effects are usually classified into two types (Abrams, 1995): (1) density-mediated indirect effects, which act through changes in species

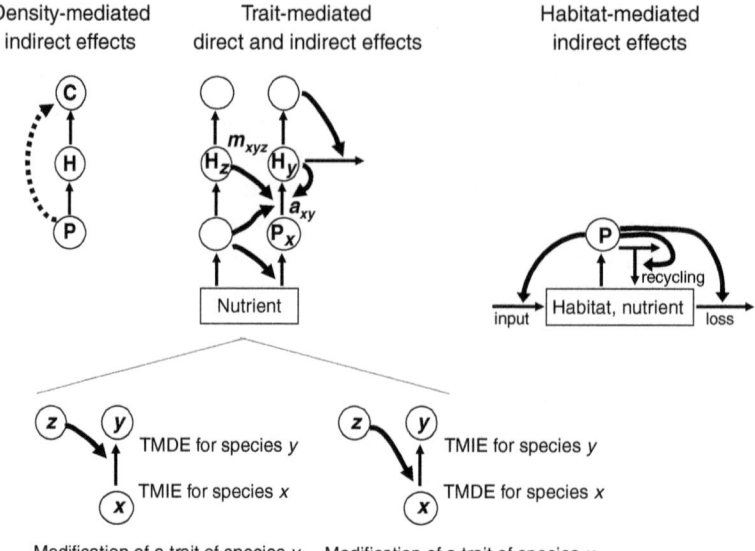

Figure 21.1 A classification of indirect effects. Examples of density-mediated indirect effects, trait-mediated direct (TMDE) and indirect (TMIE) effects, and habitat-mediated indirect effects. The latter two types of effects show the importance of interaction modifications and ecosystem engineering.

densities and propagate along a chain of direct interactions (e.g., indirect competition by exploitation of a shared resource); (2) trait-mediated indirect effects (TMIEs), which are caused by changes in species traits and may affect several species simultaneously (Fig. 21.1). TMIEs can take a variety of forms, such as non-lethal effects of a predator on its prey. A number of recent contributions have discussed methodological approaches to detect and quantify these effects and some of their potential pitfalls (Werner and Peacor 2003; Schmitz *et al.* 2004; Okuyama and Bolker 2007; Abrams 2007).[1]

Interaction modifications

Interaction modifications (Wootton 1994; Arditi *et al.* 2005) are modifications of an interaction by a species. We can consider two types of interaction modifications (Figs 21.1, 21.2): (1) the modification of an interaction between two species by a third species (for instance, a non-trophic modification of a trophic interaction) through some behavioural (e.g., interference) or chemical (e.g., allelopathy) effect; (2) the modification of an interaction between a species and an abiotic factor. Both types generate trait-mediated effects. Although interaction modifications were regarded by Wootton (1994) as a class of indirect effects, they may be viewed as either direct or indirect trait-mediated effects (Abrams 1995). If species *z* changes a

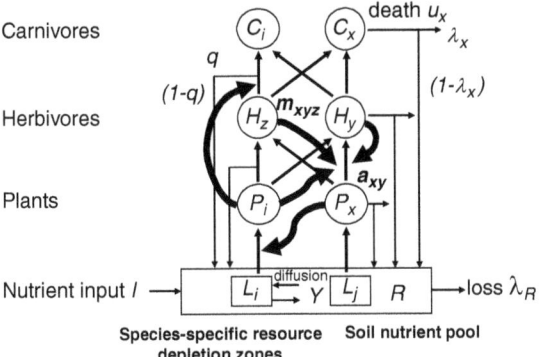

Figure 21.2 The interaction web model. Solid thin arrows represent nutrient flows. For clarity of the figure, flows of non-assimilated nutrient returned to the soil nutrient pool during consumption by carnivores and herbivores are represented only on the left trophic chain, while flows of nutrient either recycled or lost from the ecosystem following death are represented only on the right trophic chain. Thick lines represent interaction modifications, a form of trait-mediated effect. Only five examples of interaction modifications are represented here for the sake of clarity. For instance, herbivore species H_z modifies the trophic interaction between herbivore species H_y and plant species P_x with a magnitude of interaction modification m_{xyz}. The modification of the nutrient flow between plant species P_i and its species-specific resource depletion zone L_i corresponds to intraspecific competition or facilitation.

trait of species x, for instance the resistance of a plant x to predation by a herbivore y, then species z has an indirect effect on species y and an direct effect on species x (Fig. 21.1). If species z changes traits of both species x and y simultaneously, however, then species z has direct effects on both species x and y. Habitat modification always generates indirect effects; for instance, a dense habitat may allow prey to hide and avoid predators or, conversely, predators to hide and better hunt prey. Thus, a trait-mediated effect is a modification of a biotic parameter of a species, and may be either direct or indirect, whereas a habitat modification is a modification of an abiotic factor that has an indirect effect on all the species that are affected by this factor.

Density-, trait- and habitat-mediated indirect effects

Therefore it may be useful to extend Abrams's (1995) definitions to include three types of indirect effects in ecosystems (Fig. 21.1): (1) *density-mediated biotic indirect effects*, i.e., indirect effects that occur through changes in the density of a transmitter species; (2) *trait-mediated biotic indirect effects*, i.e., indirect effects that occur through changes in the traits of a transmitter species; and (3) *habitat-mediated indirect effects* i.e., indirect effects that occur through modification of the physical and chemical habitat. TMIEs are also associated with *trait-mediated direct effects*.

This extension of Abrams's definitions allows us to include *ecosystem engineers* (Jones et al. 1994) as vehicles of trait- and habitat-mediated indirect effects. By modifying their physical environment, ecosystem engineers directly or indirectly modify the availability of resources for other species, and thus create many non-trophic species interactions. Autogenic engineers change the environment by modifying their own structure, i.e., their living or dead tissues (e.g., cyanobacteria biofilms, wood structure of forest trees). Allogenic engineers change the environment by transforming living and non-living materials from one physical state to another (e.g., earthworms, bioturbation).

Species that modify trophic interactions can be regarded as ecosystem engineers because they modify a trophic resource flux, either directly through their own materials or indirectly by transforming materials. For instance, the plant cover of the invasive species *Eichhornia crassipes* creates refuges for mosquito larvae, thereby decreasing the strength of their trophic interaction with their predators. Species that modify an abiotic parameter (e.g., light intensity, recycling rate, water fluxes such as infiltration, evaporation and runoff) are also ecosystem engineers because they modify an abiotic resource flux or the control of a resource by an abiotic factor.

An interaction web model that includes non-trophic interactions and trait-mediated indirect effects
Building the model

Our interaction web model (Fig. 21.2) extends the food web model developed by Thébault and Loreau (2003) for a nutrient-limited ecosystem with three trophic levels by adding non-trophic modifications of trophic interactions. The model respects the principle of mass conservation, and allows a wide range of non-trophic interactions to occur. A complete description can be found in Goudard and Loreau (2008).

The key feature of the model is the potential for each species z to modify the trophic interaction between any two species x and y (including itself), and thereby to increase or decrease their per capita population growth rate or fitness. These interaction modifications are non-trophic modifications of trophic interactions, and are thus trait-mediated biotic effects (Fig. 21.2). The non-trophic effect of a species z on the interaction between species x and y depends on both its biomass X_z and a magnitude of interaction modification m_{xyz}. μ_{xy} is the non-trophic coefficient: it is the total non-trophic effect of all the $3S$ species in the ecosystem (S species per trophic level) on the trophic interaction between species x and y. Thus, the realized consumption rate of species y by species x, $a_{xy}\mu_{xy}$, is the product of the potential consumption rate

a_{xy}, i.e., the intensity of the trophic interaction between predator species x and prey species y ($a_{xy} \geq 0$ and $a_{xy} = -a_{yx}$), and the non-trophic coefficient μ_{xy} :

$$c_{xy} = a_{xy}\mu_{xy}$$

$$\text{where } \mu_{xy} = \exp\left(\sum_{z=1}^{3S} m_{xyz} \log\left(1 + b_z X_z\right)\right) = \prod_{z=1}^{3S}(1 + b_z X_z)^{m_{xyz}} \qquad (21.1)$$

The function that describes non-trophic effects in this equation was chosen such that it satisfies several conditions. First, it is a strictly increasing function of both the magnitude of interaction modification m_{xyz} and biomass X_z. Second, the non-trophic coefficient μ_{xy} is unchanged if either $m_{xyz} = 0$ or $X_z = 0$. Third, the magnitude of interaction modification is symmetrical ($m_{xyz} = m_{yxz}$) to maintain mass balance. Fourth, the non-trophic coefficient must be strictly positive whatever the sign of the interaction modification, so that the sign of the realized consumption rate does not change, and thus the nutrient flow between species x and species y is not reversed. Fifth, the non-trophic coefficient must be larger than 1 if the interaction modification is positive, and smaller than 1 if the interaction modification is negative. Thus, each species z can affect any two species x and y by increasing ($\mu_{xy} > 1$) or decreasing ($\mu_{xy} < 1$) the realized consumption rate of species y by species x. The coefficient b_z converts biomass X_z into a dimensionless number; we used $b_z = 1$ for simplicity in our simulations.

In the absence of trait-mediated effects, the only direct species interaction is predation, and our interaction web reduces to a food web. When interaction modifications are added, all types of species interactions are present (competition, mutualism, exploitation, commensalism, amensalism), including negative or positive intraspecific effects ($m_{xzz} \neq 0$ for species z). Our model, however, respects the principle of mass conservation since interaction modifications affect the material flow between a resource and a consumer in the same way for both species. The model also includes nutrient cycling and a volume allocation rule in the soil, allowing functional complementarity between plant species (Loreau 1998).

We constrained the model as little as possible to explore its general properties. Accordingly, we randomly assigned the various biological parameters (potential consumption rates, intensities of interaction modifications, death rates, non-recycled proportions of nutrient) to a regional pool of species from a uniform distribution within appropriate intervals, and let the local ecosystem assemble spontaneously. Each simulated ecosystem resulted from an assembly process involving successive introductions of species picked at random from the regional species pool and species eliminations as a result of local interactions. Despite constant species turnover, aggregate ecosystem properties turned out to stabilize relatively quickly in a quasi-stationary regime.

Assessing ecosystem properties and species interactions

We analysed the effects of species richness and non-trophic interactions in the regional species pool on a wide range of community and ecosystem properties in the local ecosystems resulting from the assembly process, in particular total local species richness, local species richness of each trophic level, proportions of the various types of net species effects and interactions, interaction web connectance, total biomass, biomass of each trophic level, production of each trophic level, invasion resistance, and robustness to resident extinctions. Non-trophic interactions were manipulated by varying the connectance and maximal magnitude of interaction modifications in the regional species pool. The *non-trophic connectance* of the regional species pool, defined as the probability that a species modifies the trophic interaction between any two species, measures the connectance of trait-mediated effects. Food web connectance was kept constant in our simulations. The magnitude of interaction modification, m_{xyz}, was randomly taken between a maximum value called *maximal non-trophic magnitude* and a symmetrical minimum (minus maximal non-trophic magnitude). This maximal non-trophic magnitude represents the maximal value of trait-mediated effects.

Since all species can affect other species in a large number of different ways, we defined *net species effects* (facilitation, inhibition, or no effect) and *net species interactions* (mutualism, competition, exploitation, commensalism, amensalism or neutral interaction) phenomenologically. The net species effect (sum of trophic and non-trophic effects) of species g on species i, E_{ig}, was measured by the partial derivative of the growth rate of species i with respect to the biomass of species g:

$$E_{ig} = \frac{\partial \left(\dfrac{dX_i}{dt} \right)}{\partial X_g} \qquad (21.2)$$

This measure includes trophic and non-trophic direct effects, as well as, potentially, TMIEs. If $E_{ig} > 0$, the effect of species g on species i is facilitative. If $E_{ig} > 0$ and $E_{gi} > 0$, the interaction between species i and g is mutualistic. Local *interaction web connectance* was measured as the proportion of non-neutral species interactions among all possible species interactions. We called *mean value of facilitation (inhibition)* the mean value of positive (negative) net species effects. We considered only interspecific species effects and interactions, without taking into account the effect of a species on itself (E_{ii}).

We also measured invasion resistance and robustness to resident extinctions following species introductions in the assembled local ecosystems. An invasion attempt was considered successful if the introduced species was able to increase when rare (Kokkoris et al. 1999). Invasion failure probability was computed as the ratio between the cumulated number of failed invasion attempts and the number of introductions from the beginning of the simulation. *Invasion resistance* was then measured by the short-term failure probability of an introduced species, obtained by recording the presence or absence of the introduced species 100 time steps after the introduction event. *Robustness to resident extinctions* following species introductions was measured by recording the number of resident species extinctions caused by an introduced species during the 100 time steps that followed its introduction. The smaller the number of resident extinctions, the higher the ecosystem's robustness to resident extinctions.

The interdependence between species richness and species interactions
Effect of species richness on the prevalence of species interactions

Although the relationship between trophic connectance and species richness has been well studied in food webs (Martinez 1992; Montoya and Solé 2003), we lack knowledge about the connectance of ecosystems considered as full interaction webs. Recent experimental studies suggest that the prevalence of species interactions such as facilitation may increase with species richness (e.g., Cardinale et al. 2002). Our model predicts that interaction web connectance increases with species richness and that the proportions of the various types of species effects and species interactions depend on species richness (Fig. 21.3a, b). A higher species richness increases the number of trophic links of a given species (as long as consumers are not strict specialists), which increases the probability for this species to have at least one trophic link modified by any other species in the web, and thus the probability for each species to interact with any other species. The fact that interaction web connectance tends to 100% here (Fig. 21.3a), however, is due to the assumption that consumers are generalists in our model. Other food web configurations may lead to smaller upper limits.

Effect of species richness on the strength of species interactions

The analysis of natural food webs (Neutel et al. 2002) suggests that natural ecosystems are characterized by a large number of weak interactions and a small number of strong interactions. Our model predicts that the strength of species effects (interspecific facilitation and inhibition) decreases with species

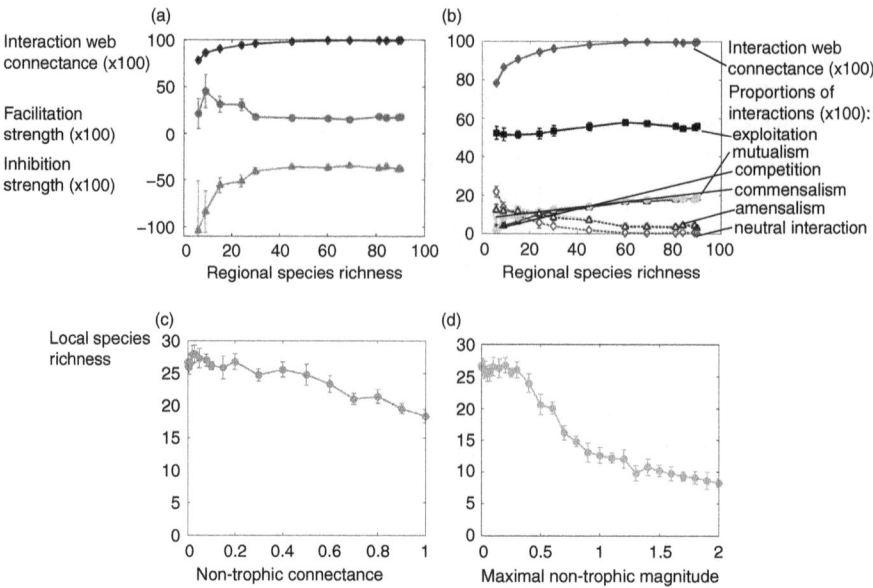

Figure 21.3 Reciprocal effects between species richness and species interactions: results from the interaction web model. Strength of species effects (a), interaction web connectance and proportions of species interactions (b) in the community as a whole in the quasi-stationary regime as functions of regional species richness (non-trophic connectance = 0.2, maximal non-trophic magnitude = 0.2) in interaction webs. Panel (a) shows means and standard deviations of facilitation strength, inhibition strength, and interaction web connectance. Panel (b) shows the means and standard deviations of interaction web connectance and the proportions of mutualism, competition, exploitation, commensalism, amensalism, neutral interactions. Total local species richness in the quasi-stationary regime as a function of non-trophic connectance (c, regional species richness = 45, maximal non-trophic magnitude = 0.2) and maximal non-trophic magnitude (d, regional species richness = 45, non-trophic connectance = 0.2).

richness (Fig. 21.3a), in agreement with competition models (Kokkoris *et al.* 1999, 2002). Thus, species-rich interaction webs are more connected but have weaker species interactions on average. Their lower interaction strength is probably what allows them to maintain a high diversity and connectance, in agreement with previous theory (May 1972; Kokkoris *et al.* 1999, 2002).

Effect of interaction modifications on the prevalence and strength of species effects

Our model shows that both non-trophic connectance and maximal non-trophic magnitude affect the prevalence, strength and variability of species effects. In particular, non-trophic connectance increases interaction web connectance and the proportions of non-trophic species interactions such as mutualism and competition (Goudard and Loreau 2008).

Effect of interaction modifications on local species richness

Our model also predicts that high levels of non-trophic connectance and maximal non-trophic magnitude have negative effects on local species richness (Fig. 21.3c, d), in particular at the plant trophic level. Interaction modifications are likely to generate strong constraints on species coexistence. The species selected during the assembly process have higher realized consumption rates on average, which makes them more efficient but also more competitive (Goudard and Loreau 2008).

These results emphasize the mutual interdependence between species richness and species interactions. Species richness affects the prevalence and strength of species interactions, just as the latter affects species richness. This interdependence makes the relationships between ecosystem structure (species richness and species interactions) and ecosystem functioning more complex (see below).

Effects of non-trophic interactions on ecosystem functioning and its relationship with biodiversity

Effects of non-trophic interactions on biomass and production

Our model predicts that biomass and production at all trophic levels tend to decrease as either non-trophic connectance or maximal non-trophic magnitude increases (Fig. 21.4a, b). Two factors explain this counterintuitive result (Goudard and Loreau 2008). First, non-trophic connectance increases the mean realized consumption rates of the various species, which eventually contributes to decrease biomass and production. Second, it also increases the proportions of inhibition and competition more than those of facilitation and mutualism in plants.

In contrast, Arditi et al. (2005) found an increasing proportion of 'super-efficient' systems as the magnitude of interaction modifications increases in another interaction web model. These contrasting predictions are likely explained by two key differences between the two models. The first difference is the level of trophic connectance among species, which was relatively low in their model and high in ours. We allowed all species to be more or less generalist consumers, and this increases the potential for resource overexploitation. The second difference concerns the way interaction modifications are represented in the two models: interaction modifications combined additively in their model, and multiplicatively in our model (Equation (21.1)). Non-trophic effects can increase resource consumption more strongly in our model, again enhancing the potential for resource overexploitation. The two models

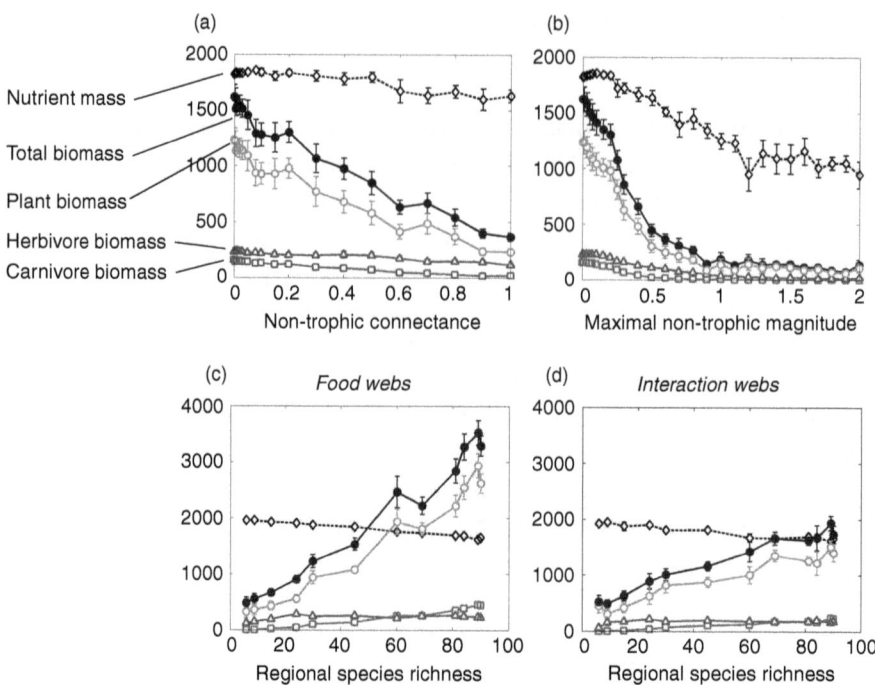

Figure 21.4 Effects of non-trophic interactions on biomass and the biodiversity–biomass relationship in food webs and interaction webs: results from the interaction web model. Biomass in the quasi-stationary regime as a function of non-trophic connectance (a, regional species richness = 45, maximal non-trophic magnitude = 0.2) and maximal non-trophic magnitude (b, regional species richness = 45, non-trophic connectance = 0.2) in interaction webs. Biomass in the quasi-stationary regime as a function of regional species richness in food webs (c, non-trophic connectance = 0, maximal non-trophic magnitude = 0) and in interaction webs (d, non-trophic connectance = 0.2, maximal non-trophic magnitude = 0.2). The various curves show the means and standard deviations of total biomass, nutrient mass, plant biomass, herbivore biomass and carnivore biomass.

highlight different potential outcomes that might occur in natural ecosystems. Experimental tests of these contrasting predictions are now needed to move forward on this topic. We still have very limited knowledge of the mechanisms and consequences of non-trophic interactions, trait-mediated effects and habitat-mediated effects in ecology. In particular, their effects on ecosystem properties such as biomass and production deserve much more attention.

Relationship between biodiversity and ecosystem functioning

Our model predicts that total biomass and biomass at each trophic level increase with regional species richness both in the absence and presence of non-trophic interactions (Fig. 21.4c and d), with a bottom-up control of plants on carnivores, a top-down control of carnivores on herbivores and a better exploitation of the limiting nutrient by plants. Production at all trophic levels increases with regional species richness (Goudard and Loreau 2008). Biomass and production, however, increase less rapidly with species richness in interaction webs with non-trophic interactions than in food webs without non-trophic interactions. This is explained again by the fact that non-trophic interactions tend to increase the average realized consumption rate, and that this effect becomes stronger as species richness increases. Thus, species become more efficient at exploiting resources, but also more competitive, and this increases the probability of observing resource overexploitation. Our model predicts positive biodiversity–ecosystem functioning relationships in both food webs and interaction webs, due to the same mechanisms, but with a strong impact of non-trophic interactions on the shape of the diversity–biomass relationship (Goudard and Loreau 2008).

Thus, our interaction web model allows a generalization of the positive biodiversity–ecosystem functioning relationship typically found in simple single-trophic-level ecosystems to complex interaction webs resulting from a long assembly process. Interestingly, we did not observe the unimodal relationships predicted under some conditions by existing theory on multitrophic ecosystems (Thébault and Loreau 2003; Ives et al. 2005; Loreau 2010). This highlights the difference between potential ecosystem configurations and those actually realized at the outcome of an assembly process.

Effects of non-trophic interactions on ecosystem responses to biological invasions
Effects of biodiversity on invasion resistance and robustness to resident extinctions

Our model predicts that species introductions induce extinctions of resident species, and that the number of these extinctions increases, while biomass and production decrease, as the frequency of species introductions increases. Thus it confirms empirical evidence for the ecological impacts of species introductions, which are widely regarded as one of the main drivers of species extinctions, and hence of loss of ecosystem services.

In turn, biodiversity affects the ability of ecosystems to resist invasions. Experimental studies conducted at local scales and controlling for abiotic extrinsic factors show that species richness increases invasion resistance in plant communities (e.g., Kennedy et al. 2002), in contrast to empirical studies at regional scales, which often show positive or negative relationships between species richness and invasion resistance (e.g., Stohlgren et al.

1999) because of covarying extrinsic factors. Models of competitive communities and food webs have usually showed positive relationships between biodiversity and invasion resistance, at least at small scales in the absence of covarying extrinsic factors (e.g., Case 1990; Law and Morton 1996; Byers and Noonburg 2003), but no theoretical or empirical study is available on the effects of other non-trophic interactions on the relationship between biodiversity and invasion resistance.

Our model also predicts that regional species richness increases invasion resistance (Fig. 21.5a), thus extending previous findings to complex interaction webs. This result is probably explained by an increased resource-use complementarity in species-rich ecosystems, which decreases the amount of resources available to invaders, and by an increased probability of including natural enemies of invaders or species with traits similar to invaders.

Few experimental (Pfisterer *et al.* 2004) or theoretical (Case 1990) studies have investigated the effect of species richness on ecosystem robustness to resident extinctions. Our model predicts that species richness increases robustness to resident extinctions in species-rich ecosystems (Fig. 21.5a). The decrease in the number of resident extinctions at a high regional species richness was mostly due to a decrease in the number of consumer extinctions (figure not shown).

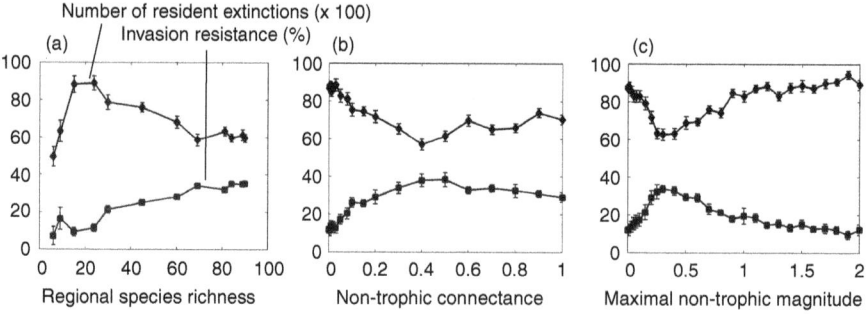

Figure 21.5 Species richness, non-trophic effects and biological invasions: results from the interaction web model. Invasion resistance and robustness to resident extinctions as functions of regional species richness (a, non-trophic connectance = 0.2, maximal non-trophic magnitude = 0.2), non-trophic connectance (b, regional species richness $3S = 45$, maximal non-trophic magnitude = 0.2) and maximal non-trophic magnitude (c, regional species richness $3S = 45$, non-trophic connectance = 0.2). Lower curves show the mean and standard deviation of invasion resistance (short-term failure probability × 100); upper curves show the mean and standard deviation of the number of resident extinctions during an interval of 100 time steps (× 100) in interaction webs.

Effects of non-trophic interactions on invasion resistance and robustness to resident extinctions

Few studies have explicitly explored the effects of species interactions on invasion resistance (Case 1990; Kokkoris *et al.* 1999) and robustness to resident extinctions. Experiments and theory suggest that trophic connectance strongly affects robustness to cascading extinctions due to primary species loss in food webs (Law and Blackford 1992; Thébault *et al.* 2007). Our interaction web model predicts that both resistance to invasions and robustness to extinctions due to invasions depend strongly on species interactions. Interestingly, intermediate values of non-trophic connectance and magnitude of interaction modifications maximize invasion resistance and robustness to resident extinctions (Fig. 21.5b and c).

The reciprocal interaction between ecosystem properties and biological invasions

Our model shows that biological invasions alter the structure and functioning of ecosystems, by inducing loss of species, biomass and production, especially when species introductions are frequent. Since ecosystem functioning depends on its structure and diversity (species richness, species interactions), invasive species can alter ecosystem functioning both directly and indirectly. But ecosystem structure – species richness and the prevalence and strength of species interactions – also conditions the probability of success of introduced species and their impacts. For instance, Mitchell *et al.* (2006) studied introduced plant species that modify interactions between native species (exploitation, competition, mutualism), and these changes in ecosystem structural properties had a feedback effect on the success of the introduced species. Thus, structural and functional ecosystem properties and biological invasions are strongly interdependent.

Our model shows that the positive relationships between biodiversity and invasion resistance and between biodiversity and robustness to resident extinctions are partly explained by the diversity-dependence of species interactions and the impacts of species interactions on invasion resistance and robustness to resident extinctions. These results suggest that species interactions, and especially non-trophic effects, trait-mediated effects and habitat-mediated effects, should receive more attention in the study of the relationship between ecosystems and biological invasions. Understanding how species interactions affect the success of biological invasions and the damage they cause is likely to improve our ability to avoid undesirable invasions and their associated costs.

The complex relationship between ecosystem structure and functioning

The structure of an ecosystem describes the elements it contains and the relationships between these elements. 'Ecosystem structure', thus, is a generic term that includes species diversity, species interactions and abiotic factors. In contrast, the term 'ecosystem functioning' denotes the various processes and properties that make the ecosystem operate as an entity; these processes and properties include biomass, production, nutrient cycling, ecosystem stability and invasion resistance.

Our results show consistently that ecosystem structure and ecosystem functioning are strongly interdependent (Fig. 21.6). They further predict strong relationships between various structural ecosystem properties as well as between various functional ecosystem properties. In particular, species diversity affects the nature and prevalence of species interactions, and, reciprocally, species interactions affect species diversity, creating a complex web of relationships between ecosystem structure and ecosystem functioning (Fig. 21.6). Biodiversity has not only direct effects on ecosystem processes, but also indirect effects through its effects on the strength and

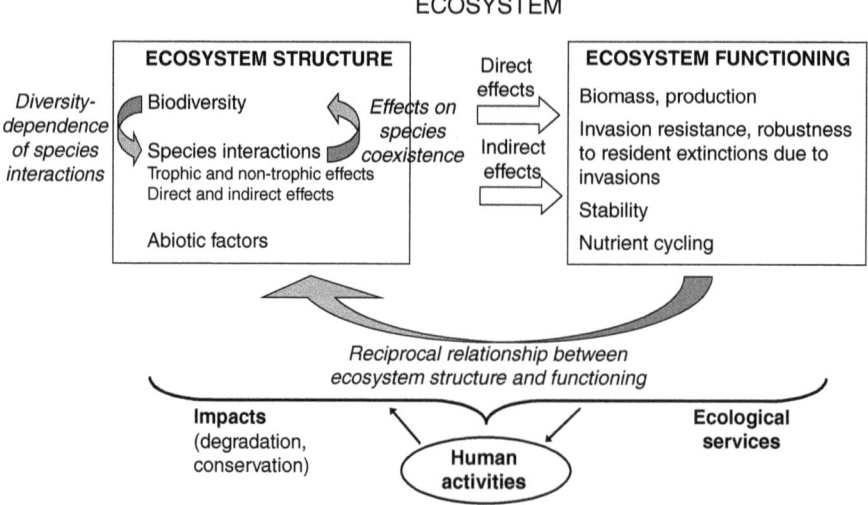

Figure 21.6 Importance of non-trophic interactions and trait-mediated effects in the relationship between ecosystem structure and functioning. White arrows show the importance of non-trophic interactions and trait-mediated effects for ecosystem functioning. Grey arrows show the complexity of the ecosystem structure–functioning relationship due to the interdependence between biodiversity and species interactions and the reciprocal effect of ecosystem functioning on ecosystem structure.

prevalence of species interactions. Similarly, species interactions have direct effects on ecosystem functioning, but also indirect effects through their effects on species coexistence. Therefore, biodiversity–ecosystem functioning relationships are much richer and more complex in complex interaction webs than in relatively simple plant communities or even in food webs, not only because they take into account non-trophic effects and trait-mediated or habitat-mediated indirect effects, but also because of the additional interdependence that these effects generate between species richness and species interactions. It would be interesting to explore further the relationships between abiotic factors, ecosystem structure and ecosystem functioning.

Incorporating non-trophic interactions, trait-mediated effects and habitat-mediated effects in theoretical ecology

Our theoretical work, together with recent empirical evidence, strongly suggests that non-trophic interactions, trait-mediated effects and habitat-mediated effects play an important role in ecosystems and should be taken into account in experimental and theoretical ecology. It also suggests specific hypotheses that would deserve to be tested experimentally. In particular, it would be interesting to test the effects of species richness predicted by our model on the strength and prevalence of species interactions in experimental ecosystems. It would also be useful to study experimentally the impacts of species interactions, and especially non-trophic, trait-mediated and habitat-mediated effects, on ecosystem processes and properties, such as biomass and production at various trophic levels, invasion resistance and robustness to resident extinctions due to biological invasions. In order to study the impacts of non-trophic effects, it would likely be easier to manipulate the number of species interactions in microcosms that have the same species richness, than the strength of these interactions.

Because our interaction web model incorporates non-trophic interactions in the form of interaction modifications, it could easily be applied or extended to the study of ecosystem engineering, either using specific forms of non-trophic modifications of trophic interactions or through modifications of abiotic parameters such as those that govern the input, recycling and loss of nutrients. Ecosystem engineers can modify ecosystem structural properties, such as species richness, composition and interactions, as well as ecosystem functional processes such as primary production (Badano *et al.* 2006; Zhu *et al.* 2006). Our model could also be applied straightforwardly to ecological studies of manipulative parasites. Manipulative parasites are perfect examples of interaction modifiers: by altering their host's behaviour, they affect the strength of the trophic links involved in their transmission as well as other trophic interactions (Lefèvre *et al.* 2009). Thus, the approach we have presented here has great potential in addressing a wide range of trait-mediated and habitat-mediated effects in theoretical ecology.

Conclusion

Our chapter highlights the importance of species interactions, in particular non-trophic interactions, in the structural and functional properties of eco-systems and in the relationships between biodiversity and ecosystem functioning. Species richness affects the nature, prevalence and strength of species interactions, and these in turn affect species richness, thus making the mechanisms of the biodiversity–ecosystem functioning relationships more complex in interaction webs than in classical food webs or competitive communities. Non-trophic interactions, trait-mediated effects and habitat-mediated effects should be given more attention in studying the relationships between ecosystem structural properties (such as species richness, species interactions and abiotic factors) and ecosystem functional properties (such as biomass, production, nutrient cycling, ecosystem stability and invasion resistance) to understand better the ecological consequences of biodiversity loss and predict the impacts of environmental changes, including biological invasions, on ecosystem services.

Note

1. Abrams (2007) also notes a gradual terminological shift from 'indirect effect' to 'indirect interaction' in the recent literature. We keep here the initial distinction between 'indirect effect' as a directional pathway of effects from one species to another and 'indirect interaction' as reciprocal action between two species through some transmitter.

References

Abrams, P. A. (1995) Implications of dynamically variable traits for identifying, classifying, and measuring direct and indirect effects in ecological communities. *American Naturalist*, **146**, 112–134.

Abrams, P. A. (2007) Defining and measuring the impact of dynamic traits on interspecific interactions. *Ecology*, **88**, 2555–2562.

Arditi, R., Michalski, J. and Hirzel, A. H. (2005) Rheagogies: modeling non-trophic effects in food webs. *Ecological Complexity*, **2**, 249–258.

Badano, E. I., Jones, C. G., Cavieres, L. A. and Wright, J. P. (2006) Assessing impacts of ecosystem engineers on community organization: a general approach illustrated by effects of a high-Andean cushion plant. *Oikos*, **115**, 369–385.

Balvanera P., Pfisterer A. B., Buchmann, N. *et al.* (2006) Quantifying the evidence for biodiversity effects on ecosystem functioning and services. *Ecology Letters*, **9**, 1146–1156.

Bolker, B., Holyoak, M., Krivan, V., Rowe, L. and Schmitz, O. (2003) Connecting theoretical and empirical studies of trait-mediated interactions. *Ecology*, **84**, 1101–1114.

Bruno, J. F., Stachowicz, J. J. and Bertness, M. D. (2003) Inclusion of facilitation into ecological theory. *Trends in Ecology and Evolution*, **18**, 119–125.

Byers, J. E. and Noonburg, E. G. (2003) Scale dependent effects of biotic resistance to biological invasion. *Ecology*, **84**, 1428–1433.

Callaway, R. M. and Walker, L. R. (1997) Competition and facilitation: a synthetic approach to interactions in plant communities. *Ecology*, **78**, 1958–1965.

Cardinale, B. J., Harvey, C. T., Gross, K. and Ives, A. R. (2002) Species diversity enhances ecosystem functioning through interspecific facilitation. *Nature*, **415**, 426–429.

Cardinale B. J., Srivastava D. S., Duffy J. E. *et al.* (2006) Effects of biodiversity on the functioning of trophic groups and ecosystems. *Nature*, **443**, 989–992.

Case, T. J. (1990) Invasion resistance arises in strongly interacting species-rich model competition communities. *Proceedings of the National Academy of Sciences of the United States of America*, **87**, 9610–9614.

Duffy, J. E., Cardinale, B. J., France, K. E. *et al.* (2007) The functional role of biodiversity in ecosystems: incorporating trophic complexity. *Ecology Letters*, **10**, 522–538.

Goudard, A. and Loreau, M. (2008) Non-trophic interactions, biodiversity and ecosystem functioning: an interaction web model. *American Naturalist*, **171**, 91–106.

Hooper, D. U., Chapin, F. S. III, Ewel, J. J. *et al.* (2005) Effects of biodiversity on ecosystem functioning: a consensus of current knowledge. *Ecological Monographs*, **75**, 3–23.

Ings, T. C., Montoya, J. M., Bascompte, J. *et al.* (2009) Ecological networks – beyond food webs. *Journal of Animal Ecology*, **78**, 253–269.

Ives, A. R., Cardinale, B. J. and Snyder, W. E. (2005) A synthesis of subdisciplines: predator–prey interactions, and biodiversity and ecosystem functioning. *Ecology Letters*, **8**, 102–116.

Jones, C. G., Lawton, J. H. and Shachak, M. (1994) Organisms as ecosystem engineers. *Oikos*, **69**, 373–386.

Kennedy, T. A., Naeem, S., Howe, K. M. *et al.* (2002) Biodiversity as a barrier to ecological invasion. *Nature*, **417**, 636–638.

Kokkoris, G. D., Jansen, V. A. A., Loreau, M. and Troumbis, A. Y. (2002) Variability in interaction strength and implications for biodiversity. *Journal of Animal Ecology*, **71**, 362–371.

Kokkoris, G. D., Troumbis, A. Y. and Lawton, J. H. (1999). Patterns of species interaction strength in assembled theoretical competition communities. *Ecology Letters*, **2**, 70–74.

Law, R. and Blackford, J. C. (1992) Self-assembling food webs: a global viewpoint of coexistence of species in Lotka–Volterra communities. *Ecology*, **73**, 567–578.

Law, R. and Morton, R. D. (1996) Permanence and the assembly of ecological communities. *Ecology*, **77**, 762–775.

Lefèvre, T., Labarenchon, C., Gauthier-Clerc, M. *et al.* (2009) The ecological significance of manipulative parasites. *Trends in Ecology and Evolution*, **24**, 41–48.

Loreau, M. (1998) Biodiversity and ecosystem functioning: a mechanistic model. *Proceedings of the National Academy of Sciences of the United States of America*, **95**, 5632–5636.

Loreau, M. (2010) *From Populations to Ecosystems: Theoretical Foundations for a New Ecological Synthesis*. Princeton, NJ: Princeton University Press.

Loreau, M., Naeem, S., Inchausti, P. *et al.* (2001) Biodiversity and ecosystem functioning: current knowledge and future challenges. *Science*, **294**, 804–808.

Martinez, N. D. (1992) Constant connectance in community food webs. *American Naturalist*, **139**, 1208–1218.

May, R. M. (1972) Will large complex systems be stable? *Nature*, **238**, 413–414.

Mitchell, C., Agrawal, A., Bever, J. *et al.* (2006) Biotic interactions and plant invasions. *Ecology Letters*, **9**, 726–740.

Montoya, J. M. and Solé, R. V. (2003) Topological properties of food webs: from real data to community assembly models. *Oikos*, **102**, 614–622.

Mooney, H. A. and Hobbs, R. J. (2000) *Invasive Species in a Changing World*. Washington DC: Island Press.

Mulder, C. P. H., Uliassi, D. D. and Doak, D. F. (2001) Physical stress and diversity-productivity relationships: the role of positive interactions. *Proceedings of the National Academy of Sciences of the United States of America*, **98**, 6704–6708.

Naeem, S., Bunker, D. E., Hector, A., Loreau, M. and Perrings, C. (2009) *Biodiversity, Ecosystem Functioning, and Human Wellbeing: an Ecological and Economic Perspective*. Oxford: Oxford University Press.

Okuyama, T. and Bolker, B. (2007) On quantitative measures of indirect interactions. *Ecology Letters*, **10**, 264–271.

Olff, H., Alonso, D., Berg, M. P. *et al.* (2009) Parallel ecological networks in ecosystems. *Philosophical Transactions of the Royal Society of London, Series B*, **364**, 1755–1779.

Neutel, A. M., Heesterbeek, J. A. P. and de Ruiter, P. C. (2002) Stability in real food webs: weak links in long loops. *Science*, **296**, 1120–1123.

Pfisterer, A. B., Joshi, J., Schmid, B. and Fischer, M. (2004) Rapid decay of diversity-productivity relationships after invasion of experimental plant communities. *Basic and Applied Ecology*, **5**, 5–14.

Ringel, M. S., Hu, H. H. and Anderson, G. (1996) The stability and the persistence of mutualisms embedded in community interactions. *Theoretical Population Biology*, **50**, 281–297.

Rixen, C. and Mulder, C. P. H. (2005) Improved water retention links high species richness with increased productivity in Arctic tundra moss communities. *Oecologia*, **146**, 287–299.

Schmitz, O. J., Krivan, V. and Ovadia, O. (2004) Trophic cascades: the primacy of trait-mediated indirect interactions. *Ecology Letters*, **7**, 153–163.

Stohlgren, T. J., Binkley, D. and Chong, G. W. (1999) Exotic plant species invade hot spots of native plant diversity. *Ecological Monographs*, **69**, 47–68.

Thébault, E. and Loreau, M. (2003) Food-web constraints on biodiversity–ecosystem functioning relationships. *Proceedings of the National Academy of Sciences of the United States of America*, **25**, 14949–14954.

Thébault, E., Huber, V. and Loreau, M. (2007) Cascading extinctions and ecosystem functioning: contrasting effects of diversity depending on food web structure. *Oikos*, **116**, 163–173.

Werner, E. E. and Peacor S. D. (2003) A review of trait-mediated indirect interactions in ecological communities. *Ecology*, **84**, 1083–1100.

Williamson, M. (1996) *Biological Invasions*. New York: Chapman and Hall.

Wootton, T. (1994) The nature and consequences of indirect effects in ecological communities. *Annual Review of Ecology and Systematics*, **25**, 443–466.

Zhu, B., Fitzgerald, D. G., Mayer, C. M., Rudstam, L. G. and Mills, E. L. (2006) Alteration of ecosystem function by zebra mussels in Oneida Lake: impacts on submerged macrophytes. *Ecosystems*, **9**, 1017–1028.

PART IV

Applied Ecology

Perspective: consequences of trait-mediated indirect interactions for biological control of plant pests

MAURICE W. SABELIS, ARNE JANSSEN and IZABELA LESNA

Institute for Biodiversity and Ecosystem Dynamics, University of Amsterdam

Introduction

Biological control of plant pests involves the use of natural enemies to control arthropods that in turn are enemies of the plant. Classic theory on biocontrol focused on density-dependent interactions between natural enemies and plant pests (Hassell 1978), but largely ignored a role for the plant other than directly affecting the growth rate of the plant pest (i.e., resulting from nutritional quality and direct anti-herbivore defence of the plant). That the plant's role goes far beyond setting the growth rate of the herbivore was realized much later. Prompted by a seminal review by Price *et al.* (1980), an avalanche of research now shows the multifarious ways in which plants influence the effectiveness of natural enemies against plant pests and the vulnerability of these pest organisms to their enemies. Such effects on the herbivore's predators may arise from constitutive investment by the plant, or from versatile investments induced by herbivory. Thus, herbivore-induced changes in the state of the plant may directly and indirectly, i.e., via predators, affect herbivores (Dicke and Sabelis 1988; Turlings *et al.* 1995; Karban and Baldwin 1997). To the best of our knowledge, predators do not change the state of plants, but they can induce changes in behaviour and life history of herbivores, thereby indirectly affecting the plant (Schmitz *et al.* 2004). Moreover, herbivorous arthropods may foresee predation risk and avoid predator-occupied plants in response to chemical alarm from herbivores under attack (Janssen *et al.* 1998; Pallini *et al.* 1999; Nomikou *et al.* 2003; Lee *et al.* 2011). Taken together, tritrophic systems of arthropods on plants are replete with trait changes induced by organisms at adjacent trophic levels and with effects cascading across all trophic levels. Such indirect effects become ever more complex in multispecies food webs because they arise whenever there are more than two species (Werner and Peacor 2003). One example is competition between plant pest species, which not only directly compete for food or by interference, but may also compete indirectly via the plant's defence response and via responses of shared and non-shared predators.

Trait-Mediated Indirect Interactions: Ecological and Evolutionary Perspectives, eds. Takayuki Ohgushi, Oswald J. Schmitz and Robert D. Holt. Published by Cambridge University Press. © Cambridge University Press 2012.

In this chapter we ask how TMIEs determine plant pest control by predators. We sketch case studies on biocontrol to illustrate the different ways in which TMIEs emerge: (1) herbivore-induced indirect plant defence, (2) predator-induced escape behaviour of herbivores, (3) predator-induced ontogenetic escape by herbivores, and (4) plant- and predator-mediated competition between herbivores. This review – though not exhaustive and largely based on our own work – aims to specify mechanisms by which biological control is affected by trait-mediated interactions and timescales at which individual and population responses become manifest.

Herbivore-induced indirect plant defence

Herbivory can induce plant responses that affect natural enemies of the herbivore. Upon attack by herbivorous arthropods, plants are systemically induced to release volatile chemicals that help arthropod predators (including parasitoids) to find their prey, i.e., herbivorous arthropods (e.g., Dicke and Sabelis 1988; Turlings *et al.* 1995; Sabelis *et al.* 2007). These responses reduce time elapsed between herbivore attack on the plant and natural enemy arrival. Such enemy response times may be short (minutes to hours) when the enemy is abundant or very mobile and then plant alarm may help only marginally, but they may also occur at long timescales (days to weeks) when the enemy is not abundant and has limited mobility. In the latter case the benefit to the plant from alarm 'calls' is magnified if the herbivore multiplies rapidly. For example, consider two-spotted spider mites on cucumber plants. They double their populations every 2 to 4 days, overexploiting the plant in a few weeks. Their most effective control agent, the predatory mite *Phytoseiulus persimilis*, has similar population doubling times, given sufficient prey for conversion into predator eggs. Since its prey consumption rate is modest (i.e., *c.* 1 per hour for young prey down to 1 per 7 hours for adult female prey), the capacity of predators to keep up with prey population growth depends on initial predator-to-prey ratios. The first arriving predators largely determine the realized predator population growth. In an experiment with small cucumber plants in open tubes on a gauze frame, removal of headspace volatiles significantly delayed first arrival of predatory mites (released 15 m downwind) from several hours up to several days (Sabelis, unpublished data). Clearly, long-distance search requires a timescale close to that of mite population growth on the plant and will thus affect pest control. In greenhouse crops, however, effects of plant volatiles may not become manifest because predatory mites are released on alternate plants in each row, making large-distance searching less vital to control.

Apart from inducing the plant to produce alarm chemicals, herbivores may also induce plant morphological changes, promoting predator access to herbivores. This mechanism is well understood for another taxon of herbivorous mites, the eriophyoid mites, the smallest arthropods on Earth (Lindquist *et al.* 1996). Their worm-like body has a cross-section diameter of c. 50 μm, at least five-times smaller than that of phytoseiid mites, one of their most significant predators (Sabelis 1996). The minute size of the eriophyoid mite is key to their ecological success, enabling them to reach places small enough to be free of predators (Sabelis 1996). Moreover, it allows them to develop a plant-parasitic lifestyle quite different from other herbivorous arthropods (Lindquist *et al.* 1996). Many eriophyoids live in plant galls they induce, but the ones of interest here have a vagrant lifestyle, frequently changing feeding sites that vary in the degree of protection against predators. In agricultural crops, such mites may easily reach pest status when predatory mites are lacking. Chemical control is often ineffective because eriophyoid mites may feed under protective structures of the plant. Hence, biological control with predatory mites may be essential. Consider tulip bulbs under attack by *Aceria tulipae*, an eriophyid mite so tiny (*c*. 0.06 mm) that it can move between bulb scales (Lesna *et al.* 2005). Herbivory induces these bulbs to modify their internal structure. The resulting changes in distance between bulb scales (from 0.1 to 0.2 mm) are microscopic, yet allow the phytoseiid predator *Neoseiulus cucumeris* (with a cross-sectional diameter of 0.2 mm) to enter the interior bulb space. The consequence of this plant 'behaviour' is dramatic: predators clear the inside of the bulb of herbivores, whereas the bulb otherwise is eaten from within. The crucial changes in bulb morphology enabling predator access are controlled by ethylene, a plant hormone released upon herbivore attack (Lesna *et al.* 2005). This plant hormone simultaneously induces the release of plant volatiles that attract predatory mites. Changes in bulb attractiveness and accessibility were demonstrated by combining chemical analysis (Aratchige 2007), olfactometry (Aratchige *et al.* 2004) and experiments on predator–prey dynamics in which the effect of ethylene was either promoted or blocked by 1-methyl-cyclopropene, a chemical occupying ethylene receptor sites (Lesna *et al.* unpublished manuscript). In climate cabinets ventilated to remove ethylene, short, 2-weekly exposure of bulbs to ethylene had major effects on the ability of the predatory mite *N. cucumeris* to control dry bulb mites in the interior of the bulb when compared to identical exposure to ethylene blockers or ambient air (Fig. 22.1). Exposure to ethylene and its blockers had no direct effect on dry bulb mites and predatory mites and at the concentrations offered ethylene exposure had similar effects on the morphology of healthy bulbs as exposure to (ethylene and other odours from) infested tulip bulbs. Taken together, these results explain why earlier experiments yielded successful biological control of dry bulb mites on tulip bulbs in closed cardboard boxes with consequently higher ethylene concentrations, but were not successful in open trays in ventilated climate rooms (Lesna *et al.* 2005).

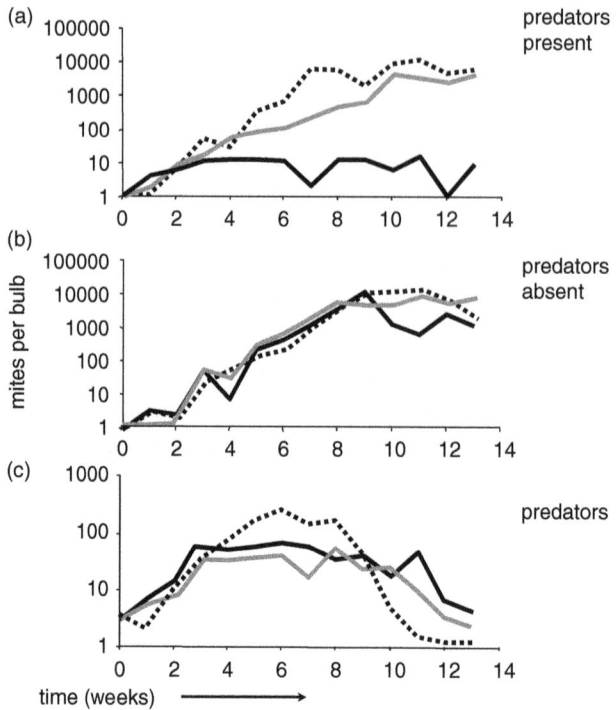

Figure 22.1 Dynamics of dry bulb mites (*Aceria tulipae*) in the interior of tulip bulbs when exposed to predatory mites (b) or not (a) and of predatory mites (*Neoseiulus cucumeris*) in the interior as well the exterior parts of the bulbs (c). Three trays, each with 100 bulbs, were assigned to one of three treatments involving 2-weekly exposure for 1 day to ambient air (grey lines), ethylene (10 ppm) (black lines) or 1-methyl-cylopropene (MCP) (1 ppm) (dashed lines). The experiments were done in climate cabinets ventilated to remove bulb-produced ethylene and kept at 20°C and 60% RH.

As another example, consider the coconut mite *Aceria guerreronis* (Aratchige 2007; Aratchige *et al.* 2007). It lives under the fruit perianths of coconut palms. The perianth tightly covers the meristematic zone of female flowers and developing coconuts thereafter. The perianth edge is closely appressed to the fruit, leaving little space for mites to enter, but due to their minute size coconut mites can do so. Some predatory mite species are small enough to move under fruit perianths and attack coconut mites. In Sri Lanka, the phytoseiid mite *Neoseiulus baraki* is commonly found in association with coconut mites and is three-times larger than its prey. It has a flat body and elongated idiosoma with short distal setae, so relative to many other phytoseiid mites it is better able to reach the narrow space under perianths of infested nuts. On uninfested nuts, however, they are hardly ever observed under the perianth. Motivated by the above tulip study, we hypothesized that the nuts change their morphology in response to damage by coconut mites thereby allowing predatory mites to enter under the perianth of infested nuts. This was tested in an experiment where we measured the distance between the perianth and the coconut fruit surface in three cultivars commonly grown in Sri Lanka (Aratchige *et al.* 2007). In uninfested nuts, this distance was large enough for coconut mites to creep under the perianth, yet too small for predators. But in nuts infested by coconut mites the perianth–fruit distance increased, such that the predatory mites could also enter. This coconut response to infestation is not instantaneous, but takes time to develop, and depends on coconut age. Very young fruits have a perianth so tightly appressed to the nut that predatory mites have much more difficulty penetrating under the perianth than do coconut mites. This gives coconut mites a head start in population buildup, leading to strongly diverging population curves as a function of fruit age. Protecting very young fruits from coconut mite colonization is therefore key for successful biological control (Negloh et al. 2010).

Like the coconut perianth, dense covers of (glandular) hairs also protect vital plant parts against many herbivorous arthropods. On tomato, for example, glandular hairs on leaves and stems protect against two-spotted spider mites (Chatzivasileiadis et al. 1999, 2001), but also hinder predatory mites (Van Haren et al. 1987). This creates a competitor-free and enemy-free space for the eriophyoid mite Aculops lycopersici being so minute that it can seek refuge and feed between the glandular hairs. A pest outbreak would then be inevitable unless tomato plants take a countermeasure. Indeed, after attack by A. lycopersici, some tomato cultivars drop their defensive glandular hairs, which in turn exposes the attackers to their predators (Glas et al., unpublished data).

Predator-induced escape behaviour of herbivores

Herbivorous arthropods do not only defend themselves against predators, but they may also exhibit avoidance behaviour in response to predators or predator-associated cues (Janssen et al. 1998). For example, western flower thrips larvae vigorously swing their abdomen to kick predators upon attack and produce anal droplets upon contact causing predators to spend considerable time in cleaning themselves (Bakker and Sabelis 1989). Moreover, thrips larvae move into refuges when exposed to predators (Pallini et al. 1998). As another example, two-spotted spider mites protect themselves against predatory mites by silken webs that cover the leaf area on which they feed and that are hard to penetrate by most (though not all) predatory mites (Sabelis and Bakker 1992). They also avoid spider-mite-infested plants when occupied by the predatory mite P. persimilis, which can penetrate defensive webs of two-spotted spider mites (Pallini et al. 1999). When western flower thrips and two-spotted spider mites co-occur on the same plant, then the presence of predatory mites attacking thrips triggers western flower thrips to seek refuge in the spider-mite web. Here, thrips larvae suffer from severe food competition with spider mites as they both feed on leaf parenchyma, but they gain protection against their own predators (as these cannot penetrate spider-mite webs) and they are not eaten by the predatory mite, P. persimilis, which is specialized on spider mites producing profuse webs. Moreover, thrips larvae can feed on spider-mite eggs in webs and utilize this food for development (Pallini et al. 1998). Consequently, biological control of western flower thrips using predators, such as Iphiseius degenerans and N. cucumeris, fails in the presence of (sufficient refuge created by) two-spotted spider mites. This phenomenon of secondary thrips outbreaks when two-spotted spider mite control fails, occurs in greenhouse practice (Bolckmans, pers. comm.) but only occasionally because *P. persimilis* is so effective in controlling two-spotted spider mites.

Herbivorous arthropods usually have multiple enemies (Sabelis 1992; Helle and Sabelis 1985; Lindquist *et al.* 1996), raising the question of whether their escape responses are tuned to predator identity (Schmitz *et al.* 2004). A fascinating example of such behaviour is described for mites inhabiting cassava in Africa. The cassava green mite (*Mononychellus tanajoa*) accidentally invaded the African continent on plant material shipped from South America in 1971. It quickly became a serious pest of cassava throughout much of its growing areas in sub-Saharan Africa (Yaninek 1988). Two exotic predator species from South America were successfully released in Africa for pest control (Yaninek and Hanna 2003; Hanna *et al.* 2005): (1) a leaf-inhabiting predatory mite, *Amblydromalus* (=*Typhlodromalus*) *manihoti* (introduced in 1989) and (2) an apex-inhabiting predatory mite, *Typhlodromalus aripo* (introduced in 1993). The apex-inhabiting predatory mite resides in the plant tips during the day, but emerges at night to forage for cassava green mites on the upper leaves (Onzo *et al.* 2003). Leaf-inhabiting predatory mites are never found in plant tips and occur on leaves day and night. In response to *T. aripo* foraging down from the apex at night, cassava green mites move down in the plant (Magalhães *et al.* 2002; Onzo *et al.* 2003) but they tend to shift up in the plant in response to *A. manihoti* on mid-stratum leaves (Magalhães *et al.* 2002). This behaviour provides temporary escape for mobile stages of cassava green mites because immobile stages (eggs and moulting stages) preoccupy the predators that would otherwise pursue mobile prey. However, when both predator species are present, cassava green mites can escape only by leaving the plant, which is very risky.

Though the two predator species share cassava green mites as prey, they exhibit microhabitat segregation, reducing intra-guild predation. Biological control is expected to fare better using both predator species instead of either one. However, the leaf-inhabiting predator *A. manihoti* established only in a small area of the cassava belt due to limited drought tolerance. The apex-inhabiting predator is more protected during the day from low humidity and incoming UV radiation (Onzo *et al.* 2010), permitting it to establish over vast areas of the cassava belt. Time series analysis of monthly mite counts in cassava fields (free of *T. manihoti*) over a period of 7 years in Benin (Hanna *et al.* 2005; Fig. 22.2a) showed that mite abundance was strongly modified by seasonal effects of rainfall and that *T. aripo* and cassava green mites exhibit predator–prey oscillations with a period of *c.* 6 months and a prey–predator phase difference of *c.* 1 month, close to that predicted from a parameterized

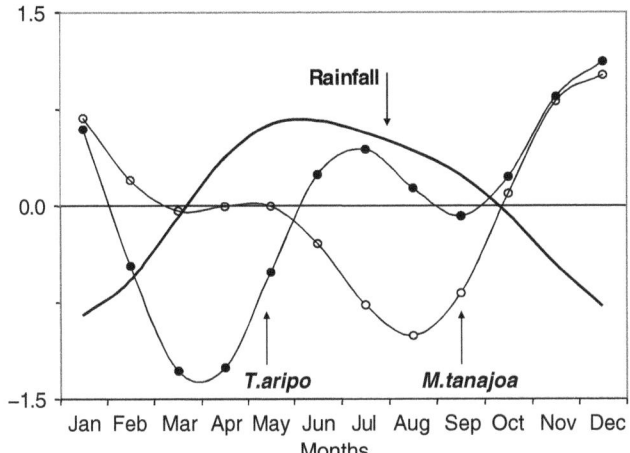

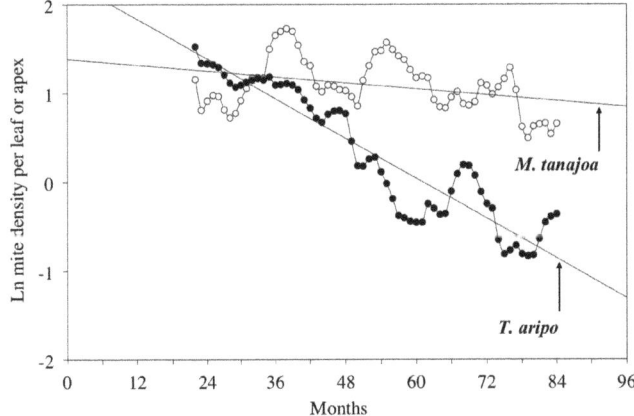

Figure 22.2 Some results from time series analysis applied to monthly estimates (from 1993 to 2001) of the natural logarithm of densities of cassava green mites (*Mononychellus tanajoa*) per leaf and of predatory mites (*Typhlodromalus aripo*) per apex (Hanna *et al.* 2005). (a) Moving annual means of log-transformed data (straight lines represent estimated regression lines) and (b) annual models of the periodical components of rainfall and of the ln densities.

Rosenzweig–MacArthur predator–prey model. Over all 7 years, the density of *T. aripo* declined steadily, whereas that of cassava green mites stayed constant and low (Hanna *et al.* 2005; Fig. 22.2b). Because the 7-year period includes 14 predator–prey cycles, there must have been a change in parameters determining predator, but not prey, equilibria of the Rosenzweig–MacArthur model. This can only be due to a gradual decrease in the prey population growth rate (or carrying capacity). Such a decrease is conceivable, if apex-inhabiting predators exert selection on cassava green mites to escape by moving deeper down in the plant to older leaves that are photosynthetically less active, conferring lower population growth rates on this prey (Hanna *et al.* 2005). Given that cassava green mites have adapted to African conditions over more than 20 years before this predator was introduced, a reasonable hypothesis is that such avoidance behaviour of the prey has been favoured by selection, initiated after predator release. Although speculative, this explanation accords with expectations from a parameterized, mechanistic predator–prey model;

changes in the food conversion rate, the predator's death rate or its functional response are implausible explanations because they determine the prey equilibrium which apparently did not change (Fig. 22.2b). The case in point is that natural selection on a prey trait – predator-induced avoidance behaviour of cassava green mites – may have caused the gradual decline of predator density over ecological timescales.

Little is known as to how herbivorous arthropods assess predation risks. One would expect them to sample their environment and gain experience before they settle on a plant. Indeed, adult females of a small herbivore, the whitefly *Bemisia tabaci*, can learn to avoid plants with predatory mites that attack only juveniles of the whiteflies, while they lay eggs on plants without predators (Nomikou *et al.* 2003). Predatory mites disperse more slowly than whiteflies as they cannot fly and therefore have to walk from plant to plant. Hence, by avoiding plants with predators, the whiteflies create a temporary refuge for their offspring. Whereas adult whiteflies avoid plants with predatory mites feeding on immature whiteflies, they did not avoid plants with predators feeding on pollen (Meng *et al.* 2006). Possibly, they use cues (e.g., alarm pheromones) from their predator-damaged congeners. The responses to such cues can have significant effects at the population level, as shown in an aphid–parasitoid system; aphid resource use changed when congeners died from parasitization, ultimately leading to aphid population decline (Fievet *et al.* 2008).

Predator-induced ontogenetic escape by herbivores

Beyond short-term behavioural responses induced by predators, herbivorous arthropods may also show long-term ontogenetic responses. For example, when exposed to coccinellid predators, more pea aphids in a colony develop into winged morphs (Weisser *et al.* 1999). The sluggishness of coccinellid consumption may permit enough time for some young aphids to escape via development of wings. Such induced ontogenetic responses are to be expected in colony-forming arthropods and occur in spider mites (Kroon *et al.* 2004, 2005, 2008). These mites are specialized to feed on leaves in spring and summer, but risk death in late summer when the leaves harbour relatively large numbers of predatory mites, their main enemies. The mites face a dilemma: stay on the leaf, eat and risk being eaten, or move away from the leaf to plant parts with no food. Female two-spotted spider mites solve this dilemma by ontogenetic change into a diapause state when exposed to predation-associated cues in the juvenile phase (Kroon *et al.* 2004, 2005, 2008). Diapause is otherwise induced above a critical night length at sufficiently low temperatures and low food quality. Apart from increased cold tolerance, diapause involves a dramatic reduction in physiological activity, promoting survival in areas without food. Moreover, diapause females change their behaviour: they disperse away from leaves to survive in winter refuges on the bark of trees or in the soil. Here, they are protected from leaf-inhabiting predatory mites. Thus, predation-associated cues on leaves cause some herbivorous mites to seek refuge elsewhere. At the population level this trait change should retard population growth of the herbivorous mites and thus also their predators; this dampens their population fluctuations, promotes herbivore persistence due to the refuge and maintains the herbivore population at a somewhat higher level than in absence of the refuge (McNair 1986, 1987; Hassell 1978, 2000; Murdoch *et al.* 2003). These expectations seem to be confirmed by a long-term experiment on integrated control of European red mites (*Panonychus ulmi*) in apple orchards in The Netherlands (Gruys 1982, 1986). Starting from their effective release, predatory mites require a time span of 3–6 years to have a significant impact on European red mite populations (Fig. 22.3). This sharply contrasts to greenhouse experiments and model simulations by Rabbinge (1976) showing that predatory mites can control European red mites within a single season over a wide range of early season predator–prey ratios. We hypothesize that European red mites changing into a diapausing state create a prey refuge effect, dampening fluctuations, but delaying successful biological control at a timescale of several years. This refuge effect reflects three processes: (1) diapause development triggered by increasing night length and lower temperatures, (2) predator-induced diapause, (3) variable, but, on average, much higher winter survival rates for European red mites than for predatory mites (Sabelis, unpublished data), giving the herbivores a head-start at the beginning of each season. Disentangling their impact on orchard mite dynamics is currently underway.

Plant- and predator-mediated competition among herbivores

Herbivore species may interact with each other not only by direct interference and exploitative competition, but also through impacts on plant defence. We recently found that the spider mite *Tetranychus evansi* downregulates plant defences in attacked leaves of tomato plants, resulting in higher rates of oviposition and population growth than on leaves of clean plants (Sarmento *et al.* 2011a). The danger of such downregulation is that attacked leaves may become a more profitable resource for heterospecific competitors, such as the two-spotted spider mite *T. urticae*. Indeed, *T. urticae* had an almost two-fold higher rate of oviposition on leaves previously fed upon by *T. evansi* (Sarmento *et al.* 2011b). In contrast, induction of direct plant defences by *T. urticae* resulted in decreased oviposition of *T. evansi*. Hence, these herbivore species affect each other through induced plant responses. However, when populations of *T. evansi* and *T. urticae* competed on the same plants without predation, populations of the latter went extinct, whereas *T. evansi* was not significantly affected by the presence of its competitor. This suggests that *T. evansi*

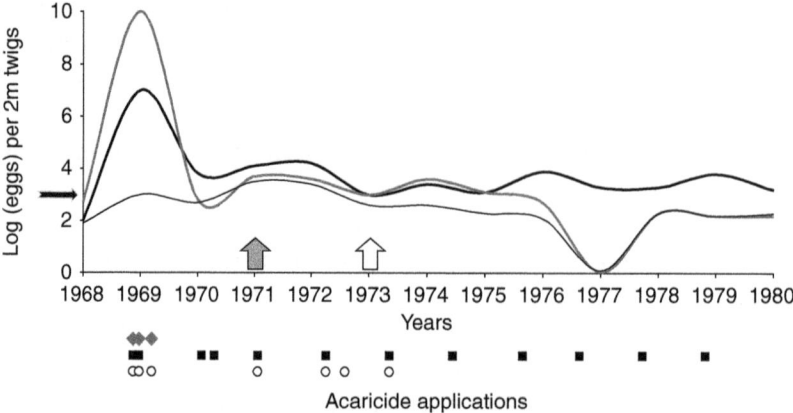

Figure 22.3 Annual dynamics of the European red mite (*Panonychus ulmi*) in the experimental orchard 'De Schuilenburg' (Lienden, The Netherlands) from 1968 to 1980 (Gruys 1986). Data (drawn as smoothed lines) concern the winter eggs (the actual overwintering stage) laid on apple twigs by diapause-induced females and they express the log-density of winter eggs per 2 m of apple twigs (sampled in February–March each year from four different apple cultivars). Winter egg densities are shown for three pest control regimes involving intensive chemical control (thick black line), moderate chemical control (thin black line) and selective chemical control (grey line). The timing of acaricide applications under the three regimes (black squares, open circles and grey diamonds respectively) is shown below the *x*-axis. Under intensive chemical control (i.e. use of broad-spectrum pesticides which as a side effect eliminate predatory mites) yearly applications of acaricides were required to prevent European red mite outbreaks. Under the other two regimes predatory mites thrived and acaricides were almost not applied (only in early years) because the use of pesticides was kept to a minimum (moderate chemical control) or because selective pesticides were used that did not harm predatory mites (selective chemical control). The vertical arrows indicate the years in which predatory mites were effectively released (presence or absence of shading refers to chemical control regime). The horizontal black arrow indicates the economic threshold, i.e. a commonly agreed density above which pest control measures are required to prevent yield reduction.

somehow prevents its competitor benefiting from downregulated plant defence. Indeed, *T. urticae* could not reach the leaf surface to feed when the leaf was covered with profuse web produced by *T. evansi* (Sarmento *et al.* 2011b). Furthermore, *T. evansi* produced even more web when exposed to damage or other cues associated with *T. urticae*. Thus, *T. evansi* protects the leaf area with downregulated plant defences by covering it with a dense silken web, preventing its competitors from feeding.

When feeding on tomato, *T. evansi* also reduced production of herbivore-induced plant volatiles (the terpenoids TMTT and β-myrcene) (Sarmento *et al.* 2011a), which are parts of tomato odours induced by *T. urticae* (Kant *et al.* 2004).

Our initial expectation was that this reduction in plant alarms would reduce attraction of predatory mites. This was tested for two predatory mite species, *Phytoseiulus macropilis* which is effective against *T. urticae* but not against *T. evansi* (due to the latter's profuse webbing) and *Phytoseiulus longipes*, a predator naturally occurring with *T. evansi* in Southern Brazil and Northern Argentina (Furtado *et al.* 2007). Contrary to expectations, olfactometer experiments showed that each of the two predator species was more attracted to infested than to uninfested tomato plants irrespective of whether they were fed upon by *T. urticae* or *T. evansi*. The predator, *P. longipes*, also did not discriminate between odours from tomato plants infested by *T. urticae* and those from plants infested by *T. evansi* when offered simultaneously. Thus, each spider mite species triggers plant alarms that attract the predator of its competitor apart from attracting their own specific predator. This mechanism may contribute to the recent emergence of *T. evansi* as an invasive pest in Africa and the Mediterranean basin, where its native predator *P. longipes* is absent (Boubou *et al.* 2011). This invasion took place into tomato crops where *T. urticae* is a pest commonly controlled by the release of *P. persimilis*. In that case, *P. persimilis* attracted to a plant infested by *T. evansi* is unable to harm this spider mite species, yet would eliminate any *T. urticae* colonizing the plant, thereby freeing *T. evansi* from its competitor. This hypothesis needs further experimental validation, but if true it represents a novel case of predator- and plant-mediated invasiveness of a herbivore (Schierenbeck *et al.* 2010).

Conclusions and perspectives

In its simplest form, biological control of plant pests concerns linear tri-trophic food chains. This not only involves interactions between adjacent trophic levels, but also between plants and predators (Sabelis *et al.* 2007) and includes density-mediated effects stemming directly from mortality, as well as a range of nonlethal effects that can lead to indirect effects on other community members due to enemies influencing victims' traits (Werner and Peacor 2003). We described case studies where nonlethal effects are manifested as altered escape or defence behaviour that likely emerges from balancing foraging opportunities against predation risks (Schmitz *et al.* 2004), and others where nonlethal effects may also emerge from ontogenetic changes into morphs better suited to defend or escape. Moreover, we provided evidence that plants can be induced to alter traits that affect the predator's ability to capture herbivores (Dicke and Sabelis 1988; Karban and Baldwin 1997; Sabelis *et al.* 2007). If this chapter has succeeded in showing that trait-mediated indirect interactions are ubiquitous in bio-control systems and potentially affect pest suppression, we have reached our goal. Moreover, we hope to have convinced the reader not only that enemy exposure can change individual traits over various timescales, but also that population responses emerge over widely different timescales.

While it is possible to show enemy-induced trait changes in the laboratory, it is notoriously difficult to demonstrate their impact in the field. If the pragmatic aim is to establish biological control, field workers tend to refrain from spending the extra efforts needed to demonstrate TMIEs. Yet, understanding the role of trait-mediated effects may well prove crucial to get insight into the mechanisms underlying biological control as well as biological invasions into agricultural systems, and this understanding may be essential for generalizing among systems. New opportunities for exploring trait-mediated effects are arising from the use of genetically modified plants and arthropods and from the use of plant hormones involved in direct and indirect plant defence (e.g., ethylene, salicylic and jasmonic acid). However, all these new experimental tools have the problem that organismal manipulations may be expressed in more complex ways than initially thought, mandating the use of mathematical models. Another general reason to use such models is the interactive nature of biological systems, where effects of a trait change cascade across all trophic levels and at each trophic level among all species that directly or indirectly compete with each other (Holt and Hochberg 2001). Finally, such models help to clarify the timescale required for individual trait changes, whether behavioural or ontogenetic, to become expressed at the population level.

References

Aratchige, N. S. (2007) Predators and the accessibility of herbivore refuges in plants. PhD Thesis, University of Amsterdam.

Aratchige, N. S., Lesna, I. and Sabelis, M. W. (2004) Belowground plant parts emit herbivore-induced volatiles: olfactory responses of a predatory mite to tulip bulbs infested by rust mites. *Experimental and Applied Acarology*, **33**, 21–30.

Aratchige, N. S., Sabelis, M. W. and Lesna, I. (2007) Plant structural changes due to herbivory: do changes in *Aceria*-infested coconut fruits allow predatory mites to move under the perianth? *Experimental and Applied Acarology*, **43**, 97–107.

Bakker, F. M. and Sabelis, M. W. (1989) How larvae of *Thrips tabaci* reduce the attack success of phytoseiid predators. *Entomologia Experimentalis et Applicata*, **50**, 47–51.

Boubou, A., Migeon, A., Roderick, G. K. and Navajas, M. (2011) Recent emergence and worldwide spread of the red tomato spider mite, *Tetranychus evansi*: genetic variation and multiple cryptic invasions. *Biological Invasions*, **13**, 81–92.

Chatzivasileiadis, E. A., Boon, J. J. and Sabelis, M. W. (1999) Accumulation and turnover of 2-tridecanone in *Tetranychus urticae* Koch and its consequences for resistance of wild and cultivated tomatoes. *Experimental and Applied Acarology*, **23**, 1011–1021.

Chatzivasileiadis, E. A., Egas, M. and Sabelis, M. W. (2001) Resistance to 2-tridecanone in *Tetranychus urticae* Koch: effects of induced resistance, cross-resistance and heritability. *Experimental and Applied Acarology*, **25**, 717–730.

Dicke, M. and Sabelis, M. W. (1988) How plants obtain predatory mites as bodyguards. *Netherlands Journal of Zoology*, **38**, 148–165.

Fievet, V., Lhomme, P. and Outreman, Y. (2008) Predation risk cues associated with killed conspecifics affect the behavior and reproduction of prey animals. *Oikos*, **117**, 1380–1385.

Furtado I. P., de Moraes, G. J., Kreiter, S., Tixier, M.-S. and Knapp, M. (2007) Potential of a Brazilian population of the predatory mite *Phytoseiulus longipes* as a biological control agent of *Tetranychus evansi* (Acari: Phytoseiidae, Tetranychidae). *Biological Control*, **42**, 139–147.

Gruys, P. (1982) Hits and misses: the ecological approach to pest control. *Entomologia Experimentalis et Applicata*, **31**, 70–87.

Gruys, P. (1986) Ontwikkeling van de fruitspintmijt en roofmijt op de Schuilenburg, 1965–1980. *Intern Report S288, Praktijkonderzoek Plant en Omgeving (PPO-WUR)*, Fruitteeltpraktijkcentrum, Randwijk, The Netherlands.

Hanna, R., Onzo, A., Lingeman, R., Yaninek, J. S. and Sabelis, M. W. (2005) Seasonal cycles and persistence in an acarine predator–prey system on cassava in Africa. *Population Ecology*, **47**, 107–117.

Hassell, M. P. (1978) *The Dynamics of Arthropod Predator–Prey Systems*. Princeton, NJ: Princeton University Press.

Hassell, M. P. (2000) *The Spatial and Temporal Dynamics of Host–Parasitoid Interactions*. Oxford: Oxford University Press.

Helle, W. and Sabelis, M. W. (1985) *Spider Mites: Their Biology, Natural Enemies and Control. World Crop Pests*, Vol. 1. Amsterdam, The Netherlands: Elsevier Science Publishers.

Holt, R. D. and Hochberg, M. E. (2001) Indirect interactions, community modules and biological control: a theoretical perspective. In E. Waijnberg, J. K. Scott and P. C. Quimby, eds., *Evaluation of Indirect Ecological Effects of Biological Control*. Wallingford, UK: CABI Publishing, pp. 13–38.

Janssen, A., Pallini, A., Venzon, M. and Sabelis, M. W. (1998) Behaviour and food

web interactions among plant inhabiting mites and thrips. *Experimental and Applied Acarology*, **22**, 497–521.

Kant M. R., Ament, K., Sabelis, M. W., Haring, M. A. and Schuurink, R. C. (2004) Differential timing of spider mite-induced direct and indirect defenses in tomato plants. *Plant Physiology*, **135**, 483–495.

Karban, R. and Baldwin, I. (1997) *Induced Responses to Herbivory*. Chicago, IL: University of Chicago Press.

Kroon, A., Veenendaal, R. L., Bruin, J., Egas, M. and Sabelis, M. W. (2004) Predation risk affects diapause induction in the spider mite *Tetranychus urticae*. *Experimental and Applied Acarology*, **34**, 307–314.

Kroon, A., Veenendaal, R. L., Bruin, J., Egas, M. and Sabelis, M. W. (2008) Sleeping with the enemy: predator-induced diapause in a mite. *Naturwissenschaften*, **95**, 1195–1198.

Kroon, A., Veenendaal, R. L., Egas, M., Bruin, J. and Sabelis, M. W. (2005) Diapause incidence in the two-spotted spider mite increases due to predator presence, not due to selective predation. *Experimental and Applied Acarology*, **35**, 73–81.

Lee, D.-H., Nyrop, J. P. and Sanderson, J. P. (2011) Avoidance of natural enemies by adult whiteflies, *Bemisia argentifolii*, and effects on host plant choice. *Biological Control*, **58**, 302–309.

Lesna, I., Conijn, C. G. M. and Sabelis, M. W. (2005) From biological control to biological insight: rust-mite induced change in bulb morphology, a new mode of indirect plant defence? In G. Weigmann, G. Alberti, A. Wohltmann and S. Ragusa, eds., *Acarine Biodiversity in the Natural and Human Sphere. Phytophaga* (Palermo), **14**, 285–291.

Lindquist, E. E., Sabelis, M. W. and Bruin, J. (1996) *Eriophyoid Mites: Their Biology, Natural Enemies and Control. World Crop Pest Series*, Vol. 6. Amsterdam, The Netherlands: Elsevier Science Publishers.

Magalhães, S., Janssen, A., Hanna, R. and Sabelis, M. W. (2002) Flexible antipredator behaviour in herbivorous mites through vertical migration in a plant. *Oecologia*, **132**, 143–149.

McNair, J. N. (1986) The effects of refuges on predator–prey interactions: a reconsideration. *Theoretical Population Biology*, **29**, 38–63.

McNair, J. N. (1987) Stability effects of prey refuges with entry–exit dynamics. *Journal of Theoretical Biology*, **125**, 449–464.

Murdoch, W. W., Briggs, C. J. and Nisbet, R. M. (2003) *Consumer–Resource Dynamics*. Princeton, NJ: Princeton University Press.

Negloh K., Hanna, R. and Schausberger, P. (2010). Season- and fruit age-dependent population dynamics of *Aceria guerreronis* and its associated predatory mite *Neoseiulus paspalivorus* on coconut in Benin. *Biological Control*, **54**, 349–358.

Nomikou, M., A. Janssen and M. W. Sabelis (2003) Herbivore host plant selection: whitefly learns to avoid host plants that are unsafe for her offspring. *Oecologia*, **136**, 484–488.

Meng, R.-X., Janssen, A., Nomikou, M., Zhang, Q.-W. and Sabelis, M. W. (2006) Previous and present diets of mite predators affect antipredator behaviour of whitefly prey. *Experimental and Applied Acarology*, **38**, 113–124.

Onzo, A., Hanna, R., Zannou, I., Sabelis, M. W. and Yaninek, J. S. (2003) Dynamics of refuge use: diurnal, vertical migration by predatory and herbivorous mites within cassava plants. *Oikos*, **101**, 59–69.

Onzo, A., Sabelis, M. W. and Hanna, R. (2010) Effects of ultraviolet radiation on predatory mites and the role of refuges in plant structures. *Environmental Entomology*, **39**, 695–701.

Pallini, A., Janssen, A. and Sabelis, M. W. (1998) Predators induce interspecific herbivore competition for food in refuge space. *Ecology Letters*, **1**, 171–176.

Pallini, A., Janssen, A. and Sabelis, M. W. (1999) Spider mites avoid plants with predators. *Experimental and Applied Ecology*, **23**, 803–815.

Price, P. W., Bouton, C. E., Gross, P. *et al.* (1980). Interactions among three trophic levels: influence of plants on interactions between insect herbivores and natural enemies. *Annual Review of Ecology, Evolution, and Systematics*, **11**, 41–65.

Rabbinge, R. (1976) Biological control of the European red mite *Panonychus ulmi*. *Simulation Monographs*. Wageningen, The Netherlands: PUDOC.

Sabelis, M. W. (1992) Arthropod predators. In M. J. Crawley, ed., *Natural Enemies: The Population Biology of Predators, Parasites and Diseases*. Oxford: Blackwell, pp. 225–264.

Sabelis, M. W. (1996) *Phytoseiidae: Their Biology, Natural Enemies and Control. World Crop Pest Series*, Vol. 6. Amsterdam, The Netherlands: Elsevier Science Publishers.

Sabelis, M. W. and Bakker, F. M. (1992) How predatory mites cope with the web of their tetranychid prey: a functional view on dorsal chaetotaxy in the Phytoseiidae. *Experimental and Applied Acarology*, **16**, 203–225.

Sabelis, M. W., Takabayashi, J., Janssen, A. *et al.* (2007) Ecology meets plant physiology: herbivore-induced plant responses and their indirect effects on arthropod communities. In T. Ohgushi, T. P. Craig and P. W. Price, eds., *Ecological Communities: Plant Mediation in Indirect Interaction Webs*. Cambridge: Cambridge University Press, pp. 188–217.

Sarmento, R. A., Lemos, F., Bleeker, P. M. *et al.* (2011a) A herbivore that manipulates plant defence. *Ecology Letters*, **14**, 229–236.

Sarmento, R. A., Lemos, F., Dias, C. R. *et al.* (2011b) A herbivorous mite down-regulates plant defence and produces web to exclude competitors. *PLoS ONE*, **6**, e23757.

Schmitz, O. J., Krivan, V. and Ovadia, O. (2004) Trophic cascades: the primacy of trait-mediated indirect interactions. *Ecology Letters*, **7**, 153–163.

Schierenbeck, K. A., Lee, C. E. and Holt, R. D. (2010) Synthesizing ecology and evolution for the study of invasive species. *Evolutionary Applications*, **3**, 96–96.

Turlings, T. C. J., Loughrin, J. H., McCall, P. J. *et al.* (1995) How caterpillar-damaged plants protect themselves by attracting parasitic wasps. *Proceedings of the National Academy of Sciences of the United States of America*, **92**, 4169–4174.

Van Haren, R. J. F., Steenhuis, M. M., Sabelis, M. W. and de Ponti, O. M. B. (1987) Tomato stem trichomes and dispersal success of *Phytoseiulus persimilis* relative to its prey *Tetranychus urticae. Experimental and Applied Acarology*, **3**, 115–121.

Weisser, W. W., Braendle, C. and Minoretti, N. (1999) Predator-induced morphological shift in the pea aphid. *Proceedings of the Royal Society of London, Series B*, **266**, 1175–1181.

Werner, E. E. and Peacor, S. D. (2003) A review of trait-mediated indirect interactions in ecological communities. *Ecology*, **84**, 1083–1100.

Yaninek, J. S. (1988) Continental dispersal of the cassava green mite, an exotic pest in Africa, and implications for biological control. *Experimental and Applied Acarology*, **4**, 211–224.

Yaninek, J. S. and Hanna, R. (2003) Cassava green mite in Africa: a unique example of successful classical biological control of a mite pest on a continental scale. In P. Neuenschwander, C. Borgemeister and L. Langewald, eds., *Biological Control in IPM Systems in Africa*. Wallingford, UK: CABI Publishing, pp. 61–75.

CHAPTER TWENTY-THREE

Natural enemy functional identity, trait-mediated interactions and biological control

TOBIN D. NORTHFIELD

Department of Zoology, University of Wisconsin-Madison

DAVID W. CROWDER

Department of Entomology, Washington State University

RANDA JABBOUR

Department of Plant, Soil, and Environmental Sciences, University of Maine

and

WILLIAM E. SNYDER

Department of Entomology, Washington State University

Introduction
Functional diversity schemes: are they useful?

Recent years have seen great interest in the importance of species richness for the functioning and stability of ecological communities (Ives and Carpenter 2007). Empirical examinations of richness effects typically vary the number of species in experimental treatments and measure resulting ecosystem functions such as biomass accumulation or resource uptake (Naeem *et al.* 2009). Across trophic levels and communities of many types, a clear pattern has emerged from these experiments: community processes (biomass accumulation, resource uptake, etc.) generally become more efficient when more species are present (Hooper *et al.* 2005; Cardinale *et al.* 2006). This pattern is generally attributed to resource partitioning among species, where species differ in ecologically significant ways such that they complement one another (Hooper *et al.* 2005). For example, in English meadow communities multiple plant species coexist, because different plant species exploit different hydrological conditions (Silvertown *et al.* 1999). The plants that dominate drought-prone areas are different from those that thrive in flood-prone areas and, presumably, total plant biomass is greatest when both plant groups (drought tolerant and flood tolerant) are present.

A remaining challenge is to effectively predict, a priori, the particular species (or groups of species) that will complement one another. One simplifying scheme that has received considerable attention is the lumping of species into 'functional groups'. In this functional-group approach, species within a

Trait-Mediated Indirect Interactions: Ecological and Evolutionary Perspectives, eds. Takayuki Ohgushi, Oswald J. Schmitz and Robert D. Holt. Published by Cambridge University Press. © Cambridge University Press 2012.

group are relatively similar to one another, and considered ecologically redundant, whereas species in different groups are distinct and complementary (Hillebrand and Matthiessen 2009). This approach gained support from studies suggesting that plant species can be classified into such functional groups (grasses, forbs, legumes and woody plants), and that the number of functional groups is a more effective predictor of ecosystem function than species richness (Diaz and Cabido 2001). For example, in savannah grasslands, plant communities that included C_3 grasses, C_4 grasses, forbs, legumes and woody plants had greater biomass and plant nitrogen accumulation, and reduced light penetration, than those communities lacking one or more of these groups (Tilman et al. 1997). These authors suggested that competition was greater within than between functional groups, consistent with niche similarity within, but niche differentiation among, groups.

The utility of functional diversity schemes, however, has been questioned (Wright et al. 2006). The central problem is that functional groups chosen arbitrarily and without ecological knowledge often capture about as many resources as functional groups defined by ecologists. Wright et al. (2006) tested the ability of a commonly used plant classification scheme (grasses, forbs, legumes and woody plants) to explain variation in biomass and nitrogen assimilation of ten experimental grasslands in Europe and the USA. Although this classification scheme has been used with some success in the past (see Diaz and Cabido 2001), the authors showed that it had no greater explanatory power than randomly assigned groupings. Indeed, a post-hoc classification scheme, using a Monte Carlo method to identify the best groupings of plant species, demonstrated that even such exceedingly well-informed classifications left significant experimental variability unexplained (Wright et al. 2006). One possible explanation for these failures is that species differ broadly across many traits, such that any single functional group classification scheme overlooks important variation across other trait axes. Perhaps, species are too consistently unique for functional groupings to be useful. In this chapter we will attempt to identify broad trends regarding mechanisms that lead to increased pest suppression by diverse communities of biological control agents, and explore whether the functional group concept is useful for understanding and predicting diversity effects in these economically important systems.

Natural enemy biodiversity and biological control

A diverse group of predators, parasitoids and pathogens contribute to the regulation of weedy and herbivorous pests of agriculture and rangeland, an ecosystem service of great economic and environmental value (McFadyen 1998). This ecosystem service is threatened by agricultural intensification, which disrupts both the number and relative abundances of biological control agents (Tylianakis et al. 2007; Crowder et al. 2010). Thus, the same factors motivating ecologists to study the importance of biodiversity in grasslands (human impacts on ecological processes) have driven agroecologists to test the effects of natural enemy biodiversity on pest control. This focus on the potential benefits of biodiversity follows a long tradition in the sub-discipline of agroecology, which is based on the idea that the biological simplification typical of modern farming systems intensifies pest outbreaks (van Emden and Williams 1974; Altieri and Whitcomb 1979). These harmful effects could perhaps be reversed through the importation of natural enemies, or by modifying farming practices to encourage greater natural enemy biodiversity, thus restoring the ecosystem services provided by natural enemies (Straub *et al.* 2008; Letourneau *et al.* 2009; Crowder *et al.* 2010).

Here, we focus our review on studies that have attempted to identify functional differences among predaceous and herbivorous biological control agents, discussing the defining characteristics of these functional groupings and their importance for control of pest communities. We then discuss whether functional classification schemes could be used to guide the importation, conservation and augmentation of biological control agents. Finally, we close by suggesting several directions for future research in this area.

Functional differences among natural enemies that promote predation by diverse communities
Enemies that complement one another
in pest species attacked

The most obvious way that natural enemies might assume functionally distinct, complementary roles is by attacking entirely different pest species. When this occurs, only multiple enemy species can control an entire pest complex. This scenario is commonly realized when an exotic crop plant species is successively colonized by a series of exotic pests, and natural enemies are located and released to attack each of these pest species. For example, the invasion of Florida citrus groves by different herbivorous pests led to releases of an equally long list of specialist natural enemies (Michaud 2002), including: egg parasitoids to control the weevil *Diaprepes abbreviatus* (Hall *et al.* 2001); ectoparasitic wasps to control the Asian citrus psyllid (*Diaphorina citri*) (Qureshi *et al.* 2009); and two parasitoid wasps to control citrus leafminers (*Phyllocnistis citrella*) and brown citrus aphids (*Toxoptera citricida*) (Hoy *et al.* 2007; Persad *et al.* 2007). In the absence of these different, complementary control agents, any single one of the citrus pests would be capable of substantially decreasing fruit yields, and so each enemy species clearly complemented the others to control the entire pest complex.

Complete feeding-niche separation, however, is not needed to promote predation by diverse communities. For example, on sugar beet plants a lady beetle (*Coccinella transversoguttata*) and a damsel bug (*Nabis americoferus*) share aphids and caterpillars as prey (Tamaki and Weeks 1972). The lady beetle concentrated its attacks on aphids, while the damsel bug focused on cater-pillars, such that a fully diverse community provided the most effective control of both pests (Tamaki and Weeks 1972). Similarly, in alfalfa crops, Cardinale *et al.* (2003) found that the lady beetle *Harmonia axyridis* preferen-tially attacked cowpea aphids (*Aphis craccivora*), whereas the parasitoid wasp *Aphidius ervi* attacked both this aphid and the pea aphid (*Acyrthosiphon pisum*). When both natural enemy species were present, lady beetles depleted cowpea aphids and *A. ervi* shifted its attacks to the more injurious pea aphids. Thus, pea aphid densities were lowest and alfalfa plants largest when a diverse natural enemy community was present over short timescales, although the presence of multiple hosts might have eventually bolstered *A. ervi* numbers over longer timescales. Similar situations occur when weeds are the control target. Gaines and Gratton (2010) examined a community of ground beetles that eat weed seeds of various sizes in potato fields. On average, smaller seed-eaters attacked smaller seeds, whereas larger species attacked larger seeds. Thus, it was suggested that both small and large seed predators were needed to control the entire weed-species complex. In all of these cases, functional groups could be defined by preferences of particular natural enemies for some pest species (or size classes) over others.

Enemies that complement one another in space

Predators can complement one another, even if they attack the same prey species, when they differ in where they forage in space. In agricultural systems, such space-use differences have been examined at a variety of scales from entire fields to single leaves. Work with predatory lady beetles attacking aphids has revealed complementary resource-use patterns at each of these scales. Smith (1971) showed that lady beetle species attacking aphids on corn (*Zea mays*) differed in where they foraged at a whole-field scale: some species foraged along field edges, whereas others congregated at field centres. Presumably, because of these differences in foraging location, aphids are attacked at both field centres and edges only when multiple lady beetle species co-occur; how-ever, controlled experiments at such large scales were not performed. Similarly, at the scale of single corn plants, the lady beetle *Adalia bipunctata* prefers to forage near the top of plants while the lady beetle *Coleomegilla maculata* prefers to forage closer to the base of plants, presumably leading to stronger predation by communities that include both predators (although this also has not been explicitly examined) (Schellhorn and Andow 1999). Finally, Straub and Snyder (2008) found spatial resource partitioning at the scale of single leaves for pred-ators attacking aphids on collard (*Brassica oleracea*) plants. Here, two species of lady beetle foraged primarily at leaf edges, whereas predatory bugs and para-sitoids foraged at leaf centres. Such space-use differences at the leaf-scale could promote predation by diverse communities (differences among species in foraging location remain in place regardless of whether other species co-occur), and indeed in this community increasing predator species richness improves aphid suppression (Snyder *et al.* 2006; Straub and Snyder 2008). In all of these cases, species could be separated into complementary functional groups based on where they forage, although resulting benefits of functional diversity for pest suppression are generally supposed rather than quantified.

Analogous, complementary space-use differences occur when weed biological control agents feed on different plant structures. For example, cinnabar moth larvae (*Tyria jacobaeae*) defoliate invasive ragwort (*Senecio jacobaea*) plants by feeding on leaves and rosettes, which complements larvae of the ragwort flea beetle (*Longitarsus jacobaeae*) that tunnel into ragwort leaf petioles and roots (McEvoy *et al.* 1991). Therefore, these two herbivores represent different functional groups, leaf/rosette versus stem/root feeders, and the combined feeding of the moth and flea beetle provides complementary control of ragwort (James *et al.* 1992). The authors suggest that lower plant fecundity in the presence of both species is driven by indirect interactions between biological control agents, where flea beetle damage inhibits the ragwort's ability to recover from defoliation and defloration by the moth. A third biological control species, the ragwort seed head fly (*Botanophila seneciella*), feeds on the developing seeds in ragwort rosettes, but is apparently functionally redundant with the cinnabar moth and is generally outcompeted (McEvoy *et al.* 1993). Therefore, in this example it is functional, rather than species, diversity that improves biological control.

Spatial separation among natural enemy species can be beneficial to pest suppression in ways beyond simple resource partitioning. When prey readily move between habitat types, predators foraging in one habitat may chase prey into the waiting clutches of a second predator in another habitat. In some cases, this may lead to 'risk enhancement', where the probability of being consumed by a multiple predator species community exceeds what would be predicted by simply multiplying the probability of consumption by each predator species alone (Sih *et al.* 1998; Schmitz 2007). For example, Losey and Denno (1998) found that lady beetles foraging in alfalfa foliage caused pea aphids to drop to the ground, where they were gobbled up by predatory ground beetles on the soil surface. Presumably, aphids disturbed by lady beetles also have their feeding disrupted, eventually leading to lower reproductive rates (e.g., Nelson *et al.* 2004). Thus, improvement in prey consumption by the diverse community stemmed from an indirect interaction between lady beetles and ground beetles that was mediated by aphid behaviour. Furthermore, predator species that forage in different habitats are less likely to encounter one another, reducing the disruptive effects of intraguild predation (e.g., Barton and Schmitz 2009). This means that in addition to increasing pest control due to a partitioning of the prey population, spatial segregation may also reduce predator interference. It is important to note, though, that spatial separation among natural enemy species does not guarantee stronger predation by more diverse predator communities. For example, a community of spiders that feed on grasshoppers in grasslands includes species feeding in the top, middle and bottom of the grassland canopy (Sokol-Hessner and Schmitz 2002). Because the grasshoppers freely move throughout the canopy, however, any grasshopper can be eaten by any spider species, and the spiders all act as a single functional group (Sokol-Hessner and Schmitz 2002; Schmitz 2007). That is, the spatial separation of the spider species from one another does not lead them to attack different subsets of the herbivore population, which would improve predation by the diverse communities, because the entire grasshopper population moves among spider habitats and so is equally available to each spider species.

Enemies that complement one another in time

Natural enemies also might complement one another by differing in their temporal activity. For example, Rathet and Hurd (1983) showed that three mantid species (*Tenodera sinensis*, *T. angustipennis* and *Mantis religiosa*) occupy the same parts of pasture vegetation and feed on similar prey, but hatch at different times in the season. Because prey selection is related to body size, staggered hatch dates may maximize prey consumption between mantid species throughout development (Rathet and Hurd 1983). Similarly, in Californian alfalfa fields, lady beetles, syrphids and lacewings are active early and late in the season when temperatures are relatively mild, whereas *Geocoris* and *Nabis* bugs are active in the middle of the summer (Neuenschwander *et al.* 1975). These authors suggested that both cold- and hot-weather predators are needed to provide complete aphid control throughout the year. Temporal niche partitioning can also occur at a much finer scale. For example, working in corn fields, Pfannenstiel and Yeargan (2002) found that the lady beetle *Coleomegilla maculata* primarily foraged during the day, whereas the predatory bug *Nabis* sp. fed primarily at night. Thus, attacks on both diurnal and nocturnal pests presumably required both predator species. In all of these cases, predators that forage earlier versus later in the year, or during the day versus at night, might form distinct and complementary functional groups.

While temporal niche partitioning can result from predator phenology, so too can this arise over the course of prey development. Wilby and Thomas (2002) explored this idea using a theoretical model, and showed that natural enemy species that attack different stages of the same pest species exerted greater control than could be achieved by any single enemy species. This occurred because some prey in each stage were protected from predation, and predators specialized on attacking different prey stages; thus, overall pest mortality was maximized only when different predator species were present to drive each stage to its refuge density. In a follow-up experiment, Wilby *et al.*

(2005) demonstrated that a caterpillar pest of rice with morphologically very different developmental stages was best controlled by a diverse predator community. In contrast, these authors found no advantage of predator diversity when the pest was a leafhopper with a simple life history, such that all herbivore life stages were morphologically similar. These differing results would be consistent with life-stage-based partitioning of prey being the key driver of predator diversity effects.

A similar temporal partitioning can occur among weed biological control agents. For example, the community of natural enemies that attack diffuse knapweed (*Centaurea diffusa*) comprises beetle and fly larvae that feed in flower heads in May through July, but also larvae of a different beetle species that burrows in roots in late summer through early autumn (Seastedt *et al.* 2007). The combination of flower-feeders and root-feeders exerts synergistic effects on the weed, but it is not clear if these effects of herbivore diversity are due to temporal separation of attacks, or simply result from differences among herbivore species in the plant part consumed (Seastedt *et al.* 2007).

As with space-use partitioning, temporal partitioning of prey can lead to super-additive effects of multiple enemy species, often through trait-mediated indirect interactions (TMIIs) between natural enemies. For example, Ramirez and Snyder (2009) examined the effects of predators and pathogens attacking the Colorado potato beetle (*Leptinotarsa decemlineata*). The predators attacked beetle eggs and larvae in plant foliage, whereas the pathogens only attacked pupating beetles in the soil. Thus, there was clear separation among enemies along both spatial and temporal axes. These authors reported that the combined mortality exerted by predators and pathogens was significantly greater than any single enemy species, apparently because the stress of nonlethal exposure to predators in early beetle stages led to a weakened immune system and heightened susceptibility to pathogens later in beetle development (this was demonstrated through a combination of field and laboratory experiments; Ramirez and Snyder 2009). Thus, early exposure to predators indirectly improved later ability of pathogens to infect the beetles. Temporal separation among enemy species also has the potential to reduce intraguild predation. For example, in corn fields, the lady beetle *C. maculata* is highly susceptible to intraguild predation by the lady beetle *Harmonia axyridis*. Intraguild predation rarely occurs, however, because *C. maculata* is active earlier in the season than *H. axyridis* (Musser and Shelton 2003). Thus, temporal separation among enemies can benefit pest control both by fostering complementary effects on shared prey, and by reducing enemy–enemy interference.

Complementary foraging behaviour

Predator communities often include some species that actively forage, and others that adopt a sit-and-wait hunting strategy. Members of these two groups might feed on different prey species, which could foster complementary impacts on herbivores (Rosenheim *et al.* 2004; Schmitz 2008). Often, actively foraging predators consume more sedentary prey while sit-and-wait predators consume mobile prey that come to them (Rosenheim *et al.* 2004). For example, in a community of predators feeding on sedentary, herbivorous spider mites (*Tetranychus cinnabarinus*), actively foraging lady beetles (*Stethorus siphonulus*) were able to suppress spider mite densities while a sit-and-wait spider capable of feeding on the spider mites (*Nesticodes rufipes*) was not (Rosenheim *et al.* 2004). In this case, however, there were no actively foraging herbivores in the community and the sit-and-wait spiders instead consumed the lady beetles, with this intraguild predation leading to increased spider mite densities. Therefore, while the combination of differently foraging predators has the potential to heighten herbivore suppression, it can also lead to greater intraguild predation and disrupted herbivore control (see also Schmitz 2007). Differences in foraging behaviour can have impacts that cascade beyond predators and their prey. For example, in one study, actively foraging spiders chased grasshoppers away from a competitively dominant plant species, increasing overall plant productivity and nitrogen mineralization (Schmitz 2008). In contrast, spiders with a sit-and-wait strategy drove grasshoppers onto the dominant plant, reducing plant community performance but increasing plant community evenness (Schmitz 2008). Thus different, and perhaps complementary, foraging strategies among these spider species directly altered the structures of herbivore communities and indirectly altered plant communities by changing the behavioural traits of grasshoppers.

Are functional diversity schemes useful in predicting enemy diversity effects?

Predator biodiversity research has generally been plagued by a poor ability to make a-priori predictions about which species will complement one another (Finke and Snyder 2010). The most comprehensive attempt to use natural functional groups to understand multi-enemy-species effects comes from the work of Schmitz and colleagues with spider predators of old-field grasshoppers (e.g., Sokol-Hessner and Schmitz 2002; Schmitz 2007, 2008). They describe the following four combinations of predator–prey traits, and the resulting effects on prey populations: (1) Prey have a broad habitat domain and predators have complementary habitat domains. In this case, each prey individual moves between predator habitats and is equally available to each predator, so predator diversity provides no benefits (either positive or negative; note that this is true only if prey do not selectively avoid habitats occupied by particular predator species). (2) Prey have a narrow habitat range, while predators have broad, overlapping habitat ranges and identical hunting modes. In this case, diversifying predator communities will lead to increased prey consumption, because prey have no refuge from predators and predators can feed on alternative prey to keep predator densities high when focal prey densities are low. (3) Prey have a broad habitat range and predators have narrow, overlapping habitat ranges, but different hunting styles. In this case, diversifying predator communities will lead to the consumption of actively foraging predators by the sit-and-wait predators, and a reduction in prey consumption (assuming that the details of species encounter rates, relative size, etc., allow intraguild predation to occur). (4) Prey and predators have the same habitat ranges, and predators have overlapping habitat domains and identical hunting modes. In this case, diversifying predator communities will lead to lower prey consumption, due to increases in interference competition between predators that defend their hunting grounds. Indeed, this framework seems to be consistent with prey-consumption patterns across a wide range of predator diversity studies (Schmitz 2007, 2008).

The Schmitz framework, however, cannot explain all results from biodiversity–biocontrol study systems. For example, on collards, as mentioned above, some predators feed on leaf edges and others in leaf cores, but aphids occur throughout the plant. A similar spatial separation is apparent among enemy functional groups on potato, where a guild of predators occurs in the foliage and a guild of insect-attacking pathogens is in the soil. In both cases the herbivores span both of these spatially distinct, functional enemy groups, matching Schmitz's scenario number 1 where substitutable enemy effects are predicted (Schmitz 2007). Yet, manipulations of species richness in both of these communities have generated clear evidence of super-additive, rather than substitutable, effects among functional groups (Ramirez and Snyder 2009; Northfield *et al.* 2010). Thus, both systems generate 'risk enhancement' of the type Schmitz predicts to occur only when predators exhibit overlapping foraging locations and hunting modes (scenario 2 above).

One reason that results in the predator–aphid–collard system (Snyder *et al.* 2006; Straub and Snyder 2008) might differ from those predicted by Schmitz (2007) is that aphids are relatively sedentary prey. If prey move randomly and frequently between predator habitats, as do Schmitz's grasshoppers, then a single prey is eventually available to any predator species and the benefit of predator diversity for prey suppression may be minimal (Fig. 23.1a). If prey rarely move between habitats occupied by different predator species, however, like aphids on collard plants, then the prey population may represent spatially isolated sub-groups only subject to attack by the predator species in each stratum. In this case, predators foraging in different locations would provide complementary effects, and increasing predator diversity would heighten aphid consumption (Fig. 23.1b). Furthermore, if predators use different habitat types, then facilitation between predators can occur whenever predator species in one habitat can only access the prey in the presence of the other predator (Fig. 23.1c). This TMII, predators affecting herbivore movement, expands the

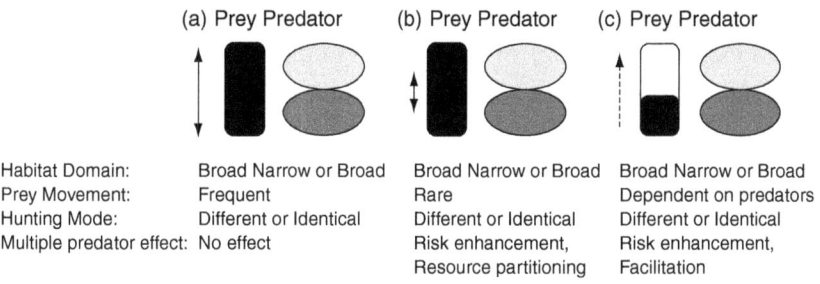

	(a) Prey Predator	(b) Prey Predator	(c) Prey Predator
Habitat Domain:	Broad Narrow or Broad	Broad Narrow or Broad	Broad Narrow or Broad
Prey Movement:	Frequent	Rare	Dependent on predators
Hunting Mode:	Different or Identical	Different or Identical	Different or Identical
Multiple predator effect:	No effect	Risk enhancement, Resource partitioning	Risk enhancement, Facilitation

Figure 23.1 Modification of Schmitz (2007) framework to include prey movement between habitat domains. Dark rounded rectangles represent prey spatiotemporal habitats, and grey ovals represent the spatiotemporal habitats of predator species. When predators have complementary habitats and prey frequently move between predator habitats predator diversity provides no benefit to prey consumption (a). When predators have complementary habitats and extend across, but prey rarely move between, predator habitat domains the two predators provide complementary roles in prey consumption (b). When predators have complementary habitats and prey habitat domain generally only overlaps with a single predator species, facilitation occurs between predators if prey switch habitats (open circle) to avoid predation (c).

Schmitz (2007) framework to include the possibility that one predator can indirectly facilitate predation by another (e.g., Losey and Denno 1998).

Can enemy functional group designations improve biological control?
Classical biological control

Classical biological control is deployed against invasive pests, by searching the native range for natural enemies that can be imported. This approach could benefit from using functional group diversity as a guiding strategy, such that species selected for importation span each group important for suppression in the pest's native range. Indeed, this philosophy has been used to guide classical biological control efforts against invasive weeds (Blossey 1995). For example, Davis *et al.* (2006) used demographic models to predict that importation of the herbivorous insects *Ceutorhynchus scrobicollis* and *C. constrictus*, which feed on rosettes and seeds, respectively, would optimize biological control of garlic mustard (*Alliaria petiolata*). This is because each herbivore targets a different, vulnerable demographic bottleneck in the plant's life history. Using a similar approach, Pedersen and Mills (2004) suggested that the importation of parasitoids that target prey in different habitats might complement one another by preventing prey from escaping into any would-be refuge (Hawkins *et al.* 1993). Of course, the value of any of these model predictions will only be known once the particular mix of complementary control agents that the models recommend is actually released.

Conservation and augmentative biological control

In conservation biological control, managed systems are manipulated to provide food, shelter, or other resources to natural enemies (Landis *et al.* 2000). Augmentative biological control is a related approach, but one where natural enemies are reared and then released en masse to control pests (Collier and van Steenwyck 2004). Both approaches could benefit from a functional group framework.

One obvious approach to promoting functional diversity would be to develop conservation strategies that benefit both generalist predators and specialist parasitoids. Conservation of generalists often stems from increasing the densities of non-pest prey, providing refuges from competition and overwintering sites, or providing a more benign temperature and moisture environment (Landis *et al.* 2000). For example, in rice crops Settle *et al.* (1996) conserved densities of detritivores and plankton feeders by avoiding early season insecticide sprays. These non-pest species provided supplemental prey to generalist predators, building predator densities and heightening their impacts on pests later in the season (Settle *et al.* 1996). In contrast, parasitoids are usually conserved through the provision of floral resources, as parasitoids use plant nectar to extend their egg-laying period and overall lifespan (Landis *et al.* 2000; Heimpel and Jervis 2005). In turn, providing flowering plants to parasitoids can improve biological control by these specialists (Wäckers 2004). So, conservation strategies could be developed to conserve both functional groups simultaneously. For example, refuge plantings could be constructed that include both bunch grasses, which provide non-pest prey and shelter for generalist predators, and flowering plants, which provide floral resources to parasitoids.

Conservation strategies could also target natural enemy species differing in their spatial or temporal activity. For example, in Midwestern-US soybean fields mulching or conservation tillage can conserve ground-active spiders, while web-builders in plant foliage can be conserved by building aboveground structures to anchor their webs (Rypstra *et al.* 1999). Similarly, Millar and Barbercheck (2001) demonstrated that no-till management preserves soil structure on farms and may conserve functional biodiversity in entomopathogenic nematode communities, as different nematode functional groups forage in different soil strata and attack different pests. Similarly, in potato crops foliar predators can be conserved by reducing insecticide applications, while soil-dwelling insect pathogens can be conserved by curtailing the use of broad-spectrum soil fumigants (Crowder *et al.* 2010).

Like conservation biological control, augmentative control might also be improved through a consideration of functional diversity. The manipulation of natural enemy community structure is perhaps easiest in enclosed environments, where managers can easily manipulate the species and functional diversity of natural enemies to be released. For example, Snyder *et al.* (2004) found that generalist lady beetles (*Harmonia axyridis*) reduced peak densities of aphids on greenhouse roses, and did not disrupt aphid control by the specialist parasitoid *Aphelinus asychis*. The authors suggest that releasing members of both the generalist and specialist functional groups would be the most effective tactic to control aphids.

Summary and future directions

Species-specific differences among natural enemy species might lead to complementary resource-use patterns along food-preference, spatial, ontogenetic or temporal niche axes, and lead to complementary impacts on pests. As an additional advantage, predator species in different functional groups appear relatively unlikely to interfere with one another. Unfortunately, complementary impacts on pests by members of different functional groups are most often supposed rather than demonstrated, leading to a clear need for more experimental research. In turn, we suggest the following research directions would be particularly fruitful.

Further refinement of our predictive abilities

Schmitz (2007) has developed a preliminary framework to predict when natural enemies are likely to complement, versus interfere with, one another. As discussed above, however, there is room for improvement in this framework. In particular, we need to predict better the occurrence of TMIIs and their effects on pest suppression. This can be achieved through a better understanding of how TMIIs drive natural enemy diversity effects. One particularly powerful approach towards this end is the experimental manipulation of flexible foraging behaviours to increase or lessen resource-use differences among enemy species (through various means), while simultaneously manipulating enemy species richness (e.g., Finke and Snyder 2008). This approach presents the opportunity for direct experimental tests of the role of TMIIs among species as the root causes of enemy-diversity effects.

Examining complementary under realistic conditions

For obvious logistical reasons, nearly all studies that evaluate the effects of predator diversity on controlling pests come from relatively small laboratory or field-cage arenas (Finke and Snyder 2010). This leaves unanswered the question of how, and how often, such processes might operate at the scale of whole fields or agricultural landscapes (Tscharntke *et al.* 2005). Advances in the use of molecular methods to track predation among free-foraging predators (e.g., Harwood *et al.* 2007) provide a particularly promising avenue for determining the diets of natural enemies, and their trophic linkages to other species, in real-world settings.

Evaluating the relationship between predator functional diversity and climate change

Global climate change can have strong effects on species interactions (Tylianakis *et al.* 2008), and therefore may impact the effects of predator functional diversity on biological control. For example, Harmon *et al.* (2009) found that two lady beetles (*Coccinella septempunctata* and *H. axyridis*) reacted to periodic heat shocks differently while feeding on pea aphids (*Acyrthosiphon pisum*). While *C. septempunctata* continued to feed on pea aphids during heat shocks, predation by *H. axyridis* dropped dramatically. Climate change can also impact predator–predator interactions. Barton and Schmitz (2009) demonstrated that in a community of spiders feeding on grasshoppers, experimental warming reduced the spatial segregation of two spider species, leading to increased intraguild predation and reduced prey consumption. Therefore, increased temperatures led to negative effects of predator diversity on prey consumption. Altogether, these results suggest that functional classifications relevant to current climate conditions may need to be altered to accommodate climate changes.

References

Altieri, M. A. and Whitcomb, W. H. (1979) The potential use of weeds in the manipulation of beneficial insects. *Horticultural Science*, **14**, 12–18.

Barton, B. T. and Schmitz, O. J. (2009) Experimental warming transforms multiple predator effects in a grassland food web. *Ecology Letters*, **12**, 1317–1325.

Blossey, B. (1995) A comparison of various approaches for evaluating potential biological-control agents using insects on *Lythrum salicaria*. *Biological Control*, **5**, 113–122.

Cardinale, B. J., Harvey, C. T., Gross, K. and Ives, A. R. (2003) Biodiversity and biocontrol: emergent impacts of a multi-enemy assemblage on pest suppression and crop yield in an agroecosystem. *Ecology Letters*, **6**, 857–865.

Cardinale, B. J., Srivastava, D. S., Duffy, J. E., Wright, J. P., Downing, A. L., Sankaran, M. and Jouseau, C. (2006) Effects of biodiversity on the functioning of trophic groups and ecosystems. *Nature*, **443**, 989–992.

Collier, T. and van Steenwyk, R. (2004) A critical evaluation of augmentative biological control. *Biological Control*, **31**, 245–256.

Crowder, D. W., Northfield, T. D., Strand, M. R. and Snyder, W. E. (2010) Organic agriculture promotes evenness and natural pest control. *Nature*, **466**, 109–112.

Davis, A. S., Landis, D. A., Nuzzo, V. *et al.* (2006) Demographic models inform selection of biocontrol agents for garlic mustard (*Alliaria petiolata*). *Ecological Applications*, **16**, 2399–2410.

Diaz, S. and Cabido, M. (2001) Vive la difference: plant functional diversity matters to ecosystem processes. *Trends in Ecology and Evolution*, **16**, 646–655.

Finke, D. L. and Snyder, W. E. (2008) Niche partitioning increases resource exploitation by diverse communities. *Science*, **321**, 1488–1490.

Finke, D. L. and Snyder, W. E. (2010) Conserving the benefits of predator biodiversity. *Biological Conservation*, **143**, 2260–2269.

Gaines, H. R. and Gratton, C. (2010) Seed predation increases with ground beetle diversity in a Wisconsin (USA) potato agroecosystem. *Agriculture, Ecosystems and Environment*, **137**, 329–336.

Hall, D. G., Pena, J., Franqui, R. *et al.* (2001) Status of biological control by egg parasitoids of *Diaprepes abbreviatus* (Coleoptera: Curculionidae) in citrus in Florida and Puerto Rico. *BioControl*, **46**, 61–70.

Harmon, J. P., Moran, N. A. and Ives, A. R. (2009) Species response to environmental change: impacts of food web interactions and evolution. *Science*, **323**, 1347–1350.

Harwood, J. D., Desneux, N., Yoo, H. Y. S. *et al.* (2007) Tracking the role of alternative prey in soybean aphid predation by *Orius insidiosus*: a molecular approach. *Molecular Ecology*, **16**, 390–4400.

Hawkins, B. A., Thomas, M. B. and Hochberg, M. E. (1993) Refuge theory and biological control. *Science*, **262**, 1429–1432.

Heimpel, G. E. and Jervis, M. A. (2005). Does floral nectar improve biological control by parasitoids? In F. L. Wacker, P. C. J. van Rijn and J. Bruin, eds., *Plant-Provided Food and Herbivore-Carnivore Interactions*. New York: Cambridge University Press, pp. 267–304.

Hillebrand, H. and Matthiessen, B. (2009) Biodiversity in a complex world: consolidation and progress in functional biodiversity research. *Ecology Letters*, **12**, 1405–1419.

Hooper, D. U., Chapin, F. S., Ewel, J. J. *et al.* (2005) Effects of biodiversity on ecosystem functioning: a consensus of current knowledge. *Ecological Monographs*, **75**, 3–35.

Hoy, M. A., Singh, R. and Rogers, M. E. (2007) Citrus leafminer, *Phyllocnistis citrella* (Lepidoptera: Gracillariidae), and natural enemy dynamics in Central Florida during 2005. *Florida Entomologist*, **90**, 358–369.

Ives, A. R. and Carpenter, S. R. (2007) Stability and diversity of ecosystems. *Science*, **317**, 58–62.

James, R. R., McEvoy, P. B. and Cox, C. S. (1992) Combining the cinnabar moth (*Tyria jacobaeae*) and the ragwort flea beetle (*Longitarsus jacobaeae*) for control of ragwort (*Senecio jacobaea*): an experimental analysis. *Journal of Applied Ecology*, **29**, 589–596.

Landis, D. A., Wratten, S. D. and Gurr, G. M. (2000) Habitat management to conserve natural enemies of arthropod pests in agriculture. *Annual Review of Entomology*, **45**, 175–201.

Letourneau, D. K., Jedlicka, J. A., Bothwell, S. G. and Moreno, C. R. (2009) Effects of natural enemy biodiversity on the suppression of arthropod herbivores in terrestrial ecosystems. *Annual Review of Ecology, Evolution, and Systematics*, **40**, 573–592.

Losey, J. E. and Denno, R. F. (1998) Positive predator-predator interactions: enhanced predation rates and synergistic suppression of aphid populations. *Ecology*, **79**, 2143–2152.

McEvoy, P., Cox, C. and Coombs, E. (1991) Successful biological control of ragwort, *Senecio jacobaea*, by introduced insects in Oregon. *Ecological Applications*, **1**, 430–442.

McEvoy, P. B., Rudd, N. T., Cox, C. S. and Huso, M. (1993) Disturbance, competition, and herbivory effects on ragwort *Senecio jacobaea* populations. *Ecological Monographs*, **63**, 55–75.

McFadyen, R. E. C. (1998) Biological control of weeds. *Annual Review of Entomology*, **43**, 369–393.

Michaud, J. P. (2002) Classical biological control: A critical review of recent programs against citrus pests in Florida. *Annals of the Entomological Society of America*, **95**, 531–540.

Millar, L. C. and Barbercheck, M. E. (2001) Interaction between endemic and introduced entomopathogenic nematodes in conventional-till and no-till corn. *Biological Control*, **22**, 235–245.

Musser, F. R. and Shelton, A. M. (2003) Factors altering the temporal and within-plant distribution of coccinellids in corn and their impact on potential intra-guild predation. *Environmental Entomology*, **32**, 575–583.

Naeem, S., Bunker, D. E., Hector, A., Loreau, M. and Perrings, C., eds., (2009) *Biodiversity, Ecosystem Functioning, and Human Wellbeing: An Ecological and Economic Perspective*. Oxford: Oxford University Press.

Nelson, E. H., Matthews, C. E. and Rosenheim, J. A. (2004) Predators reduce prey population growth by inducing changes in prey behavior. *Ecology*, **85**, 1853–1858.

Neuenschwander, P., Hagen, K. S. and Smith, R. F. (1975) Predation of aphids in California's alfalfa fields. *Hilgardia* **43**, 53–78.

Northfield, T. D., Snyder, G. B., Ives, A. R. and Snyder, W. E. (2010) Niche saturation reveals resource partitioning among consumers. *Ecology Letters*, **13**, 338–348.

Pedersen, B. S. and Mills, N. J. (2004) Single vs. multiple introduction in biological control: the roles of parasitoid efficiency, antagonism and niche overlap. *Journal of Applied Ecology*, **41**, 973–984.

Persad, A. B., Hoy, M. A. and Nguyen, R. (2007) Establishment of *Lipolexis oregmae* (Hymenoptera: Aphidiidae) in a classical biological control program directed against the brown citrus aphid (Homoptera: Aphididae) in Florida. *Florida Entomologist*, **90**, 204–213.

Pfannenstiel, R. S. and Yeargan, K. V. (2002) Identification and diel activity patterns of predators attacking *Helicoverpa zea* (Lepidoptera: Noctuidae) eggs in soybean and sweet corn. *Environmental Entomology*, **31**, 232–241.

Qureshi, J. A., Rogers, M. E., Hall, D. G. and Stansly, P. A. (2009) Incidence of invasive *Diaphorina citri* (Hemiptera: Psyllidae) and its introduced parasitoid *Tamarixia radiata* (Hymenoptera: Eulophidae) in Florida citrus. *Journal of Economic Entomology*, **102**, 247–256.

Ramirez, R. A. and Snyder, W. E. (2009) Scared sick? Predator–pathogen facilitation enhances exploitation of a shared resource. *Ecology*, **90**, 2832–2839.

Rathet, I. H. and Hurd, L. E. (1983) Ecological relationships of 3 co-occurring mantids, *Tenodera sinensis* (Saussure), *Tenodera angustipennis* (Saussure), and *Mantis religiosa* (Linnaeus). *American Midland Naturalist*, **110**, 240–248.

Rosenheim, J. A., Glik, T. E., Goeriz, R. E. and Ramert, B. (2004) Linking a predator's foraging behavior with its effects on herbivore population suppression. *Ecology*, **85**, 3362–3372.

Rypstra, A. L., Carter, P. E., Balfour, R. A. and Marshall, S. D. (1999) Architectural features of agricultural habitats and their impact on the spider inhabitants. *Journal of Arachnology*, **27**, 371–377.

Schellhorn, N. A. and Andow, D. A. (1999) Cannibalism and interspecific predation: role of oviposition behavior. *Ecological Applications*, **9**, 418–428.

Schmitz, O. J. (2007) Predator diversity and trophic interactions. *Ecology*, **88**, 2415–2426.

Schmitz, O. J. (2008) Effects of predator hunting mode on grassland ecosystem function. *Science*, **319**, 952–954.

Seastedt, T. R., Knochel, D. G., Garmoe, M. and Shosky, S. A. (2007) Interactions and effects of multiple biological control insects on diffuse and spotted knapweed in the front range of Colorado. *Biological Control*, **42**, 345–354.

Settle, W. H., Ariawan, H., Astuti, E. T. *et al.* (1996) Managing tropical rice pests through conservation of generalist natural enemies and alternative prey. *Ecology*, **77**, 1975–1988.

Sih, A., Englund, G. and Wooster, D. (1998) Emergent impacts of multiple predators on prey. *Trends in Ecology and Evolution*, **13**, 350–355.

Silvertown, J., Dodd, M. E., Gowing, D. J. G. and Mountford, J. O. (1999) Hydrologically defined niches reveal a basis for species richness in plant communities. *Nature*, **400**, 61–63.

Smith, B. C. (1971) Effects of various factors on the local distribution and density of coccinellid adults on corn. *Canadian Entomologist*, **103**, 1115–1120.

Snyder, W. E., Ballard, S. N., Yang, S. *et al.* (2004) Complementary biocontrol of aphids by the ladybird beetle *Harmonia axyridis* and the parasitoid *Aphelinus asychis* on greenhouse roses. *Biological Control*, **30**, 229–235.

Snyder, W. E., Snyder, G. B., Finke, D. L. and Straub, C. S. (2006) Predator biodiversity strengthens herbivore suppression. *Ecology Letters*, **9**, 789–796.

Sokol-Hessner, L. and Schmitz, O. J. (2002) Aggregate effects of multiple predator species on a shared prey. *Ecology*, **83**, 2367–2372.

Straub, C. S., Finke, D. L. and Snyder, W. E. (2008) Are the conservation of natural enemy biodiversity and biological control compatible goals? *Biological Control*, **45**, 225–237.

Straub, C. S. and Snyder, W. E. (2008) Increasing enemy biodiversity strengthens herbivore suppression on two plant species. *Ecology*, **89**, 1605–1615.

Tamaki, G. and Weeks, R. E. (1972) Efficiency of three predators, *Geocoris bullatus, Nabis americoferus*, and *Coccinella transversoguttata*, used alone or in combination against three insect prey species, *Myzus persicae, Ceramica picta*, and *Mamestra configurata*, in a greenhouse study. *Environmental Entomology*, **1**, 258–263.

Tilman, D., Knops, J., Wedin, D. *et al.* (1997) The influence of functional diversity and composition on ecosystem processes. *Science*, **277**, 1300–1302.

Tscharntke, T., Klein, A. M., Kruess, A., Steffan-Dewenter, I. and Thies, C. (2005) Landscape perspectives on agricultural intensification and biodiversity: ecosystem service management. *Ecology Letters*, **8**, 857–874.

Tylianakis, J. M., Didham, R. K., Bascompte, J. and Wardle, D. A. (2008) Global change and species interactions in terrestrial ecosystems. *Ecology Letters*, **11**, 1351–1363.

Tylianakis, J. M., Tscharntke, T. and Lewis, O. T. (2007) Habitat modification alters the structure of tropical host–parasitoid food webs. *Nature*, **445**, 202–205.

van Emden, H. F. and Williams, G. F. (1974) Insect stability and diversity in agro-ecosystems. *Annual Review of Entomology*, **19**, 455–475.

Wäckers, F. L. (2004) Assessing the suitability of flowering herbs as parasitoid food sources: flower attractiveness and nectar accessibility. *Biological Control*, **29**, 307–314.

Wilby, A. and Thomas, M. B. (2002) Natural enemy diversity and pest control: patterns of pest emergence with agricultural intensification. *Ecology Letters*, **5**, 353–360.

Wilby, A., Villareal, S. C., Lan, L. P., Heong, K. L. and Thomas, M. B. (2005) Functional benefits of predator species diversity depend on prey identity. *Ecological Entomology*, **30**, 497–501.

Wright, J. P., Naeem, S., Hector, A. *et al.* (2006) Conventional functional classification schemes underestimate the relationship with ecosystem functioning. *Ecology Letters*, **9**, 111–120.

CHAPTER TWENTY-FOUR

Trait-mediated effects modify patch-size density relationships in insect herbivores and parasitoids

PETER A. HAMBÄCK and PETTER ANDERSSON

Department of Botany, Stockholm University

and

TIBOR BUKOVINSZKY

Department of Terrestrial Ecology, Netherlands Institute of Ecology

Introduction

The spatial distribution of individuals is influenced by bottom-up and top-down effects, and by processes both in the local habitat and at larger landscape scales (Strong 1979; Tscharntke and Brandl 2004; Ryall and Fahrig 2006). Patterns of resource distributions are often primary determinants of consumer abundance. For example, herbivore densities vary with the size of host plant patches, with the density of plant individuals and with the presence of other, non-host, plant species in the neighbourhood (Andow 1991; Bender *et al.* 1998; Bowman *et al.* 2002; Hambäck and Beckerman 2003). Based on observations of these spatial patterns, Richard Root formulated the resource concentration hypothesis (RCH) more than 35 years ago (Root 1973) and predicted that herbivore densities were higher in areas with high resource concentrations. Subsequent research has revealed a more complex picture, where herbivore densities may be either higher or lower in resource dense areas and where patterns may vary with the spatial scale of the study (Bowers and Matter 1997; Bender *et al.* 1998; Bowman *et al.* 2002; Bommarco and Banks 2003; Hambäck and Englund 2005). The limited predictive capacity of RCH can be explained by the incompleteness in the formulation of underlying mechanisms (Bukovinszky *et al.* 2005; Hambäck and Englund 2005).

Root's argument was that herbivore densities would be higher in large patches, because immigration rates were higher into and emigration rates were lower from large areas. This argument defines immigration rate as the number of immigrating individuals, not correcting for the fact that a larger number of individuals in large patches will also be distributed across a larger area. A recent paper by Hambäck and Englund (2005) examined RCH

Trait-Mediated Indirect Interactions: Ecological and Evolutionary Perspectives, eds. Takayuki Ohgushi, Oswald J. Schmitz and Robert D. Holt. Published by Cambridge University Press. © Cambridge University Press 2012.

mathematically and found a much wider range of predicted density–area relationships. The analysis showed that the main drivers explaining density variation along patch size gradients are (1) the relative role of local versus regional processes and (2) individual search mode. The basic message was straightforward, though hardly surprising, that patterns in large patches are determined by local processes whereas patterns in small patches are determined by regional processes. However, patterns in small patches should also vary with search mode, or specifically with the relative scaling of immigration and emigration rates in relation to patch size. The model was more successful than RCH in explaining empirical patterns, showing differences for species with passive dispersal (e.g., aphids) for olfactory searchers (e.g., many moth species) and for visual searchers (e.g, many butterflies) (Bukovinszky *et al.* 2005; Hambäck and Englund 2005). Patterns also changed with the mobility of the organisms, as suggested by the model, for both moths (Hambäck *et al.* 2007a) and butterflies (Hambäck *et al.* 2010a). These observations were promising, but obviously much variation is left to explain. In this chapter, we explore two additional factors affecting spatial patterns in herbivore densities: induced volatiles and natural enemies.

Herbivores and natural enemies both use damage-induced volatiles to speed up host location and to avoid competition, and volatiles from the primary host plant may therefore cause positive or negative feedbacks on density distributions (Turchin 1989). Positive feedbacks are known from many field systems, including important forest pest species (e.g., Raffa 1991), and may modify the density distribution in spatially heterogeneous environments (Bukovinszky *et al.* 2010). Natural enemies impose strong effects on spatial patterns in herbivore densities, but they are also affected by the spatial distribution of plants and herbivores. Field data often show variation in natural enemy density and predation rates along patch size gradients and in response to other patterns of plant distributions (Sheehan and Shelton 1989; Zabel and Tscharntke 1998; Thies and Tscharntke 1999; Holt *et al.* 1999; Cronin and Reeve 2005; Bolger *et al.* 2008). Possible ecological mechanisms for these patterns include nectar feeding in adult parasitoids (Siekmann *et al.* 2004), search mode in parasitoids (host location mechanism), avoidance of interference competitors and intraguild predation, or availability of alternative hosts (Holt and Lawton 1993). However, few studies have explored how these processes interact with patch size or have attempted to derive predictions for density–area distributions in multitrophic systems.

In this chapter, we extend a basic population model for herbivore density distributions to examine how feedbacks, such as behavioural responses, modify the spatial distribution of both herbivores and their natural enemies. We first describe the basic model with patch-size dependent migration rates, and examine how trait modifications in plants induced by herbivore feeding affect subsequent attraction by herbivores. Such trait-mediated effects are overlooked in population models. There are plenty of studies examining the molecular or behavioural side of secondary attraction, but the consequences for density distributions are seldom explored (Schoonhoven et al. 2006). Induced changes in plant nutritional chemistry may also affect residence time of herbivores in patches (Underwood et al. 2005; Poelman et al. 2008; Viswanathan et al. 2008), but such trait modifications are beyond the scope of this chapter. In the following models, we include natural enemies and particularly the role of host dependent immigration and emigration rates in host–parasitoid systems. Finally, we discuss some applied consequences based on our modelling approach.

A basic population model

As a baseline for studying the effect of patch size on density distributions, we use the model of Hambäck and Englund (2005). In this model, local density is a function both of movement processes (immigration and emigration) and local growth. There is no explicit spatial structure in the model, and mean immigration rates are not assumed to vary depending on the relative position of patches. Simulations suggest that this assumption is reasonable for reasonably large patch networks (Englund and Hambäck 2007). The assumption essentially implies that patch choice is independent of comparisons between patches. For this reason, the model is most relevant to arthropods with limited spatial cognitive capacity. Specifically, the model assumes habitat patches where the rate of change in the local density (n) in a patch with area A is described as

$$\frac{dn}{dt} = g(n)n - En + I = rn - \varepsilon A^{-\beta}n + iA^{-\varsigma}. \tag{24.1}$$

In the model, $g(n)$ is a function describing local growth, I is the number of individuals arriving in the patch, per unit area, and E is the mean probability that an individual in the patch will leave. For simplicity, and because the focus of the analysis is to examine effects due to movement processes, we assume no local density dependence and $g(n) = r$. The novel feature about the model was the assumption about migration rates and how these may depend on patch size. The particular formulation was based on both theoretical arguments and empirical data (Englund and Hambäck 2004, 2007). In the model, immigration and emigration rates are modelled as power functions of patch size with parameters determining both absolute movement rates (i, ε) and how rates change with patch size (β, ς). To explain the scaling factors, assume that individuals searching for host plant patches move in straight lines. The probability of intersecting and entering a host plant patch is proportional to the patch diameter D. For circular and square patches, $D \propto A^{1/2}$, and the number of individuals arriving in the patch (I) is therefore proportional to $A^{1/2}$. Individuals entering a patch are diluted across the patch area, and the effect on local density is

$$I = iA^{-0.5}. \tag{24.2}$$

The same relationship applies when individual foragers detect patches from a distance (Bowman *et al.* 2002). In contrast, when individuals move according to Brownian motion, and do not discover patches until they literally sit in them, immigration rates rather depend on the patch perimeter. For such movement behaviours, ζ is proportional to the fractal dimension of the patch edge and for a typical natural patch $\zeta \approx 0.33$ (Englund and Hambäck 2007). A similar reasoning can explain variation in emigration rates among patches of different size, depending on insect movement behaviour, and in the size of β (Englund and Hambäck 2004). The model in Equation (24.1) can be solved analytically, predicting that the density distribution across a range of patch sizes is

$$n(t) = \frac{iA^{\beta-\zeta}\left(1-e^{t\left(r-\varepsilon A^{-\beta}\right)}\right)}{\varepsilon - rA^{\beta}} \xrightarrow{t\to\infty} \frac{iA^{\beta-\zeta}}{\varepsilon - rA^{\beta}}, \tag{24.3}$$

where t is the time since patches were established. The quantitative prediction relies mainly on two factors; the relative size of local growth and absolute emigration rates (r versus ε) on one hand and the relative scaling of immigration and emigration rates (β versus ζ) on the other hand. When migration completely dominates local densities ($\varepsilon \gg r$) then $n \propto A^{\beta-\zeta}$ and $\log(n) \propto (\beta-\zeta)*\log(A)$ (Hambäck and Englund 2005). In other words, the slope in the relationship between density and area (DAR_{slope}) should equal the relative scalings of immigration and emigration rates when migration dominates local density. Because the model predicts a linear relationship on a log–log scale, a logical model for estimating parameters for the DAR_{slope} is Poisson regression with log-link.

Englund and Hambäck (2004, 2007) estimated β and ζ for a wide range of insect herbivores, showing that the variation among species can be traced to differences in search mode among species. For one group, butterflies, data were sufficient for calculating the predicted slope for the density–area relationship ($DAR_{slope} = \beta - \zeta = 0.17 - 0.54 = -0.37$). To test the prediction, Hambäck *et al.* (2010a) calculated the DAR_{slope} for 60 butterfly species from 11 studies (273 independent slope estimates). Because model predictions are only relevant when migration dominates local densities, species were grouped based on mobility. The logic behind this grouping was that migration should dominate local density for species with a high mobility. In the analysis, two mobility estimates were used; wing span and expert mobility rankings. The analysis provided good agreement with predictions: $DAR_{slope} = -0.42$ for large (mobile) species and $DAR_{slope} = -0.40$ for species with a high mobility ranking. This test is however not ideal because the scalings of migration rates and DAR_{slope} were not estimated for the same species. Such a data set is available for the butterfly

Iolanta iolas by Rabasa *et al.* (2007, 2008). The calculated scaling of net migration rates for *I. iolas* ($\beta - \zeta = -0.70$), estimated with the virtual migration model, was somewhat steeper than the observed DAR_{slope} (range -0.31 to -0.59). Partly, this discrepancy may be explained by an inherent bias when estimating immigration rates using mark–recapture data (Englund and Hambäck 2007), but it could also indicate the importance of feedback mechanisms not covered by the model.

While our model predicts density distributions based on patch size variations in migration rates, and how these interact with local growth, similar patterns may also arise for other reasons (Bowers and Matter 1997; Matter 1997; Bowers and Dooley 1999; Gaston and Matter 2002). A common dichotomy in describing fragmentation effects on abundance and diversity is whether observed effects are due to edge or area per se. Habitat edges affect individuals and populations in several ways, and these most commonly involve edge-mediated movement behaviour or differences in vital rates along edges and inside patches (Fagan *et al.* 1999; Cantrell *et al.* 2002; Fletcher *et al.* 2007; Ries and Sisk 2008). The definition of area effects per se is even less clear, and the dichotomy between area and edge effects is often poorly defined in the literature. A common method for identifying edge effects is to quantify population density from patch edges inwards. However, density gradients may arise through both movement processes and effects on vital rates. A more fruitful dichotomy than edge versus area effects is to separate effects through animal movements and effects through vital rates (reproduction and mortality), as in our model. We assume that reproductive or mortality rates are independent of patch size. However, as variation in vital rates along patch size gradients is often observed, it represents an alternative hypothesis to a movement-based hypothesis for density–area relationships. Separating hypotheses for a specific system necessitates that vital rates and movement rates are estimated along the same patch size gradients. Vital rates should be estimated based on somatic growth and reproduction (of individuals) and not on population growth as the latter is subject to changes by migration rates.

Adding induced attraction

Migration rates and density distributions of insects living on plants do not depend only on plant density, but plants may vary in their attractiveness due to individual variation in traits such as volatile emission. For instance, herbivore damage induces both qualitative and quantitative changes in volatile emissions (Vet and Dicke 1992; Raffa 2001), causing differences between the primary attraction towards undamaged plants and secondary attraction towards damaged plants, in both herbivores and their natural enemies. Secondary attraction implies some type of density dependence in search and patch immigration rates, and field data suggest that positive feedbacks caused by secondary attraction may disproportionally increase densities in large patches and cause the DAR_{slope} to shift from negative to positive (Hambäck and Englund 2005). For instance, several studies suggest that the DAR_{slope} for *Phyllotreta* spp. (Coleoptera: Chrysomelidae) shifts from negative to positive with time (Kareiva 1985; Bukovinszky *et al.* 2010) and that these beetles are known to be strongly attracted to plant damage (Peng and Weiss 1992, see also Bukovinszky *et al.* 2010).

The specific formulation for the density-dependence depends on insect search behaviour, and the scaling of olfactory search, and how damage induced volatiles change the odour distribution. Olfactory search is complicated by the fact that odour molecules are patchily distributed in odour plumes, and that odour concentration gradients are not readily apparent (Murlis *et al.* 1992). Most insects using olfactory cues for long-distance attraction therefore use anemotaxis, where individuals fly up-wind when stimulated by odour molecules. The key feature for understanding the relative attraction to patches of different size, or with different numbers of odour sources, is therefore the area where the odour concentration is above the detection threshold for the searching insect. Modelling odour plumes, however, is tricky due to their filamentous nature (Lof *et al.* 2007) but theoretical studies and field data suggest that the lengths of odour plumes are roughly proportional to the number of odour sources (Andersson *et al.*, in prep.). Thus, the probability of detecting a patch by olfactory search is proportional to S^w, where S = the number of odour sources, and this requires a reformulation of the model. Assume that insects detect patches both as a function of the number of odour sources and the patch diameter, but that primary search is diameter dependent and secondary search is dependent on the number of odour sources (nA):

$$\frac{dn}{dt} = -\varepsilon A^{-\beta} n + i\left(vn^{\omega} A^{\omega-1} + A^{-\zeta}\right). \tag{24.4}$$

The parameter ω is factor scaling immigration rate to the number of odour sources and v is a factor describing the relative strength of primary and secondary search. Equation (24.4) cannot be analytically integrated over time, but to interpret the role of secondary attraction we derive the stable density distribution when secondary attraction is much stronger than primary attraction:

$$n^* = \left(\frac{ivA^{\beta-1+\omega}}{\varepsilon}\right)^{\frac{1}{1-\omega}}, \tag{24.5}$$

and the DAR_{slope} is

$$\frac{d\ln(n)}{d\ln(A)} = \frac{\beta - 1 + \omega}{1 - \omega}. \tag{24.6}$$

The comparison of this slope with the slope from the basic model, where primary attraction determines DAR_{slope} $(= \beta - \zeta)$, suggests the importance of the scaling parameter ω. Recent work in our group has estimated that olfactory attraction scales with the square-root of the number of odour sources, suggesting that $\omega \approx 0.5$ (Andersson et al. in prep.; see also Bossert and Wilson 1963). While this estimate is based on average plume values, the size of the parameter suggests that the DAR_{slope} should in most cases be negative, because $\beta \leq 0.5$. This finding implies that the previous assumption, that induced attraction shifts the slope from negative to positive, is incorrect. Further studies are clearly needed to explain this pattern, but it is possible that a different scaling of olfactory information might be involved.

Adding natural enemies

Herbivore density is not only affected by plant distributions but also by natural enemies and, in field studies, natural enemy densities, predation and parasitism rates often vary with patch size. Predation can be included in the model but this extension necessarily increases model complexity, not only because predators consume prey but also because migration rates of both trophic levels may be interdependent. It has been suggested that prey movement rates often increase in response to predator contact, and indirectly increase patch emigration rates (Englund 1997; Cronin et al. 2004). Predators may similarly be attracted to patches with high prey densities or reduce emigration rates from such patches (Sheehan and Shelton 1989; Geervliet et al. 1998). A general dynamic model describing the dynamics in predator (p) and prey (n) densities, following the reasoning above, would be

$$\frac{dn}{dt} = g_n(n, p)n - E_n(n, p)n + I_n(n, p)$$

$$\frac{dp}{dt} = g_p(n, p)p - E_p(n, p)p + I_p(n, p), \tag{24.7}$$

where all rates may be functions of predator and prey densities, as well as of patch size. The analysis here focuses on short-term dynamics of hosts and parasitoids, mimicking an experiment where patches are established de novo (e.g., Bach 1988), or for a broader set of cases when $E_i \gg g_i$. We can for these cases ignore local growth rates and focus on movements to and from patches. We explore cases with both generalist and specialist natural enemies.

Larval/pupal parasitoids and host attraction

Larval and pupal parasitoids attack hosts at a stage when the spatial distribution of host individuals is already established. When modelling spatial effects on parasitism rates by these parasitoids, we may therefore use equation (24.3) or (24.5) for describing the initial host distribution. We define $0 < t < t_n$ as the window in time for host egg laying, and $n_0 = n(t_n)$ as the distribution of hosts available for parasitoid search. Parasitoid search and parasitism can then be modelled by sequentially following host distributions, in a similar fashion to earlier models of univoltine host–parasitoid systems (Hassell 1978). As discussed for the single species model, search mode in parasitoids also affects the spatial distribution of parasitism. Search cues for parasitoids are variable and include volatiles from both undamaged and herbivore-induced plants, or cues directly associated with the presence of the host (i.e., frass, silk, faeces). Cues directly related to the host are typically more reliable, but detectable at a close range only, whereas herbivore-induced cues are easily detectable from a distance but are less reliable predictors of host identity (Vet and Dicke 1992). We initially focus on olfactory searchers using either host odour or damage-induced cues for long-range attraction. The attraction radius is then proportional to host number, and relates to patch size, similar to induced plant cues. Parasitoid densities can then be described as

$$\frac{dp}{dt} = -\varepsilon_p A^{-\chi} p + i_p \frac{S^\omega}{A} = -\varepsilon_p A^{-\chi} p + i_p \frac{(nA)^\omega}{A}, \tag{24.8}$$

where ε_p and i_p are absolute emigration and immigration rates respectively for parasitoids, and where χ and ω are the scaling coefficients for emigration and immigration rates. The temporal change in parasitoid density can be found by integrating Equation (24.7) over time as (assuming that the initial parasitoid density in the patch is zero):

$$p(t) = \frac{i_p n^\omega A^{\chi+\omega-1}\left(1-e^{-t\left(\varepsilon_p A^{-\chi}\right)}\right)}{\varepsilon_p} \xrightarrow{t\to\infty} \frac{i_p n^\omega A^{\chi+\omega-1}}{\varepsilon_p}. \tag{24.9}$$

This expression describes the observed density of parasitoids at either time t or when the system has reached a steady state. To translate this density into observed parasitism rate, we make the same assumptions about the parasitism process as the Nicholson–Bailey model. Two important assumptions are that host finding within a patch is roughly proportional to host density and that parasitoid females are unable to differentiate between parasitized and unparasitized hosts from a distance, which are both reasonable assumptions for several species (e.g., Geervliet et al. 1998; Shiojiri et al. 2001). From these assumptions follow that encounter rates with unparasitized hosts within a patch will be proportional to the density of such hosts. If we also assume a small handling time, a simple model describing the density of unparasitized hosts (n_u) with time will be

$$n_u(t) = e^{-ap(t)t}n_0, \tag{24.10}$$

where α is the parasitoid search efficiency. Many hosts are not immediately killed but rather hijacked (i.e., koinobiont parasitic lifestyle), and larval hosts continue feeding and release damage-induced volatiles after parasitism. Consequently, parasitism will not affect long distance attraction. To find the proportion of unparasitized hosts (or the host survival probability = 1 – parasitism rate) for this model, enter Equations (24.3) and (24.9) into Equation (24.10):

$$n_{prop} = \frac{n_u(t_p)}{n_0} = e^{a\frac{i_p}{\varepsilon_p}\left(\frac{i_n}{\varepsilon_n}\right)^{\omega}t_p\left(A^{2\chi+\beta-\zeta+\omega-1}\left(1-e^{-\varepsilon_n A^{-\beta}t_n}\right)\left(e^{-\varepsilon_p A^{-\chi}t_p}-1\right)^{\omega}\right)^{\omega}}, \tag{24.11}$$

where t_x is the length of the host (t_n) and parasitoid (t_p) egg-laying period, respectively. The next step will be to derive a relationship between patch size and survival probability, similar to the DAR_{slope}. Interestingly, some reshuffling suggests that a linear relationship (hereafter PAR_{slop} = slope in the relationship between ln(proportion of unparasitized hosts) and ln(area)) can be derived by a complementary log transformation:

$$\frac{d\ln\left(-\ln\left(n_{prop}\right)\right)}{d\ln(A)} = \omega(\beta-\zeta) + (\chi+\omega-1) - \frac{\omega\beta\varepsilon_n A^{-\beta_n}t_n}{e^{\varepsilon_n t_n A^{-\beta_{tn}}}-1} + \frac{\chi\varepsilon_p A^{-\chi_p}t_p}{e^{\varepsilon_p t_p A^{-\chi_p}}-1}. \tag{24.12}$$

This expression may seem complex but note that the latter parts involve the temporal trajectory. This part will have small effects on the predicted slope when the redistribution of individuals has stabilized. The first, methodological, insight is that a generalized linear model with binomial error distribution and complementary log-link can be used for estimating the slope between the survival probability and log(area), similar to the Poisson regression for DAR_{slope}. It is important that the PAR_{slope} is estimated on the proportion surviving rather than on the proportion parasitized as the latter have a more complex relation to patch size. The second insight is that the spatial variation in survival probabilities from parasitism depends on a combination of host and parasitoid female search modes. Four parameters determine this variation: the immigration and emigration scaling factors for hosts and parasitoids. However, the scaling of parasitoid migration rates weighs more heavily in the overall scaling of survival probabilities because the effect from host distribution is modified by ω (typically less than 0.5). Thus far, it is not possible to evaluate theoretical predictions because no one has estimated migration and parasitism rates along the same patch size gradient. One may wonder how the model relates to all studies showing that emigration rates of parasitoids depend on local host density, and we return to this question.

The only study, known to us, that evaluated patch size variations in both immigration and emigration rates for parasitoids is that of Sheehan and Shelton (1989), which examined the aphid parasitoids (*Diaeretiella rapae*). Even though the model is not perfectly relevant for aphid systems because aphid reproduction and parasitism occur simultaneously, the study nevertheless gives important insights into the scaling of migration rates for parasitoids. The scaling parameter ω was estimated from Sheehan and Shelton by regressing the number of wasps arriving per plant per day in relation to patch size (i.e., slope = $\omega - 1$, Equation (24.7)). Surprisingly, the scaling of immigration rates for *D. rapae* relative to patch size was close to zero (≈ -0.04) translating to $\omega \approx 0.96$. Such patch size invariance in immigration rates has mostly been observed for aphids and has been explained through their passive dispersal strategy (Hambäck and Englund 2005). Sheehan and Shelton speculated that the patch size invariance may be due to an insensitivity of *D. rapae* to damage-induced volatiles (isothiocyanates, but see Bradburne and Mithen 2000) and argue that parasitoids rather find hosts through random movements coupled with short range detection. This explanation is, however, not satisfactory as such movement behaviour should lead to a perimeter dependence and negative scaling for immigration rates (Hambäck and Englund 2005). Sheehan and Shelton also estimated emigration rates along the same patch size gradient (Fig. 24.1) and the calculated negative slope ($\chi \approx 0.43$, range 0.21–0.73) is within the predicted range. The data also showed that emigration rates changed with aphid density, with no interaction to patch size, but we return to this pattern in the next section. The corresponding scaling coefficients are not known for the aphid (*Brevicoryne brassicae*) but an earlier study has estimated the $DAR_{slope} \approx 0.05$ (Grez and Gonzales 1995; Banks 1998). Please note, however, that theoretical and empirical studies suggest that the density of aphids, based on aphid movement behaviour, may be a hump-shaped relationship of patch size (Hambäck and Englund 2005; Hambäck *et al.* 2007b). We ignore this possibility, and the predicted patch size dependent variation in the survival probability for this system is PAR_{slope} [slope in the relationship between ln(proportion of unparasitized hosts) and ln(area)] = $\omega(\beta - \zeta) + (\chi + \omega - 1) = 0.96 \times 0.05 + (0.43 - 0.04) \approx 0.44$. Notice that χ, the scaling of parasitoid emigration rates, basically determines the relationship for this example. To appreciate the effect of the linearization through complementary log transformation, assume two patches where one patch is 10-times larger than the other patch. The effect will then vary depending on the absolute size of parasitism rates which is logical as the effect cannot be as large when rates are close to one as when rates are close to zero. For the estimated parameters

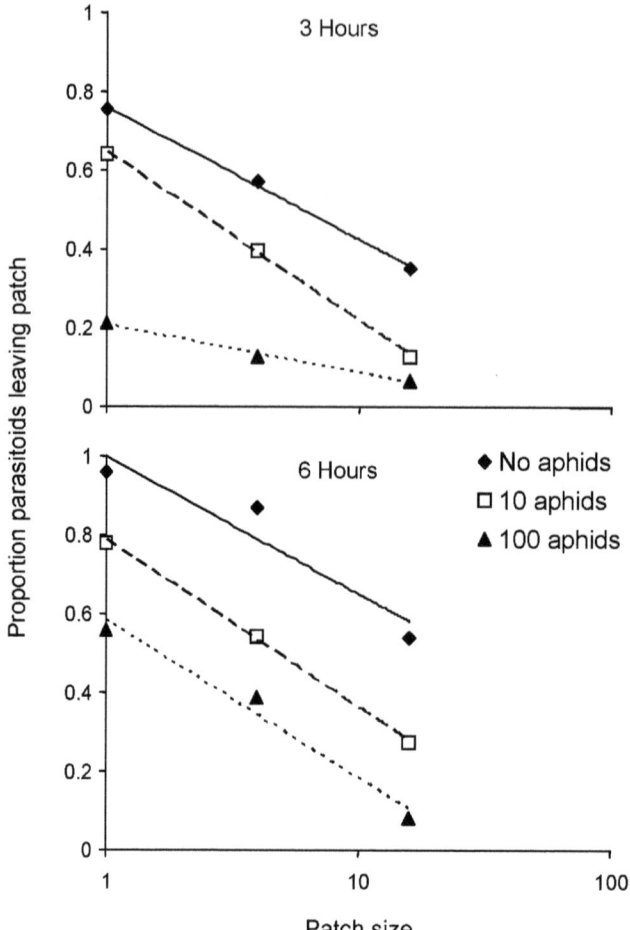

Figure 24.1 Estimated emigration rates for *Diaretiella rapae* in a patch size gradient. The figure is redrawn from Sheehan and Shelton (1989) with permission from the Ecological Society of America.

in the *D. rapae–B. brassicae* system, the survival probability will be 7.5 times higher in the larger patch than in the small when parasitism rates are 0.99 but survival probabilities are virtually identical when parasitism rates are 0.01 (Fig. 24.2).

Density-dependent parasitoid emigration rates

The study system of Sheehan and Shelton evidently displayed a different density dependence than host attraction. We now turn to the case when emigration rates, and not immigration rates, depend on host density, similar to what was observed by Sheehan and Shelton and in several other parasitoid studies (e.g., French and Travis 2001). This necessitates that E becomes a

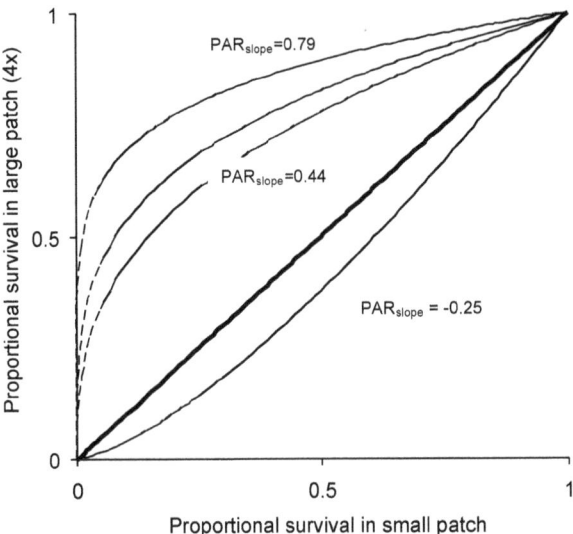

Figure 24.2 Relationship between mean host survival probability for two patches with different size, for the patch size relationship 10:1.

function of n. The shape of this function may vary, but we follow Sheehan and Shelton and use $\ln(E_p) = \ln(\varepsilon_p) - \chi\ln(A) - \kappa\ln(n)$ or $E = \varepsilon_p A^{-\chi} n^{-\kappa}$. When we enter this function in Equation (24.3), we get

$$\frac{dp}{dt} = -\varepsilon_p n^{-\kappa} A^{-\chi} p + i_p A^{(\omega-1)}. \tag{24.13}$$

Equation (24.12) can be integrated over time and the temporal change in local parasitoid density is

$$p(t) = \frac{i_p n^{\kappa} A^{\chi+\omega-1}\left(1-e^{-t\left(\varepsilon_p n^{-\kappa} A^{-\chi}\right)}\right)}{\varepsilon_p} \xrightarrow{t\to\infty} \frac{i_p n^{\kappa} A^{\chi+\omega-1}}{\varepsilon_p}. \tag{24.14}$$

Using the same analytical approach and assumption as for host attraction, the proportion of unparasitized host is

$$n_{prop} = \frac{n_u(t)}{n}$$

$$= e^{a\frac{i_p}{\varepsilon_p}\left(\frac{i_n}{\varepsilon_n}\right)^{\kappa} t_p\left(A^{(\beta-\zeta)+\kappa(\chi+\omega-1)}\left(1-e^{-\varepsilon_n A^{-\beta}t_n}\right)\right)^{\kappa}\left(e^{\varepsilon_p\left(\frac{i_n}{\varepsilon_n}\right)^{-\kappa} t_\zeta\left(A^{(\beta-\zeta)-\kappa\chi}\left(e^{-\varepsilon_n A^{-\beta}t_n}-1\right)\right)^{-\kappa}}-1\right)}.$$

$$\tag{24.15}$$

This expression is fundamentally similar to Equation (24.10) and the complementary log-transformation provides the slope, at sufficiently large A:

$$\frac{d \ln(-\ln(n_{prop}))}{d \ln(A)} = \kappa(\beta-\zeta) + (\chi + \omega - 1) - (time\_trajectory). \qquad (24.16)$$

The time trajectory is more complex for this model and the full expression is not shown. However, we can parameterize the model at stable density distributions to appreciate the difference relative to the effect of host attraction. In the previous exercise we estimated χ and ω, and the missing scaling relationship (κ) can also be estimated from Sheehan and Shelton ($\kappa \approx 0.35$; Fig. 24.1). The predicted slope for these parameters is very similar to the assumption without density dependent emigration rates, $PAR_{slope} = \kappa(\beta - \zeta) + (\chi + \omega - 1) = 0.35 \times 0.05 + (0.43 - 0.04) = 0.42$.

A similar model could in principle be used for generalist natural enemies, such as spiders and carabids. Generalist natural enemies are seldom attracted to prey from a distance and immigration rates therefore only depend on general patch characteristics. For instance, carabids and spiders are known to reduce movement rates, and are arrested, in areas with a more dense vegetation or in response to local prey density (Bommarco and Fagan 2001). Such behaviour will cause a similar scaling of emigration rates along a patch size gradient as for D. rapae. However, in contrast to most parasitoids, spiders and carabids directly kill prey and therefore affect prey densities directly.

Predator-induced prey emigration

Prey commonly respond to predator presence by increasing their movement rates, even though cases of freezing also occur. Englund and co-workers have in several papers estimated how predator induced emigration rates affect estimates of prey mortality in enclosure experiments (Englund 1997; Englund and Cooper 2003). Besides, prey responses to predators will also affect emigration rates from natural patches (Cronin et al. 2004). Using our model framework, predator dependent emigration rates can be added to Equation (24.6) as

$$E_n = \varepsilon_n (1 + p^\eta) A^{-\beta}. \qquad (24.17)$$

This formulation makes the assumption that predators increase the basic emigration rate but that this effect is similar for small and large patches. By entering equation (24.17) into the prey model, we derive the following relationship for prey density distributions

$$n(t) = \frac{i_n A^{\beta-\zeta}\left(1 - e^{-(1+p^\eta)t_{\varepsilon_n} A^{-\beta}}\right)}{(1 + p^\eta)\varepsilon_n} \xrightarrow{t \to \infty} \frac{i_n A^{\beta-\zeta}}{(1 + p^\eta)\varepsilon_n}. \qquad (24.18)$$

Now assume that predators distribute themselves according to habitat characteristics such as patch size and not according to a particular prey. Predation risk, and predator induced emigration rates in prey, can then be predicted for a patch size gradient using the predator density distribution

$$p = \frac{i_p A^\chi A^{\omega-1}}{\varepsilon_p}. \tag{24.19}$$

It is now possible to derive the predicted prey DAR_{slope} for sufficiently large A as

$$\frac{d\ln(n)}{d\ln(A)} = (\beta-\zeta)-\eta(\chi+\omega-1)-(time\_trajectory). \tag{24.20}$$

Notice the similarity to Equations (24.12) and (24.16), but that the scale modifier is now attached to the predator migration scaling. The effect from the predator induced emigration will depend on the shape of this effect but also on the relative scalings of prey and predator migration rates. This finding is important because it argues that future studies on trait-mediated effects on prey abundance should consider not only the effect from the induced behaviour but also how this effect interacts with search traits of both prey and predator.

Field data

There are many suggestions that natural enemy density and parasitism rates vary with patch size, but the data are seldom quantified in a way where parameter values of scaling relationships can be estimated. The study by Sheehan and Shelton is the only study estimating immigration and emigration rates for parasitoids, and we found additional studies that had estimated natural enemy densities or parasitism rates in patch size gradients (Table 24.1). Published studies have used both experimentally established patches, where study length was shorter than the reproductive period, and surveys of natural patches. Strictly speaking, our predictions are developed for cases when $r \ll E$, which should apply to all experimental studies but they are less relevant for studies involving natural enemy densities. As an example of an experimental study, Hambäck et al. (2010b) quantified the density of *Diadegma* spp. (two species), and their host *Plutella xylostella*, in cabbage patches of different size. This study lasted about 2 months and no reproduction occurred in this period. The estimated density–area relationship for *P. xylostella* was fairly shallow, but negative ($DAR_{slope} = -0.12$). The estimated PAR_{slope} for *Diadegma* spp. was actually quite similar, although of opposite sign. We estimated the PAR in two ways; both with a binomial regression with complementary log-link on the raw data for unparasitized pupae and with mean values similar to the other studies, where we did not have access to the original data. The slope estimated on the raw data was 0.24 and the estimate for mean values was 0.21 (Table 24.1). Recall that the positive PAR_{slope} implies that survival probabilities are higher and the parasitism rates are lower in large patches.

Table 24.1 *Field estimates of the relationship between patch size and host survival probability (a), or natural enemy density (b, c). PAR$_{slope}$ was estimated with a binomial regression with complementary log-link on mean values for each patch size (extracted from figures in the original paper). DAR$_{slope}$ was estimated using regression on log-transformed data. Studies include both experimentally established patch networks (type = exp) and natural patches (type = nat).*

Parasitoid species (family)	PAR$_{slope}$/ DAR$_{slope}$	Host species	Host plant species	Type	Reference
(a) Host survival probability					
Diadegma spp. (Ichneumonidae)	0.21	Plutella xylostella	Brassica oleracea	exp	Hambäck et al. (2010b)
Anagrus columbi (Mymaridae)	0.002	Prokelisia crocea	Spartina pectinata	nat	Cronin (2003a)
Ophion pteridis (Ichneumonidae)	0.07	Hadena bicruris	Silene latifolia	exp	Elzinga et al. (2005)
Bracon variator (Braconidae)	0.04	Hadena bicruris	Silene latifolia	exp	Elzinga et al. (2005)
Microplitis tristis (Braconidae)	0.03	Hadena bicruris	Silene latifolia	exp	Elzinga et al. (2005)
Campoletis sp. (Ichneumonidae)	−0.02	Itame andersoni	Dryas drummondi	nat	Doak (2000)
Aleiodes sp. (Braconidae)	0.01	Itame andersoni	Dryas drummondi	nat	Doak (2000)
Phyrxe pecosensis (Tachinidae)	0.02	Itame andersoni	Dryas drummondi	nat	Doak (2000)
Multiple species	−0.11	Eupithecia immundata	Actaea spicata	nat	von Zeipel et al. (2006)
(b) Enemy density (ind/plant)					
Cotesia glomerata (Braconidae)	0.02	Pieris brassicae	Brassica nigra	exp	Bezemer et al. (2010)
Chrysopidae	0.07			exp	Prasifka et al. (2005)
Coccinellidae	−0.27			exp	Prasifka et al. (2005)
(c) Enemy density (ind/unit area)					
Anagrus spp. (Mymaridae)	0.45	Delphacodes scolochloa	Scolochloa festucacea	nat	Cronin (2009)
Coccinella septempunctata (Coccinellidae)	−0.16	Uroleucon nigrotuberculatum	Solidago canadensis	exp	Kareiva (1987)
Cursorial spiders	−0.42			nat	Cronin et al. (2004)

Web spiders	−0.73	nat	Cronin *et al.* (2004)
Sit-and-wait spiders	−0.25	nat	Cronin *et al.* (2004)
Lycosid spiders	−0.29 (non-isolated islands)	nat	Östman *et al.* (2009)
	0.25 (isolated islands)		
Parasitic wasps	−0.40 (non-isolated islands)	nat	Östman *et al.* (2009)
	0.18 (isolated islands)		
Chilopoda spp.	−0.25	exp	Prasifka *et al.* (2005)
Dolichopodidae	−0.35	exp	Prasifka *et al.* (2005)
Non-lycosid spiders	−0.04	exp	Prasifka *et al.* (2005)
Orius insidiosus (Anthocoridae)	−0.25	exp	Prasifka *et al.* (2005)

In the published data, it is apparent that PAR_{slopes} are generally positive, although weakly so in most cases, whereas DAR_{slopes} are generally negative, similar to the DAR_{slopes} for other insect groups. An interesting case was described by Östman *et al.* (2009), where slopes for lycosid spiders and parasitic wasps were negative on islands situated close to a mainland and positive for more isolated islands. A similar shift was not seen for other groups. The authors speculate that the sign shift may be caused by a shift in the relative importance of local growth and migration processes for density–area relationships.

The work by Cronin and coworkers is probably the most detailed study on the role of spatial heterogeneity for host–parasitoid and natural enemy interactions. Their study system consists of discrete grass patches (*Spartina pectinata* or *Scolochloa festucacea*), planthoppers (*Prokelisia crocea* or *Delphacodes scholochloa*) and their natural enemies (spiders and parasitoids (*Anagrus* spp.)). Immigration rates (as defined in our study) of planthoppers (*P. crocea*) were invariant in relation to patch size whereas emigration rates were nonlinear functions of patch size (Cronin 2003b). Cronin argued that the nonlinear emigration rates are caused by a lower patch quality in large patches. The mechanism seems to be different than in our model assumptions because patch sizes in their system correlate with age, caused by plant spatial spread. For the cordgrass system, plant quality seems to decrease with age, indirectly affecting planthopper emigration rates. Quantifications of parasitoid densities and parasitism rates on *P. crocea* suggested that the immigration rate increased with patch size whereas the parasitism rate was independent of patch size. This apparent discrepancy may be caused by parasitoid interference (Cronin 2003a). Quantifications of other natural enemies suggested, in contrast to parasitoids, that web-building and cursorial spiders had higher density in small patches (Cronin *et al.* 2004). The authors tested whether this variation in spider densities could affect planthopper densities. The data showed such an effect, but the main mechanism was not mortality but rather an increased planthopper emigration rate caused by spider presence.

Applied aspects

Including trait-mediated aspects in the density responses of herbivores and their natural enemies to plant patch size has implications for the understanding and design of pest suppressive habitat diversification strategies. Habitat diversification strategies, like mixed cropping systems, are being implemented with the intention to suppress populations of herbivore populations. The two main hypotheses, the resource concentration and the enemies hypothesis, apparently are unable to explain all variation in the effects of mixed cropping systems on host plant finding, acceptance and survival of herbivorous pests. It has been previously argued that much of this variation originates from the fact that both the resource concentration and the enemies hypotheses are based on population responses, while the underlying behavioural responses remain largely unaccounted for (Bommarco and Banks 2003; Hambäck and Beckerman 2003). Meta-analyses show that, in general, mixed cropping systems support lower densities of herbivores and higher abundances of natural enemies (Tonhasca and Byrne 1994; Bommarco and Banks 2003; Langellotto and Denno 2004). However, literature data also typically show large unexplained variation in the responses of both trophic levels to mixed cropping systems. Part of this variation is likely explained by the fact that the outcome may depend on the type and scale of crop diversification strategy that is being implemented (e.g., intercropping, trapcropping, push–pull, etc.) (Bommarco and Banks 2003).

Recent studies suggest that the impact of vegetation heterogeneity on density responses of different herbivore species may be in part determined by how patch size and (constitutive and induced) traits of plants interact with herbivore aggregation behaviour (Bukovinszky et al. 2010). Herbivore-induced plant responses can cause herbivores to interact indirectly, e.g., by altering plant nutritional traits (i.e., primary and allelochemistry), which may be assessed by searching herbivores at a distance (Van Zandt and Agrawal 2004; Poelman et al. 2008). This way, phenotypic changes in plant defence chemistry for example can influence colonization and community assembly. In a spatial context, habitat characteristics such as patch size may affect the history of colonization of patches, resulting in priority effects and altered patch size–density relationships. Thus, variation found in density responses to mixed cropping systems may reflect trait-mediated interactions between plant-derived cues and insect movement behaviour.

The effects of mixed cropping systems on densities of natural enemies show an even greater variation than that of herbivores. A likely explanation for this is that most of the studies compare parasitism rates, while leaving host densities not standardized, confounding the effects of mixed cropping systems with the functional responses of natural enemies. Few studies examining patch responses of parasitoids under standardized conditions found variable results. A recent release–recapture study showed that more *Cotesia glomerata* (a parasitoid of *Pieris* butterflies) individuals have located hosts when they were in patches of four plants than on individual plants (Bezemer et al. 2010). Also, host density was positively correlated with the number of parasitoids locating the host patch, which might indicate the role of olfactory cues for host finding in the field. Surrounding vegetation (bare soil, short and tall vegetation), however, had a strong effect on the number of parasitoids successfully locating the plant-host patches. In the context of mixed cropping systems this indicates that patch size, plant infochemicals and vegetation context all likely affect migration behaviour of parasitoids and top-down control of herbivores. Many crop diversification strategies rely on infochemical, visual or structural aspects of plant traits interacting with herbivore movement behaviour (e.g., push–pull, trap cropping, undersowing). The model predictions here as well as recent empirical data suggest that considering the role of trait-mediated effects in movement behaviour of insect herbivores and their natural enemies will help us better understand the dynamics of host–parasitoid interactions in spatially heterogeneous habitats.

Conclusion

Most population models, if not all, treat the information about resource distributions as constant, whereas empirical studies clearly show that both positive and negative feedbacks are not only common but also substantially affect density distributions and species interactions. The models presented in this chapter are in many ways too simple for describing the complexity of insect responses to plant patches and to host densities, and how these responses are affected by the shape of the information landscape. Nevertheless, important insights do arise from simple models such as these and in order to understand other processes affecting density distributions we have to account for the direct effects of host distributions on immigration and emigration rates. As argued elsewhere (Hambäck and Englund 2005), models with scale-dependent migration rates provide a base-line for studies on the role of patch size on insect density akin to the use of allometric relationships in evolutionary studies. Differential migration rates from small and large patches are evident and cannot be ignored but it is also clear that other processes, particularly strong density dependence or strong local growth, can reduce the importance of patch size dependent migration rates. The relative role of local growth versus migration has been explored in previous papers and was found to depend on both patch structure and on animal mobility. Our intention with this chapter is to argue that the future understanding of density distributions in arthropod predator–prey systems not only depends on a close development between general population models and data (see also Cronin and Reeve 2005), but also that future analyses should consider both direct effects caused by predation and indirect effects caused by changes in movement patterns. Such a spatial dimension is currently ignored in the recent literature on consumptive and non-consumptive effects of predators on prey. Including the scaling of population processes is also necessary for making conclusions at larger spatial scales from processes at smaller spatial scales and useful future extensions from the modelling approach could be for a wider set of metapopulation and metacommunity models.

References

Andow, D. A. (1991) Vegetational diversity and arthropod population response. *Annual Review of Entomology*, **36**, 561–586.

Bach, C. E. (1988) Effects of host plant patch size on herbivore density: patterns. *Ecology*, **69**, 1090–1102.

Banks, J. E. (1998) The scale of landscape fragmentation affects herbivore response to vegetation heterogeneity. *Oecologia*, **117**, 239–246.

Bender, D. J., Contreras, T. A. and Fahrig, L. (1998) Habitat loss and population decline: a meta-analysis of the patch size effect. *Ecology*, **79**, 517–533.

Bezemer, T. M., Harvey, J. A., Kamp, A. F. D. *et al.* (2010) Behaviour of male and female parasitoids in the field: influence of patch size, host density, and habitat complexity. *Ecological Entomology*, **35**, 341–351.

Bolger, D. T., Beard, K. H., Suarez, A. V. and Case, T. J. (2008) Increased abundance of native and non-native spiders with habitat fragmentation. *Diversity and Distributions*, **14**, 655–665.

Bommarco, R. and Banks, J. E. (2003) Scale as modifier in vegetation diversity experiments: effects on herbivores and predators. *Oikos*, **102**, 440–448.

Bommarco, R. and Fagan, W. F. (2001) Influence of crop edges on movement of generalist predators: a diffusion approach. *Agricultural and Forest Entomology*, **3**, 1–11.

Bossert, W. H. and Wilson, E. O. (1963) The analysis of olfactory communication among animals. *Journal of Theoretical Biology*, **5**, 443–469.

Bowers, M. A. and Dooley, J. L. (1999) A controlled, hierarchical study of habitat fragmentation: responses at the individual, patch, and landscape scales. *Landscape Ecology*, **14**, 381–389.

Bowers, M. A. and Matter, S. F. (1997) Landscape ecology of mammals: relationships between density and patch-size. *Journal of Mammalogy*, **78**, 999–1013.

Bowman, J., Cappuccino, N. and Fahrig, L. (2002) Patch size and population density: the effect of immigration behavior. *Conservation Ecology*, **6**, Art. No. 9.

Bradburne, R. P. and Mithen, R. (2000) Glucosinolate genetics and the attraction of the aphid parasitoid *Diaeretiella rapae* to Brassica. *Proceedings of the Royal Society of London, Series B*, **267**, 89–95.

Bukovinszky, T., Gols, R., Kamp, A. *et al.* (2010) Combined effects of patch size and plant nutritional quality on local densities of insect populations. *Basic and Applied Ecology*, **11**, 396–405.

Bukovinszky, T., Potting, R. P. J., Clough, Y., van Lenteren, J. C. and Vet, L. E. M. (2005) The role of pre- and post-alighting detection mechanisms in the responses to patch size by specialist herbivores. *Oikos*, **109**, 435–446.

Cantrell, R. S., Cosner, C. and Fagan, W. F. (2002) Habitat edges and predator–prey interactions: effects on critical patch size. *Mathematical Biosciences*, **175**, 31–55.

Cronin, J. T. (2003a) Patch structure, oviposition behavior, and the distribution of parasitism risk. *Ecological Monographs*, **73**, 283–300.

Cronin, J. T. (2003b) Movement and spatial population structure of a prairie planthopper. *Ecology*, **84**, 1179–1188.

Cronin, J. T. (2009) Habitat edges, within-patch dispersion of hosts, and parasitoid oviposition behavior. *Ecology*, **90**, 196–207.

Cronin, J. T. and Reeve, J. D. (2005) Host–parasitoid spatial ecology: a plea for a landscape-level synthesis. *Proceedings of the Royal Society of London, Series B*, **272**, 2225–2235.

Cronin, J. T., Haynes, K. J. and Dillemuth, F. (2004) Spider effects on planthopper mortality, dispersal, and spatial population dynamics. *Ecology*, **85**, 2134–2143.

Doak, P. (2000) The effects of plant dispersion and prey density on parasitism rates in a naturally patchy habitat. *Oecologia*, **122**, 556–567.

Elzinga, J.A., Turin, H., van Damme, J.M.M. and Biere, A. (2005) Plant population size and isolation affect herbivory of *Silene latifolia* by the specialist herbivore *Hadena bicruris* and parasitism of the herbivore by parasitoids. *Oecologia*, **144**, 416–426.

Englund, G. (1997) Importance of spatial scale and prey movements in predator caging experiments. *Ecology*, **78**, 2316–2325.

Englund, G. and Cooper, S.D. (2003) Scale effects and extrapolation in ecological experiments. *Advances in Ecological Research*, **33**, 161–213.

Englund, G. and Hambäck, P.A. (2004) Scale-dependence of emigration rates. *Ecology*, **85**, 320–327.

Englund, G. and Hambäck, P.A. (2007) Scale dependence of immigration rates: models, metrics, and data. *Journal of Animal Ecology*, **76**, 30–35.

Fagan, W.F., Cantrell, R.S. and Cosner, C. (1999) How habitat edges change species interactions. *American Naturalist*, **153**, 165–182.

Fletcher, R.J., Ries, L., Battin, J. and Chalfoun, A.D. (2007) The role of habitat area and edge in fragmented landscapes: definitively distinct or inevitably intertwined? *Canadian Journal of Zoology*, **85**, 1017–1030.

French, D.R. and Travis, J.M.J. (2001) Density-dependent dispersal in host–parasitoid assemblages. *Oikos*, **95**, 125–135.

Gaston, K.J. and Matter, S.F. (2002) Individuals-area relationships: comment. *Ecology*, **83**, 288–293.

Geervliet, J.B.F., Ariëns, S., Dicke, M. and Vet, L.E.M. (1998) Long-distance assessment of patch profitability through volatile infochemicals by the parasitoids *Cotesia glomerata* and *C. rubecula* (Hymenoptera: Braconidae). *Biological Control*, **11**, 113–121.

Grez, A.A. and Gonzales, R.H. (1995) Resource concentration hypothesis: effect of host plant patch size on density of herbivorous insects. *Oecologia*, **103**, 471–474.

Hambäck, P.A. and Beckerman, A.P. (2003) Herbivory and plant resource competition: a review of two interacting interactions. *Oikos*, **101**, 26–37.

Hambäck, P.A. and Englund, G. (2005) Patch area, population density and the scaling of migration rates: the resource concentration hypothesis revisited. *Ecology Letters*, **8**, 1057–1065.

Hambäck, P.A., Bergman, K.-O., Bommarco, R. et al. (2010a) Allometric density responses in butterflies: the response to small and large patches by small and large species. *Ecography*, **33**(6), 1149–1156.

Hambäck, P.A., Björkman, M. and Hopkins, R.J. (2010b) Patch size effects are more important than genetic diversity for plant-herbivore interactions in *Brassica* crops. *Ecological Entomology*, **35**, 299–306.

Hambäck, P.A., Summerville, K.S., Steffen-Dewenter, I. et al. (2007a) Habitat specialisation, body-size and phylogeny explains density-area relationships in Lepidoptera: a cross-continental comparison. *Proceedings of the National Academy of Sciences of the United States of America*, **104**, 8368–8373.

Hambäck, P.A., Vogt, M., Tscharntke, T., Thies, C. and Englund, G. (2007b) Spatiotemporal dynamics of cereal aphids: testing a scaling theory for local density. *Oikos*, **116**, 1996–2006.

Hassell, M.P. (1978) *Arthropod Predator–Prey Systems*. Princeton, NJ: Princeton University Press.

Holt, R.D. and Lawton, J.H. (1993) Apparent competition and enemy-free space in insect host–parasitoid communities. *American Naturalist*, **142**, 623–645.

Holt, R.D., Lawton, J.H., Polis, G.A. and Martinez, N.D. (1999) Trophic rank and the species-area relationship. *Ecology*, **80**, 1495–1504.

Kareiva, P. (1985) Finding and losing host plants by *Phyllotreta*: patch size and surrounding habitat. *Ecology*, **66**, 1809–1816.

Kareiva, P. (1987) Habitat fragmentation and the stability of predator–prey interactions. *Nature*, **326**, 388–390.

Langellotto, G. A. and Denno, R. F. (2004) Responses of invertebrate natural enemies to complex-structured habitats: a meta-analytical synthesis. *Oecologia*, **139**, 1–10.

Lof, M., Hemerik, L. and de Gee, M. (2007) Chemical communication: does odor plume shape matter. *Proceedings of the Netherlands Entomological Society Meetings*, **18**, 61–70.

Matter, S. F. (1997) Population density and area: the role of between- and within-patch processes. *Oecologia*, **110**, 533–538.

Murlis, J., Elkinton, J. S. and Cardé, R. T. (1992) Odor plumes and how insects use them. *Annual Review of Entomology*, **37**, 505–532.

Östman, Ö., Mellbrand, K. and Hambäck, P. A. (2009) Edge or dispersal effects – Their relative importance on arthropod densities on small islands. *Basic and Applied Ecology*, **10**, 475–484.

Peng, C. and Weiss, M. J. (1992) Evidence of an aggregation pheromone in the flea beetle, *Phyllotreta cruciferae* (Goeze) (Coleoptera: Chrysomelidae). *Journal of Chemical Ecology*, **18**, 875–884.

Poelman, E. H., Broekgaarden, C., Van Loon, J. J. A. and Dicke, M. (2008) Early season herbivore differentially affects plant defence responses to subsequently colonizing herbivores and their abundance in the field. *Molecular Ecology*, **17**, 3352–3365.

Prasifka, J. R., Hellmich, R. L., Dively, G. P. and Lewis, L. C. (2005) Assessing the effects of pest management on nontarget arthropods: the influence of plot size and isolation. *Environmental Entomology*, **34**, 1181–1192.

Rabasa, S. G., Gutiérrez, D. and Escudero, A. (2007) Metapopulation structure and habitat quality in modelling dispersal in the butterfly *Iolana iolas*. *Oikos*, **116**, 793–806.

Rabasa, S. G., Gutierrez, D. and Escudero, A. (2008) Relative importance of host plant patch geometry and habitat quality on the patterns of occupancy, extinction and density of the monophagous butterfly *Iolana iolas*. *Oecologia*, **156**, 491–503.

Raffa, K. F. (1991) Induced defensive reactions in conifer-bark beetle systems. In D. W. Tallamy and M. J. Raupp, eds., *Phytochemical Induction by Herbivores*. New York: John Wiley and Sons, pp. 245–276.

Raffa, K. F. (2001) Mixed messages across multiple trophic levels: the ecology of bark beetle chemical communication systems. *Chemoecology*, **11**, 49–65.

Ries, L. and Sisk, T. D. (2008) Butterfly edge effects are predicted by a simple model in a complex landscape. *Oecologia*, **156**, 75–86.

Root, R. B. (1973) Organization of a plant-arthropod association in simple and diverse habitats: the fauna of collards (*Brassica oleracea*). *Ecological Monographs*, **43**, 95–124.

Ryall, K. L. and Fahrig, L. (2006) Response of predators to loss and fragmentation of prey habitat: a review of theory. *Ecology*, **87**, 1086–1093.

Schoonhoven, L. M., Van Loon, J. J. A. and Dicke, M. (2006) *Insect-Plant Biology*. Oxford: Oxford University Press.

Sheehan, W. and Shelton, A. M. (1989) Parasitoid response to concentration of herbivore food plants: finding and leaving plants. *Ecology*, **70**, 993–998.

Shiojiri, K., Takabayashi, J., Yano, S. and Takafuji, A. (2001) Infochemically mediated tritrophic interaction webs on cabbage plants. *Population Ecology*, **43**, 23–29.

Siekmann, G., Keller, M. A. and Tenhumberg, B. (2004) The sweet tooth of adult parasitoid *Cotesia rubecula*: ignoring hosts for nectar? *Journal of Insect Behavior*, **17**, 459–476.

Strong, D. R. (1979) Biogeographic dynamics of insect–host plant communities. *Annual Review of Entomology*, **24**, 89–119.

Thies, C. and Tscharntke, T. (1999) Landscape structure and biological control in agroecosystems. *Science*, **285**, 893–895.

Tonhasca, A. and Byrne, D. N. (1994) The effects of crop diversification on herbivorous insects: a meta-analysis approach. *Ecological Entomology*, **19**, 239–244.

Tscharntke, T. and Brandl, R. (2004) Plant–insect interactions in fragmented landscapes. *Annual Review of Entomology*, **49**, 405–430.

Turchin, P. (1989) Population consequences of aggregative movement. *Journal of Animal Ecology*, **58**, 75–100.

Underwood, N., Anderson, K. and Inouye, B. D. (2005) Induced vs. constitutive resistance and the spatial distribution of insect herbivores among plants. *Ecology*, **86**, 594–602.

Van Zandt, P. A. and Agrawal, A. A. (2004) Specificity of induced plant responses to specialist herbivores of the common milkweed *Asclepias syriaca*. *Oikos*, **104**, 401–409.

Vet, L. E. M. and Dicke, M. (1992) Ecology of infochemical use by natural enemies in a tritrophic context. *Annual Review of Entomology*, **37**, 141–172.

Viswanathan, D. V., McNickle, G. and Thaler, J. S. (2008) Heterogeneity of plant phenotypes caused by herbivore-specific induced responses influences the spatial distribution of herbivores. *Ecological Entomology*, **33**, 86–94.

von Zeipel, H., Eriksson, O. and Ehrlen, J. (2006) Host plant population size determines cascading effects in a plant-herbivore-parasitoid system. *Basic and Applied Ecology*, **7**, 191–200.

Zabel, J. and Tscharntke, T. (1998) Does fragmentation of *Urtica* habitats affect phytophagous and predatory insects differentially? *Oecologia*, **116**, 419–425.

CHAPTER TWENTY-FIVE

Plasticity and trait-mediated indirect interactions among plants

ERIK T. ASCHEHOUG and RAGAN M. CALLAWAY

Division of Biological Sciences, University of Montana

Introduction

Ecologists have long recognized the importance of phenotypic plasticity as a mechanism by which organisms acclimate and adapt to local environments. Phenotypic plasticity is commonly defined as variation in the morphological or physiological phenotype of a given genotype in response to the abiotic and biotic environment (Bradshaw 1965). Plants are particularly plastic organisms because they must solve the fundamental problems of resource acquisition, competition and herbivore attack without mobility (Sultan 1987, 2000). Plasticity has been demonstrated in the morphological, developmental, physiological and biochemical traits of plant species, with many traits showing flexibility in expression both between and within individual plant species (Novoplansky 2002; de Kroon *et al.* 2005; Valladares *et al.* 2006). The plastic responses vary not only in their form, but also in their permanence. Responses may be permanent for the lifetime of an individual, fixed for long periods of time (e.g., a growing season), or dynamic at the scale of minutes (Metlen *et al.* 2009).

Research on the nature of plasticity and its potential to broaden the ecological niches of species in shifting abiotic and biotic conditions has historically focused on the morphological responses of plants. However, morphological responses tend to be slow and are largely irreversible – two parameters that are not favoured by selection (Valladares *et al.* 2007). In contrast to changes in morphology, plants can also respond via biochemistry. These responses can be exceptionally rapid and highly ephemeral (Metlen *et al.* 2009); traits that lend themselves to adaptive value and are favoured by natural selection, but biochemical traits have been studied very little in the explicit context of plasticity.

Plasticity and trait variation have been studied extensively in plants, but the ecological consequences of such phenotypic variation are poorly understood (Miner *et al.* 2005). Experimental settings commonly overestimate the degree to which plants can exhibit plastic responses to changes in natural conditions. This may be because of the continually changing conditions in nature which

Trait-Mediated Indirect Interactions: Ecological and Evolutionary Perspectives, eds. Takayuki Ohgushi, Oswald J. Schmitz and Robert D. Holt. Published by Cambridge University Press. © Cambridge University Press 2012.

makes optimization of some plastic responses difficult, especially if plastic responses have long lag times and are largely irreversible. In addition, the diffuse nature of competition among plants within communities may limit a plant's ability to exhibit its maximum potential plasticity in response to biotic interactions. Thus there appear to be greater limits and constraints to plasticity in plants in natural systems than would be predicted from responses to controlled environments (Valladares *et al.* 2007). However, investigations that focus on trait responses that can mediate multiple environmental stimuli in plants should have a higher potential for adaptive value. For example, secondary biochemistry responses to nutrient stress may also mediate competitive or facilitative interactions (Tharayil *et al.* 2009) and therefore can potentially provide a more stable cue for plant plasticity responses in natural systems.

It is surprising that the consequences of plasticity for interactions among plants have not been more deeply explored since high phenotypic plasticity in plants is thought to be a characteristic of 'good competitors' (Grime 2001), and good competitors can have powerful effects on communities (Connell 1983). Species classified as good competitors generally show more rapid responses to variation in their environment, such as adjusting root:shoot ratios, leaf specific area, proportions of fine to coarse roots and diversity of biochemistry, than species that are poor competitors (Grime 2001; Callaway *et al.* 2003). Despite the high degree of plasticity expressed by plants, and the potential for this plasticity to affect the way a species might interact with its neighbours (known as trait-mediated interaction or TMI), very few studies have been conducted with plants that focus on plasticity and interactions. However, by re-examining earlier work with a focus on plasticity we can piece together direct evidence for TMIs.

For example, Callaway (1990) found that the root architecture of *Quercus douglasii* seedlings demonstrated plasticity to variation in water source. *Quercus douglasii* seedlings with experimentally restricted access to deep stores of water produced roughly twice as many fine lateral roots and more than 5× the lateral root mass as seedlings with access to a deep water source. This phenotypic plasticity demonstrated by seedlings in controlled experiments corresponded with apparent plasticity in the field, where mature trees without access to deep water possessed very dense surface lateral root systems while trees with deep water access did not (Callaway *et al.* 1991). Plasticity in root architecture in the field appeared to create a TMI as trees with abundant shallow roots strongly suppressed understory productivity; whereas trees without abundant shallow roots had strong facilitative effects.

More recently, ecologists have extended the view of TMIs from direct interactions between plants to indirect interactions (TMIIs) among species. While we know of no studies of plant–plant interactions that test for the presence of TMIIs, conceptually all of the necessary components to produce

TMIIs have been studied, making the next step ripe for empirical research. For example, in the scenario of the TMI apparently mediated by plasticity in the root architecture of *Quercus douglasii*, this plasticity also correlated with different understory community compositions. The abundance of the native *Nassella* (nee *Stipa*) *pulchra* was higher under trees with abundant shallow roots (Callaway *et al.* 1991). Simultaneously the abundance of European annuals (primarily *Avena fatua* and *Bromus diandrus*), which can competitively exclude natives such as *Nassella*, was lower. This pattern suggests the occurrence of a TMII but does not demonstrate it. However, a relatively simple experiment could explore this spatial pattern in the context of TMIIs, and similar experiments could be used to study TMII in other systems. The key would be to determine whether shallow root architecture simply promoted *Nassella* directly, or altered the competitive effects of the European annuals in ways that indirectly promoted *Nassella* (see Rice and Nagy 2000). Using *Nassella* as a target species, TMIIs would be demonstrated if experimental treatments in which European annuals were removed from around *Nassella* under trees without shallow root architecture improved the growth or fitness of *Nassella*.

This link between plastic responses to environment and its affect on plant–plant interactions (direct and indirect) represents a major gap in our understanding of how plant communities assemble. TMIs and TMIIs have the potential to create tremendous variation, or conditionality, in the outcomes of interactions among competing species, and thus have important implications for how competitors might coexist (Chesson and Rosenzweig 1991). In other words, we know that indirect interactions among groups of competitors can promote coexistence among species that would otherwise be likely to exclude each other competitively (Miller 1994; Callaway and Howard 2007), thus plasticity among species can greatly enhance the potential for indirect interactions to sustain coexistence among competing species and thus increase community diversity.

Interactions among plants

Negative direct interactions among plants appear to derive primarily from the need to acquire basic resources such as light, water and nutrients, which are often in limited supply (Goldberg 1990; Miller and Travis 1996). Because plants are sessile, resource competition between individuals can be intense, potentially making coexistence difficult when essential resources are scarce (Tilman 1982). In addition, allelopathy, the negative biochemical effects of neighbours on each other (Turlings *et al.* 1990; Williamson 1990; Mahall and Callaway 1992; Schenck *et al.* 1999), can also be a mechanism by which plant species inhibit each other.

Positive interactions among plants, or facilitation, occur when the presence of one plant enhances the growth, survival or reproduction of a neighbour

(Callaway 2007). But it is important to note that facilitation by one species on another may correspond with reciprocal negative, positive or neutral responses. Direct positive interactions may incorporate a wider range of mechanisms than direct negative interactions (Callaway 2007). Like competition, facilitation may occur through resource effects, one species increasing nutrient, water or light availability to another, or through chemical effects (Metlen *et al.* 2009). However, facilitation can also be driven by non-resource processes. Most commonly, species that are physically tolerant to stresses such as cold, heat, wind, salinity and disturbance buffer other species from these abiotic conditions.

Indirect interactions among plants can be derived from direct resource competition, allelopathy or facilitation (Pages and Michalet 2003; Callaway 2007; Callaway and Howard 2007), but these have received far less attention than the direct impacts that plants have on one another. This may be because plants are generally embedded within a matrix of many other plants, all of which require the same basic resources of light, water and nutrients; thereby creating an environment in which direct interactions appear to be assured. However, the highly aggregate nature of plant communities also sets the stage for common and strong indirect interactions – situations in which the direct interaction between two species is caused or altered by simultaneous interactions with additional species (Miller 1994; Levine 1999; Callaway and Pennings 2000; Callaway and Howard 2007; Cuesta *et al.* 2010).

Ecological consequences of plant interactions

Competition and facilitation among plants are the basic processes through which TMIs and TMIIs can operate, and these interactions can be powerful organizing forces in structuring plant communities (Allen and Forman 1976; Grime 1977; Connell 1983; Tilman 1985; Ortega and Pearson 2005; Callaway 2007; Cavieres and Badano 2009). Because of this, evaluating species' inherent competitive abilities can likely provide some insight into how they will perform in a community context. However, assessing the relative competitive strengths of species is difficult in anything other than simple pairwise or 'bioassay' experiments; and it now is becoming apparent that such experiments do not accurately predict how individuals may respond when subjected to the diffuse nature of interactions found in plant communities (Callaway and Howard 2007; Perkins *et al.* 2007; Engel and Weltzin 2008; Schmidtke *et al.* 2010; Aschehoug 2011). Even more, rankings of competitive effects and responses may not be complete indicators of individual competitive abilities when plants are in real communities (Wang *et al.* 2010). Thus theory for how plant communities assemble that fails to incorporate indirect interactions is probably incomplete, and this has very important implications for studying TMIIs.

There are two general, but contrasting, theories for how plant species may assemble into communities under equilibrium conditions as a result of competition among plants. The first does not incorporate indirect interactions, and thus does not have the potential to integrate TMIIs, and poses that plant communities are competitively transitive in nature (Goldsmith 1978; Mitchley and Grubb 1986; Keddy and Shipley 1989). In other words, all species in a given pool, or community, can be ranked in a linear competitive hierarchy. The strict 'pecking order' that results from hierarchical competitive abilities provides a predictive tool for community organization. A transitive, or hierarchical, perspective on assembly rules assumes that communities will consistently move towards dominance by the best competitor in the hierarchy in a homogeneous abiotic environment, and this can potentially lead to the development of monocultures. An important theoretical consequence of not allowing TMIIs among plants in such transitive or hierarchical communities is that weak competitors will be competitively excluded given enough time and the absence of non-equilibrium processes. In this paradigm only non-equilibrium forces, such as fire, herbivory (which can establish TMIIs) or abiotic heterogeneity, can prevent the dominance of a small number of species or the formation of monocultures in a local community.

In contrast to hierarchical assembly rules is the theory that plants exhibit non-transitive or non-hierarchical competitive properties as they form communities (Jackson and Buss 1975; May and Leonard 1975; Petraitis 1979). Whereas hierarchical organization is best described mathematically as A > B > C, non-hierarchical organization occurs when loops form in the hierarchy such as A > B, B > C, but C > A. In other words, species C indirectly benefits species B by having a direct negative impact on species A, which creates the opportunity for TMIIs to be included in conceptual models. Given the right starting point, a simple loop within a suite of competing species can result in a perpetually shifting state in which all three species coexist indefinitely (Buss and Jackson 1979). This coexistence is based entirely on the balance of direct competitive interactions, but leads to the formation of complex networks of species interactions which may be mediated directly or indirectly via plant plasticity. Proponents of non-transitive competitive processes note that because plants interact with many other species simultaneously, clear pecking orders are likely rare. In addition, sporadic reversals of dominance among species can create powerful and facilitative indirect effects among competitors which can transform overall community structure.

When community members interact in complex 'networks' of interactions, competitive exclusion is much less likely. In addition, coexistence may be maintained among large pools of species in the absence of abiotic heterogeneity or non-equilibrium processes. Mathematical evaluations of such interactions predict that such indirect interactions among competitors can allow large communities of species to coexist (Karlson and Jackson 1981; Laird and Schamp 2006; Laird and Schamp 2008).

Non-transitive theory requires quite specific combinations or sequences of interactions among species to produce indirect interactions, and thus TMIIs. However, groups of plant species appear to compete in ways that produce indirect interactions, but without the competitive 'loops' required for non-transitive theory. In other words, some species appear to 'modify' interactions among other species without establishing the classic non-transitive combinations of competitive dominance (Callaway and Pennings 2000; Metlen 2010).

A key commonality of the transitive and non-transitive theories of plant community assembly is the requirement of strong, species-specific, direct negative effects. In a hierarchical system, the direct effect of each species on another is linear; a single dominant species that exerts primary control of community wide species diversity via competitive interactions, and subdominant species that exert lesser degrees of control in direct proportion to their place in the hierarchy. In non-hierarchical systems, there is the requirement of at least one of the weaker competitors directly outcompeting a species of higher competitive ranking. In other words, a species that loses most of its interactions with other community members must be able to outcompete a species that wins most of its interactions in the community, an unlikely scenario without the presence of TMIs. In fact, the conditionality of competitive outcomes may be explained in large part by the plastic response of plants to competition (Cahill *et al.* 2010).

Transitive and non-transitive processes occur in communities but plants also experience 'diffuse' competition in a community context. Instead of a distinct interaction with another single species, for which ranks might be determined, a plant experiences the additive effects (positive or negative) of many species interacting in space and time. Such diffuse interactions may have profound impacts on the species composition of plant communities (Davidson 1980; Vandermeer 1980; Wilson and Keddy 1986; Miller 1994; Li and Wilson 1998; Levine 1999; Callaway and Pennings 2000) and provide tremendous potential, conceptually if not logistically, to study TMIIs.

There is a substantial escalation in the complexity of assembly rules as we move from transitive to non-transitive models and as we add interaction modifications and diffuse interactions to competitive loops. And this complexity increases again once we consider that most current thinking is built on the construct, or at least the implicit assumption, that plants are fixed in their competitive abilities. However, we know that plants vary in their competitive abilities both within and between populations. For example, Grøndahl and Ehlers (2008) found that genotypic variation in the production of different terpenes by ecotypes of *Thymus pulegioides* and *T. serpyllum* altered the effects of the *Thymus* species on co-occurring plant species. The ecological effects demonstrated for *Thymus* species (also see Ehlers and Thompson 2004; Jensen and Ehlers 2010) corresponded with selective effects of *Thymus* on their neighbours. Plants that came from sites where they co-occurred naturally with a carvacrol (a terpene)-producing ecotype of *Thymus* also performed better on soil treated with carvacrol. This example of how genotypic variation can affect the competitive effects of a species derives from at least two general ways that plants interact directly, facilitation and allelopathy. If relatively subtle differences in 'fixed' competitive interactions can have such a large impact on community formation, then *phenotypic* shifts that lead to changes in interaction outcomes (TMIs and TMIIs) have the potential to be very powerful in determining how plant communities assemble.

Plasticity and direct interactions

In both transitive and non-transitive models, the intensity of direct interactions determines the degree to which plants can coexist. Therefore understanding how plasticity affects interaction intensities can greatly improve our ability to predict plant coexistence in communities. Much is known about how plant traits such as morphology, growth rates, final size, reproduction, qualitative and quantitative biochemical traits, and biomass allocation can vary widely for a given genotype (Sultan 1987, 2000; Metlen *et al.* 2009), which creates exceptional opportunities for exploring how phenotypic plasticity within an individual species can influence interactions with other species and the subsequent effects on the structure of plant communities.

A substantial component of the way that plants affect each other (either negatively or positively) is based on plant size and growth rate. For example, Brooker *et al.* (2005) reanalysed data from Reader *et al.* (1994) to compare the intensity of the competitive effect of neighbours on *Poa pratensis* to the importance of the competitive effect. Among grasslands that varied in productivity, both components of competition were significantly affected by total neighbour biomass. Plant size can also affect facilitative interactions. Tewksbury and Lloyd (2001) found that larger *Olneya testota* trees in the Sonoran Desert supported higher numbers of beneficiary species and larger beneficiary perennials than small canopies. Because of the importance of the size of individual plants for competitive and facilitative effects, phenotypic plasticity in size may have substantial effects on interaction outcomes.

Morphological plasticity as a response to abiotic conditions, however, is often slow and costly, which may limit the ability of plants to respond when subjected to intense competition (Novoplansky 2002). Biochemical responses, such as the release of secondary metabolites that increase nutrient availability in the rhizosphere, are less costly and more ephemeral responses that can have immediate impacts on plant performance (Metlen *et al.* 2009) and potentially effects on neighbouring competitors. Li *et al.* (2007) found that the cluster root forming species *Vicia faba* increased phosphorus availability in the soil rhizosphere via the release of acidifying chemicals (citrate and malate). The biochemical response of *V. faba* to phosphorus deficiency is also exceptionally fast; laboratory tests show it reducing the pH of nutrient agar by ~2 units in 6 hours. Such changes in soil acidity can result in 10-fold changes in phosphorus availability. In field experiments, the increase in phosphorus availability resulted in an overyielding of 26% for *V. faba*. In addition, *V. faba* directly facilitated *Zea mays* through the shared increase in phosphorus availability leading to an overyielding of 43% by *Z. mays*. This example demonstrates the strong potential for biochemical plasticity to be a model system for understanding how TMIs and TMIIs can impact both competitive and facilitative interactions and the organization of communities.

Plasticity and indirect interactions

We know of no examples in which phenotypic plasticity in a plant trait has been shown to alter *indirect* interactions among other plant species. Conceptually, however, all of the component pieces of TMIIs can be examined from existing empirical studies. What is lacking is a comprehensive set of experiments that explicitly link plasticity and indirect interactions. Ideally, such studies would entail an experiment in which species 'A' demonstrated two or more phenotypes (e.g., A_{Ph1} and A_{Ph2}), and then the indirect effects of these two phenotypes would be tested in experiments involving two or more other neighbours (Fig. 25.1). For example, in Fig. 25.1a, the hypothetical A_{Ph1} has weak competitive effects on species B, and species B has strong competitive effects on species C. Thus the indirect effects of A_{Ph1} on C are weak. In contrast, the hypothetical A_{Ph2} has strong competitive effects on species B, and thus strong indirect facilitative effects on species C. Figure 25.2 illustrates how TMIIs might occur when there is plasticity in a facilitative benefactor (species A) or a beneficiary (species B).

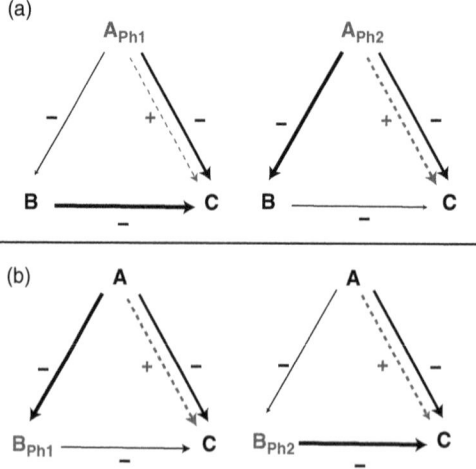

Figure 25.1 (a) A conceptual model of how a phenotypic shift in species A can alter the competitive *effect* (solid line) of species A on species B, resulting in an increase in indirect (dashed line) benefit to species C. (b) A model of how a phenotypic change in species B in *response* to competition by species A can result in a change in the indirect effect of species A on species C. Both models represent TMIIs as a result of competitive interactions. See colour plate section.

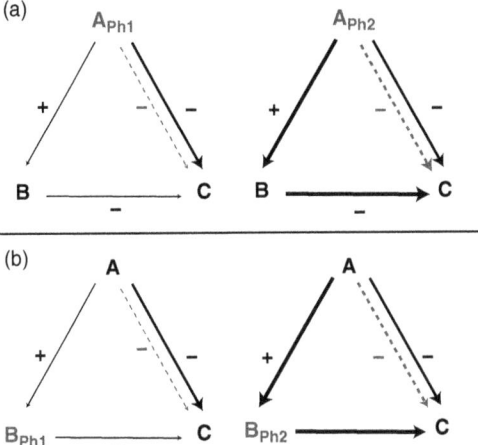

Figure 25.2 (a) A conceptual model of how a phenotypic shift in species A can alter the facilitative *effect* (solid line) of species A on species B, resulting in an increase in indirect (dashed line) negative effect on species C. (b) A model of how a phenotypic change in species B in *response* to facilitation by species A can result in a change in the indirect effect of species A on species C. Both models represent TMIIs as a result of facilitative interactions. See colour plate section.

Next we suggest potential scenarios in nature in which these TMIIs might occur, with the goal of stimulating ideas for how such studies might be approached in the future. An intriguing scenario may exist for *Q. agrifola*, the native perennial herb *Pholistima auritum*, and European annual grasses in California grasslands. *Pholistima* can form near monocultures directly beneath the canopies of some *Q. agrifolia* trees, but is much less abundant in the grassland directly adjacent to the canopies where European annuals dominate (Parker and Muller 1982). However, if *Pholistima* is not present under the oaks, European grass species are intensely facilitated (as *Pholistima* itself appears to be) by *Q. agrifolia* which suggests that the low abundance of annual grass species in the understory is not due to the direct effects of the oaks. Instead, it appears that once facilitated, *Pholistima* excludes the European annual grasses through the inhibitory effects of its litter and leachates. In field experiments, fresh *Pholistima* litter reduced *Bromus* germination by 73% and *Avena* by 96%. However, when experiments were conducted with *Pholistima* litter that had been leached, at least 92% of seeds germinated in every treatment. *Q. agrifolia* appears to have powerful negative indirect effects on grasses, and the stage is set for the next necessary step for TMIIs – if *Q. agrifolia* demonstrates plasticity in some way that affects its facilitative effect on *Pholistima*, the indirect effects of the tree on annual grasses are likely to change as well.

Q. douglasii and *Q. agrifolia* may help us understand the *effects* of plasticity on indirect interactions (Fig. 25.2a), but plants show a great deal of plasticity in *response* to neighbours as well (Callaway *et al.* 2003; Cahill *et al.* 2010). We know of no examples in which plastic responses have been connected to cascading indirect interactions with other species, but much like the *Quercus* example above, we can speculate about how the plastic response to competition may lead to indirect interactions.

Cahill *et al.* (2010) found that plants altered their root foraging strategies based on the amount and distribution of resources in the soil and the presence of competitors. When grown alone, *Abutilon theophrasti* had broadly developed root systems regardless of whether resources were uniformly or patchily distributed within the soil. However, when in competition with a conspecific, *A. theophrasti* was highly plastic in root distribution and distance from stem depending on the distribution of resources, suggesting that plants are capable of altering the plastic response of roots to nutrients depending on the presence or absence of a competitor, a good example of a TMI. While this example highlights the plastic response of *A. theophrasti* to competition, this particular set of experiments was limited to pairwise competition, which does not test TMIIs. Nevertheless, the alteration of root system morphology based on the presence of competitors and variable resources suggests that when placed in a multispecies context, the intensity of competition that *A. theophrasti* experiences (stronger or weaker) is highly dependent on the plastic response. A test of this TMII response could easily be carried out by replicating the experiments with multiple competitors. TMIIs would be demonstrated if competition intensities differed between single and multiple competitor experiments.

Plasticity and exotic invaders

Plasticity expressed by any plant species provides an opportunity to explore TMIIs in novel ways, but exotic invaders might provide unusually good opportunities because they are thought to be unusually plastic (Richards *et al.* 2006; Hulme 2008) and unusually strong competitors (Maron and Marler 2008). High phenotypic plasticity has been suggested as a good predictor of invasiveness (Mal and Lovett-Doust 2005; Chun *et al.* 2007, but see Bossdorf *et al.* 2005; Hulme 2008). Unlike our focus here on plasticity and TMII, both Richards *et al.* (2006) and Hulme (2008) focus on the potential role of plasticity in allowing an invader to express advantageous phenotypes as they colonize a broad range of environments (Bradshaw 1965; Van Valen 1965; Whitlock 1996; Sultan *et al.* 1998a, b; Donohue *et al.* 2001; Richards *et al.* 2005). While likely true, exotic 'invasion' is only defined in part by colonization by exotic species. Callaway and Maron (2006) and Hierro *et al.* (2005) note that exotic 'invasions' involve biogeographic shifts in the fundamental ecology of a species; generally much higher abundances and stronger apparent impacts in non-native ranges than in native ranges. Thus the essence of an 'invader' as opposed to an 'exotic' is not just in the processes involved in colonization, but in the processes involved in the attainment of very high densities, biomass and impacts on other species. It is in the context of plasticity within these processes and impacts that invaders provide the best opportunities to study TMIIs.

Strong competitive interactions are likely to play an important role in establishing the dominance of some invaders (D'Antonio and Mahall 1991; Levine *et al.* 2003; Vila and Weiner 2004; Maron and Marler 2008; Munshaw and Lortie 2010). Release from specialist herbivores or pathogens may allow plants to be more competitive in non-native ranges (Keane and Crawley 2002), or successful invaders may possess competitive advantages because they come from a more competitive species pool, or happen to possess inherent traits that give them an advantage relative to their new neighbours. Invaders may have strong competitive effects in their non-native ranges through their ability to attain higher biomass, or because of novel traits that confer more subtle competitive advantages (Callaway and Pennings 2000; Callaway and Ridenour 2004). There have been quantitative biogeographic comparisons of productivity, biomass or density in both the native and non-native ranges of invasive plant species (Woodburn and Sheppard 1996; Grigulis *et al.* 2001; Paynter *et al.* 2003; Jakobs *et al.* 2004; Beckmann *et al.* 2009), and many studies have clearly documented strong negative impacts of invaders in their non-native ranges; apparently much stronger than most if not all native species (e.g., Bruce *et al.* 1997; Ridenour and Callaway 2001; Lu and Ma 2005; Ortega and Pearson 2005; Hejda *et al.* 2009). One study has quantified the impact of an invasive species on the productivity or diversity of its neighbours in the field in both its native and non-native ranges. Inderjit *et al.* (2011) found that the canopies of *Ageratina adenophora*, a widespread and aggressive subtropical invader, had facilitative effects on other species in its native Mexico but highly inhibitory effects in its non-native ranges in China and India.

Despite the wealth of information on the plasticity of invaders and the powerful impacts they have in their non-native ranges, we know nothing about the phenotypic plasticity of invaders within the context of TMIs or TMIIs. We can only speculate again on likely scenarios and ways in which we might experimentally explore TMIIs produced by the phenotypic plasticity expressed by exotic invasive species. For example, when the California native shrubs *Haplopappus ericoides* and *H. venetus* var. *seloides* grow in the absence of competition their root systems are concentrated near the soil surface (D'Antonio and Mahall 1991). However, when competing with the exotic *Carpobrotus edulis* the root systems of *Haplopappus* shift to a much deeper morphology as they are displaced by the mat-forming exotic (D'Antonio and Mahall 1991). This change in rooting depth by *Haplopappus* suggests that neighbouring species can exert strong control over the phenotype of competitors – in this case inducing a change that may result in a trade-off in access to nutrients and water (Ho *et al.* 2005). Further, a change in *Haplopappus* rooting depth may decrease the intensity of competition between *Haplopappus* and *Carpobrotus* but may increase the intensity of competition with other species that utilize deeper soil sections, which would represent a TMII.

Carpobrotus invades different abiotic habitats (D'Antonio 1993) which is likely to elicit plastic responses by *Carpobrotus* (Weber and D'Antonio 1999). Plasticity expressed by *Carpobrotus* may change its effects on the root architecture of *Haplopappus*, creating a complex suite of plastic effects and responses between the two species. It would be intriguing to explore the next step by experimentally subjecting *Haplopappus* to competition with other species while it is experiencing at least two different manifestations of plasticity in *Carpobrotus*.

Chemically mediated interactions among plants, such as allelopathy, can also have strong impacts on the organization of communities and represent a promising area in the search for TMIIs. *Centaurea stoebe*, an European invader in North America, exudes the compound (±)-catechin from its roots (Tharayil and Triebwasser 2010), which can inhibit the growth of neighbouring competing plants (Callaway *et al.* 2005; Inderjit *et al.* 2008a, b; Simoes *et al.* 2008; He *et al.* 2009; Pollock *et al.* 2009; but see Blair *et al.* 2006; Duke *et al.* 2009). In addition to inhibiting neighbour performance, (±)-catechin is also a chelator, the addition of which makes phosphorus available in soils where it is bound by calcium (Thorpe *et al.* 2006; Tharayil *et al.* 2008, 2009) which can improve the performance of *C. stoebe* in phosphorus deficient soils. Native species vary a great deal in their susceptibility to (±)-catechin (Thorpe *et al.* 2009). Weir *et al.* (2006) found that two good competitors with *C. stoebe*, *Lupinus sericeus* and *Gaillardia grandiflora*, produced levels of oxalate in their root exudates that were more than an order of magnitude higher than that of three poor competitors. They also found that oxalic acid reduces the oxidative damage generated by (±)-catechin. Furthermore, exposure to (±)-catechin increased the exudation of oxalate by *G. grandifolia* four-fold and *L. sericeus* by 50-fold. This suggests that some native plants may respond to competition with *C. stoebe* in a plastic way, which is a demonstration of a TMI. This response creates the opportunity for a TMII involving the amelioration of (±)-catechin effects on co-occurring species. Interestingly, native grasses are highly spatially associated with *L. sericeus* in communities invaded by *C. stoebe* and field experiments show that *L. sericeus* indirectly facilitates native grasses in vegetation dominated by *C. stoebe*. This facilitation was correlated with the presence of oxalic acid in the soil in the field. When oxalic acid was applied to the roots of native grasses it alleviated the allelopathic effects of (±)-catechin, indicating that root secreted oxalic acid may act as a chemical facilitator for plant species that do not produce the chemical. Again, this example is not an explicit test of TMIIs, but it does suggest that the chemically mediated suite of indirect interactions derives from the plastic response of some species to the presence of a novel chemical in the soil rhizosphere.

Conclusion

Although we know of no examples in which researchers have specifically investigated the effects of plasticity on indirect interactions among plants, the requisite component pieces of TMIIs in plants are well understood. Because of both the highly plastic nature of plants and a myriad of probable indirect interactions in plant communities, TMIIs among plants are clearly an important future research direction. But beyond linking existing ideas about plasticity and interactions, we also have considered how to use TMIIs to provide fundamental insight into broader ecological questions, such as how plant communities assemble, or how invasive species can act as powerful reorganizing forces in communities.

Among the more promising lines of research, the biochemical plasticity of plants (Metlen *et al.* 2009) has the potential to provide highly dynamic and inducible phenotypic shifts in plants that may also have strong allelopathic effects on some, but not all, neighbours (e.g., Thorpe *et al.* 2009), and in some, but not all abiotic contexts (Pollock *et al.* 2009). Because plant secondary biochemistry can also be specialized in purpose and unique to a family, genus, or even an individual species, the potential for plasticity, and thus TMIIs, via plant biochemistry is nearly endless. In addition, the cascading effects of induced biochemical plasticity could also be facilitative as it can provide associational defence (Pfister and Hay 1988) and possibly alert other species to the presence of herbivores (Karban *et al.* 2006).

The absence of studies of TMIIs among plants may be due in part to the daunting matter of experimenting with highly diffuse interactions occurring among multiple species. But diffuse interactions are the product, in part, of the immobility of plants, and immobility in multispecies complexes may be why plants are so unusually plastic and provide such exceptional opportunities for studying TMIs and TMIIs. Exploring how shifts in phenotypes respond to changing abiotic and biotic conditions, and in turn affect interactions with multispecies complexes, may yield major advances towards a more mechanistic understanding of the distributions and abundances of plant species.

Acknowledgements

E. T. Aschehoug thanks the NSF Graduate Research Fellowship Program for support and R. M. Callaway thanks the NSF (DEB 0614406) and the Andrew W. Mellon Foundation for support.

References

Allen, E. and Forman, R. (1976) Plant species removals and old-field community structure and stability. *Ecology*, **57**, 1233–1243.

Aschehoug, E. T. (2011) Indirect interactions and plant community structure. PhD Thesis, University of Montana.

Beckmann, M., Erfmeier, A. and Bruelheide, H. (2009) A comparison of native and invasive populations of three clonal plant species in Germany and New Zealand. *Journal of Biogeography*, **36**, 865–878.

Blair, A., Nissen, S., Brunk, G. and Hufbauer, R. (2006) A lack of evidence for an ecological role of the putative allelochemical (±)-catechin in spotted knapweed invasion success. *Journal of Chemical Ecology*, **32**, 2327–2331.

Bossdorf, O., Auge, H., Lafuma, L. *et al.* (2005) Phenotypic and genetic differentiation between native and introduced plant populations. *Oecologia*, **144**, 1–11.

Bradshaw, A. (1965) Evolutionary significance of phenotypic plasticity in plants. In E. Caspari, ed., *Advances in Genetics*. London: Academic Press, pp. 115–151.

Brooker, R., Kikvidze, Z., Pugnaire, F. I. *et al.* (2005) The importance of importance. *Oikos*, **111**, 208–208.

Bruce, K., Cameron, G., Harcombe, P. and Jubinsky, G. (1997) Introduction, impact on native habitats, and management of a woody invader, the Chinese tallow tree, *Sapium sebiferum (l.) roxb. Natural Areas Journal*, **17**, 255–260.

Buss, L. W. and Jackson, J. B. C. (1979) Competitive networks: nontransitive competitive relationships in cryptic coral reef environments. *American Naturalist*, **113**, 223–234.

Cahill Jr, J., McNickle, G., Haag, J. *et al.* (2010) Plants integrate information about nutrients and neighbors. *Science*, **328**, 1657.

Callaway, R. and Howard, T. (2007) Competitive networks, indirect interactions, and allelopathy: a microbial viewpoint on plant communities. In U. L. K. Esser, W. Beyschlag and Jin Murata, eds., *Progress in Botany*. Berlin: Springer, pp. 317–335.

Callaway, R., Ridenour, W., Laboski, T., Weir, T. and Vivanco, J. (2005) Natural selection for resistance to the allelopathic effects of invasive plants. *Ecology*, **93**, 576–583.

Callaway, R. M. (1990) Effects of soil-water distribution on the lateral root development of three species of California oaks. *American Journal of Botany*, **77**, 1469–1475.

Callaway, R. M. (2007) *Positive Interactions and Interdependence in Plant Communities*. Dordrecht, The Netherlands: Springer.

Callaway, R. M. and Maron, J. L. (2006) What have exotic plant invasions taught us over the past 20 years? *Trends in Ecology and Evolution*, **21**, 369–374.

Callaway, R. M. and Pennings, S. C. (2000) Facilitation may buffer competitive effects: indirect and diffuse interactions among salt marsh plants. *American Naturalist*, **156**, 416–424.

Callaway, R. M. and Ridenour, W. M. (2004) Novel weapons: invasive success and the evolution of increased competitive ability. *Frontiers in Ecology and the Environment*, **2**, 436–443.

Callaway, R. M. Nadkarni, N. M. and Mahall, B. E. (1991) Facilitation and interference of *Quercus douglasii* on understory productivity in central California. *Ecology*, **72**, 1484–1499.

Callaway, R. M., Pennings, S. C. and Richards, C. L. (2003) Phenotypic plasticity and interactions among plants. *Ecology*, **84**, 1115–1128.

Cavieres, L. A. and Badano, E. I. (2009) Do facilitative interactions increase species richness at the entire community level? *Journal of Ecology*, **97**, 1181–1191.

Chesson, P. and Rosenzweig, M. (1991) Behavior, heterogeneity, and the dynamics of interacting species. *Ecology*, **72**, 1187–1195.

Chun, Y. J., Collyer, M. L., Moloney, K. A. and Nason, J. D. (2007) Phenotypic plasticity of native vs. invasive purple loosestrife: a two-state multivariate approach. *Ecology*, **88**, 1499–1512.

Connell, J. H. (1983) On the prevalence and relative importance of interspecific competition: evidence from field experiments. *American Naturalist*, **122**, 661–696.

Cuesta, B., Villar-Salvador, P., Puertolas, J., Benayas, J. M. R. and Michalet, R. (2010) Facilitation of *Quercus ilex* in Mediterranean shrubland is explained by both direct and indirect interactions mediated by herbs. *Journal of Ecology*, **98**, 687–696.

D'Antonio, C. (1993) Mechanisms controlling invasion of coastal plant communities by the alien succulent *Carpobrotus edulis*. *Ecology*, **74**, 83–95.

D'Antonio, C. M. and Mahall, B. E. (1991) Root profiles and competition between the invasive, exotic perennial, *Carpobrotus edulis*, and two native shrub species in California coastal scrub. *American Journal of Botany*, **78**, 885–894.

Davidson, D. W. (1980) Some consequences of diffuse competition in a desert ant community. *American Naturalist*, **116**, 92–105.

de Kroon, H., Huber, H., Stuefer, J. F. and van Groenendael, J. M. (2005) A modular concept of phenotypic plasticity in plants. *New Phytologist*, **166**, 73–82.

Donohue, K., Pyle, E. H., Messiqua, D., Heschel, M. S. and Schmitt, J. (2001) Adaptive divergence in plasticity in natural populations of *Impatiens capensis* and its consequences for performance in novel habitats. *Evolution*, **55**, 692–702.

Duke, S., Blair, A., Dayan, F. *et al.* (2009) Is (-)-catechin a novel weapon of spotted knapweed (*Centaurea stoebe*)? *Journal of Chemical Ecology*, **35**, 141–153.

Ehlers, B. K. and Thompson, J. (2004) Do co-occurring plant species adapt to one another? The response of *Bromus erectus* to the presence of different *Thymus vulgaris* chemotypes. *Oecologia*, **141**, 511–518.

Engel, E. C. and Weltzin, J. F. (2008) Can community composition be predicted from pairwise species interactions? *Plant Ecology*, **195**, 77–85.

Goldberg, D. E. (1990) Components of resource competition in plant communities. In J. Grace and D. Tilman, eds., *Perspectives on Plant Competition*. San Diego, CA: Academic Press, pp. 27–50.

Goldsmith, F. B. (1978) Interaction (competition) studies as a step towards the synthesis of sea-cliff vegetation. *Journal of Ecology*, **66**, 921–931.

Grigulis, K., Sheppard, A. W., Ash, J. E. and Groves, R. H. (2001) The comparative demography of the pasture weed *Echium plantagineum* between its native and invaded ranges. *Journal of Applied Ecology*, **38**, 281–290.

Grime, J. P. (1977) Evidence for existence of three primary strategies in plants and its relevance to ecological and evolutionary theory. *American Naturalist*, **111**, 1169–1194.

Grime, J. P. (2001) *Plant Strategies, Vegetation Processes, and Ecosystem Properties*. Chichester, UK: John Wiley and Sons.

Grøndahl, E. and Ehlers, B. K. (2008) Local adaptation to biotic factors: reciprocal transplants of four species associated with aromatic *Thymus pulegioides* and *T. serpyllum*. *Journal of Ecology*, **96**, 981–992.

He, W., Feng, Y., Ridenour, W. *et al.* (2009) Novel weapons and invasion: biogeographic differences in the competitive effects of *Centaurea maculosa* and its root exudate (±)-catechin. *Oecologia*, **159**, 803–815.

Hejda, M., Pysek, P. and Jarosik, V. (2009) Impact of invasive plants on the species richness, diversity and composition of invaded communities. *Journal of Ecology*, **97**, 393–403.

Hierro, J. L., Maron, J. L. and Callaway, R. M. (2005) A biogeographical approach to plant invasions: the importance of studying exotics in their introduced and native range. *Journal of Ecology*, **93**, 5–15.

Ho, M., Rosas, J., Brown, K. and Lynch, J. (2005) Root architectural trade-offs for water and phosphorus acquisition. *Functional Plant Biology*, **32**, 737–748.

Hulme, P. E. (2008) Phenotypic plasticity and plant invasions: is it all jack? *Functional Ecology*, **22**, 3–7.

Inderjit, Evans, H., Crocoll, C. *et al.* (2011) Volatile chemicals from leaf litter are associated with invasiveness of a neotropical weed in Asia. *Ecology*, **92**, 316–324.

Inderjit, Pollock, J. L., Callaway, R. M. and Holben, W. (2008a) Phytotoxic effects of (±)-catechin in vitro, in soil, and in the field. *PLoS ONE*, **3**, e2536.

Inderjit, Seastedt, T. R., Callaway, R. M. and Kaur, J. (2008b) Allelopathy and plant invasions: traditional, congeneric, and biogeographical approaches. *Biological Invasions*, **10**, 875–890.

Jackson, J. B. C. and Buss, L. (1975) Allelopathy and spatial competition among coral-reef invertebrates. *Proceedings of the National Academy of Sciences of the United States of America*, **72**, 5160–5163.

Jakobs, G., Weber, E. and Edwards, P. (2004) Introduced plants of the invasive *Solidago gigantea* (Asteraceae) are larger and grow denser than conspecifics in the native range. *Diversity and Distributions*, **10**, 11–19.

Jensen, C. G. and Ehlers, B. K. (2010) Genetic variation for sensitivity to a thyme monoterpene in associated plant species. *Oecologia*, **162**, 1017–1025.

Karban, R., Shiojiri, K., Huntzinger, M. and McCall, A. (2006) Damage-induced resistance in sagebrush: volatiles are key to intra- and interplant communication. *Ecology*, **87**, 922–930.

Karlson, R. H. and Jackson, J. B. C. (1981) Competitive networks and community structure: a simulation study. *Ecology*, **62**, 670–678.

Keane, R. M. and Crawley, M. J. (2002) Exotic plant invasions and the enemy release hypothesis. *Trends in Ecology and Evolution*, **17**, 164–170.

Keddy, P. A. and Shipley, B. (1989) Competitive hierarchies in herbaceous plant communities. *Oikos*, **54**, 234–241.

Laird, R. A. and Schamp, B. S. (2006) Competitive intransitivity promotes species coexistence. *American Naturalist*, **168**, 182–193.

Laird, R. A. and Schamp, B. S. (2008) Does local competition increase the coexistence of species in intransitive networks? *Ecology*, **89**, 237–247.

Levine, J. M. (1999) Indirect facilitation: evidence and predictions from a riparian community. *Ecology*, **80**, 1762–1769.

Levine, J. M., Vila, M., D'Antonio, C. M. *et al.* (2003) Mechanisms underlying the impacts of exotic plant invasions. *Proceedings of the Royal Society of London, Series B*, **270**, 775–781.

Li, L., Li, S.-M., Sun, J.-H. *et al.* (2007) Diversity enhances agricultural productivity via rhizosphere phosphorus facilitation on phosphorus-deficient soils. *Proceedings of the National Academy of Sciences of the United States of America*, **104**, 11192–11196.

Li, X. D. and Wilson, S. D. (1998) Facilitation among woody plants establishing in an old field. *Ecology*, **79**, 2694–2705.

Lu, Z. J. and Ma, K. P. (2005) Scale dependent relationships between native plant diversity and the invasion of croftonweed (*Eupatorium adenophorum*) in southwest China. *Weed Science*, **53**, 600–604.

Mahall, B. E. and Callaway, R. M. (1992) Root communication mechanisms and intracommunity distributions of two Mojave desert shrubs. *Ecology*, **73**, 2145–2151.

Mal, T. K. and Lovett-Doust, J. (2005) Phenotypic plasticity in vegetative and reproductive traits in an invasive weed, *Lythrum salicaria* (Lythraceae) in response to soil moisture. *American Journal of Botany*, **92**, 819–825.

Maron, J. L. and Marler, M. (2008) Field-based competitive impacts between invaders and natives at varying resource supply. *Journal of Ecology*, **96**, 1187–1197.

May, R. M. and Leonard, W. J. (1975) Nonlinear aspects of competition between three species. *Siam Journal on Applied Mathematics*, **29**, 243–253.

Metlen, K. L. (2010) Using patchy plant invasions to understand how diffuse interactions modify facilitation and competition.

PhD Thesis, The University of Montana, Missoula, MT, USA.

Metlen, K. L., Aschehoug, E. T. and Callaway, R. M. (2009) Plant behavioural ecology: dynamic plasticity in secondary metabolites. *Plant Cell and Environment*, **32**, 641–653.

Miller, T. E. (1994) Direct and indirect species interactions in an early old-field plant community. *American Naturalist*, **143**, 1007–1025.

Miller, T. E. and Travis, J. (1996) The evolutionary role of indirect effects in communities. *Ecology*, **77**, 1329–1335.

Miner, B. G., Sultan, S. E., Morgan, S. G., Padilla, D. K. and Relyea, R. A. (2005) Ecological consequences of phenotypic plasticity. *Trends in Ecology and Evolution*, **20**, 685–692.

Mitchley, J. and Grubb, P. (1986) Control of relative abundance of perennials in chalk grassland in southern England: I. Constancy of rank order and results of pot- and field-experiments on the role of interference. *Journal of Ecology*, **74**, 1139–1166.

Munshaw, M. and Lortie, C. (2010) Back to the basics: using density series to test regulation versus limitation for invasive plants. *Plant Ecology*, **211**, 1–5.

Novoplansky, A. (2002) Developmental plasticity in plants: implications of non-cognitive behavior. *Evolutionary Ecology*, **16**, 177–188.

Ortega, Y. K. and Pearson, D. E. (2005) Weak vs. strong invaders of natural plant communities: assessing invasibility and impact. *Ecological Applications*, **15**, 651–661.

Pages, J. P. and Michalet, R. (2003) A test of the indirect facilitation model in a temperate hardwood forest of the northern French Alps. *Journal of Ecology*, **91**, 932–940.

Parker, V. T. and Muller, C. H. (1982) Vegetational and environmental-changes beneath isolated live oak trees (*Quercus agrifolia*) in a California annual grassland. *American Midland Naturalist*, **107**, 69–81.

Paynter, Q., Downey, P. O. and Sheppard, A. W. (2003) Age structure and growth of the woody legume weed *Cytisus scoparius* in native and exotic habitats: implications for control. *Journal of Applied Ecology*, **40**, 470–480.

Perkins, T. A., Holmes, W. R. and Weltzin, J. F. (2007) Multi-species interactions in competitive hierarchies: new methods and empirical test. *Journal of Vegetation Science*, **18**, 685–692.

Petraitis, P. S. (1979) Competitive networks and measures of intransitivity. *American Naturalist*, **114**, 921–925.

Pfister, C. A. and Hay, M. E. (1988) Associational plant refuges: convergent patterns in marine and terrestrial communities result from differing mechanisms. *Oecologia*, **77**, 118–129.

Pollock, J., Callaway, R., Thelen, G. and Holben, W. (2009) Catechin–metal interactions as a mechanism for conditional allelopathy by the invasive plant *Centaurea maculosa*. *Journal of Ecology*, **97**, 1234–1242.

Reader, R. J., Wilson, S. D., Belcher, J. W. *et al.* (1994) Plant competition in relation to neighbor biomass: an intercontinental study with *Poa pratensis*. *Ecology*, **75**, 1753–1760.

Rice, K. and Nagy, E. (2000) Oak canopy effects on the distribution patterns of two annual grasses: the role of competition and soil nutrients. *American Journal of Botany*, **87**, 1699.

Richards, C. L., Bossdorf, O., Muth, N. Z., Gurevitch, J. and Pigliucci, M. (2006) Jack of all trades, master of some? On the role of phenotypic plasticity in plant invasions. *Ecology Letters*, **9**, 981–993.

Richards, C. L., Pennings, S. C. and Donovan, L. A. (2005) Habitat range and phenotypic variation in salt marsh plants. *Plant Ecology*, **176**, 263–273.

Ridenour, W. M. and Callaway, R. M. (2001) The relative importance of allelopathy in interference: the effects of an invasive weed on a native bunchgrass. *Oecologia*, **126**, 444–450.

Schenck, J., Mahall, B. and Callaway, R. (1999) Spatial segregation of roots. *Advances in Ecology*, **28**, 145–180.

Schmidtke, A., Rottstock, T., Gaedke, U. and Fischer, M. (2010) Plant community diversity and composition affect individual plant performance. *Oecologia*, **164**, 1–13.

Simoes, K., Du, J., Kretzschmar, F. S. *et al.* (2008) Phytotoxic catechin leached by seeds of the tropical weed *Sesbania virgata*. *Journal of Chemical Ecology*, **34**, 681–687.

Sultan, S. E. (1987) Evolutionary implications of phenotypic plasticity in plants. *Evolutionary Biology*, **21**, 127–178.

Sultan, S. E. (2000) Phenotypic plasticity for plant development, function and life history. *Trends in Plant Science*, **5**, 537–542.

Sultan, S. E., Wilczek, A. M., Bell, D. L. and Hand, G. (1998a) Physiological response to complex environments in annual *Polygonum* species of contrasting ecological breadth. *Oecologia*, **115**, 564–578.

Sultan, S. E., Wilczek, A. M., Hann, S. D. and Brosi, B. J. (1998b) Contrasting ecological breadth of co-occurring annual *Polygonum* species. *Journal of Ecology*, **86**, 363–383.

Tewksbury, J. and Lloyd, J. (2001) Positive interactions under nurse-plants: spatial scale, stress gradients and benefactor size. *Oecologia*, **127**, 425–434.

Tharayil, N. and Triebwasser, D. (2010) Elucidation of a diurnal pattern of catechin exudation by *Centaurea stoebe*. *Journal of Chemical Ecology*, **36**, 200–204.

Tharayil, N., Bhowmik, P., Alpert, P. *et al.* (2009) Dual purpose secondary compounds: phytotoxin of *Centaurea diffusa* also facilitates nutrient uptake. *New Phytologist*, **181**, 424–434.

Tharayil, N., Bhowmik, P. and Xing, B. (2008) Bioavailability of allelochemicals as affected by companion compounds in soil matrices. *Journal of Agricultural and Food Chemistry*, **56**, 3706–3713.

Thorpe, A., Archer, V. and DeLuca, T. (2006) The invasive forb, *Centaurea maculosa*, increases phosphorus availability in Montana grasslands. *Applied Soil Ecology*, **32**, 118–122.

Thorpe, A. S., Thelen, G. C., Diaconu, A. and Callaway, R. M. (2009) Root exudate is allelopathic in invaded community but not in native community: field evidence for the novel weapons hypothesis. *Journal of Ecology*, **97**, 641–645.

Tilman, D. (1982) *Resource Competition and Community Structure*. Princeton, NJ: Princeton University Press.

Tilman, D. (1985) The resource-ratio hypothesis of plant succession. *American Naturalist*, **125**, 827–852.

Turlings, T. C. J., Tumlinson, J. H. and Lewis, W. J. (1990) Exploitation of herbivore-induced plant odors by host-seeking parasitic wasps. *Science*, **250**, 1251–1253.

Valladares, F., Gianoli, E. and Gómez, J. M. (2007) Ecological limits to plant phenotypic plasticity. *New Phytologist*, **176**, 749–763.

Valladares, F., Sanchez, D. and Zavala, M. (2006) Quantitative estimation of phenotypic plasticity: bridging the gap between the evolutionary concept and its ecological applications. *Ecology*, **94**, 1103–1116.

Van Valen, L. (1965) Morphological variation and width of ecological niche. *American Naturalist*, **99**, 377–390.

Vandermeer, J. (1980) Indirect mutualism: variations on a theme by Stephen Levine. *American Naturalist*, **116**, 441–448.

Vila, M. and Weiner, J. (2004) Are invasive plant species better competitors than native plant species? Evidence from pair-wise experiments. *Oikos*, **105**, 229–238.

Wang, P., Stieglitz, T., Zhou, D. W. and Cahill Jr, J. F. (2010) Are competitive effect and response two sides of the same coin, or fundamentally different? *Functional Ecology*, **24**, 196–207.

Weber, E. and D'Antonio, C. (1999) Phenotypic plasticity in hybridizing *Carpobrotus* spp. (Aizoaceae) from coastal California and its role in plant invasion. *Botany*, **77**, 1411–1418.

Weir, T. L., Bais, H. P., Stull, V. J. *et al.* (2006) Oxalate contributes to the resistance of *Gaillardia grandiflora* and *Lupinus sericeas* to a phytotoxin produced by *Centaurea maculosa*. *Planta*, **223**, 785–795.

Whitlock, M. C. (1996) The red queen beats the jack-of-all-trades: the limitations on the evolution of phenotypic plasticity and niche breadth. *American Naturalist*, **148**, S65–S77.

Williamson, G. (1990) Allelopathy, Koch's postulates, and the neck riddle. In J. B. Grace and D. Tilman, eds., *Perspectives in Plant Competition*. San Diego, CA: Academic Press, pp. 143–162.

Wilson, S. D. and Keddy, P. A. (1986) Measuring diffuse competition along an environmental gradient: results from a shoreline plant community. *American Naturalist*, **127**, 862–869.

Woodburn, T. and Sheppard, A. (1996) The demography of *Carduus nutans* as a native and an alien weed. *Plant Protection Quarterly*, **11**, 236–238.

CHAPTER TWENTY-SIX

Climate change, phenology and the nature of consumer–resource interactions: advancing the match/mismatch hypothesis

JEFFREY T. KERBY

Department of Biology, Pennsylvania State University

CHRISTOPHER C. WILMERS

Environmental Studies Department, University of California-Santa Cruz

and

ERIC POST

Department of Biology, Pennsylvania State University

Introduction

Understanding how species cope with ecological and environmental variation is a fundamental concern of ecology. Over the course of their lives, many organisms alter their phenotypes in response to biotic and abiotic pressures (Miner *et al.* 2005), responses that cascade through the food web to, in turn, affect the dynamics of species interactions. These effects, called trait-mediated effects, are pervasive in ecological communities, and their study has offered new insights into community ecology, a subject previously dominated by a density-mediated understanding of species interactions (Werner and Peacor 2003). Most analyses of trait-mediated effects take a top-down perspective where variation in consumer traits causes phenotypic responses by prey species. These phenotypic responses include behavioural, morphological and/or physiological plasticity that have ramifying consequences for the food web by influencing how predators and prey interact (Werner and Peacor 2003). This top-down perspective on the influence of traits in communities suggests that it is consumers that determine the nature and strength of the mediated effects.

Climate change is an ongoing global perturbation that also affects the densities and traits of interacting species, although these effects are not necessarily related to food web trade-offs. Cohesive shifts in phenology – the timing of periodic biological events, such as migration, flowering or mating – reveal the global scale of climate change's influence on species' traits (Parmesan and Yohe 2003; Root *et al.* 2003). These phenological changes affect conditions that influence

Trait-Mediated Indirect Interactions: Ecological and Evolutionary Perspectives, eds. Takayuki Ohgushi, Oswald J. Schmitz and Robert D. Holt. Published by Cambridge University Press. © Cambridge University Press 2012.

the relative fitness contributions of life-history traits, traits such as age-structured growth, reproductive timing or developmental rates. For some species, these traits are plastic to fitness trade-offs created by phenological shifts. In this way, climate change can affect the expression of traits that have an overwhelming influence on species interactions. Unlike the top-down influence of consumers, this non-trophic forcing can affect food webs via bottom-up processes. Phenology not only affects the nature and timing of species interactions, but also influences the very likelihood that two species will interact at all. In this manner, it can conflate or confound prey trait responses to immediate food web trade-offs, like those mediated by predators. Climate-driven phenological variability provides new context for understanding the interaction between trophic and non-trophic traits and how this influences overall food web dynamics.

The consequences of phenological shifts for consumer–resource interactions have been most clearly documented when interacting species experience a differential response in time and/or space to a shared change in climate (Parmesan 2006). Phenological asynchrony related to climate change has been identified among trophic levels (Thackery *et al.* 2010), species (Visser and Both 2005) and even within species (Høye *et al.* 2007). These differential shifts reveal frequent changes in consumer–resource interactions that many communities are likely to experience (Walther *et al.* 2002; Post *et al.* 2009). Emergent inter-trophic asynchrony can trigger demographic changes that affect the entire food web by affecting interactions that structure communities (Costello *et al.* 2006; Borcherding *et al.* 2010). For example, differential climate-driven phenological shifts have been implicated in the collapse of avian population cycles at high latitudes (Ludwig *et al.* 2006). Traditionally, these population irruptions have provided periodic flushes of nutrients that support predator populations while simultaneously regulating plant successional dynamics (Ims *et al.* 2008).

Analyses of the consequences of climate-driven shifts in phenological traits trace their conceptual origins to the 'match/mismatch hypothesis' – a simple framework that links climate-driven trophic mismatch with population and community-level consequences.

Origins of the match/mismatch hypothesis

The match/mismatch hypothesis emerged in the 1970s from the marine fisheries literature to explain the extreme variation in population recruitment of economically important fish stocks of cod (*Gadus* spp.) and herring (*Clupea* spp.) in the North Atlantic (Cushing 1974). It proposes that in seasonal waters, fish recruitment is determined by the degree of temporal overlap between a 'critical period' of fish larval development, a period marked by high food- and predator-mediated mortality, and the timing of the peak abundance of their food resource, pelagic zooplankton. The magnitude of this overlap, conceptualized by two bell curves of species abundance resting on a temporal axis

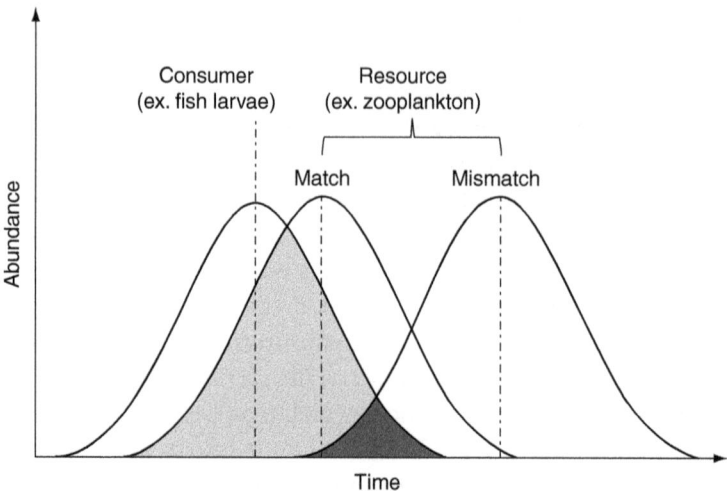

Figure 26.1 The temporal match/mismatch hypothesis. The abundances of consumers and resources are shown as distributions in time. The relative overlap between consumer and resources varies with resource phenology and results either in a match and high consumer population recruitment (light shading), or a mismatch with low consumer population recruitment (dark shading). Modified from Durant *et al.* (2007).

(Fig. 26.1), conceivably results in either a temporal overlap of resource demand and availability, a trophic match, and thus high fish population recruitment, or a temporal disjunction between resource demand and availability, a trophic mismatch, with low fish population recruitment.

Over the course of several decades, David Cushing, a British fisheries biologist, expanded on hypotheses proposed in the early twentieth century (Hjort 1914) and developed what he named the 'match/mismatch hypothesis' (Cushing 1974, 1982, 1990). Cushing observed that the mean timing of peak fish spawning and, by extension, the phenology of the critical period for the majority of fish larvae, was relatively fixed from year to year, whereas the appearance of zooplankton populations was regulated from the bottom-up by stochastic climatic processes (Cushing 1990). Earlier hypotheses had assumed that the critical period for these fishes was brief, lasting only from the time of hatching to first feeding (Hjort 1914). Cushing broadened the application of the critical period to include all of larval development and just beyond (Cushing 1990). By relaxing this assumption, Cushing's hypothesis emphasized an outlook where the per capita effects of life-history stage transitions were regarded as processes rather than fixed events. Cushing's match/ mismatch hypothesis also emphasizes that climate variability plays a decisive, but indirect role in species interactions by affecting the expression of species' life-history traits via its influence on their reproductive phenology. When generalized, the match/mismatch hypothesis proposes that nascent consumers are unable to track consistently variability in the reproductive phenology of lower trophic levels, and that this failure has disproportionately large consequences on population recruitment relative to other instances of interspecific interaction throughout their ontogeny.

Empirical support for Cushing's match/mismatch hypothesis has been somewhat equivocal; however, this has often been the result of data limitations and the model's simplification of complex multitrophic dynamics (Leggett and Deblois 1994; Durant *et al.* 2007). Despite this, in the fisheries literature alone, the match/mismatch hypothesis has spawned decades of research, numerous allied hypotheses, and encouraged ongoing debate about the mechanisms of bottom-up community regulation in marine systems (reviewed in Durant *et al.* 2007).

Climate change and the match/mismatch hypothesis

Cushing's simple framework is not conceptually bound to marine systems, and has proven readily adaptable for the study of the consequences of differential phenological responses to climate change across several systems. In recent decades, the scientific community has drawn increasing attention to the ecological consequences of climate change (Walther *et al.* 2002; Forchhammer and Post 2004, Fig. 2; IPCC 2007; Post *et. al* 2009). Phenological shifts relative to calendar dates (Fitter and Fitter 2002), and more recently phenological shifts relative to other species' phenologies (Visser and Both 2005), have emerged as foci for climate ecology research. Cushing himself perceived the relevance of his framework for addressing questions related to climate change (Cushing 1982); however, the first applications of this framework beyond the North Atlantic system focused on the mistimed reproduction in great tits (*Parus major*) in the Netherlands (Visser *et al.* 1998).

Since this advance in the late 1990s, population-level effects of trophic mismatch caused by differential phenological shifts among species have been documented in detail across diverse consumer–resource pairings, including interactions between birds and invertebrates (Visser *et al.* 1998; Hipfner 2008; Both *et al.* 2009), birds and fish (Durant *et al.* 2005; Gremillet *et al.* 2008), vertebrate herbivores and plants (Post and Forchhammer 2008; Post *et al.* 2008a), invertebrate herbivores and plants (Visser and Holleman 2001), pollinators and plants (Memmott *et al.* 2007; Hegeland *et al.* 2009) and marine and freshwater fishes and invertebrates (Edwards and Richardson 2004; Winder and Schindler 2004). Trophic mismatch may occur at any level in a food web, or even in multiple levels simultaneously, from primary producers to apex predators (Grebmeier *et al.* 2006; Both *et al.* 2009; Gremillet *et al.* 2008; Primack *et al.* 2009; Montes-Hugo *et al.* 2009; Thackeray *et al.* 2010). Cushing's match/mismatch hypothesis is the progenitor of these studies, but several key conceptual advances, some of which are discussed below, have granted this framework broader relevance to the abovementioned and future investigations of the ecological consequences of climate change.

Accounting for abundance, temporal variance and adaptation

Resource and/or consumer abundance can influence the strength of a trophic match/mismatch by decreasing or increasing the likelihood that consumers will encounter resources at the 'tails' of their temporal distributions (Cushing 1982; Durant *et al.* 2005). While the original match/mismatch hypothesis focused on the mean timing of peak abundances, it is clear that the magnitude of either resource or consumer abundance, represented by a narrower or more highly dispersed distribution (Fig. 26.2a), can influence the degree of temporal matching during the critical period by increasing the area of potential overlap between the consumer and resource curves (Durant *et al.* 2005). The relative effects of resource timing versus resource abundance can be separated from one another using time series analyses (Durant *et al.* 2005); however, the prevalence and significance of these relationships across diverse systems remains relatively underreported and, at times, equivocal (Hipfner 2008).

The extent to which the abundance curves of interacting species overlap is also determined by their temporal variance (Fig. 26.2a, b). Warming manipulations of two Arctic shrub and one forb species in Greenland demonstrate that in addition to shifts in the timing of phenological events, the duration of phenological life-history periods, or phenophases, may also be sensitive to climatic factors (Post *et al.* 2008b). Most match/mismatch studies have focused on the timing of the first or mean date of phenological processes, whereas few have explored the prevalence and consequences of shifts in phenological duration (but see Both and Visser 2001, 2005). Despite this, differential shifts in the duration of phenophases in response to climate change could conceivably give rise to match/mismatch conditions similar to, but independent of, those linked with the mean date of peak abundance.

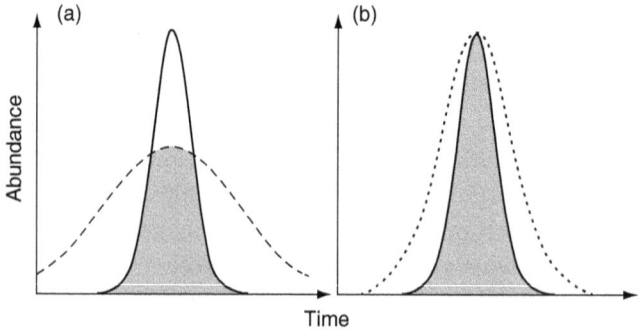

Figure 26.2 Factors which may affect the degree of match/mismatch independent of a shift in the peak timing of resource abundance. A shift in consumer or resource abundance (a), or temporal variance (a, b) about the peak can conceptually limit or magnify the effects of temporal trophic mismatch. Reproduced from Miller-Rushing *et al.* (2010).

To understand the consequences, both observed and expected, of phenological shifts requires, at the minimum, a coarse understanding of the evolutionary context of the development of each interacting species' phenological trait plasticity. For example, fish–zooplankton interactions in the North Atlantic, the focus of Cushing's original hypothesis, presumably coevolved in an environment that commonly experiences variable climatic conditions. In these fish populations, a mismatch between the timing of the peak in larval food requirements and the timing of peak food availability can clearly limit population recruitment. A complete mismatch between consumers and resources in these populations is, however, unlikely, because the duration of fish spawning throughout the season may occur for well over a month, albeit at low levels, before and after the 'fixed' peak date of spawning (Cushing 1982, 1990). This wide temporal variance about the peak spawning date ensures that at least some individuals of each year class will experience high-quality resource conditions and presumably thrive in 'mismatch' years (Cushing 1982, 1990). Such a prolonged period of spawning represents a bet-hedging strategy (Slatkin 1974); one that emerged from the evolutionary context of selective pressures that existed while this community was formed.

In many regions, climate is changing at rates that exceed those under which existing communities have been formed and maintained – a situation predicted to become increasingly commonplace in the coming decades (IPCC 2007). Without ecological precedent of such climatic pressures, species will not have evolved the adaptive plasticity necessary to hedge against emergent mismatches (Williams 1966). In some cases, species will have sufficiently plastic traits capable of tracking climatic and ecological shifts merely by chance, thus minimizing the potential for mismatches. Conversely, other species will be unable to respond at a sufficient rate to remain functional members of interaction webs under the selection pressures brought on by climate change (Visser 2010). This element of chance makes predicting future instances of mismatch more difficult.

Spatial mismatch

The spatial dimension of trophic mismatch (Post *et al.* 2008a) can also influence the magnitude and type of consumer–resource interactions in ecological communities. While many factors, including species interactions, combine with environmental conditions ultimately to determine spatial patterns of consumers and resources within and among trophic levels (Hutchinson 1957), the influence of climate change on these patterns has recently been the subject of increasing research and debate (Pearson and Dawson 2003; Levinsky *et al.* 2007). As with phenological trends, mean distributional shifts in response to climate change have been documented across numerous taxa around the globe (Parmesan and Yohe 2003; Root *et al.* 2003). Of particular significance to community ecology is how species' distributions covary in response to shared climate change. The match/mismatch hypothesis can again act as a framework for this line of study, by focusing on the consequences of trophic asynchrony using new methods to overcome the complexities associated with spatial analyses.

Until recently, temporal mismatch, that which occurs at a single point in space, was the primary focus of research related to the match/mismatch hypothesis. Unlike temporal processes that occur in one ordinal dimension, spatial changes in the same processes can occur in three; these additional factors, combined with variable interpretations of the term 'spatial mismatch', complicate studies that seek to account for spatiotemporal components of trophic match or mismatch. The term 'spatial mismatch' has been used in several, often complementary, mechanistic explanations of trophic asynchrony that arise from spatiotemporal variability, some examples of which are discussed below.

One usage of 'spatial mismatch' refers to predicting how distributions of interaction-paired species will differentially respond to climate change, by using bioclimatic-niche models (Levinsky et al. 2007). The methodologies of these models are diverse, but their basic aim is to project a species' realized niche onto a map and explore how this niche space will respond in a geographic sense to predicted changes in niche-limiting variables. Comparisons between the predicted niche spaces of interacting species under various climate models often reveal niche divergence. For example, it has been suggested that climate-linked niche divergence may cause a spatial mismatch between a monophagous butterfly (*Boloria titania*) and its larval host plant (*Polygonum bistorta*) in Europe (Schweiger et al. 2008). Using a combination of climate, soil and land-cover variables, the authors of that study suggest that the potential northward expansion of these butterflies may outstrip the dispersal ability of their larval host plant over the next 70 years, resulting in a reduced and increasingly fragmented consumer niche-space (Schweiger et al. 2008). In some areas this could lead to a complete extirpation of this interaction pairing, and therefore, all components of the interaction web that stem from it. Interaction diversity is an essential component of biodiversity (Thompson 1996; Price 2002), and the loss of interactions to mismatch, potentially independent of immediate changes in taxonomic diversity, may presage future taxonomic losses, yet this area of research continues to be relatively under-emphasized by conservation scientists.

Despite predictions of complete niche divergence, there are few empirical examples of this that can be directly linked to climate change. This paucity of empirical evidence may be the result of many factors, including difficulties in defining niche space. To some extent, this difficulty may also owe to conflation of the concepts of 'niche' and 'habitat', which is one of the most easily quantified and described niche components. Beyond this conceptual hurdle are the empirical challenges of measuring dynamic changes in niche-limiting factors across large geographic areas. Broad-scale phenological monitoring networks, such as the USA National Phenology Network (NPN) and the European Phenology Network (EPN), may be able to ease some of the data limitations that plague coarse-scale modelling approaches. Because many species traits that influence niche space are plastic with respect to both climatic and ecological influences, realistic parameterization of bioclimatic niche models is difficult. Furthermore, niche modelling studies are currently unable to incorporate the possibility of the emergence of new species interactions, which may be particularly important for so called 'specialist' species such as the butterflies described above. As a resource becomes rare or disappears, it is unclear whether or not 'specialist' consumers will express latent plasticity in their ability to respond to these pressures (Miller-Rushing *et al.* 2010), and if not, this raises questions about the evolutionary advantages for specialization in what are inherently dynamic environments. Future studies will need to clarify how species interactions emerge from rapidly changing community milieus across a continuum of spatial scales (see Araujo and Luoto 2007).

The term spatial mismatch is also used to describe how the *strength* of consumer–resource interactions is affected by climate-sensitive distance relationships (Durant *et al.* 2007; Gremillet *et al.* 2008). In this case, the effects of differential shifts of species distributions in space are analogous to the effects of phenological mismatch. A conceptually simple model of this scenario might arise for central place foragers if the mean distance between the forager and its resource varies with climate or other pressures (Durant *et al.* 2007). Greater distance between resources and reproductive sites can lead to tradeoffs of increased travel and/or search time, which translate to decreased efficiency in provisioning young, a situation that could have serious repercussions during an energetically demanding 'critical period' around reproduction (Durant *et al.* 2007). For example, Cape gannets (*Morus capensis*) are central place foraging sea birds that have recently experienced this type of spatial mismatch with their primary prey – sardines and anchovies (Gremillet *et al.* 2008). These large seabirds nest along the Atlantic coastlines of South Africa and Namibia, but make long foraging flights out to marine regions of high primary productivity, regions that traditionally have been linked with abundant stocks of their preferred food (Gremillet *et al.* 2008). Spatial mismatch between the distributions of copepods and fish, potentially caused by a combination of climate factors and direct anthropogenic influences, has contributed to a strong decline in Cape gannet prey in these foraging regions, decreasing the efficiency with which Cape gannets can find and acquire resources needed to provision their chicks (Gremilllet *et al.* 2008). This type of spatial mismatch, which arises from linear distance–time relationships between resources and consumers, may be widespread, although it is not widely reported outside of the context of apex marine predators (Veit *et al.* 1997; Grebmeier *et al.* 2006; Montes-Hugo *et al.* 2009).

Many species rely on environmental cues to inform them of current or future ecological conditions and they respond to this information by altering their phenotype to address perceived trade-offs (Miner *et al.* 2005). If the relationship between a cue and an associated environmental factor changes, *and* if these changes occur at a rate that exceeds a species' ability to adapt their decision making to these changes, species' responses to these cues may be poorly informed and lead to trophic asynchrony or even ecological traps (Visser 2010). Climate change is capable of influencing the relationship between cues and environmental conditions in several ways. For example, photoperiod and mean expected temperature may diverge with climate change because only temperature is affected by current global climate forcings. If species were to make decisions that rely on one to inform about the other, they may experience a decoupling between the type of phenotypic trait plasticity they express and the type of phenotypic plasticity that might be best suited to actual conditions (Visser *et al.* 1998; Phillimore *et al.* 2010). A prominent spatial dimension to these decouplings can arise because climate change occurs unevenly in space (IPCC 2007). As distance increases between two ecosystems, they are increasingly unlikely to experience similar climatic change as the result of variability in regional biosphere–atmosphere interactions. The potential for trophic mismatch in migratory species is therefore heightened relative to residents. These animals experience this temporal variability across a spatial continuum, not just at a single point, and can be particularly vulnerable to mismatch if they rely upon cues in one location to inform about another. Said another way, match or mismatch may arise from differential species response at any one point in space, and/or from the influence of differential climate change at multiple locations. This forms the basis for another usage of the term 'spatial mismatch'.

In Europe, many long-distance migrant bird populations are in decline relative to non-migrants (Sanderson *et al.* 2006), part of which can be explained by the above type of spatial mismatch between wintering and breeding grounds (Both *et al.* 2010; Jones and Cresswell 2010). One study of Palearctic passerines found evidence to support the 'distance hypothesis' – that long-distance migrants are more likely to experience population declines associated with mismatch than shorter-distance migrants or range residents because the probability of mismatch occurring at any one location along the migration route increases with migratory distance. However, this was only supported by empirical evidence when distance was considered in context with the seasonality of the migrant's breeding ground, where seasonality was defined as the temporal variance about the mean peak in consumer resources (Both *et al.* 2010). Long-distance migrants that bred primarily in more seasonal forest

habitats, with a narrow window of food abundance, experienced significantly sharper population declines than long-distance migrants that bred
in less seasonal marshy areas (Both *et al.* 2010). Irrespective of seasonality,
resident and shorter-distance migratory species that lived in both areas were
comparatively less affected than long-distance migrants (Both *et al.* 2010).
Another study, that did not incorporate a seasonal variance component, but
included migratory birds from both hemispheres, also found evidence that
suggested absolute migration distance could be a factor in bird population
declines, and that overall, migratory birds were more likely to experience
mismatch conditions and population declines than were residents (Jones
and Cresswell 2010). In both of these studies, the great distances between
wintering and breeding grounds imply an increasingly likely probability
that ineffective migratory cues will result from divergent climate regimes
(Both *et al.* 2010; Jones and Cresswell 2010). Studies at these broad scales
require simplified assumptions about abiotic influences on trait plasticity
that inevitably accompany low-resolution phenological data. However, that
these studies were still able to detect effects of migratory distance and
divergent climates in spite of these limitations raises important questions
about how spatiotemporal components of species interactions that occur
over continental scales will be affected by climate change.

The pattern of resource distribution at the landscape scale may also vary
with changing climate conditions. For example, by differentially affecting
the timing of plant emergence – a phenophase with high nutrition and low
digestive costs for herbivores – climate is capable of affecting spatial *patterns*
of resource quality across a wide array of scales from thousands of
kilometres to less than one metre (Chen *et al.* 2005; Post and Stenseth
1999; Post and Forchhammer 2008; Post *et al.* 2008b). This variation in spatial
patterning can have repercussions for consumer foraging decisions and
manifest itself as another type of spatial mismatch. At the landscape scale,
a spatial continuum of temporal shifts in resource availability and/or quality
is expressed as spatiotemporal resource heterogeneity, an important factor
in population dynamics (Roughgarden 1974; Levin 1976). Consumers have
evolved foraging strategies to cope with and even rely on heterogeneous
distributions of high quality resources. In seasonal environments, migratory
ungulates take advantage of spatiotemporal resource heterogeneity by following the early/mid phases of plant phenology through the landscape. This
can effectively prolong their access to high quality resources (Senft *et al.*
1987). Because climate change alters the pattern of resource quality
expressed in a landscape by affecting plant phenology, it can impact the
efficiency of herbivore foraging strategies designed to maximize high quality forage intake required to offset the high costs of reproduction. For
example, in highly seasonal West Greenland, reproductive success of

migratory caribou (*Rangifer tarandus*) depends on their ability to arrive and give birth at their calving site around the mean temporal peak in resource abundance (Post and Forchhammer 2008), but also on their ability to track spatial phenological heterogeneity along a local forage horizon during and around the calving period (Post *et al.* 2008b). These studies are an early step towards unifying the spatial match/mismatch hypothesis with landscape ecology concepts (e.g., Turner 2005), but future investigations will need to explore a more complete range of climatic effects on species interactions in relation to resource patterning, rather than just timing, across a hierarchy of spatial scales.

Migratory species may offer a clear insight into how differential spatio-temporal shifts in the distributional patterns of resources can be expected to influence trophic interactions. In some situations, migration itself may become an ineffective strategy as a result of what might be termed a spatial mismatch. This would be spatial mismatch in the sense that changes in the spatial patterning at one trophic level would negatively influence foraging success in higher trophic levels and result in a trophic mismatch, potentially independent of the mean timing of resource availability throughout the study area. Ungulate migration has been studied for decades and may provide a good starting point for these investigations. If migration is the result of spatial patterns of resource distribution, as is predicted by the forage maturation hypothesis (Fryxell 1991), spatial compression (i.e., homogeneity) of plant phenology along the migratory route, as has been observed at local scales in West Greenland (Post *et al.* 2008b), could conceivably alter selection coefficients between non-migratory and migratory members of populations. For instance, elk (*Cervus elaphaus*) populations in the Canadian Rockies are composed of both migratory and non-migratory individuals (Hebblewhite *et al.* 2008). A 3-year observational study found that, on average, migrant individuals of these populations were exposed to more nutritious and digestible food resources than residents, owing to their strategy of exploiting heterogeneous spatial patterns of plant phenology during migration (Hebblewhite *et al.* 2008). The advantages that migration confers on individuals could diminish in this population should spatial compression of resource phenology occur here. In seasonal environments, even slight shifts in foraging efficiency can have dramatic impacts on reproductive success (White 1983). Large herbivores are often important interactors in ecological communities, and their removal from interaction networks has been shown to induce significant community restructuring (Pringle *et al.* 2007; Post and Pedersen 2008). While the potential for this type of spatial mismatch to influence migratory species' population dynamics is clear, future studies will be required to verify to what extent these concepts may apply to empirical situations.

Integrating match/mismatch with life-history strategies

Trophic mismatch is most widely documented in seasonal environments where food resources are limited throughout much of the year however, even within these environments, consumer sensitivity to temporal resource limitation will vary among species as a function of, among other factors, variation in their life-history strategies. Income breeders, for example, require a continuous influx of energy to offset the high costs of reproduction, and thus are likely to have a critical period clearly related to food acquisition around the timing of their reproductive efforts. Conversely, capital breeders build up an energy surplus throughout the year that they later expend during reproduction, giving their reproductive effort relatively more independence from immediate food resource conditions (Drent et al. 2006). Capital breeders may thus prove less sensitive, but by no means immune, to climate-driven fluctuations in resources.

Traditionally, infant or juvenile mortality associated with what is assumed to be a fixed and intuitively described critical period of breeding phenology has been the sole effect reported by match/mismatch investigations (Durant et al. 2007). While this may be the most tractable metric of mismatch, it has almost certainly drawn attention away from efforts to document other potential consequences of a mismatch. For some insect species, ecological and environmental conditions during an early critical stage of ontogeny may have a delayed influence on adult body size and fertility (Prout and McChesney 1985) – both of which are life-history linked traits that greatly contribute to fitness. These delayed effects on traits could also have ramifications for the strength and type of interactions species experience throughout their development (Yang and Rudolf 2010). Identifying a broader range of direct effects that trophic mismatch can have on populations is a pressing, but presently poorly documented component of the demographic consequences of mismatch (Miller-Rushing et al. 2010). This approach will, however, challenge the traditional interpretation of the 'critical period' concept.

In a general sense, a species' 'critical period' is the product of interactions from within a hierarchy of biological sensitivities, integrating individual's traits from embryology, neurobiology and/or behaviour (Browman 1989). Ecological or environmental factors can induce trait plasticity within each level of this hierarchy, and by extension, affect how biological sensitivities interact to be expressed as a critical period of the life history of an organism. In contrast to this perspective, most analyses treat the critical period of a species' life history as an intrinsic property, i.e., as though it were a fixed trait (Visser and Both 2005). This assumption may limit our understanding of trophic decoupling in some species, whereby new critical periods will emerge as a consequence of novel ecological forcings associated with phenological shifts of life-history traits.

In many communities, species interact over their respective lifespans. These interactions can change in intensity or type depending on the timing of one or both interacting species' stage specific development and/or body size (Fig. 26.3a, b) (Werner and Gilliam 1984; Osenberg *et al.* 1988; Yang and Rudolf 2010). For example, fish eggs of one species may initially be prey items for fish or larvae of another, but once hatched, these larvae may compete with or even prey upon their former predators before eventually becoming generalist predators that consume prey from several trophic levels. When species interactions are stage structured in this manner, as might be expected in many invertebrate and/or aquatic ecosystems, the timing, duration and physical traits associated with life-history stages can determine the magnitude and type of interactions a species experiences (Werner and Gilliam 1984; Osenberg *et al.* 1988; Yang and Rudolf 2010). Similarly in plant–herbivore interactions, invertebrate herbivores may transition from predators to pollinators depending on the timing and duration of both species' ontogeny (Bronstein *et al.* 2009). Because shifts in the timing or duration of ontogeny may be variable in response to climate change (Werner and Gilliam 1984; Post *et al.* 2008b; Yang and Rudolf 2010), there is potential for species interactions to change, decouple or even strengthen over the entire span of their respective life histories. Instead of limiting focus to the differential shifts in phenological events to a traditionally critical period of development (Visser and Both 2005), some match/mismatch studies will benefit by focusing on stage-specific per capita effects of species interactions throughout the entire span of their trophic coupling (Yang and Rudolf 2010). A focus on per capita interactions throughout aggregate life history may clarify how trait plasticity and ecological sensitivity interact to affect population fitness, even via delayed responses to mismatch. This may be an effective approach better to understand a broader suite of the ecological consequences of climate change, even in less seasonal environments where sensitivities to resource limitation may be harder to predict and are certainly less well documented.

Conclusions

As empirical evidence of climate change's perturbing effects on ecological communities mounts, phenology has emerged as an essential component of species trait-responses to these emergent forcings. Trait-mediated ecological effects are increasingly the subject of community ecology research. However, the effects of non-trophic forcings on species traits must not be overlooked. Differential phenological shifts will continue to affect trophic interactions as climate regimes change by influencing not only the timing of species interactions, but also their very nature.

The match/mismatch hypothesis has been a popular framework for analyses documenting the immediate consequences of these climatic

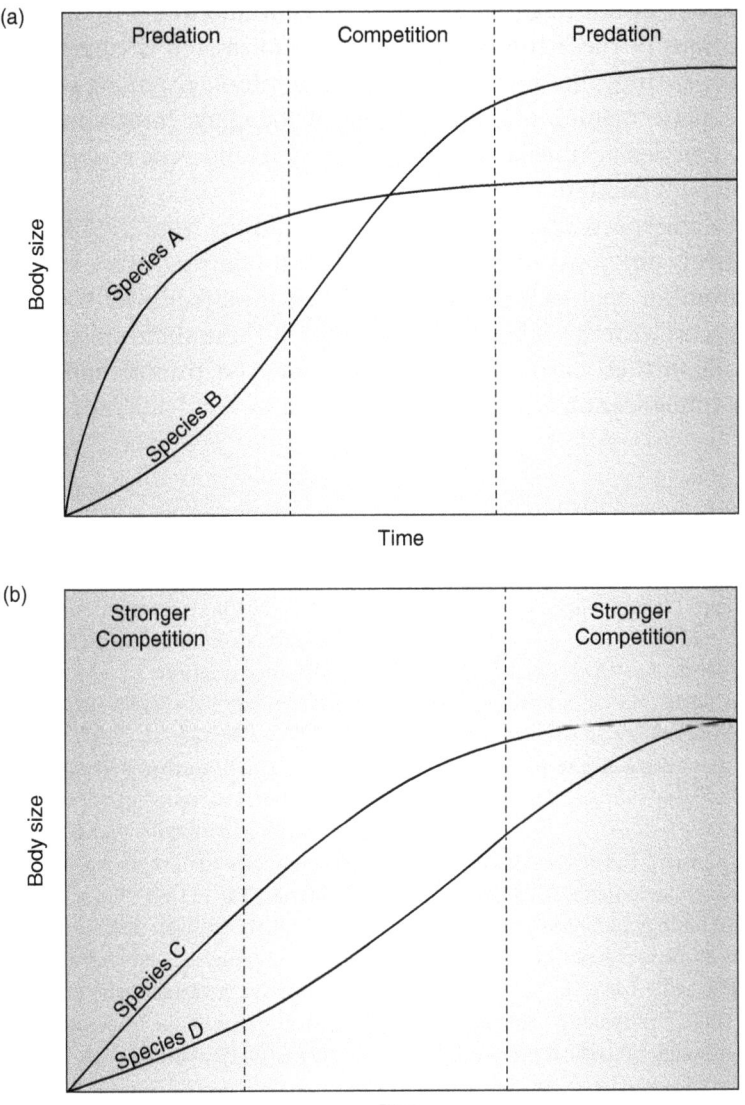

Figure 26.3 Conceptual diagram demonstrating how both the type (a) and strength (b) of species interactions may change over time as a consequence of differential shifts in the phenology of ontogeny. Shading represents the transition in type (a) or strength (b) of an interaction. For example, a hypothetical species A may switch from consumer of species B to its competitor, or even a prey item for species B as a function of the relative timing of growth between the two species (a). Similarly, a differential timing of growth between interacting species C and D can alternately ease or strengthen competition (b). Reproduced from Yang and Rudolf (2010).

perturbations, but future research will need to continue to clarify further spatial dimensions of these consequences. Spatial mismatch is currently a popular subject of study, but inconsistent use of terminology among studies has made a concise definition of the concept challenging. Integrating the study of trait- and density-mediated interactions with landscape ecology is a logical next step for community ecology research.

Furthermore, the spectrum of possible effects stemming from trophic mismatch is relatively unknown apart from recruitment failures. Future studies will benefit from an approach that links shifts in the life-history traits of interacting species with the delayed consequences of these shifts on species' traits. This will further clarify the interaction between trophic and non-trophic traits' influences on ecological communities.

References

Araujo, M.B. and Luoto, M. (2007) The importance of biotic interactions for modelling species distributions under climate change. *Global Ecology and Biogeography*, **16**, 743–753

Borcherding, J., Beeck, P., DeAngelis, D.L. and Scharf, W.R. (2010) Match or mismatch: the influence of phenology on size-dependent life history and divergence in population structure. *Journal of Animal Ecology*, **79**, 1101–1112.

Bronstein, J.L., Huxman, T., Horvath, B., Farabee, M. and Davidowitz, G. (2009) Reproductive biology of *Datura wrightii*: the benefits of a herbivorous pollinator. *Annals of Botany*, **103**, 1435–1443.

Browman, H.I. (1989) Embryology, ethology, and ecology of ontogenetic critical periods in fish. *Brain Behavioural Ecology*, **34**, 5–12.

Both, C. and Visser, M.E. (2001) Adjustment to climate change is constrained by arrival date in a long-distance bird. *Nature*, **411**, 296–298.

Both, C. and Visser, M.E. (2005) The effect of climate change on the correlation between avian life history traits. *Global Change Biology*, **11**, 1606–1613.

Both, C., van Asch, M., Bijlsma, R.B., van den Burg, A.B. and Visser, M.E. (2009) Climate change and unequal phenological changes across four trophic levels: constraints or adaptations? *Journal of Animal Ecology*, **78**, 73–83.

Both, C., Van Turnhout, C.A, M., Bijlsma, R.B. et al. (2010) Avian population consequences of climate change are most severe for long-distance migrants in seasonal habitats. *Proceedings of the Royal Society of London, Series B*, **277**, 1259–1266.

Chen, X., Hu, B. and Yu, R. (2005) Spatial and temporal variation of phenological growing season and climate change impacts in temperate eastern China. *Global Change Biology*, **11**, 1118–1130.

Costello, J.H., Sullivan, B.K. and Gifford, D.J. (2006) A physical-biological interaction underlying variable phenological responses to climate change by coastal zooplankton. *Journal of Plankton Research*, **28**, 1099–1105.

Cushing, D.H. (1974) The natural regulation of fish populations. In F.R. Harden Jones, ed., *Sea Fisheries Research*. London: Elek Science, pp. 399–412.

Cushing, D.H. (1982) *Climate and Fisheries*. London: Academic Press.

Cushing, D.H. (1990) Plankton production and year-class strength in fish populations: an update of the match/mismatch hypothesis. *Advances in Marine Biology*, **26**, 249–293.

Drent, R.H., Fox, A.D. and Stahl, J. (2006) Travelling to breed. *Journal of Ornithology*, **147**, 122–134.

Durant, J. M., Hjermann, D. Ø., Anker-Nilssen, T. et al. (2005) Timing and abundance as key mechanisms affecting trophic interactions in variable environments. *Ecology Letters*, **8**, 952–958.

Durant, J. M., Hjermann, D. Ø., Otterson, G. and Stenseth, N. C. (2007) Climate and the match or mismatch between predator requirements and resource availability. *Climate Research*, **33**, 271–283.

Edwards, M. and Richardson, A. J. (2004) Impact of climate change on marine pelagic phenology and trophic mismatch. *Nature*, **430**, 881–884.

Elton, C. S. (1958) *The Ecology of Invasions by Plants and Animals*. London: Methuen.

Fitter, A. H. and Fitter, R. S. R. (2002) Rapid changes in flowering time in British plants. *Science*, **296**, 1689–1691.

Forchammer, M. C. and Post, E. (2004) Using large-scale climate indices in climate change ecology studies. *Population Ecology*, **46**, 1–12.

Fryxell, J. M. (1991) Forage quality and aggregation by large herbivores. *American Naturalist*, **138**, 478–498.

Grebmeier, J. M., Overland, J. E., Moore, S. E. et al. (2006) A major ecosystem shift in the northern Bering Sea. *Science*, **311**, 1461–1464.

Gremillet, D., Lewis, S., Drapeau, L. et al. (2008) Spatial match–mismatch in the Benguela upwelling zone: should we expect chlorophyll and sea-surface temperature to predict marine predator distributions? *Journal of Applied Ecology*, **45**, 610–621.

Hebblewhite, M., Merrill, E. and McDermid, G. (2008) A multi-scale test of the forage maturation hypothesis in a partially migratory ungulate population. *Ecological Monographs*, **76**, 141–166.

Hipfner, J. M. (2008) Matches and mismatches: ocean climate, prey phenology and breeding success in a zooplanktivorous seabird. *Marine Ecology Progress Series*, **368**, 295–304.

Hjort, J. (1914) Fluctuations in the Great Fisheries of Northern Europe viewed in the light of biological research. *Rapports et Procès-Verbaux des Réunions, Conseil International pour l'Exploration de la Mer*, **20**, 124–169.

Høye, T. T., Post, E., Meltofte, H., Schmidt, N. M. and Forchhammer, M. C. (2007) Rapid advancement of spring in the High Arctic. *Current Biology*, **17**, R449–R451.

Hutchinson, G. E. (1957) Concluding remarks. *Ecology*, **22**, 415–427.

Ims, R. A., Henden, J-A. and Killengreen, S. T. (2008) Collapsing population cycles. *Trends in Ecology and Evolution*, **23**, 79–86.

IPCC (2007) *Climate Change 2007: The Physical Science Basis*. S. Solomon, D. Qin, M. Manning et al., eds. New York: Cambridge University Press.

Jones, T. and Cresswell, W. (2010) The phenology mismatch hypothesis: are declines of migrant birds linked to uneven global climate change? *Journal of Animal Ecology*, **79**, 98–108.

Leggett, W. C. and Deblois, E. (1994) Recruitment in marine fishes: is it regulated by starvation and predation in the egg and larval stages? *Netherlands Journal of Sea Research*, **32**, 119–134.

Levin, S. A. (1976) Population dynamic models in heterogeneous environments. *Annual Review of Ecology and Systematics*, **7**, 287–310.

Levinksky, I., Skov, F., Svenning, J.-C. and Rahbek, C. (2007) Potential impacts of climate change on the distributions and diversity patterns of European mammals. *Biodiversity Conservation*, **16**, 3803–3816.

Ludwig, G. X., Alatalo, R. V., Helle, P. et al. (2006) Short- and long-term population dynamical consequences of asymmetric climate change in black grouse. *Proceedings of the Royal Society of London, Series B*, **273**, 2009–2016.

Memmott, J., Craze, P. G., Waser, N. M. and Price, M. V. (2007) Global warming and the disruptions of plant-pollinator interactions. *Ecology Letters*, **10**, 710–717.

Miller-Rushing, A. J., Høye, T. T., Inouye, D. W. and Post, E. (2010) The effects of phenological mismatches on demography. *Philosophical Transactions of the Royal Society of London, Series B*, **365**, 3177–3186.

Miner, B.G., Sultan, S.E., Morgan, S.G., Padilla, D.K. and Relyea, R.A. (2005) Ecological consequences of phenotypic plasticity. *Trends in Ecology and Evolution*, **20**, 685–692.

Montes-Hugo, M., Doney, S.C., Ducklow, H.W. *et al.* (2009) Recent changes in phytoplankton communities associated with rapid regional climate change along the western Antarctic peninsula. *Science*, **323**, 1470–1473.

Osenberg, C.W., Werner, E.E., Mittelbach, G.G. and Hall, D.J. (1988) Growth patterns in bluegill (*Lepomis macrochirus*) and pumpkinseed (*L. gibbosus*) sunfish: environmental variation and the importance of ontogenetic niche shifts. *Canadian Journal of Fisheries and Aquatic Science*, **45**, 17–26.

Parmesan, C. (2006) Ecological and evolutionary responses to recent climate change. *Annual Review of Ecology, Evolution, and Systematics*, **37**, 637–669.

Parmesan, C. and Yohe, G. (2003) A globally coherent fingerprint of climate change impacts across natural systems. *Nature*, **421**, 37–42.

Pearson, R.G. and Dawson, T.P. (2003) Predicting the impacts of climate change on the distribution of species: are bioclimate envelope models useful? *Global Ecology and Biogeography*, **12**, 361–371.

Phillimore, A.B., Hadfield, J.D., Jones, O.R. and Smithers, R.J. (2010) Differences in spawning date between populations of common frog reveal local adaptation. *Proceedings of the National Academy of Sciences of the United States of America*, **107**, 8292–8297.

Post, E. and Forchhammer, M.C. (2008) Climate change reduces reproductive success of an Arctic herbivore through trophic mismatch. *Philosophical Transactions of the Royal Society of London, Series B*, **363**, 2367–2373.

Post, E. and Pederson, C. (2008) Opposing plant community responses to warming with and without herbivores. *Proceedings of the National*

Academy of Sciences of the United States of America, **105**, 12353–12358.

Post, E. and Stenseth, N.C. (1999) Climatic variability, plant phenology, and northern ungulates. *Ecology*, **80**, 1322–1339.

Post, E., Forchhammer, M.C., Bret-Harte, M.S. *et al.* (2009) Ecological dynamics across the Arctic associated with recent climate change. *Science*, **325**, 1355–1358.

Post, E., Pedersen, C., Wilmers, C.C. and Forchhammer, M.C. (2008a) Warming, plant phenology and the spatial dimension of trophic mismatch for large herbivores. *Proceedings of the Royal Society of London, Series B*, **275**, 2005–2013

Post, E., Pederson, C., Wilmers, C.C. and Forchhammer, M.C. (2008b) Phenological sequences reveal aggregate life history response to climate warming. *Ecology*, **89**, 363–370.

Primack, R.B., Ibanez, I., Higuchi, H. *et al.* (2009) Spatial and interspecific variability in phenological responses to warming temperatures. *Biological Conservation*, **142**, 2569–2577.

Pringle, R.M., Young, T.P., Rubenstein, D.I. and McCauley, D.J. (2007) Herbivore-initiated interaction cascades and their modulation by productivity in an African savanna. *Proceedings of the National Academy of Sciences of the United States of America*, **104**, 193–197.

Price, P.W. (2002) Species interactions and the evolution of biodiversity. In C.M. Herrera and O. Pellmyr, eds., *Plant–Animal Interactions: An Evolutionary Approach*. Oxford: Blackwell Science, pp. 3–25.

Prout, T. and McChesney, F. (1985) Competition among immatures affects their adult fertility: population dynamics. *American Naturalist*, **126**, 521–558.

Root, T.L., Price, J.T., Hall, K.R. *et al.* (2003) Fingerprints of global warming on wild animals and plants. *Nature*, **421**, 57–60.

Roughgarden, J. (1974) Population dynamics in a spatially varying environment: how population size 'tracks' spatial variation in carrying capacity. *American Naturalist*, **108**, 649–664.

Sanderson F. J., Donald, P. F., Pain, D. J., Burfield, I. J. and van Bommel, F. P. J. (2006) Long-term population declines in Afro-Palearctic migrant birds. *Biological Conservation*, **131**, 93–105.

Schweiger, O., Settele, J., Kudrna, O., Klotz, S. and Kuhn, I. (2008) Climate change can cause spatial mismatch of trophically interacting species. *Ecology*, **12**, 3472–3479.

Senft, R. L., Coughenour, M. B., Bailey, D. W. *et al.* (1987) Large herbivore foraging and ecological hierarchies. *BioScience*, **11**, 789–795.

Slatkin, M. (1974) Hedging one's evolutionary bets. *Nature*, **250**, 704–705.

Thackeray, S. J., Sparks, T. H., Frederiksen *et al.* (2010) Trophic level asynchrony in rates of phenological change for marine, freshwater, and terrestrial environments. *Global Change Biology*, **16**, 3304–3313.

Thompson, J. N. (1996) Evolutionary ecology and the conservation of biodiversity. *Trends in Ecology and Evolution*, **11**, 300–303.

Turner, M. (2005) Landscape ecology: what is the state of the science? *Annual Review of Ecology, Evolution, and Systematics*, **36**, 319–344.

Veit, R. R., McGowan, J. A., Ainley, D. G., Wahl, T. R. and Pyle, P. (1997) Apex marine predator declines ninety percent in association with changing oceanic climate. *Global Change Biology*, **3**, 23–28.

Visser, M. E. (2010) Keeping up with a warming world; assessing the rate of adaptation to climate change. *Proceedings of the Royal Society of London, Series B*, **275**, 649–659.

Visser, M. E. and Both, C. (2005) Shifts in phenology due to global climate change: the need for a yardstick. *Proceedings of the Royal Society of London, Series B*, **272**, 2561–2569.

Visser, M. E. and Holleman, L. J. M. (2001) Warmer springs disrupt the synchrony of oak and winter moth phenology. *Proceedings of the Royal Society of London, Series B*, **268**, 289–294.

Visser, M. E., Both, C. and Lambrechts, M. M. (2004) Global climate change leads to mistimed avian reproduction. *Advances in Ecological Research*, **35**, 89–110.

Visser, M. E., van Noordwink, A. J., Tinbergen, J. M. and Lessells, C. M. (1998) Warmer springs lead to mistimed reproduction in great tits (*Parus major*). *Proceedings of the Royal Society of London, Series B*, **265**, 1867–1870.

Walther, G-R., Post, E., Convey, P. *et al.* (2002) Ecological responses to recent climate change. *Nature*, **416**, 389–395.

White, R. G. (1983) Foraging patterns and their multiplier effects on productivity of northern ungulates. *Oikos*, **40**, 377–384.

Williams, G. C. (1966) *Adaptation and Natural Selection*. Princeton, NJ: Princeton University Press.

Winder, M. and Schindler, D. E. (2004) Climatic effects on the phenology of lake processes. *Global Change Biology*, **10**, 1844–1856.

Werner, E. E. and Gilliam, J. F. (1984) The ontogenetic niche and species interactions in size-structured populations. *Annual Review of Ecology and Systematics*, **15**, 393–425.

Werner, E. E. and Peacor, S. D. (2003) A review of trait-mediated indirect interactions in ecological communities. *Ecology*, **84**, 1083–1100.

Yang, L. H. and Rudolf, V. H. W. (2010) Phenology, ontogeny and the effects of climate change on the timing of species interactions. *Ecology Letters*, **13**, 1–10.

CHAPTER TWENTY-SEVEN

Coda

TAKAYUKI OHGUSHI

Center for Ecological Research, Kyoto University

OSWALD J. SCHMITZ

School of Forestry and Environmental Studies, Yale University

and

ROBERT D. HOLT

Department of Biology, University of Florida

Community section

Phenotypic plasticity is a ubiquitous phenomenon in nature, and provides a basis for trait-mediated indirect interactions (TMIIs) among species in ecological communities. Since trait-mediated indirect effects (TMIEs) are replete across a wide range of ecosystems, it is becoming increasingly apparent that phenotypic plasticity in response to interacting species can play an important role in determining community organization and dynamics. Below we highlight the major findings of community consequences of TMIIs in this volume.

TMIIs are common and can determine trophic structure in marine pelagic and insect host–parasitoid systems, both of which have been little explored (Chapters 3 and 4).

TMIEs in prey–predator systems should be taken into consideration in terms of non-trophic links (Chapter 2), size- and age-structure of a population (Chapter 5) and density-dependence (Chapter 6).

Herbivore-caused phenotypic plasticity and/or genetic variations of plants have significant, indirect impacts on diversity and abundance of predators, and prey–predator interactions by bottom-up cascading effects (Chapters 7 and 9).

TMIEs caused by species coexistence, size-structured populations and non-host behaviour can stabilize prey–predator or host–parasitoid dynamics (Chapters 3, 5 and 8).

Enemy-induced morphological traits for dispersal could influence meta-community dynamics (Chapter 3).

The connection of top-down and bottom-up effects is a challenging issue, because they interact with each other via changes in any combination of traits and abundance of plants, herbivore prey or predators (Chapter 9).

Trait-Mediated Indirect Interactions: Ecological and Evolutionary Perspectives, eds. Takayuki Ohgushi, Oswald J. Schmitz and Robert D. Holt. Published by Cambridge University Press. © Cambridge University Press 2012.

Expanding temporal and spatial scales are needed to capture the long-term and large spatial community consequences of TMIIs (Chapters 2–4, 6, 8–10).

We need an integrated approach, which combines theory, laboratory studies and field studies, because of a large gap between theoretical and empirical work (Chapters 6–7, 10).

Coevolution section

Community ecology has been enriched by the concept of 'ecological engineering', which reflects how organisms mediate changes in the 'distribution, abundance and composition of energy and materials in the environment', for reasons other than the direct assimilation of resources (Jones and Gutierrez 2007, p. 7). It has become increasingly clear that understanding community structure requires an evolutionary perspective, ranging from elucidating the consequences of within-species genetic variation for interactions to assessing the determinants of speciation rates. We suggest that the evolutionary dimension of TMIIs spanned by the chapters in this volume constitutes a kind of 'evolutionary engineering', in that the traits of one species may alter how another pair (or larger set) of species interacts via reciprocal selection over evolutionary timescales. The chapters in the coevolution section of this volume provide a rich array of examples, and conceptual perspectives, on how such evolutionary engineering may arise.

TMIIs dovetail nicely with traditional issues in the study of coevolution, such as diffuse coevolution and the geographic mosaic theory, but most studies lack crucial details needed to nail down definitively the importance of such interactions (Chapter 11).

Biological invasions can provide a cascade of evolutionary indirect effects, including those mediated by phenotypic plasticity in traits (Chapter 12).

In a food web context, the evolutionary implications of trait-mediated interactions can extend even to distantly related taxa (Chapter 13).

Classes of interactions that have traditionally been studied in isolation, such as dispersal mutualisms, pollination mutualisms and seed predation, can actually be strongly linked via TMIEs (Chapters 14 and 15).

Ecosystem section

Collectively, the ecosystem section chapters expand our understanding of the factors shaping bottom-up and top-down control of ecosystem structure and functioning in ways that once and for all begin to bridge the divide between the evolutionary ecology of species interactions and ecosystem functioning.

All species are component parts of ecosystems and can play important functional roles by exerting some control over ecosystem properties and functioning. It stands to reason that ecosystem function overall must be in some way related to the diversity of species comprising a system (Chapters 16 and 21).

Focusing on species diversity alone is insufficient to understand the suite of functional roles that species assume. Species' impacts in ecosystems can be linked to their morphological, physiological and behavioural traits that determine the way they function (Chapters 17 and 18).

Trait variation among genotypes can have large and predictable effects on community and ecosystem processes indicating that diversity within a species may be as important as species diversity (Chapters 16, 18 and 19).

The fear of being eaten by predators causes profound changes in species physiology that alter ecosystem nutrient fluxes and pool sizes, and hence control over nutrient and elemental cycling and trophic transfer efficiencies (Chapter 17).

Trait-mediated effects may operate within the microbial community helping to demystify one of the largest unknowns in ecosystem ecology that is typically treated as a 'black box' in ecosystem analysis (Chapter 20; see also Chapters 9 and 16).

A complete understanding of ecosystem functioning requires perspectives that consider two dimensions of diversity: species and trait diversity within trophic levels and trophic level diversity in ecosystems (Chapter 21).

Applied ecology section

There is a large body of scientific evidence that trait-mediated effects matter in communities and ecosystems. Accordingly, the field has matured sufficiently that we can now entertain the application of the scientific findings to environmental management (Chapter 22).

Trait-mediated effects theory helps us to be more strategic about choosing predator species and combining them in 'communities' in order to make biological control of pest species more efficient within an agricultural system (Chapters 22 and 23) as well as managing movement and dynamics of pest species across landscapes (Chapter 24).

Advances in dealing with invasive species are currently constrained by the lack of a general conceptual framework for predicting invasion success. Consideration of TMIIs can help us understand the community contexts that encourage invasions and community contexts that help to resist invasions (Chapter 25).

Global change is a pervasive influence on ecological communities because it stands to disassemble and reconfigure species interdependencies in communities. A TMIEs perspective helps us understand how global change, especially climate change, can alter the expression of species phenotypes (e.g., timing of life-history strategies) that ultimately influence consumer-resource interactions and indirect effects in ecological food chains (Chapter 26).

Reference

Jones, C. G. and Gutierrez, J. L. (2007) On the purpose, meaning, and usage of the physical ecosystem engineering concept. In K. Cuddington, J. E. Byers, W. G. Wilson and A. Hastings, eds., *Ecosystem Engineers: Plants to Protists.* Burlington, MA: Academic Press, Elsevier, pp. 3–24.

Index

Abies spp. (firs), 282
Abutilon theophrasti, 498
Acanthina angelica (whelk), 62
Acanthinucella spirata (predatory snail), 50
Aceria guerreronis (coconut mite), 438–439
Aceria tulipae (dry bulb mite), 437
Acremonium strictum, 263
Aculops lycopersici (mite), 439
Acyrthosiphon pisum (pea aphid), 31–36, 232,
    399, 401, 453, 462
Adalia bipunctata (lady beetle), 453
adaptive predator behavior, 135
adaptive prey trait modification (APTM), 140
    and environmental stochasticity, 145–147
    future needs for APTM theory, 145–153
    gap between theory and experiment,
        150–153
    in food web theory, 140–143
    in larger food web systems, 142–143
    in one predator–two prey webs, 142
    in predator–prey models, 140
    in tritrophic food chains, 141
    in two predator–one prey webs, 141–142
    influence on system dynamics, 143–144
    prey response to system dynamics, 147–148
    question of inclusion in ecological theory,
        144–145
    role in ecological theory, 132–133
    scaling up insights to large webs, 148–150
    terminology, 135
    theoretical issues, 144
    See also modeling adaptive prey trait
        modification.
Aeshna umbrosa (dragonfly), 75, 76–77
Ageratina adenophora, 499
Aleppo pine (*Pinus halepensis*), 284
Alexandrium minutum (microalga), 56–57
alfalfa, 453, 455
algae (marine), 50
algal toxins, 56–57
Allee effects, 91
Alliaria petiolata (garlic mustard), 459
Amblydromalus manihoti (predatory mite),
    440–442
amphipods, 18–19, 51
Amphiprion percula (clownfish), 59–60

Anagrus spp. (parasitoids), 482
Anax junius (dragonfly), 76–77
anglerfishes, 404
Anopheles gambiae (mosquito), 401
antibiotic compounds in corals, 400
ants, 112
    effects of plants on, 15, 18
    Formica japonica, 169–170
    parasitoids, 23
    Pheidole diversipilosa, 36–37
Aphelinus asychis (parasitoid), 461
Aphidius ervi (parasitoid), 31–36, 401, 453
aphid–parasitoid system, 40
    trait-mediated trophic cascades, 38–40
aphids
    Acyrthosiphon pisum, 31–36, 232, 401, 453
    Aphis craccivora, 232, 453
    Brassica oleracea food webs, 41–42
    Brevicoryne brassicae, 112, 475
    cowpea aphid, 232, 453
    development of winged morph, 15, 442
    effects of parasitoid predators, 14, 23
    effects of parasitoid presence, 15
    influences on body size, 15
    Megoura viciae, 31–36
    Myzocallis asclepiadis, 112
    Myzus persicae, 112
    natural enemies, 452–454, 461, 462
    pea aphid, 31–36, 232, 399, 401, 453
    Pemphigus betae, 304–305, 306–307,
        375, 382
    predation by coccinellids, 20
    secondary symbionts, 401
    Toxoptera citricida, 452
    Uroleucon nigrotuberculatum, 169–170
    Uroleucon rudbeckiae, 122–124
Aphis craccivora (cowpea aphid), 232, 453
Apocephalus 'sp.8' (parasitoid), 36–37
apparent competition, 10, 19–20, 140
    between prey, 142
applied ecology, 528
    summary of consequences of TMIIs, 528
aquatic ecosystems, 400
    microbially mediated TMIIs, 399–400
    protective symbioses, 399–400
Arabadopsis thaliana, 383

arbuscular-mycorrhizal fungi, 261–264
arthropod communities, 167
  effects of herbivore-initiated bottom-up
    cascades, 162–167
  interspecific indirect genetic effects (IIGEs),
    308–309
Ascophyllum nodosum (seaweed), 58
Asian citrus psyllid (*Diaphorina citri*), 452
aspen (*Populus* spp.)
  effects of wolf reintroduction, 13, 37
  genotypes, 373
augmentative biological control, 461
  use of enemy functional diversity, 460–461
Avena fatua, 491

Baccharis, 296
Baccharis salicifolia, 122–124
barnacles, 50
  (*Semibalanus balanoides*), 330, 331
  effects of whelk predation, 15
  predator-induced morophological
    change, 62
bass habitat shift, 22
beaver (*Castor canadensis*), 287, 300, 303
  interspecific indirect genetic effects (IIGEs),
    308–309
beetles, 37, 179
  Galerucella calmariensis, 37
  Galerucella tenella, 37
  ground beetles, 453
  Ips typographus, 349
  Mordellistena convicta, 245, 247–248, 251–252
  Trirhabda virgata, 348
behavioural flexiblity
  influence on community dynamics, 42–43
behavioural plasticity, 11–12
  trait-change mechanism, 16
behaviourally mediated indirect effects, 135
Bemisia tabaci (whitefly), 442
Bifidobacterium, 402
biodiversity
  and ecosystem functioning, 414–415,
    424–425
  and resistance to invasion, 425–426
  and robustness to resident extinctions,
    425–426
  consequences of herbivore-initiated
    bottom-up cascades, 178–179
  importance of non-trophic interactions,
    414–415
  importance of TMIIs, 414–415
biological control
  and natural enemy biodiversity, 451–452
  augmentative biological control, 460–461
  classical biological control, 459
  conservation biological control, 460–461
  herbivore-induced indirect plant defence,
    436–439
  influence of trait-mediated effects, 445–446
  mathematical models, 446
  plant-mediated competition among
    herbivores, 443–445

predator-induced escape behaviour of
    herbivores, 439–442
predator-induced ontogenetic escape by
    herbivores, 442–443
predator-mediated competition among
    herbivores, 443–445
trait changes in tritrophic systems,
    435–436
use of enemy functional diversity, 459–461
weeds, 454, 456
  See also natural enemies.
biological invasions. *See* invasive species
black-capped chickadee (*Parus atricapillus*),
    245, 247, 347, 349
black locust (*Robinia pseudoacacia*), 264
blue crab (*Callinectes sapidus*), 54, 57–58
body size, 15
  indirect effects on, 1, 15
Boloria titania (butterfly), 514
Botanophila seneciella (ragwort seed head
    fly), 454
bottlenose dolphins, 50
bottom-up cascading effects, 181
  future research directions, 180–181
  observed trends, 180–181
bottom-up trophic cascades
  herbivore initiation, 162–164
  initiated by a stem borer in a willow
    system, 167–168
  initiated by aphids in a goldenrod system,
    169–170
  initiated by belowground microbe in a
    soybean system, 170–171
  initiated by microbial symbionts, 170–171
Brassica oleracea aphid–parasitoid food webs,
    41–42
Brassica oleracea var. *gemmifera* (Brussels
    sprouts), 112
Brevicoryne brassicae (aphid), 112, 475
broad-sense community heritability, 311
Bromus diandrus, 491
Bromus tectorum (grass), 232
brown citrus aphids (*Toxoptera citricida*), 452
Brussels sprouts (*Brassica oleracea* var.
    gemmifera), 112
bryostatins, 399
Bugula neritina (bryozoan), 399
bullfrogs, 22
bur oak system, 349–350
Burkholderia spp., 403
Busycon carica (knobbed whelk), 57–58
butterflies, 470
  Boloria titania, 514
  Iolanta iolas, 470
  Pieris spp., 483

C:N:P cycling in ecosystems, 332–333
Callinectes sapidus (blue crab), 54, 57–58
Cancer spp. (crabs), 58–59
candidate genes, 297, 302–304
  definition, 316
cannibalistic conspecifics, 74–75, 81

Cape gannet (*Morus capensis*), 515–516
Carcinus maenus (green crab), 50, 52, 55, 57,
    58–59, 330, 331
caribou (*Rangifer tarandus*), 518
Carpobrotus edulis, 499–500
carvacrol, 495
cassava green mite (*Mononychellus tanajoa*),
    440–442
Castor canadensis (beaver), 287, 300, 303
    interspecific indirect genetic effects (IIGEs),
       308–309
Centaurea diffusa (diffuse knapweed), 456
Centaurea stoebe, 500–501
Cerastoderma edule (cockle), 57
Cervus elaphus (elk), 13, 18, 37, 518
Ceutorhynchus constrictus, 459
Ceutorhynchus scrobicollis, 459
Chamerion angustifolium (fireweed), 261
cinnabar moth (*Tyria jacobaeae*), 454
citrus leafminers (*Phyllocnistis citrella*), 452
citrus pests
    range of specialist natural enemies, 452
clams, 50
    Macoma balthica, 57
    Mulinia lateralis, 57
Clark's nutcracker (*Nucifraga columbiana*),
    279–281, 282–283
Clavicipitaceae (endophytic fungi), 402
climate change, 462
    and the match/mismatch hypothesis,
       511–512
    driver of phenological trait shifts, 508–509
    future research directions, 520–522
    impacts on migratory species, 516–518
    impacts on natural enemy functional
       diversity, 461–462
    resource abundance variation, 512–513
    spatial mismatch in consumer-resource
       interactions, 513–518
Clupea spp. (herring), 59, 509
Coccinella septempunctata (lady beetle), 462
Coccinella transversoguttata (lady beetle), 452
coccinellids, 20
cockle (*Cerastoderma edule*), 57
coconut mite (*Aceria guerreronis*), 438–439
cod (*Gadus* spp.), 509
coevolution, 207
    broad definition, 207
    diffuse coevolution, 208
    geographic mosaic theory, 208
    Janzen's definition, 207–208
    origin of the term, 207
    pairwise coevolution, 207–208
    summary of consequences of TMIIs, 527
coevolutionary process and TMIIs, 218
    conceptual and theoretical importance,
       217–218
    future research directions, 217–218
coevolutionary theory and TMIIs, 209–217
    diffuse coevolution, 217
    geographic mosaic theory and TMIIs,
       214–217

    hot and cold spots, 217
    selection mosaics, 217
    TMIIs and pairwise vs. diffuse interactions,
       211–212
coevolutionary TMIIs, 208
    examples of influence of TMIIs, 209–210
    origins of, 207–208
    requirements for pairwise coevolution,
       208–209
Coleomegilla maculata (lady beetle), 453,
    455, 456
Coleoptera, 28
collard (*Brassica oleracea*), 453
Collembola, 357
Colorado potato beetle (*Leptinotarsa
    decemlineata*), 456
common periwinkle (*Littorina littorea*), 13, 61
communities
    effects of individual plant genotypes,
       371–377
    importance of trait- vs. density-mediated
       indirect effects, 9–10
    non-additive effects of plant genotype
       diversity, 382–384
    summary of consequences of TMIIs,
       526–527
    taxonomic framework for TMIIs, 10–25
community, 316
    definition, 316
community composition, 296, 316
community diversity, 296, 316
    effects of phenotypic plasticity, 491
community dynamics
    density interactions, 2–3
    influence of behavioural flexiblity, 42–43
community ecology
    potential contribution of APTM, 153–154
    recognition of indirect effects, 1–2
    traditional pairwise approach to
       interactions, 131
community evolution
    evolutionary indirect interactions, 253–255
community genetics, 295–297, 300, 316
community genomics, 301–302, 316
community heritability, 311, 316
    quantification, 300
community interactions and IIGEs, 304–307
community-level selection, 311–314, 317,
    452, 455
    simulation approach, 311–314
community phenotypes, 305, 310–311, 316
community properties
    influence of TMIIs, 295–297
community stability, 296, 317
community structure
    effects of microbially mediated TMIIs,
       404–405
    effects of trait-mediated interactions, 40–42
comparative genomics, 315, 317
cones
    trait evolution, 278–279
    variations in DMIIs and TMIIs, 281–284

conifers
  disc loading of seeds, 285
  evolutionary consequences of TMIIs, 279–281
  reproductive trait evolution, 278–279, 285–287
  role of DMIIs and TMIIs in trait evolution, 281–284
  selection pressures on serotiny, 285–287
connectance, 42
  and species diversity, 41–42
conservation biological control
  use of enemy functional diversity, 460–461
conspecific cannibalism, 74–75
consumer–resource interactions
  impacts of phenological asynchrony, 509
  phenological shift and spatial mismatch, 513–518
consumptive competition, 19–20
context-dependent effects, 2
copepods, 56–57
coral probiotic hypothesis, 400
coral reefs, 59
cordgrass
  induced defenses, 16
cordgrass (Spartina alterniflora), 61
core species, 296
corn (Zea mays), 453, 496
Cotesia glomerata (parasitoid), 30–31, 483
cottonwoods (Populus spp.), 279, 287
  genomic sequencing, 301–302
  interspecific indirect genetic effects (IIGEs), 308–309
cowpea aphid (Aphis craccivora), 232, 453
crab predation, 21
crabs (Cancer spp.), 58–59
crayfish, 21
crickets, 37
crossbills, 216–217
Cytisus scoparius, 230

damage-induced volatiles
  effects on herbivore densities, 467
damsel bug (Nabis americoferus), 452
Darwin, Charles, 207
decomposers, 340
  effects of herbivore TMIEs, 352–353
defense induction in plants, 349–351
defoliation-induced root exudation of labile C, 346
Delphacodes scholochloa (planthopper), 482
demography, 89
  influence of traits, 89
density dependence
  Allee effects, 91
  analysis of trait-mediated effects, 101–103
  and trait-mediated interactions, 90–94
  discrete-time model of trait-mediated effects, 100–101
  influence of trait plasticity, 101
  influence on stability of ecosystems, 89
  positive density dependence, 91

density interactions, 2–3
density-mediated biotic indirect effects, 417
density-mediated effects
  in marine systems, 47–48
density-mediated indirect effects (DMIEs), 12
  comparison with trait-mediated indirect effects, 9–10, 415–416
  definition, 135, 237
Desmarestia ligulata (seaweed), 63
detritivores, 340
  effects of herbivore TMIEs, 353–354
detritus-based food chains, 325
developmental plasticity, 11–12
  trait-change mechanism, 16–17
developmental stage variation within species, 70
Diadegma spp., 479
Diaeretiella rapae (parasitoid), 475
diamond back moth (Plutella xylostella), 30–31, 479
diapause in spider mites, 442–443
Diaphorina citri (Asian citrus psyllid), 452
Diaprepes abbreviatus (weevil), 452
diffuse coevolution, 208, 211–212, 217
diffuse knapweed (Centaurea diffusa), 456
dimethyl sulfide (DMS)
  release by grazing zooplankton, 57
dimethyl sulfioproponate (DMSP), 57
dinoflagellates, 400
Dioryctria albovittella (stem-boring moth), 349
Diptera, 28
direct density dependence
  modelling trait-mediated effects, 95–100
discrete-time model
  trait-mediated density dependence, 100–101
dog whelk (Nucella lapillus), 52, 55
dominant (foundation) species, 279, 296
downy woodpecker (Poecile pubescens), 245, 247
dragonflies
  Aeshna umbrosa, 75–77
  Anax junius, 76–77
  Plathemis lydia, 76
  size-structured interactions among larvae, 76
  stage-structured mutual predation, 76–77
dry bulb mite (Aceria tulipae), 437
dugongs, 50

eastern hemlock, 301
eco-evolutionary feedback, 237
ecological engineers, 135
ecological theory
  adaptive prey trait modification (APTM), 132–133
  food web theory, 133–135
  foraging theory, 132
  higher-order interactions (HOIs), 131–135, ₋
  incorporating habitat-mediated effects, 429
  incorporating non-trophic interactions, 429

ecological theory (cont.)
  incorporating the interaction web model,
    429
  incorporating trait-mediated effects, 429
  potential contribution of APTM, 153–154
  traditional pairwise approach to
    interactions, 131
ecological traps, 516
ecosystem
  definition, 317
ecosystem engineering
  application of the interaction web model,
    429
  herbivore-induced plant phenotypes,
    171–174
ecosystem engineers, 18, 170, 296, 418
ecosystem functioning
  and biodiversity, 414–415, 424–425
  interaction web model, 415, 418–420,
    421–423
  kinds of indirect effects, 415–418
  link between species richness and species
    interactions, 421–423
  relationship to ecosystem structure,
    428–429
  role of non-trophic interactions, 414–415
  role of TMIIs, 414–415
ecosystem genetics, 300, 316
ecosystem genomics, 301–302, 316
ecosystem heritability, 316
ecosystem phenotypes, 306, 316
ecosystem processes, 307
  and IIGEs, 304–307
  defence induction in plants, 349–351
  effects of green-fall, 345
  effects of hebivore cadavers, 344–345
  effects of herbivore TMIEs in soil systems,
    352–354
  effects of herbivore faeces and urine, 343–344
  effects of predator-induced fear in prey,
    326–328
  effects of premature leaf abscission, 345
  effects of selective foraging, 346–349
  effects of through-fall, 345
  effects on soil microclimate, 351–352
  effects on soil resources, 352
  fast-cycle effects, 341–343, 351–352
  herbivore-induced root exudation of labile
    C, 346, 495
  herbivore influences, 339–341
  herbivore influences on nutrient recycling,
    349–351
  herbivore TMIE impacts in soil systems,
    354–355
  impact of individual plant genotypes,
    377–381
  influences on litter decomposition,
    349–351
  mechanisms of herbivore influence, 341
  Quercus (oak) ecosystems herbivore TMIEs
    study, 355–357
  slow-cycle effects, 341–343

  slow-cycle pathways, 346–352
ecosystem properties, 297
  influence of TMIIs, 295–297
  interaction with invasive species, 427
ecosystem structure
  relationship to ecosystem functioning,
    428–429
ecosystems
  consequences of energy flow in food
    chains, 328–330
  consequences of plant interactions,
    492–495
  consequences of TMIIs, 279–281
  effects of individual plant genotypes,
    371–375
  effects of microbially mediated TMIIs,
    405–406
  factors affecting flows of energy and
    materials, 333–334
  non-additive effects of plant genotype
    diversity, 382–384
  nutrient constraints on energy transfer,
    332–333
  summary of consequences of TMIIs,
    527–528
Eichhornia crassipes, 418
elk (Cervus elaphus), 13, 18, 37, 518
Elymus multisetus (grass), 232
Endobugula sertula (bacterium), 399
Endoclita excrescence (stem-boring moth),
    167–168
endophytic bacteria, 22
endophytic fungi
  interaction with mycorrhizae, 264
  symbioses, 402–403
energy flow in food chains, 328–330
  factors affecting, 333–334
energy transfer in ecosystems
  nutrient constraints, 332–333
Epinephelus striatus (Nassau grouper), 15, 59
eriophyoid mites, 436–437
Eucalyptus spp., 301, 372
Eucosma recissoriana (lodgepole pine cone
    borer moth), 282
Euplotidium ciliates, 399
European Phenology Network (EPN), 515
European red mite (Panonychus ulmi), 443
Eurosta host races
  web of indirect interactions, 248–250
Eurosta solidaginis (gall fly), 244–245,
    251–252
  mediation of enemy indirect interactions,
    247–248
Eurycea cirrigera (salamander), 72
Eurytoma gigantea (parasitoid), 245, 247–248,
    251–252
evolution, 12
  as trait-change mechanism, 11–12
  trait-change mechanisms, 16
evolution of increased competitive ability
    (EICA), 222, 229–230, 237
evolutionary consequences of TMIIs, 279–281

evolutionary indirect effects, 221
  change in invaders and natives, 221–223
  definition, 237–238
  evolution of increased competitive ability
    (EICA), 229–230
  future research directions, 232–235
  increased anti-herbivore defences in
    natives, 231–232
  increased competitive ability in natives,
    232
  indirect ecological effects of invading
    species, 223–225
  insights from study of invasive species,
    235–237
  loss of mutualism in invaders, 230–231
  potential consequences of invading
    species, 225–232
  study of biological invasions, 221
evolutionary indirect interactions, 244
  community evolution, 253–255
  Eurosta solidaginis mediates enemy
    interactions, 247–248
  measuring indirect interactions,
    245–253
  reciprocal transplant experiments,
    252–253
  selection experiments, 253
  testing indirect selection assumptions, 247
  using geographic variation to test indirect
    selection, 251–252
  web of indirect interactions of Eurosta host
    races, 248–250
  web of indirect interactions of Solidago
    species, 248–250
exploitative competition between predators,
    141–142
expressed sequence tags (ESTs), 302, 317
extra-floral nectaries, 263, 265

fear of predation
  effects at ecosystem level, 326–328
feeding traits, 13
fire ants, 23
fireweed (Chamerion angustifolium), 261
firs (Abies), 282
fish
  predation on salamanders, 20
fish stocks
  match/mismatch hypothesis, 509–511
flexible traits, 187
  consideration in community modelling,
    186–187
food chains
  C:N:P cycling, 332–333
  connections between the bottom and the
    middle, 330–331
  connections between the top and the
    middle, 331
  consequences of foraging decisions in the
    middle, 326–328
  detritus-based, 325
  energy flow, 328–330

factors affecting flows through, 333–334
factors affecting length of, 327–328
length of, 327–328
predator-induced fear in prey, 326–328
trophic control from the middle,
    325–331
food web interactions, 3
food web theory, 69
  adaptive prey trait modification (APTM),
    140–143
  higher-order interactions (HOIs), 133–135
food webs
  diversity of species in the middle, 326
  interactions in the middle, 326
foraging theory, 132
forest tent caterpillars, 382
Formica japonica (ant), 169–170
foundation species, 279, 296, 297, 302, 317
  as mediators of IIGEs, 301–302
  influence of individual plant genotype,
    371–373
four species webs, 21–23
fucoid algae, 50
Fucus sp. (seaweed), 17, 61
functional diversity schemes
  limitations of, 450–451
  predicting enemy diversity effects,
    457–459
functional genomics, 315, 317
Fundulus heteroclitus (killifish), 57

Gadus spp. (cod), 509
Gaillardia grandiflora, 500–501
Galeocerdo cuvier (tiger shark), 50
Galerucella calmariensis (beetle), 37
Galerucella tenella (beetle), 37
gall fly (Eurosta solidaginis), 244–245, 247–248,
    251–252
gall-forming sawflies (Phyllocolpa spp.), 308
garlic mustard (Alliaria petiolata), 459
genetic basis for trophic interactions,
    304–307
genetic basis of TMIIs, 297–301
genetic fingerprinting, 302
genetic maps, 297
  organization of genomic information, 297
  Populus spp., 302–303
genetic similarity rule, 373
genetic traits, 11
genetic type, 108
genetic variation terminology, 108
genetics
  community and ecosystem, 316
  community genetics, 295–297
genomic information
  organization of, 297
genomic sequencing
  cottonwoods (Populus spp.), 302
genomics
  community and ecosystem, 301–302, 316
  comparative genomics, 317
  functional genomics, 317

genomics research, 316
    relevance of interspecific indirect genetic
        effects (IIGEs), 314–316
genotype, 108
Geocoris spp., 455
geographic mosaic theory, 208, 214–217
geographic variation
    testing the indirect selection hypothesis,
        251–252
Geranium sylvaticum, 261
gerbils, 22
giant kelp (Macrocystis pyrifera), 63
Glomus hoi, 261
goldenrod (Solidago altissima), 169–170, 379,
    383, 384
goldenrod (Solidago rugosa), 330
goldenrod (Solidago spp.), 244–245, 296,
    306, 348
    reciprocal transplant experiments, 253
    web of indirect interactions, 248–250, 252
goldenrod–herbivore–natural enemy
    interactions, 244–245
grass shrimp (Palaemonetes pugio), 57
grasses
    Bromus tectorum, 232
    Elymus multisetus, 232
    Paspalum dilatatum, 351
    Poa pratensis, 495
    Scolochloa festucacea, 482
    Spartina pectinata, 482
grasshoppers, 9, 13, 348, 455, 457, 462
great tit (Parus major), 511
green crab (Carcinus maenus), 50, 52, 55, 57,
    58–59, 330, 331
green-fall
    effects on ecosystem processes, 345
groupers, 15
guppy (Poecilia reticulata), 287
gut microbacteria in vertebrates, 401–402
gut microbes in humans, 373
gypsy moth, 382
Gyrinophilus porphyriticus (salamander), 72

habitat-mediated effects
    incorporation into theoretical ecology, 429
habitat-mediated indirect effects,, 417
habitat selection traits, 13–14
haddock, 59
hairy woodpecker (Picoides villosus), 283
Hamiltonella defensa, 399, 401
Haplopappus ericoides, 499–500
Haplopappus venetus var. seloides, 499–500
harbour seals, 50
hard clam (Mercenaria mercenaria), 52–53, 57–58
Harmonia axyridis (lady beetle), 36, 453, 456,
    461, 462
herbivore density and patch size
    resource concentration hypothesis (Root),
        466–467
herbivore density distributions
    effects of damage-induced volatiles, 467
    influence of natural enemies, 467

herbivore density distributions model,
    467–468
    adding induced attraction to the model,
        470–472
    adding natural enemies to the model,
        472–479
    applied aspects, 482–484
    basic population model, 468–470
    density-dependent parasitoid emigration
        rates, 476–478
    field data, 479–482
    future research directions, 484
    insights from models, 484
    larval/pupal parasitoids and host
        attraction, 473–476
    predator-induced prey emigration, 478–479
herbivore–enemy interactions, 127
    tritrophic perspective, 107–108
    See also plant effects on herbivore–enemy
        interactions.
herbivore-induced indirect plant defence, 15,
    23, 31, 36, 37, 436–439
herbivore-induced phenotypic plasticity in
    plants, 161–162, 230, 231
herbivore-induced plant defences, 10, 13, 15,
    16–17
herbivore-induced plant phenotypes, 245
    changes in plant nutritional quality, 172
    damage-induced regrowth, 173
    ecosystem engineering, 173–174
    herbivore responses to, 174–176
    predator responses to changes in
        herbivores, 176–178
    resistance mediated by secondary
        metabolites, 171–172
    spatial and temporal resource mosaics,
        179–180
    susceptibility mediated by secondary
        metabolites, 171–172
herbivore-induced plant volatile chemicals,
    20–, 20, 30–31
herbivore-initiated bottom-up cascades, 231
    biodiversity consequences, 178–179
    effects on arthropod communities,
        164–167
    future research directions, 180–181
    observed trends, 180–181
herbivore suites, 223, 225, 234, 237
herbivore TMIEs
    case study with Quercus (oak) ecosystems,
        355–357
    defence induction in plants, 349–351
    ecosystem effects of cadavers, 344–345
    ecosystem effects of faeces and urine,
        343–344
    ecosystem effects of green-fall, 345
    ecosystem effects of premature leaf
        abscission, 345
    ecosystem effects of through-fall, 345
    effects on decomposers (microbes),
        352–353
    effects on detritivores, 353–354

effects on soil microclimate, 351–352
effects on soil resources, 352
fast-cycle effects, 341–343, 351–352
fast-cycle pathways, 343–346
future research directions, 360–361
importance for soil systems, 354–355
induced root exudation of labile C, 346
influence on litter decomposition rates,
    349–351
influence on nutrient recycling, 349–351
relative importance of, 354–355
selective foraging, 346–349
slow-cycle effects, 341–343
slow-cycle pathways, 346–352
herbivores
influences on ecosystem processes,
    339–341
mechanisms of influence on ecosystem
    processes, 341
herring (*Clupea* spp.), 59, 509
higher-order interactions (HOIs)
alternative terminology, 135
importance in ecological communities,
    131–133
in food web theory, 133–135
Holling type II functional response
    model, 186
homeostatic adjustments, 17
host–parasitoid interactions
effects of non-host species, 29–37
stability of, 29–37
trait-mediated trophic cascades, 37–40
human gut microbes, 373
human influences on marine systems, 58–60
Human Microbiome Project, 402
Hydrilla verticillata, 303–304
Hymenoptera, 28
Hynobios retardatus (salamander), 73
Hypera brunneipennis, 222–223, 231–232
Hypericum perforatum, 230–231
Hypnea sp. (seaweed), 63–64

indirect ecological effects, 225
biological invasions, 223–225
indirect effects
definition, 237
nature of, 1–2
indirect genetic effects (IGEs), 299, 300, 317
indirect selection hypothesis
testing assumptions, 247
using geographic variation to test, 251–252
induced plant defences, 16–17
induced plant volatile chemicals, 30–31
induced responses
in plankton, 56–57
induced volatiles
effects on herbivore densities, 467
inducible defences
effects of ocean acidification, 59–60
inducible responses
in marine systems, 47
interaction modifications, 135, 416–417

interaction web model, 415
application to ecological theory, 429
application to ecosystem engineering,
    429
assessing ecosystem properties and species
    interactions, 420
biodiversity and ecosystem functioning,
    424–425
biodiversity and resistance to invasion,
    425–426
biodiversity and robustness to resident
    extinctions, 425–426
building the model, 418–420
ecosystem structure and functioning,
    428–429
effects of interaction modifications,
    422–423
effects of non-trophic interactions on
    biomass and production, 423–424
interaction between ecosystem properties
    and invasions, 427
link between species richness and species
    interactions, 421–423
non-trophic interactions and resistance to
    invasion, 427
non-trophic interactions and robustness to
    resident extinctions, 427
potential extension of applications, 429
species richness and connectance, 421
species richness and prevalence of
    interactions, 421
species richness and strength of
    interactions, 421–422
interaction-web topologies, 17–23
consumptive competition/apparent
    competition, 19–20
future research directions, 24
taxonomic framework, 12
three-species web with non-trophic links,
    20–21
tritrophic cascades, 17–19
webs with four or more species, 21–23
interspecific indirect genetic effects
    (IIGEs), 301
and TMIIs, 297–301
candidate gene approach, 302–304
community interactions and ecosystem
    processes, 304–307
cottonwoods, beavers and arthropod
    communities, 308–309
definition, 317
genetic and genomic basis for
    identification, 301–302
mediation by foundation species, 301–302
plant genotypes, 384–385
quantitative trait loci (QTL) analysis,
    302–304
relevance for future research, 314–316
selection at community level, 311–314
selection within a community context,
    310–311
interspecific variation in plant traits, 125–126

intraspecific genetic variation in plant
    traits, 125
intraspecific variation, 69–70
    impacts of individual plant genotypes,
        384–385
introduced species
    in marine systems, 58–59
invasive species
    and plant phenotypic plasticity, 498–501
    biodiversity and resistance to invasion,
        425–426
    direct and indirect effects, 221
    effects of non-trophic reactions on
        ecosystem responses, 425–427
    evolution of increased competitive ability
        (EICA), 222
    evolutionary change in natives and
        invaders, 221–223
    impact on biodiversity, 414–415
    in marine systems, 58–59
    indirect ecological effects, 223–225
    insights into evolutionary indirect effects,
        235–237
    interaction with ecosystem properties, 427
    limiting factors, 230–231
    mutualist-limited spread, 230–231
    non-trophic interactions and resistance to
        invasion, 427
    potential evolutionary indirect effects,
        225–232
    robustness to resident extinctions,
        425–426, 427
invasive species indirect effects
    evolution of increased anti-herbivore
        defences in natives, 231–232
    evolution of increased competitive ability
        (EICA), 229–230
    evolutionary loss of mutualism, 230–231, .
    future research directions, 232–235
    increased competitive ability in natives, 232
invertebrates
    protective symbioses, 400–401
Iolanta iolas (butterfly), 470
Iphiseius degenerans, 439
Ips typographus (beetle), 349
isopods, 20

Jacobian matrix, 93–94
Japanese brown frog (Rana pirica), 73
jimsonweed, 111

keystone species, 296
killifish (Fundulus heteroclitus), 57
kinds of traits, 13–15
    feeding, 13
    future research directions, 23–24
    life-history traits, 15
    morphological traits, 15
    physiological traits, 14–15
    space use/habitat selection, 13–14
    taxonomic framework, 11
knobbed whelk (Busycon carica), 57–58

lace bugs, 349–350
lacewing larvae, 111
lacewings, 455
lady beetles
    Adalia bipunctata, 453
    as biological controls, 453–454, 455
    Coccinella septempunctata, 462
    Coccinella transversoguttata, 452
    Coleomegilla maculata, 453, 455, 456
    Harmonia axyridis, 36, 453, 456, 461, 462
    Stethorus siphonulus, 457
leaf-rolling moth, 15, 22
leaf trichomes, 108
Leptinotarsa decemlineata (Colorado potato
    beetle), 456
Leucanthemum vulgare, 263
life-history strategies
    and the match/mismatch hypothesis,
        519–520
life-history traits, 14, 15
lima bean (Phaseolus lunatus), 265
limber pine (Pinus flexilis), 279–281, 285
Listeria monocytogenes, 401
listeriosis, 401
Lithophragma parviflorum, 214–215
litter decomposition rates
    influence of herbivores, 349–351
Littoraria irrorata (marsh periwinkle), 61
Littorina littorea (common periwinkle), 13, 61
Littorina obtusata (smooth periwinkle), 58
lizards, 38
locust (Schistocerca sp.), 400
lodgepole pine (Pinus contorta latifolia),
    216–217, 279, 283
lodgepole pine cone borer moth (Eucosma
    recissoriana), 282
Longitarsus jacobaeae (ragwort flea
    beetle), 454
loop analysis of trait-mediated effects, 102
Lotka-Volterra equations and extensions, 69
Lotka-Volterra model, 93, 95
Lotus wrangelianus, 222–223, 231–232
Loxia curvirostra (red crossbill), 283–284
Lupinus sericeus, 500–501
Lycopersicon esculentum (tomato), 263, 439,
    443–445
lygaeid bugs, 111

Macoma balthica (clam), 57
Macrocystis pyrifera (giant kelp), 63
Manduca sexta, 51
mangrove, 301
mantid species, 455
mantis (Tenodera angustipennis), 455
mantis (Tenodera sinensis), 455
Mantis religiosa, 455
marine sponges, 400
marine system TMIIs
    cascading effects of predator avoidance,
        54–58
    cascading effects of predator avoidance
        beyond three species, 50–52

consumer-induced TMIIs between basal species, 62
context-dependency, 52–54
effects of predator cues, 55–56
future research directions, 64
human influences on indirect interactions, 58–60
prey-induced TMIIs between prey species, 63–64
trait-mediated grazer–grazer interactions, 61–62
types of experimental design, 49–50
wider effects of TMIIs, 60–64
marine systems
density-mediated vs trait-mediated effects, 47–48
inducible responses and TMIIs, 48
phyletic diversity, 48
proportion of generalist consumers, 48
range and consequences of TMIIs, 47
top predator avoidance effects, 59
variety of inducible responses, 47
marker–trait association studies, 302
marsh periwinkle (*Littoraria irrorata*), 61
match/mismatch hypothesis
and climate change, 511–512
future research directions, 520–522
integrating with life-history strategies, 519–520
origins of, 509–511
resource abundance variation, 512–513
spatial mismatch, 513–518
temporal variance and adaptation, 512–513
mathematical model, 467
herbivore density and patch size, 466–467
mayflies, 14, 19
measuring indirect interactions, 247
structured equation modelling, 246–247
testing indirect selection assumptions, 247
Medicago polymorpha, 222–223, 231–232
Megoura viciae (aphid), 31–36
Mercenaria mercenaria (hard clam), 52–53, 57–58
metabolic theory of ecology, 333
microbial symbionts, 171
initiation of bottom-up trophic cascades, 170–171
microbially mediated TMIIs
aquatic ecosystems, 399–400
defining, 391–393
detecting, 391–393
distinction from DMIIs, 391–393
ecosystem-level consequences, 405–406
effects on community structure and assembly, 404–405
examples of protective symbiosis, 393–404
future research directions, 408
pairwise species interactions, 404
predicting direction and strength, 406–408
protective symbioses in invertebrates, 400–401

protective symbioses in vertebrates, 401–402
terrestrial systems, 400–404
migratory species
impacts of climate change, 516–518
potential for trophic mismatch, 516–518
milkweed, 14, 112
minnows, 22
mite (*Aculops lycopersici*), 439
model-based analysis of response surface designs, 193–200
characterizing flexible trait models, 193–196
extrapolation and estimation, 197–199
modelling adaptive prey trait modification
incorporating trait modification into models, 138–139
incorporating traits into models, 135–138
modelling the dynamics of trait change, 139–140
modelling communities
common experimental design, 188–189
consideration of flexible traits, 186–187
consideration of TMIIs, 186–187
problems with the common experimental design, 189–191
static-trait communities, 191–193
models. *See* herbivore density distributions model
monarch butterfly, 14
Mononychellus tanajoa (cassava green mite), 440–442
Mordellistena convicta (beetle), 245, 247–248, 251–252
morphological traits, 15
Morus capensis (Cape gannet), 515–516
mosquito (*Anopheles gambiae*), 401
moth (*Greya politella*), 214–215
mud crab (*Panopeus herbstii*), 52–53
Mulinia lateralis (clam), 57
multispecies mutualisms, 258
categories of trait-mediated indirect effects, 260–266
future research directions, 272–273
impacts on plant ecology and evolution, 257–258
mechanisms for effects on hosts, 258–259
mediated by DMIIs, 258
mediated by TMIIs, 258–259
nutritional–nutritional mutualisms, 264
nutritional–protection mutualisms, 263
nutritional–transport mutualisms, 261–262
pollination and seed dispersal mutualism interactions (case study), 266–272
pollination mutualisms, 265–266
protection–protection mutualisms, 265
protection–transport mutualisms, 264–265
seed dispersal mutualisms, 266
transport–transport mutualisms, 265–266

mussel (*Mytilus edulis*), 50
mutualism
    evolutionary loss of, 230–231
    limitation on spread of invasive species,
        230–231
    See also multispecies mutualisms.
mutualistic ants, 18
mycorrhizae, 230–231, 258, 259, 263, 264, 306
Mytilus edulis (mussel), 50, 331
Myzocallis asclepiadis (aphid), 112
Myzus persicae (aphid), 112

Nabis americoferus (damsel bug), 452
Nabis spp., 455
Nassau grouper (*Epinephelus striatus*), 15, 59
Nassella pulchra, 491
natural enemies
    influence on herbivore density
        distributions, 467
natural enemy biodiversity
    and biological control, 451–452
natural enemy diversity effects
    predictive use of functional diversity,
        457–459
natural enemy functional diversity
    biological control applications, 459–461
    complementary foraging behaviour,
        456–457
    complementary roles in pest species
        attacked, 452–453
    complementary roles in space, 453–455
    complementary roles in time, 455–456
    future research directions, 461–462
    impacts of climate change, 461–462
natural enemy–herbivore interactions
    tritrophic perspective, 107–108
natural systems
    tritrophic interactions, 107–108
nematodes, 357
    Steinernema spp., 404
Neoseiulus baraki (predatory mite), 438–439
Neoseiulus cucumeris (predatory mite), 437, 439
Neotyphodium endophyte, 404, 405
Nesticodes rufipes (spider mite), 457
next generation sequencing, 315, 317
niche divergence, 514
Nicholson-Bailey model, 473
nitrogen-fixing bacteria, 22
    interaction with mycorrhizae, 264
non-additive outcomes
    community effects of plant genotype
        diversity (mixtures), 382–384
nonconsumptive effects (NCEs), 135
nonlethal effects, 135
non-trophic interactions
    and resistance to invasion, 427
    and robustness to resident extinctions, 427
    effects on biomass and production,
        423–424
    importance for biodiversity, 414–415
    incorporating into ecological theory, 429
    role in ecosystem functioning, 414–415

non-trophic links
    three-species web, 20–21
non-trophic responses
    effects on ecosystem responses to
        biological invasions, 425–427
Nucella lapillus (dog whelk), 55
Nucella lapillus (snail), 330, 331
Nucifraga columbiana (Clark's nutcracker),
    279–281, 282–283
nutrient constraints in ecosystems, 332–333
nutrient recyling
    influences of herbivores, 349–351
nutritional-nutritional mutualisms, 264
nutritional-protection mutualisms, 263
nutritional-transport mutualisms, 261–262

oak (*Quercus douglasii*), 490
oak (*Quercus leavis*), 378
oak (*Quercus* spp.) ecosystems, 360
    herbivore TMIEs case study, 355–357
ocean acidification, 59–60
Oenothera biennis, 296, 383
Olneya testota, 495
one predator-two prey webs
    adaptive prey trait modification (APTM),
        142
Opsanus tau (toadfish), 52–53
optimal foraging theory, 41
owls
    effects on gerbil prey, 22

pairwise coevolution, 207–209, 211–212
Palaemonetes pugio (grass shrimp), 57
Panonychus ulmi (European red mite), 443
Panopeus herbstii (mud crab), 52–53
paradox of enrichment, 140, 141
parasites, 13
    effects on host fitness, 13
parasitoid–aphid system
    trait-mediated trophic cascades, 38–40
parasitoids, 23
    Anagrus spp., 482
    Aphelinus asychis, 461
    Aphidius ervi, 31–36, 401, 453
    Apocephalus 'sp.8', 36–37
    Brassica oleracea food webs, 41–42
    Cotesia glomerata, 30–31, 483
    density-dependent emigration rates,
        476–478
    Diaeretiella rapae, 475
    effects of non-hosts on foraging, 13
    effects of trait-mediated interactions, 40–42
    effects on aphid prey, 14, 15, 23
    Eurytoma gigantea, 245, 247–248, 251–252
    host attraction for larval/pupal parasitoids,
        473–476
    influences on body size, 15
    secondary, 23
parsnip web worm, 217
Parus atricapillus (black-capped chickadee),
    245, 247
Parus major (great tit), 511

Paspalum dilatatum (grass), 351
pea aphid (*Acyrthosiphon pisum*), 31–36, 232, 399, 401, 453, 462
Pemphigus betae (aphid), 304–305, 306–307, 375, 382
perch
    cannibalistic conspecifics, 81
    effects of habitat shift, 20
Periclista (sawfly), 355–357
Phaeocystis globosa (heteromorphic phytoplankton), 56
Phaseolus lunatus (lime bean), 265
Pheidole diversipilosa (ant), 36–37
phenological asynchrony
    consequences for consumer-resource interactions, 509
phenological shifts
    consequence of climate change, 512–513
    future research directions, 520–522
    resource abundance variation, 512–513
    spatial mismatch in consumer-resource interactions, 513–518
phenological traits
    shifts driven by climate change, 508–509
phenotypes
    community and ecosystem levels, 316
phenotypic plasticity, 10
    and community diversity, 491
    nature of plastic responses, 489
    trait-change mechanisms, 11–12
phenotypic plasticity in plants, 161
    and direct interactions, 495–496
    and exotic invaders, 498–501
    and indirect interactions, 496–498
    bottom-up tophic cascades, 162–164
    future research directions, 501
    herbivore-induced effects, 171–174
    induced by herbivores, 161–162
    influence on community diversity, 161
    plant-based reorce variation, 161
    plastic responses of plants, 489–491
    range of herbivore-induced effects, 164–167
Pholistima auritum, 497–498
phorid fly, 23
Phyllocnistis citrella (citrus leafminers), 452
Phyllocolpa spp.(gall-forming sawflies), 308
Phyllotreta spp., 471
physiological traits, 14–15
phytoplankton, 56
    Phaeocystis globosa, 56
phytoseiid mites, 437
Phytoseiulus longipes (predatory mite), 444–445
Phytoseiulus macropilis (predatory mite), 444–445
Phytoseiulus persimilis (predatory mite), 436, 439–440
Picoides villosus (hairy woodpecker), 283
Pieris rapae (small white butterfly), 30–31
Pieris spp. (butterflies), 483
pine squirrels (*Tamiasciurus* spp.), 279–284, 285–287

pinfish, 18–19, 51, 54
Pinus albicaulis (whitebark pine), 285
Pinus contorta latifolia (lodgepole pine), 216–217, 279, 283
Pinus flexilis (limber pine), 279–281, 285
Pinus halepensis (Aleppo pine), 284
Pinus lambertiana (sugar pine), 285
Pinus ponderosa (ponderosa pine), 284
Pinus sylvestris (Scots pine), 283
pinyon pine, 306
Pisaster ochraceus (sea star), 50–51
plankton, 49
    predator-induced responses, 56–57
    release of dimethyl sulfide (DMS), 57
    zooplankton ciliates and flagellates, 56
    zooplankton vertical migrations, 55–56
plant defence guilds, 225, 238
plant effects on herbivore-enemy interactions, 108
    case studies, 118–124
    classification scheme, 110–113
    criteria for DMIIs, 108–109
    criteria for TMIIs, 108–109
    definitions and terminology, 108–109
    experimental approaches, 109–110
    future directions for research, 126–127
    interspecific variation in plant traits, 125–126
    interspecific variation in predator-herbivore interactions, 118–121
    intraspecific genetic variation in plant traits, 125
    intraspecific variation in predator-herbivore interactions, 121–124
    mechanisms, 110–113
    tritrophic forest food web, 118–121
    tritrophic perspective, 107–108
plant genotype diversity (mixtures)
    non-additive community and ecosystem outcomes, 382–384
plant genotypes
    effects of variations in foundation species, 371–373
    effects on communities and ecosystems, 371–375
    genotype-mediated linkages, 381–382
    importance of intraspecific variation, 384–385
    individual genotypes and communities, 375–377
    individual genotypes and ecosystem processes, 377–381
    interspecific indirect genetic effects, 384–385
plant–herbivore systems, 3, 12
plant interactions
    ecological consequences, 492–495
    indirect interactions, 492
    mechanisms of multispecies mutualist effects, 258–259
    negative direct interactions, 491
    plasticity and direct interactions, 495–496

plant interactions (cont.)
  plasticity and exotic invaders, 498–501
  plasticity and indirect interactions,
    496–498
  positive interactions (facilitation), 491–492
plant-mediated competition among
    herbivores, 14, 476
plant phenotypes
  damage-induced regrowth, 173
  herbivore responses to induced
    phenotypes, 174–176
  induced changes in nutritional quality, 172
  predator responses to changes in
    herbivores, 176–178
  resistance mediated by secondary
    metabolites, 171–172
  susceptibility mediated by secondary
    metabolites, 171–172
  See also phenotypic plasticity in plants.
plant symbioses
  aboveground, 402–403
  belowground, 403
plant trichomes, 111
planthoppers, 16, 61
  Delphacodes scholochloa, 482
  Prokelisia crocea, 482
  Prokelisia sp., 61
Plasmodium, 401
plasticity in traits, 2
Plathemis lydia (dragonfly), 76
Plutella xylostella (diamond back moth), 30–31,
    479
Poa pratensis (grass), 495
Poecile pubescens (downy woodpecker), 245, 247
Poecilia reticulata (guppy), 287
pollination 265–272
  effects of mycorrhizae, 261–262
pollination mutualisms, 262
Polygonum bistorta, 514
ponderosa pine (Pinus ponderosa), 284
Populus angustifolia, 111, 304, 306, 308–309,
    371–373, 376, 379, 382
Populus angustifolia hybrids, 377
Populus angustifolia × P.fremontii hybrids, 111
Populus fremontii, 304, 308–309, 371–373, 377
Populus spp., 372, 374
  aspen, 13, 37
  aspen genotypes, 373
  cottonwoods, 279, 287, 301–302, 308–309
  ecosystem impacts of individual genotypes,
    377–381
  effects of genotype mixtures, 383
  genetic maps, 302–303
  hybrids, 111, 304–307, 308–309, 371–373,
    377
  impacts of individual genotypes, 376–377
  interspecific indirect genetic effects (IIGEs),
    304–307
Populus tremuloides, 379, 384
positive density dependence, 91
predator avoidance
  effects in marine systems, 59

predator cues
  effects in marine systems, 55–56
predator-induced escape behaviour of
    herbivores, 439–442
predator-induced fear in prey, 326–328
predator-induced ontogenetic escape by
    herbivores, 442–443
predator-induced prey emigration, 478–479
predator-mediated competition among
    herbivores, 443–445
predator–prey interactions
  indirect effects, 9
  size-structured TMIIs, 71–72
predator–prey models
  adaptive prey trait modification
    (APTM), 140
predators
  conspecific cannibalism, 74–75
  intimidation effects on prey, 13
  responses to changes in herbivores,
    176–178
  size-structured interactions, 72–75
  stage-structured mutual predation, 76–77
predatory birds, 38
predatory mites
  Amblydromalus manihoti, 440–442
  Neoseiulus baraki, 438–439
  Neoseiulus cucumeris, 437, 439
  Phytoseiulus longipes, 444–445
  Phytoseiulus macropilis, 444–445
  Phytoseiulus persimilis, 436, 439–440
  Typhlodromalus aripo, 440–442
premature leaf abscission
  effects on ecosystem processes, 345
prey
  size-structured interactions, 76
prey-predator systems, 3
probiotic bacteria, 400, 402
Prokelisia crocea (planthopper), 482
Prokelisia sp. (planthopper), 61
protection-protection mutualisms, 265
protection-transport mutualisms, 264–265
protective symbioses
  aquatic ecosystems, 399–400
  in invertebrates, 400–401
  in plants (aboveground), 402–403
  in plants (belowground), 403
  in vertebrates, 401–402
  terrestrial systems, 400–404
Pseudocardinia (actinobacterium), 401
Pycnopodia helianthoides (sea star), 53

qualitative analysis of trait-mediated effects,
    102–103
quantifying TMIIs
  common experimental design, 188–189
  model-based analysis of response surface
    designs, 193–200
  parallels with static-trait communities,
    191–193
  problems with the common experimental
    design, 189–191

quantitative trait loci (QTL), 295, 297, 302–304
quantitative trait loci (QTL) analyses, 317
quantitative trait loci (QTL) linkage maps, 302
Quercus (oak) ecosystems
    herbivore TMIEs case study, 355–357
Quercus agrifolia (oak), 497–498
Quercus douglasii (oak), 490–491
Quercus leavis (oak), 378
Quercus spp., 372

ragwort (*Senecio jacobaea*), 454
ragwort flea beetle (*Longitarsus jacobaeae*), 454
ragwort seed head fly (*Botanophila seneciella*), 454
Rana pirica (Japanese brown frog), 73
Rangifer tarandus (caribou), 518
reciprocal transplant experiments, 252–253
red crossbill (*Loxia curvirostra*), 283–284
red-eyed treefrog, 84
Regiella insecticola, 401
reindeer, 347
resource abundance variation
    effects of climate change, 512–513
resource concentration hypothesis (Root), 466–467
resource mosaics, 179–180
rhizobia, 230, 264
    trophic effects in a soybean system, 170–171
Rhizopus microsporus, 403
Rickettsiella, 401
Robinia pseudoacacia (black locust), 264
roots
    herbivore-induced exudation of labile C, 346

salamanders
    Eurycea cirrigera, 72
    Gyrinophilus porphyriticus, 72
    Hynobios retardatus, 73
    predation by fish, 20
    size-specific interactions, 72
Salix eriocarpa (willow), 167–168
Salix gilgiana (willow), 167–168
Salix serissaefolia (willow), 167–168
Salix spp., 372
salticid spiders, 111
Sargassum filipendula (seaweed), 51, 63–64
Sargassum sp. (seaweed), 18–19
sawfly (*Periclista*), 355–357
scallops, 54
Schistocerca sp. (locust), 400
Sciurus spp. (tree squirrels), 279–281
Scolochloa festucacea (grass), 482
Scots pine (*Pinus sylvestris*), 283
sculpin, 14
sea stars, 50–51
    Pycnopodia helianthoides, 53
sea turtles, 50
sea urchins, 63
    Strongylocentrotus franciscanus, 53–54
    Strongylocentrotus purpuratus, 53–54

seaweed, 400
    Ascophyllum nodosum, 58
    Desmarestia ligulata, 63
    Fucus sp., 17, 61
    Hypnea sp., 63–64
    Sargassum filipendula, 51, 63–64
    Sargassum sp., 18–19
seed dispersal mutualisms, 266–272
seeds
    trait evolution, 278–279
    variations in DMIIs and TMIIs, 281–284
selection at community level, 311–314
    simulation approach, 311–314
selection experiments
    measuring indirect interactions, 253
selective foraging
    ecosystem effects, 346–349
Semibalanus balanoides (barnacle), 330, 331
Senecio jacobaea (ragwort), 454
serotiny in conifers
    selection pressures on, 285–287
Serratia symbiotica, 399
sharks, 59
    effects of loss of top predators, 59
simple sequence repeats (SSRs), 302, 317
single nucleotide polymorphisms (SNPs), 297, 302, 317
size
    variation within species, 70
size-structured populations, 15
size-structured predators, 72–75
size-structured TMIIs, 70–85
    conspecific cannibalism, 74–75
    effects on long-term dynamics, 80–82
    effects on predator–prey interactions, 71–72
    effects on short-term dynamics, 79–80
    expanding the TMII concept, 82–83
    future research directions, 83–85
    intraspecific variation, 69–70
    one-species system, 70
    size and developmental variation within species, 69–70
    size classes as distinct functional groups, 71–72
    size-structured mutual predation, 76–77
    size-structured prey, 76
    structural vs. numerical changes, 77–79
    two-species system, 70
slow-growth/high-mortality hypothesis, 112
small white butterfly (*Pieris rapae*), 30–31
smooth periwinkle (*Littorina obtusata*), 58, 61
snails, 50
    Acanthinucella spirata, 50
    Nucella lapillus, 330, 331
soil fauna
    effects of herbivore TMIEs, 353–354
soil microbes
    effects of herbivore TMIEs, 352–353
soil microbial communities
    impacts of *Populus* genotypes, 376–377
soil microclimate
    effects of herbivores, 351–352

soil resources
  effects on herbivores, 352
soil systems
  effects of herbivore TMIEs, 352–354
  herbivore TMIEs study in *Quercus* (oak)
    ecosystems, 355–357
  importance of herbivore TMIEs, 354–355
Solanum ptychanthum, 51
Solidago altissima (goldenrod), 169–170, 379,
  383, 384
Solidago rugosa (goldenrod), 330
Solidago spp. (goldenrod), 244–245, 296, 306,
  348
  reciprocal transplant experiments, 253
  web of indirect interactions, 248–250, 252
soybean
  trophic effects of microbial symbionts,
    170–171
space use/habitat selection traits, 13–14
Spartina alterniflora (cordgrass), 61
Spartina pectinata (grass), 482
spatial mismatch in consumer-resource
  interactions, 513–518
species diversity
  and connectance, 41–42
species-level variation, 69–70
spider mites
  Nesticodes rufipes, 457
  Tetranychus cinnabarinus, 457
  Tetranychus evansi, 443–445
  Tetranychus urticae, 436, 439–440, 442–443,
    443–445
spiders, 9, 13, 18, 37, 455, 457, 462, 482
squid, 400
squirrels, 216–217
stability of ecosystems
  influence of adaptive prey trait
    modification (APTM), 143–144
  influence of density dependence, 89
  influence of trait plasticity, 90–94
stage-structured indirect interactions, 72–77
Steinernema spp. (nematodes), 404
stem-boring moths
  Dioryctria albovittella, 349
  effects on willows, 15, 22
  Endoclita excrescence, 167–168
Stethorus siphonulus (lady beetle), 457
stoneflies, 14
strangler figs, 230
Strongylocentrotus franciscanus (sea urchin), 53–54
Strongylocentrotus purpuratus (sea urchin),
  53–54
structured equation modelling, 246–247
sugar pine (*Pinus lambertiana*), 285
symbiosis
  plants (aboveground symbioses), 402–403
  plants (belowground symbioses), 403
  protective symbioses in invertebrates,
    400–401
  protective symbioses in vertebrates,
    401–402
syrphids, 455

tadpoles
  effects of predator presence, 18, 19
  effects of predators on, 22
  response to predator cues, 73
Tamiasciurus spp. (pine squirrels), 279–284,
  285–287
taxonomic framework for TMIIs, 10–25
  future research directions, 23–25
  interaction-web topologies, 12, 17–23
  kinds of traits, 11, 13–15
  trait-change mechanisms, 11–12, 16–17
Tenodera angustipennis (mantis), 455
Tenodera sinensis (mantis), 455
terpenes, 494
terrestrial systems
  protective symbioses, 400–404
Tetranychus cinnabarinus (spider mite), 457
Tetranychus evansi (spider mite), 443–445
Tetranychus urticae (two-spotted spider mite),
  436, 439–440, 442–443, 443–445
three-species web with non-trophic links, 20–21
through-fall
  effects on ecosystem processes, 345
Thymus pulegioides, 494
Thymus serpyllum, 494
tiger shark (*Galeocerdo cuvier*), 50
  effects of loss of, 59
time scales of trait-mediated effects, 94–95
TMIEs. *See* trait-mediated indirect effects
TMIIs. *See* trait-mediated indirect
    interactions
toadfish (*Opsanus tau*), 52–53
tomato (*Lycopersicon esculentum*), 263, 439,
  443–445
Toxoptera citricida (brown citrus aphids), 452
trait, 2
  definition, 2
trait cascades, 18–19
trait-change mechanisms, 15, 16–17
  behavioural plasticity, 16
  developmental plasticity, 16–17
  evolution, 16
  future research directions, 24
  taxonomic framework, 11–12
  within-generation phenotype selection, 16
trait evolution
  community and ecosystem consequences,
    287–288
  consequences of TMIIs, 279–281
  examples and consequences of indirect
    interactions, 285–287
  indirect genetic effects (IGEs), 299
  influence of multispecies interactions, 278
  influence on ecosystem interactions, 278
  reproductive traits in conifers, 278–279
  selection pressures from other species,
    287–288
  variations in DMIIs and TMIIs, 281–284
trait invariance, 2
trait-mediated biotic indirect effects, 417
trait-mediated density dependence
  discrete-time model, 100–101

trait-mediated direct effects, 418
trait-mediated effects
  direct density dependence model, 95–100
  in marine systems, 47–48
  incorporation into theoretical ecology, 429
  methods of analysis, 101–103
  timescales, 94–95
trait-mediated indirect effects (TMIEs)
  alternative terminology, 135
  comparison with density effects, 9–10
  comparison with DMIEs, 415–416
  definition, 237
  nature of, 1–2
trait-mediated indirect interactions (TMIIs)
  alternative terminology, 135
  conditions required for, 3
  consideration in community modelling,
    186–187
  definition, 3, 48, 317
  expanding the concept, 82–83
  extent of influence in communities, 3
  implications for ecological studies, 3–4
  requirements for, 19
  types of effects, 3
trait-mediated trophic cascades
  host–parasitoid interactions, 37–40
trait plasticity, 2
  influence on density dependence, 101
  influence on ecosystem stability, 90–94
traits, 11
  feeding traits, 13
  influence on demography, 89
  life-history traits, 15
  morphological traits, 15
  physiological traits, 14–15
  space use/habitat selection, 13–14
  taxonomic framework, 11
transport-transport mutualisms, 265–266
tree squirrels (Sciurus spp.), 279–281
Trillium erectum, 266–272
Trirhabda virgata (beetle), 348
tritrophic cascades, 17–19
tritrophic food chains
  adaptive prey trait modification (APTM),
    141
tritrophic forest food web, 118–121
tritrophic interactions
  in natural systems, 107–108
tritrophic perspective, 107–108
trophic cascades, 3, 140
  herbivore-initiated bottom-up cascades,
    162–164
  trait-mediated, 37–40
  tritrophic food chains, 141
trophic control
  bottom-up view, 325
  middle of food chains, 325–331
  top-down view, 324–325

trophic interactions, 317
  genetic basis, 304–307
tunicates, 400
two predator–one prey webs, 142
  adaptive prey trait modification (APTM),
    141–142
two-spotted spider mite (Tetranychus urticae),
    436, 439–440, 442–443, 443–445
Typhlodromalus aripo (predatory mite),
    440–442
Tyria jacobaeae (cinnabar moth), 454

Uroleucon nigrotuberculatum (aphid), 169–170
Uroleucon rudbeckiae (aphid), 122–124
USA National Phenology Network (NPN), 515

vertebrates, 402
  protective symbioses, 401–402
Vicia faba, 31–36, 263, 495–496
volatile compounds produced by plants, 263,
    265, 467
  dimethyl sulfide (DMS), 57

water fleas, 10
webs with four or more species, 21–23
weeds, 454
  biological controls, 454, 456
weevil (Diaprepes abbreviatus), 452
western flower thrips, 439–440
whelks
  Acanthina angelica, 62
  effects on barnacle prey, 15
  responses to introduced species, 58–59
white clover, 230
whitebark pine (Pinus albicaulis), 285
whitefly (Bemisia tabaci), 442
wild parsnip, 217
willow galls, 111
willows, 15
  effects of stem-boring moth, 15, 22
  Salix eriocarpa, 167–168
  Salix gilgiana, 167–168
  Salix serissaefolia, 167–168
within-generation phenotype selection,
    11–12, 16
wolves
  effects of reintroduction, 13, 37

Yellowstone National Park
  effects of wolf reintroduction, 13, 37
Yersinia pestis, 401

Zea mays (corn), 453, 496
zooplankton, 10, 22
  ciliates and flagellates, 56
  release of dimethyl sulfide (DMS), 57
  vertical migrations, 55–56
Zostera marina, 383

For EU product safety concerns, contact us at Calle de José Abascal, 56–1°, 28003 Madrid, Spain or eugpsr@cambridge.org.

www.ingramcontent.com/pod-product-compliance
Ingram Content Group UK Ltd.
Pitfield, Milton Keynes, MK11 3LW, UK
UKHW060333090126
466816UK00014B/262